ncRNAs in Abiotic Stress Tolerance in Plants

ncRNAs in Abiotic Stress Tolerance in Plants offers a timely and comprehensive examination of non-coding RNAs (ncRNAs) and their pivotal role in regulating plant responses to biotic and abiotic stresses. As research continues to uncover the functional versatility of small RNAs, long non-coding RNAs and circular RNAs in plant genomes, this volume provides a focused and integrative perspective on how ncRNAs orchestrate gene expression under stress conditions, including drought, salinity, temperature extremes, pathogen attack and nutrient deficiencies.

Covering both foundational concepts and cutting-edge research, this book explores the biogenesis, classification and molecular functions of ncRNAs, while highlighting the experimental and computational approaches used to study them. It further examines the cross-talk between ncRNAs and other regulatory layers such as transcription factors, epigenetic marks and hormonal signalling pathways – offering insights into the multilayered regulatory networks that underlie plant resilience.

Written by leading experts from diverse research institutions, this volume bridges molecular biology, plant physiology and stress genomics. It serves as a valuable reference for plant molecular biologists, agronomists, biotechnologists and graduate students working at the interface of RNA biology and crop improvement. As climate change and global food insecurity intensify, *ncRNAs in Abiotic Stress Tolerance in Plants* is an essential resource for developing innovative, RNA-based strategies to enhance crop resilience and sustainability.

ncRNAs in Abiotic Stress Tolerance in Plants

Edited by
Muhammad Waseem, Sajid Fiaz and Pingwu Liu

CRC Press
Taylor & Francis Group
Boca Raton London New York

CRC Press is an imprint of the
Taylor & Francis Group, an **informa** business

Designed cover image: Shutterstock

First edition published 2026
by CRC Press
2385 NW Executive Center Drive, Suite 320, Boca Raton FL 33431

and by CRC Press
4 Park Square, Milton Park, Abingdon, Oxon, OX14 4RN

CRC Press is an imprint of Taylor & Francis Group, LLC

ISBN: 978-1-032-95114-0 (hbk)
ISBN: 978-1-032-95115-7 (pbk)
ISBN: 978-1-003-58325-7 (ebk)

DOI: 10.1201/9781003583257

Typeset in Times
by KnowledgeWorks Global Ltd.

Contents

About the Editors

Muhammad Waseem is Associate Professor/Researcher at College of Tropical Agriculture and Forestry, Hainan University, Haikou, Hainan, China. He received a doctoral degree (PhD, specialisation in Plant Molecular Biology) from the School of Life Sciences, Chongqing University, Chongqing, PR China. He has five years of research experience, including postdoctoral. His interests include transcriptional regulation of flowering time regulation, fruit ripening and development, and abiotic stresses such as salinity and drought. In addition, he has hands-on experience in next-generation sequencing data analysis, transcriptomics and data visualisation. Currently, Dr. Waseem is actively in investigating flowering time regulation in *Brassica* crops. He has been awarded with a National Science Foundation Project, RFIS-I in 2023. In brief, he has published more than 58 research articles, five review articles and nine book chapters in various high-impact factor publications, including *New Phytologist, International Journal of Macromolecules, Planta, GM Crops & Food, Environmental and Experimental Botany, Physiologia Plantarum* and *Frontiers in Plant Science.* He has also edited books with Springer publisher.

Sajid Fiaz, PhD, is Assistant Professor in Institute of Molecular Biology & Biotechnology at The University of Lahore, Lahore. He has scientific collaboration with King Saud University, Saudi Arabia, Princess Nourah bint Abdulrahman University, Saudi Arabia, and Gansu Agricultural University, Lanzhou, China. Dr. Fiaz received his PhD in Crop Genetics and Breeding from Chinese Academy of Agricultural Sciences. Dr. Fiaz's research interests include marker-assisted selection for agronomic traits, mutation breeding for biotic and abiotic stress resistance, QTL mapping and genome editing for yield and quality traits in cereals. Dr. Fiaz is actively supervising under-graduate and post-graduate students. He has been awarded a Start-Up Research Grant Project from Higher Education Commission of Pakistan worth 1.0 million PKR. He has published more than 135 research articles, ten book chapters and ten review articles with prestigious scientific magazines like *International Journal of Molecular Sciences, Frontiers in Plant Sciences, Plant Physiology and Biochemistry* and *Saudi Journal of Biological Sciences.* He is working as academic editor for *PLOS ONE*, guest associate editor for *Frontiers in Plant Sciences*, review editor for *Frontiers in Genetics* and editorial board member for GM Crops and Food. He has recently edited three books with Springer and one book with Wiley publishers. Currently, his lab group is working to explore genetic factors controlling traits with economic importance in cereals.

Pingwu Liu is Full Professor at the College of Tropical Agriculture and Forestry, Hainan University, where he also serves as a master's supervisor in horticultural sciences. He holds both a Bachelor's and PhD in Agronomy from Huazhong Agricultural University, completing a combined-degree programme between 1995 and 2004. Professor Liu's research expertise spans *Brassicaceae* vegetable germplasm innovation, protected horticulture techniques and biological control strategies. He has led multiple competitive research projects funded by provincial and national programmes, including key R&D projects in ornamental rapeseed breeding and gene mining for agronomically important traits. His work integrates molecular breeding, plant physiology and applied crop improvement, with a strong focus on tropical agricultural systems. He has published extensively in leading peer-reviewed journals, contributing to the understanding of non-coding RNAs in stress responses, QTL mapping in sugarcane and molecular diversity in vegetable crops. His most recent publications address genome-wide analysis of long non-coding RNAs in *Brassica napus* under abiotic stress and marker-assisted breeding strategies. As an editor, Professor Liu brings a wealth of academic experience, a deep understanding of plant molecular biology and a commitment to bridging fundamental research with practical agricultural applications. His editorial contributions aim to advance knowledge in the areas of plant stress biology, molecular breeding and sustainable crop production in the context of climate-resilient agriculture.

Contributors

Banzeer Ahsan Abbasi
Department of Botany
Rawalpindi Women University
Rawalpindi, Pakistan

Zainul Abideen
Dr. Muhammad Ajmal Khan Institute of Sustainable, Halophyte Utilization
University of Karachi
Karachi, Pakistan

Priyanka Aggarwal
School of Biological and Environmental Sciences
Shoolini University
Solan, India

Ishtiaq Ali
Department of Biological and Health Sciences
Pak-Austria Fachhochschule: Institute of Applied Sciences and Technology
Haripur, Pakistan

Nasir Ali
Department of Biological and Health Sciences
Pak-Austria Fachhochschule: Institute of Applied Sciences and Technology
Haripur, Pakistan

Rahmat Ali
Department of Biological Sciences
Texas Tech University
Lubbock, TX, USA

Talib Ali
Department of Plant Sciences, Faculty of Biological Sciences
Quaid-i-Azam University
Islamabad, Pakistan

Muhammad Anas
Department of Soil Science & Plant Nutrition
Selçuk University
Konya, Türkiye

Kumari Anjani
Department of Agricultural Biotechnology and Molecular Biology
Dr. Rajendra Prasad Central Agricultural University
Bihar, India

Hirdayesh Anuragi
ICAR-Central Agroforestry Research Institute
Jhansi, India

Muhammad Anwar
School of Tropical Agriculture and Forestry
Hainan University
Haikou, China

Ayyanadar Arunachalam
ICAR-Central Agroforestry Research Institute
Jhansi, India

Anosha Asghar
Department of Plant Sciences, Faculty of Biological Sciences
Quaid-i-Azam University
Islamabad, Pakistan

Alyan Ashraf
Pakistan Environmental Protection Agency
Ministry of Climate Change & Environmental Coordination
Islamabad, Pakistan

Muhammad Arslan Ashraf
Department of Botany
Government College University
Faisalabad, Pakistan

Muhammad Asif
Department of Plant Breeding and Genetics
B. Z. University
Multan, Pakistan

Arslan Asim
Pakistan Environmental Protection Agency
Ministry of Climate Change & Environmental Coordination
Islamabad, Pakistan

Rehan Aslam
Department of Natural and Basic Science
University of Turbat
Turbat, Pakistan

Chetan Awasthi
Centre for Nanoscience and Nanotechnology
Jamia Millia Islamia
New Delhi, India

Karma Landup Bhutia
Department of Agricultural Biotechnology and Molecular Biology
Dr. Rajendra Prasad Central Agricultural University
Bihar, India

Saboor Muarij Bunny
Pakistan Environmental Protection Agency
Ministry of Climate Change & Environmental Coordination
Islamabad, Pakistan

Vibhuti Chandel
School of Biological and Environmental Sciences
Shoolini University
Solan, India

Usman Khalid Chaudhry
Pakistan Environmental Protection Agency
Ministry of Climate Change & Environmental Coordination
Islamabad, Pakistan

Charupriya Chauhan
Department of Genetics and Plant Breeding
Acharya Narendra Deva University of Agriculture and Technology
Ayodhya, India

Vijay Chauhan
School of Biological and Environmental Sciences
Shoolini University
Solan, India

Subramani Divakar
Department of Plant Biotechnology
Tamil Nadu Agricultural University
Coimbatore, India

Shah Fahad
Department of Biology and Microbiology
South Dakota State University
Brookings, SD, USA

Sirmast Faiz
National Institute of Virology, Dr. Panjwani Center for Molecular Medicine and Drug Research, International Center of Chemical and Biological Sciences
University of Karachi
Karachi, Pakistan

Ummey Farwa Hassan
Dr. Muhammad Ajmal Khan Institute of Sustainable, Halophyte Utilization
University of Karachi
Karachi, Pakistan

Noor Fatima
Dr. Muhammad Ajmal Khan Institute of Sustainable, Halophyte Utilization
University of Karachi
Karachi, Pakistan

Vishnu Bapurao Gore
Division of Biochemistry
Indian Agricultural Research Institute
New Delhi, India

Arun Kumar Handa
ICAR-Central Agroforestry Research Institute
Jhansi, India

Areeba Ijaz
Department of Botany, Division of Science and Technology
University of Education
Lahore, Pakistan

Naila Ijaz
Nano Composite Lab, Department of Physics
University of Agriculture
Faisalabad, Pakistan

Shumaila Ijaz
Department of Burn and Plastic Surgery, Shenzhen Institute of Translational Medicine, The First Affiliated Hospital of Shenzhen University, School of Biomedical Engineering, Shenzhen University Medical School
Shenzhen University
Shenzhen, China

Muhammad Ikram
Jamil ur Rehman Center for Genome Research, PCMD, International Center for Chemical and Biological Sciences
University of Karachi
Karachi, Pakistan

Javed Iqbal
Department of Botany
Bacha Khan University
Charsadda, Pakistan

Junaid Iqbal
Pakistan Environmental Protection Agency
Ministry of Climate Change & Environmental Coordination
Islamabad, Pakistan

Rashid Iqbal
Department of Agronomy, Faculty of Agriculture and Environment
The Islamia University of Bahawalpur
Bahawalpur, Pakistan

Rachna Shashikant Jadhav
Faculty of Applied Sciences and Biotechnology
Shoolini University
Bajhol, India

Priyanka Jaiswal
Department of Genetics and Plant Breeding
Acharya Narendra Deva University of Agriculture and Technology
Ayodhya, India

Banshidhar Jha
Department of Genetics and Plant Breeding
Acharya Narendra Deva University of Agriculture and Technology
Ayodhya, India

Nitish Joshi
Department of Agronomy, Horticulture and Plant Science
South Dakota State University
Brookings, SD, USA

Anil Namdev Kale
Department of Agricultural Biotechnology and Molecular Biology
Rajendra Prasad Central Agricultural University
Samastipur, India

Muhammad Kaleem
Department of Botany, Division of Science and Technology
University of Education
Township Campus, Lahore, Pakistan

Imran Khan
Center for Life Science Ventures
Cornell University
Ithaca, NY, USA

Saurabh Kulshrestha
Faculty of Applied Sciences and Biotechnology
Shoolini University
Bajhol, India

Nitin Kumar
CSSS PG College Machhra
Meerut, India

Santosh Kumar
Department of Genetics and Plant Breeding
Acharya Narendra Deva University of Agriculture and Technology
Ayodhya, India

Chandresh Kumari
Faculty of Applied Sciences and Biotechnology
Shoolini University
Bajhol, India

Riffat Latif
Department of Pharmaceutical Sciences
South Dakota State University
Brookings, SD, USA

Ayush Madan
Department of Biotechnology, School of Research & Technology
People's University
Bhopal, India

Gnanasekaran Madhavan
Department of Agronomy
Tamil Nadu Agricultural University
Coimbatore, India

Tariq Mahmood
Department of Plant Sciences, Faculty of Biological Sciences
Quaid-i-Azam University
Islamabad, Pakistan

Sayanti Mandal
Department of Chemistry & Biochemistry
Sharda School of Engineering & Sciences
Sharda University
Greater Noida, India

Udit Nandan Mishra
Faculty of Agriculture
Sri-Sri University
Cuttack, India

Faiza Munir
Department of Agricultural Sciences and Technology, Atta-ur-Rahman School of Applied Biosciences (ASAB)
National University of Sciences and Technology (NUST)
Islamabad, Pakistan

Ghulam Murtaza
School of Agriculture
Yunnan University
Kunming, China

Yogesh Dashrath Naik
Department of Agricultural Biotechnology and Molecular Biology
Dr. Rajendra Prasad Central Agricultural University
Samastipur, India

Bhupathi Nandhini
Department of Plant Pathology
University of Agricultural Sciences
Dharwad, India

Muhammad Nauman
School of Tropical Agriculture and Forestry
Hainan University
Haikou, China

Taufiq Nawaz
Department of Biology and Microbiology
South Dakota State University
Brookings, SD, USA

Pragati
Institute of Plant and Microbial Biology
Academia Sinica
Taipei, Taiwan

Saleem Ur Rahman
Department of Biological and Health Sciences
Pak-Austria Fachhochschule: Institute of Applied Sciences and Technology
Haripur, Pakistan

Rizwan Rasheed
Department of Botany
Government College University
Faisalabad, Pakistan

Nadia Riaz
Department of Botany
Lahore College for Women University
Lahore, Pakistan

Muhammad Asif Saleem
Department of Plant Breeding and Genetics
B. Z. University
Multan, Pakistan

Shah Saud
College of Life Science
Linyi University
Linyi, China

Ehtisham Shafique
Department of Biological and Health Sciences
Pak-Austria Fachhochschule: Institute of Applied Sciences and Technology
Haripur, Pakistan

Anis Ali Shah
Department of Botany, Division of Science and Technology
University of Education
Lahore, Pakistan

Tayyaba Shahzadi
Department of Chemistry, Government College for Women
University of Sialkot
Sialkot, Pakistan

Nadia Sharif
Department of Biotechnology
Women University Mardan
Mardan, Pakistan

Himanshu Sharma
Amity School of Biological Sciences
Amity University Punjab
Mohali, Punjab, India

Vinay Kumar Sharma
Department of Agricultural Biotechnology and Molecular Biology
Dr. Rajendra Prasad Central Agricultural University
Samastipur, India

Vinay Sharma
Center of Excellence in Genomics and Systems Biology
International Crops Research Institute for the Semi-Arid Tropics (ICRISAT)
Hyderabad, India

Sanjay Kumar Sheshma
Department of Agricultural Biotechnology and Molecular Biology
Dr. Rajendra Prasad Central Agricultural University
Samastipur, India

Mohammad Shoeb
Department of Plant Breeding and Genetics
B. Z. University
Multan, Pakistan

Pradeep Singh
Agri-Biotechnology Division
National Agri-Food Biotechnology Institute
Mohali, India

Rajesh Kumar Singhal
ICAR-Indian Grassland and Fodder Research Institute
Jhansi, India

Seweta Srivastava
Department of Plant Pathology, School of Agriculture
Lovely Professional University
Phagwara, India

Nimra Tariq
Department of Botany, Division of Science and Technology
University of Education Township Campus
Lahore, Pakistan

Anamish Tyagi
Department of Genetics and Plant Breeding
Acharya Narendra Deva University of Agriculture and Technology
Ayodhya, India

Muhammad Muneeb Ullah
Pakistan Environmental Protection Agency
Ministry of Climate Change & Environmental Coordination
Islamabad, Pakistan

Zakir Ullah
Department of Plant Sciences, Faculty of Biological Sciences
Quaid-i-Azam University
Islamabad, Pakistan

Rooma Waqar
Department of Plant Sciences, Faculty of Biological Sciences
Quaid-i-Azam University
Islamabad, Pakistan

Tabassum Yaseen
Department of Botany
Bacha Khan University
Charsadda, Pakistan

Afifa Younas
Department of Botany, Division of Science and Technology
University of Education Township Campus
Lahore, Pakistan

Zubaida Yousaf
Department of Botany
Lahore College for Women University
Lahore, Pakistan

Sadia Zafar
Department of Botany, Division of Science and Technology
University of Education
Lahore, Pakistan

Ambreen Zari
Department of Biotechnology
Women University Mardan
Mardan, Pakistan

1 Abiotic Stress Mitigation in Plants
The Critical Role of ncRNAs

Gnanasekaran Madhavan, Subramani Divakar, Vishnu Bapurao Gore, Bhupathi Nandhini and Sayanti Mandal

1.1 INTRODUCTION

The transcription of RNA is extensive in the eukaryotic genome, yet approximately **90% of the eukaryotic genome is transcribed, with only 1–2% of these transcripts encoding proteins, while the vast majority are transcribed as non-coding RNAs (ncRNAs)** (Rai et al., 2019; Wang et al., 2017). The non-protein-coding transcriptome encompasses ncRNAs derived from regions once deemed biologically inert, including intergenic zones, pseudogenes repetitive sequences and transposons. Initially, these non-coding-transcripts were dismissed as mere transcriptional error, largely because of their negligible or absent protein-coding potential and their lacking conserved sequences. However, their emergence challenges this outdated perspective, highlighting their potential functional significance (Pauli et al., 2011; Wagh et al., 2025). NcRNAs are functional RNA molecules with low protein-coding potential. They fall into three categories: medium-sized ncRNAs (31–200 nt), long-ncRNAs (more than 200 nt) and tiny ncRNAs (sRNAs) (18–30 nt) (Figure 1.1). PIWI interacting RNAs (piRNAs) and plant microRNAs (miRNAs), typically 21-nt sRNAs, guide mRNA target degradation and/or inhibit translation, thereby suppressing target gene expression (D'Ario et al., 2017; Rajagopalan et al., 2006); small-nuclear RNA (snRNA), transfer-RNA (tRNA), small-nucleolar RNA (snoRNA) and ribosomal-RNA (rRNA) are all members of structural class of non-coding RNAs (Zhu et al., 2011). Although these regulatory ncRNAs are produced from DNA, they cannot be translated into proteins. Despite this limitation, they are indispensable for regulating plant growth, plant progress and resilience to environmental abiotic stress, exerting their influence through both transcriptional and post-transcriptional mechanisms (Brosnan & Voinnet, 2009; D'Ario et al., 2017). In current eras, vital improvement is established on finding out the complex and dynamic linkages of plant responses to climate adversaries. Researchers have focused on recognising genes for particular stress with their protein interaction networks. The reciprocations of plants are intricately governed, tightly controlled by the expression of gene and gene regulations. These processes operate through diverse and sophisticated mechanisms at multiple levels, including both transcriptional and post-transcriptional regulations (Ingram & Bartels, 1996; Origin, 2009; Sunkar & Zhu, 2004). Most research focuses on how coding proteins work, but recent studies show that ncRNAs are pivotal in growth of the plant against adverse conditions by controlling the corresponding genes (Campalans et al., 2004).

To withstand various environmental stresses, plants have evolved intricate mechanisms, including the development of transcriptional memory. This stress-induced memory enhances plant adaptability when faced with recurring or novel environmental challenges. Investigating the molecular pathways involved in generating transcriptional memory in response to plant stress could provide valuable insights for breeding stress-resistant crops. Climate change has led to a significant increase in the frequency of abiotic stressors, escalating the threat to crop yields and global food security. In this context, non-coding RNAs (ncRNAs) play a critical role in mediating plant responses to both abiotic and biotic stresses (Zuo et al., 2023).

1.2 CANONICAL BIOGENESIS

1.2.1 Characterisation of Several Plant ncRNAs

Non-coding RNAs (ncRNAs) exhibit diverse origins and functions. Emergence of distinct ncRNAs, viz., piRNAs, miRNAs, lncRNAs and siRNAs, is explained by a number of theories. The classification of non-coding RNAs is discussed in Figure 1.1. Pseudogenisation of protein-coding sequences, originating from pre-existing transposable elements (TEs), gene duplication, random hairpin structure development, RNA virus replication and double-stranded RNAs (dsRNAs) from heterochromatin areas and DNA repeats are some of these hypotheses (Allen et al., 2004; Nozawa et al., 2010; Origin, 2009) (Tables 1.1 and 1.2).

1.2.2 Biogenesis of miRNA

MicroRNAs, typically 20–22 nucleotides long, are generated from extended primary transcripts known as primary miRNAs (pri-miRNAs). RNA polymerase II (RNA pol II) initiates this process (Borges & Martienssen, 2015), which is further facilitated by Dicer-like protein *1(DCL1), SERRATE(SE), and HYPONASTIC LEAVES(HYL1)* (Axtell & Meyers, 2018; Rogers & Chen, 2012). Following its transfer to the cytoplasm, the resultant miRNA–miRNA+ duplex joins the RNA-induced-silencing-complex (RISC) with *ARGONAUTE 1(AGO1)* proteins, facilitates the silencing of the gene by cleaving mRNA that induces translational repression, are present in this complex (Gu et al., 2012; Pareek et al., 2015).

DOI: 10.1201/9781003583257-1

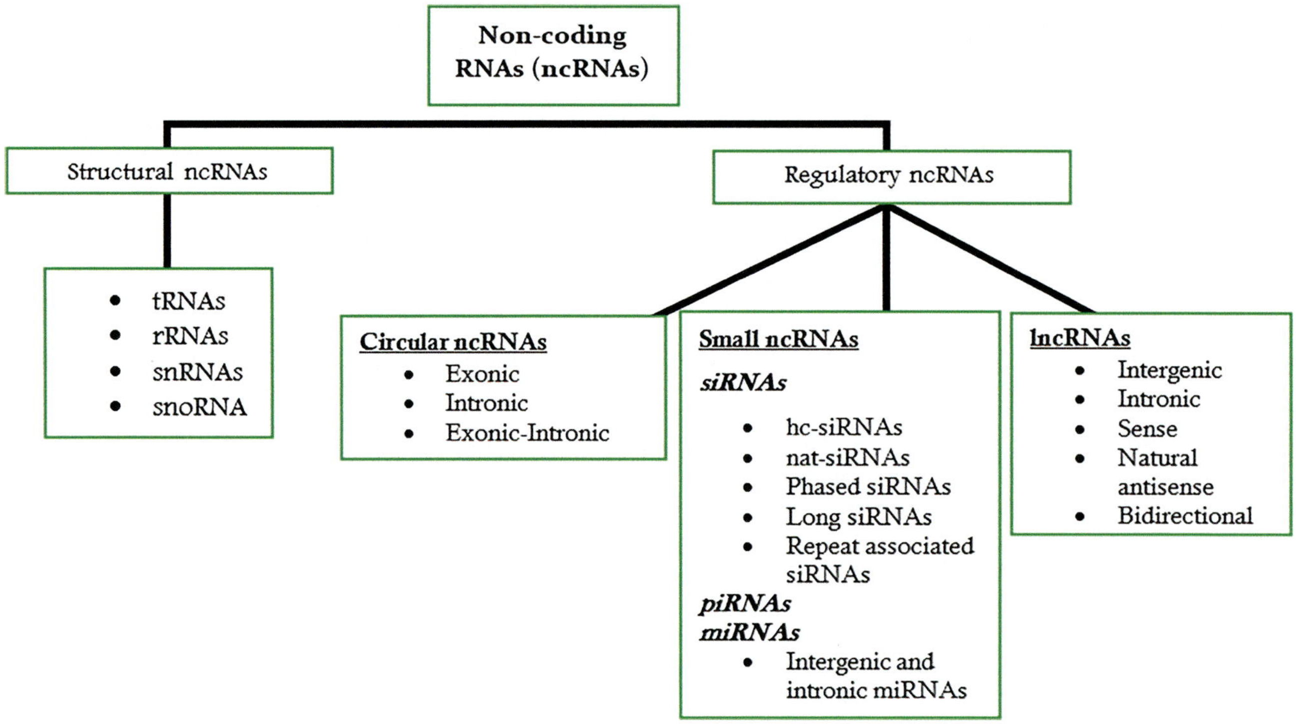

FIGURE 1.1 The classification of different non-coding RNAs.

1.2.3 Biogenesis of siRNA

Small interfering RNAs (siRNAs) can be either exogenous or endogenous, with the latter originating from complementary long double-stranded RNAs (dsRNAs) through the action of *RNA-DEPENDENT RNA POLYMERASE (RDR)* and subsequent cleavage by DCL1 proteins. Endogenous secondary siRNAs encompass various categories, including nat-siRNAs (natural antisense transcript-derived siRNAs), phasiRNAs (phased siRNAs), tasiRNAs or ta-siRNAs (trans-acting siRNAs) and hc-siRNAs (heterochromatic siRNAs). tasiRNAs function similarly to miRNAs, operating in cis- and affecting loci separate from their production region.

In contrast, hc-siRNAs predominantly function as cis-, targeting the homologous regions due to the alignment between their origin and target sites. nat-siRNAs are categorised into two such as cis- and trans-nat-siRNAs. Cis-nat-siRNAs are generated when overlapping RNAs from opposing strands at the same genetic location are transcribed, whereas trans-nat-siRNAs result from coding-transcripts that originate from distinct loci. The biogenesis of siRNAs is orchestrated through the formation of double-stranded RNAs (dsRNAs), the enzymatic activities of Dicer-like (DCL) proteins and the RISC that includes Argonaute (AGO) proteins, enabling regulation at post-transcriptional or transcriptional levels. The generation of ta-siRNAs relies on the coordinated action of SUPPRESSOR OF GENE-SILENCING3 (SGS3), DCL4 and RNA-dependent RNA polymerase 6 (RDR6)

TABLE 1.1
The Different Regulatory Roles of Long Non-Coding RNAs

Scaffolds	Guides	Signals	Decoys/Mimics
Through complementary secondary/tertiary structures, or sequence motifs, lncRNAs interact with proteins, promoting the assembly of larger functional units from smaller protein complexes.	Through interactions with chromatin-regulatory-complexes or transcriptional co-regulators, guide lncRNAs lead these molecules to targeted genome regions, influencing the expression of specific genes.	By binding to DNA-associated proteins, including histone-modifying complexes and transcription factors (TFs), lncRNAs can modify transcription, splicing, and translational processes to control the expression of gene.	By preventing transcriptional regulators from binding to their designated sites, decoy lncRNAs play a key role in regulating gene expression. In plants, lncRNAs harbour miRNA recognition regions that mimic the miRNA targets, effectively hindering miRNA activity by binding to the miRNAs and blocking their interaction with authentic targets.

TABLE 1.2
The Classification of Different Non-Coding RNAs

Non-Coding RNAs	Mode of Action	Target Enzymes	Function/ Properties	Examples	References
Micro RNAs	Inhibit target gene expression through mRNA cleavage and translational suppression.	MicroRNAs (miRNAs) recognise their target messenger RNAs (mRNAs) through complementary sequences, directing AGO1 to precisely cleave the target mRNA between the 10th and 11th nucleotides of the miRNA.	Shoot apical meristem, regulates leaf polarity, floral transition, tillering.	– miR156 targets SPL3. – miR165/166 regulate REVOLUTA. – miR172 modulates APETALA2. – miR398 governs COPPER/ZINC SUPEROXIDE DISMUTASE 1 (CSD1) and CSD2.	D'Ario et al. (2017), Silva et al. (2019), Yu et al. (2017)
TE (transposable-element)-derived-small-interfering RNAs	Transcriptional gene silencing through RNA-directed-NA-methylation.	*CLASSY* chromatin remodelling factors recruit RNA Pol IV to generate ss siRNA precursors. RDR2 along with DCL3 convert into dsRNAs thereby forms the complex and promotes the methylation.	Plant reproduction and hybridisation i.e., hybrid vigour and novel phenotypes.	In Arabidopsis, RdDM-deficient mutants often exhibit transposon de-repression, although transposition occurs only at a limited number of loci.	Das et al. (2015), Li et al. (2012), Martin et al. (2010), Swarup and Denyer (2019)
Phased siRNAs	SGS3 attaches to the 5′ or 3′ cleavage fragments following AGO-mediated slicing, recruiting RDR6 to convert the single-stranded cleavage fragments into double-stranded RNAs (dsRNAs). These dsRNAs are then processed into a series of 21- or 24-nucleotide small interfering RNAs (siRNAs) by DCL proteins.	In dicotyledonous plants, phasiRNAs are generated from protein-coding genes, such as NUCLEOTIDE-BINDING LEUCINE-RICH REPEAT (NB-LRR) and PENTATRICOPEPTIDE REPEAT (PPR) genes. In contrast, monocotyledonous plants produce phasiRNAs from long non-coding RNAs (lncRNAs) produced at PHAS (phasiRNA-generating) loci.	Regulate the plant thermotolerance, regulate the biological process of the plant.	A mutation in the OUTER CELL LAYER 4 (OCL4) gene in maize leads to defects in anther development.	Chen et al. (2010), Yu et al. (2017), Zhai et al. (2015)
Long non-coding RNAs	Alters their target genes' transcription, translation or epigenetic modifications.	Functioning as molecular scaffolds, it directs regulatory factors like chromatin remodellers to precise gene loci, thereby modulating gene expression. Acting as decoys, it binds proteins, diverting them from their intended targets to control the expression of gene.	These molecules regulate responses to various stressors, including cold, drought, heat, salinity and abscisic acid stress.	The tomato lncRNA16397, which responds to *Phytophthora infestans* infection, is a natural antisense transcript (NAT) of GLUTAREDOXIN 22. It acts in cis- to enhance GLUTAREDOXIN 22 expression, thereby increasing resistance to the pathogen.	Cui et al. (2017), Origin (2009), Plaza et al. (2017)

(Gasciolli et al., 2005; Peragine et al., 2004). In contrast, nat-siRNA production is dependent on SGS3, RDR6, DCL1, DCL2 and a plant-specific RNA polymerase (Pol II/Pol IV), particularly NRPD1A (Borsani et al., 2005). The synthesis of hc-siRNAs is driven by the activity of DCL3, Pol IV, NRPD1A, 5S rRNA genes, RDR2 (Mallory & Vaucheret, 2006; Zilberman et al., 2004). These distinct pathways underscore the intricate and highly specialised mechanisms of siRNA biogenesis and functionality.

1.2.4 Biogenesis of lncRNA

Long non-coding RNAs (lncRNAs) are identified by their extended length, exceeding 200 base pairs and their absence of protein-coding ability. Most lncRNAs, synthesised by RNA polymerase II (Pol II), possess a 5′ cap and a 3′ end that is either non-adenylated/poly-adenylated. Additional RNA polymerases, including Pol IV and Pol V, are similarly integral to lncRNA production (Wierzbicki et al., 2009). The lncRNAs generated by the enzymes Pol IV and Pol V are distinct in that they lack poly-A tails and are produced in smaller quantities compared to their Pol II counterparts. These particular lncRNAs are instrumental in facilitating RNA-directed-DNA-methylation (RdDM). Long non-coding RNAs (lncRNAs) modulate gene expression during both transcriptional and post-transcriptional phases through various mechanisms. These molecules exert their regulatory functions by acting as scaffolds, guides, signals or decoys in multiple biological processes. They operate as structural platforms by engaging with chromatin regulatory proteins, function as decoys for miRNAs (also referred to as sponges or target mimics) and contribute to epigenetic silencing processes (Table 1.1) (Wang et al., 2018).

1.2.5 Biogenesis of circRNAs

Circular RNAs (circ-RNAs) are a distinct category of non-coding RNAs (ncRNAs) that form closed, single-stranded loops. These molecules are found in various organisms, including eukaryotes, archaea and prokaryotes (Wang et al., 2018). First discovered in tomato viroids during the 1970s (Yu et al., 2017), unlike linear ncRNAs, circ-RNAs are generated through a process called back splicing during post-transcriptional modification of pre-mRNA, rather than forward splicing (Silva et al., 2019). The structure of circ-RNAs is characterised by a continuous loop that is covalently closed, lacking both the 5′ cap and the 3′ poly-A tail typically found in linear RNAs. This unique configuration grants circ-RNAs exceptional stability and resistance to degradation by Ribonuclease R (RNase R). This unique structure confers circRNAs with significantly longer half-lives compared to their linear counterparts (Chen et al., 2010). While the majority of circRNAs are derived from exons, some are also generated from introns or intronic sequences. Based on their genomic origin, circ-RNAs can be differentiated into several categories: exonic-circRNAs (EcircRNAs), exon-intron circRNAs (ElciRNAs), intergenic circRNAs, circular intronic RNAs (ciRNAs), intergene-intron circRNAs, intergene-exon circRNAs (Cui et al., 2017; Zhai et al., 2015). Additionally, circular RNAs (circ-RNAs) have been identified as microRNA (miRNA) sponges, interacting with miRNAs and competing against messenger RNAs (mRNAs) for miRNA binding sites. This interaction mitigates the binding of miRNAs to their target mRNAs, effectively enhancing the expression levels of the mRNAs. In the plant species Arabidopsis, researchers have discovered a circRNA derived from exon 6 of the SEPALLATA3 (SEP3) gene. This circRNA has been demonstrated to influence exon skipping by creating a circRNA:DNA R-loop. This process results in the suppression of SEP3 transcription and promotes the production of an alternatively spliced variant, SEP3.3, which is characterised by the absence of exon 6 (Chekanova, 2015).

1.3 THE REGULATORY FUNCTIONS OF NON-CODING RNAs IN PLANT DEVELOPMENT AND GROWTH

Non-coding RNAs (ncRNAs) such as siRNAs, siRNAs, and lncRNAs play essential regulatory roles in plant growth and development across various complex networks (Yu et al., 2019) (Table 1.3). These roles are particularly significant during crucial growth stages, including the transition from seed germination to vegetative and reproductive phases. Plant miRNAs regulate diverse biological processes by targeting various transcription factors (TFs) and corresponding genes involved in different pathways. Certain miRNA-target modules are essential regulators of plant developmental phase transitions, including *miR159-MYELOBLASTOSIS (MYB)*, *miR156-SQUAMOSA PROMOTER BINDING PROTEIN-LIKE (SPL)* and *miR172-APETALA 2 (AP2)*. These modules facilitate phase transitions by acting as either positive or negative regulators (Swarup & Denyer, 2019). Furthermore, miRNAs create regulatory networks that control the dormancy and germination processes in plants by interacting with plant hormones including gibberellic acid (GA) and abscisic acid (ABA) (Martin et al., 2010).

Long non-coding RNAs (lncRNAs) play a vital role in various plant developmental processes by modulating the expression of genes in close proximity through cis-acting mechanisms and distant genes via trans-acting modes (Mandal et al., 2023; Yu et al., 2019). These regulatory mechanisms include chromatin/histone modifications via protein recruitment, miRNA mimicry, transcriptional control, silencing and post-translational modifications.

1.4 RESPONSE OF ncRNA TO PLANT STRESS RESPONSE

To adapt to environmental stressors, plants have evolved diverse mechanisms for adjusting their growth and development. Recent research has revealed that non-coding RNAs (ncRNAs) show differential expression under adverse conditions (Figure 1.2). These ncRNAs are crucial in directly responding to stress or fine-tuning gene expression within complex cellular networks. Both biotic and abiotic stressors activate various regulatory ncRNAs, which coordinate the modulation of key stress-responsive pathways through targeted transcripts. MicroRNA-driven regulation of gene expression has been widely documented across a range of stress conditions, including drought, heat, salinity, cold, nutrient deficiencies and pathogen infections (Table 1.4). This dynamic regulation has been observed in model organisms and significant crop plants like Arabidopsis, barley, maize, rice and wheat (Ferdous et al., 2017; He et al., 2019; Hua et al., 2019; Sailaja et al., 2014).

TABLE 1.3
The Examples of miRNAs Involved in the Plant Growth and Development

miRNAs	Target	Response	Reference
miR156-SPL	Acts as a negative regulator of plant growth transitions, where the reduced levels cause SPL expression to quicken the transition	Plant growth transition from germination to vegetative to reproductive stages	Swarup and Denyer (2019)
miR172-AP2	Where the miR172 acts in reverse as the increased levels decreases the AP2 expression and helps transition		
miR159	Disrupts the equilibrium between ABA and GA hormones, asserting control over positive signalling pathways via MYB transcription factors like MYB33 and MYB101	Controls the seed dormancy and germination	Martin et al. (2010)
miR160	Inhibits the expression of *AUXIN-RESPONSE-FACTORS (ARFs)* in rice	Controls seed germination	Das et al. (2015)
TAS3-tasiRNA resultant from miR390-AGO7	ARF family factors	Embryo growth, shoot apical meristem, leaf morphology, root structure	Yu et al. (2019)
miR164	NAC1 in maize and wheat NAC262 in potato	Promotes the lateral root development	Li et al. (2012, 2017)
miR394	Suppression of FLC by positively regulating FT and SOC1 in Arabidopsis	Flowering time	Bernardi et al. (2022)
miR840	Knocking down PPR and WHY3 in Arabidopsis	Regulates leaf senescence	Ren et al. (2022)
TAS4-tasiRNA triggered by miR828	MYB-genes (*PRODUCTION OF ANTHOCYANIN-PIGMENT 1* (PAP1), PAP2, and MYB113	Regulation of anthocyanin pathway	Zhou et al. (2013)
PINOID (PID), an auxin transporter gene	Through *AUXIN-REGULATED PROMOTER LOOP (APOLO) and* lincRNA (long-intergenic non-coding RNA)	Root growth in Arabidopsis	Ariel et al. (2015)
COOLAIR COLDAIR	Suppression of FLC by these lncRNAs recruiting (*PHD-PRC2*) to assist in histone modification by forming intergenic chromatin loop	Regulates flowering time and seed dormancy	Chen and Penfield (2018)
HID 1 ***Hidden Treasure 1***	Regulates the expression of phytochrome-interacting-factor 3 (PIF3) to inhibit the green coloration of cotyledons	Promotes photomorphogenesis	Wang et al. (2014)

1.4.1 Drought

Drought severely impacts plant growth and development by altering water relations, reducing photosynthetic efficiency, accumulating reactive oxygen species (ROS), and changing enzyme activities. Drought induces osmotic stress in plants, regulating inorganic and organic metabolites, which results in ionic toxicity, ROS accumulation, plasma membrane disruption and cell wall damage (Praveen et al., 2023).

For example, in potatoes under drought conditions, stu-miR398b-3p negatively controls the antioxidant enzyme *Cu/Zn superoxide-dismutase* (SOD) (Figure 1.3). SOD, which scavenges active oxygen during drought, shows reduced activity in lines overexpressing stu-miR398b-3p, causing quicker wilting compared to wild-type plants without irrigation (Qin et al., 2017; Zheng et al., 2023). In Arabidopsis, researchers discovered the long non-coding RNA (lncRNA) DRIR, which enhances responses to salt and drought stress. DRIR, typically expressed at low levels, is significantly upregulated during stress and ABA treatment, underscoring its role in stress response. Additionally, *lncRNA35557* serves as an endogenous target mimic (eTM) for miRNA tae-miR6206, preventing its cleavage of the NAC transcription factor *TaNAC018*, thereby enhancing drought tolerance by increasing TaNAC018 transcript levels (Xu et al., 2024).

1.4.2 Heat

Heat stress is progressively undermining plant growth and development due to the escalating impacts of global warming and climate change. Heat stress causes protein denaturation and triggers the expression of heat shock proteins (HSPs) in various plant compartments, including the cytoplasm, endoplasmic reticulum and mitochondria (Figure 1.3). Many HSPs function as molecular chaperones, preventing protein denaturation and maintaining proteostasis (Lin et al., 2001; Scharf et al., 2012).

In Arabidopsis, the miR165/166-PHB (PHABULOSA) pathway is crucial in regulating HSFA1, a key factor in heat stress response (Zheng et al., 2023). Plants overexpressing miR165/166 showed markedly improved heat tolerance. MiR165/166 suppresses PHB, which in turn reduces the transcription of HSFA1, the primary regulator of plant heat stress defence. Additionally, overexpressing cucumber (*Cucumis sativus*) miR9748 improved heat resistance in Arabidopsis. CsbZIP2 (BASIC-LEUCINE ZIPPER 2) activates miR9748 expression by binding to its promoter, while miR9748 cleaves CsNPF4.4 (NITRATE TRANSPORTER 1/PEPTIDE TRANSPORTER 4.4), which negatively impacts heat tolerance by inhibiting the jasmonic acid (JA) signalling pathway (Li et al., 2022).

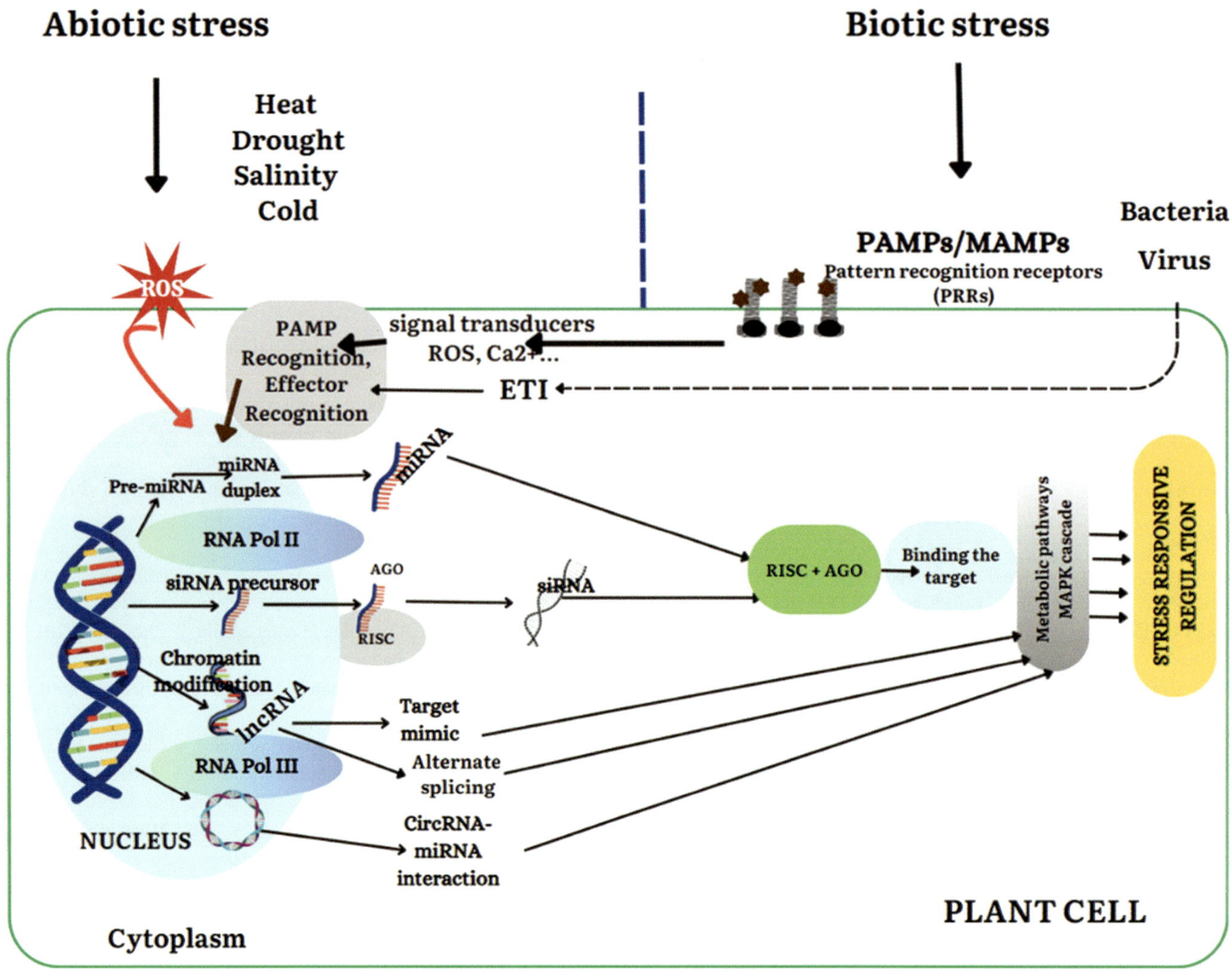

FIGURE 1.2 The general model of mechanism of abiotic and biotic stress in the plant cell. Plants respond to abiotic and biotic stresses by generating reactive oxygen species (ROS) and recognising pattern-associated-molecular-patterns (PAMPs). As a defence mechanism, various non-coding RNAs (ncRNAs) are synthesised in response to ROS and PAMPs. Stress-induced miRNAs and small-ncRNAs combine with Argonaute (AGO) proteins to form RNA-induced-silencing-complexes (RISC), which regulate gene expression. Long non-coding RNAs (lncRNAs) influence gene expression through miRNA target mimicry, alternative splicing, or chromatin modification. Circular RNAs (circRNAs) function as miRNA sponges to modulate gene expression. (The illustration is made through Canva.)

1.4.3 Cold

Frost and low temperatures threaten plant survival, growth and yield. To combat these challenges, plants utilise a variety of sophisticated adaptive strategies, including the regulation of ncRNAs (Figure 1.3). ncRNAs regulate gene expression in response to low temperatures. For instance, miR319 enhances cold hinderance by targeting *OsPCF6 and OsTCP21* in rice (Wang et al., 2014). Similarly, miR1320 overexpression boosts cold tolerance in rice by targeting *AP2/ERF TF OsERF096*, which represses the JA-mediated signalling pathway (Sun et al., 2022). The significant lncRNAs were identified in various plants. In Arabidopsis, lncRNA SVALKA negatively influences freezing tolerance by controlling CBF1 gene expression, with CBFs being conserved TFs that enhance cold tolerance across plant species (Kindgren et al., 2018).

1.4.4 Salinity

Globally, salinity hampers plant growth and fertility by causing hyperosmotic stress, disrupting ion balance and generating oxidative stress (Kumar et al., 2018). Plants under high salinity and drought stress experience hyperosmotic conditions, either directly or indirectly. These stressed plants accumulate increased organic and inorganic substances, resulting in higher solute concentrations (Sun et al., 2022; Wang et al., 2014) (Figure 1.3). Osmoregulation involves reducing osmotic potential to retain water. This process utilises two types of substances: inorganic ions (K^+, Na^+ and Ca^{2+}) and organic solutes (proline, betaine, SSs and polyols) (Munns et al., 2020). During stress, plants produce adversity proteins, also called osmoregulatory proteins, which help lower cellular osmotic potential and prevent dehydration, thereby

TABLE 1.4
The List of Non-Coding RNAs Involved in the Plant Stress Response

Stress	Crop	Non-Coding RNA	Function	Reference
Abiotic	Rice	*miR319*	Cold stress	Wang et al. (2014)
	Rice	*miR166*	Drought stress	Salvador-Guirao et al. (2018)
	Rice	*miR162a, miR390, miR171b, miR166m*	Heavy metal (cadmium stress)	Ding et al. (2011)
	Rice	miR169o	Nitrogen use efficiency	Yu et al. (2018)
	Wheat	*miR156, miR160*	Heat and salinity stress	Pandey et al. (2014)
	Soybean	*miR172c*	Drought and saline responsive	Li et al. (2016)
	Soybean	*miR167d, miR169d, miR4369*	Drought and saline stress responsive	Li et al. (2011)
	Arabidopsis (thale cress)	miR397	Salt stress tolerance	Nguyen et al. (2020)
	Arabidopsis (thale cress)	*miR393*	Salinity stress	Denver and Ullah (2019)
	Arabidopsis (thale cress)	*lncRNA SVALKA*	Cold tolerance	Kindgren et al. (2018)
	Wheat	*lncRNA117, lncRNA616*	Cold tolerance	Lu et al. (2020)
Biotic stress	Rice	*miR535*	Against *Magnaporthe oryzae*	Zhang et al. (2022)
	Pigeon pea	*CcmiR3695, CcmiR8182, CcmiR1214*	Fusarium udam infection (Fusarium wilt)	Hussain et al. (2018)
	Rapeseed	*miR156b*	Pseudomonas Brassicaceae infection	Verma et al. (2014)
	Sweet potato	miR408	Against pest *Spodoptera litura*	Kuo et al. (2019)
	Tomato	*lncRNA16397*	Resistance against *Phytophthora infestans*	Cui et al. (2019)
	Cotton	*miR156*	Resistance to bollworm	Mao et al. (2017)

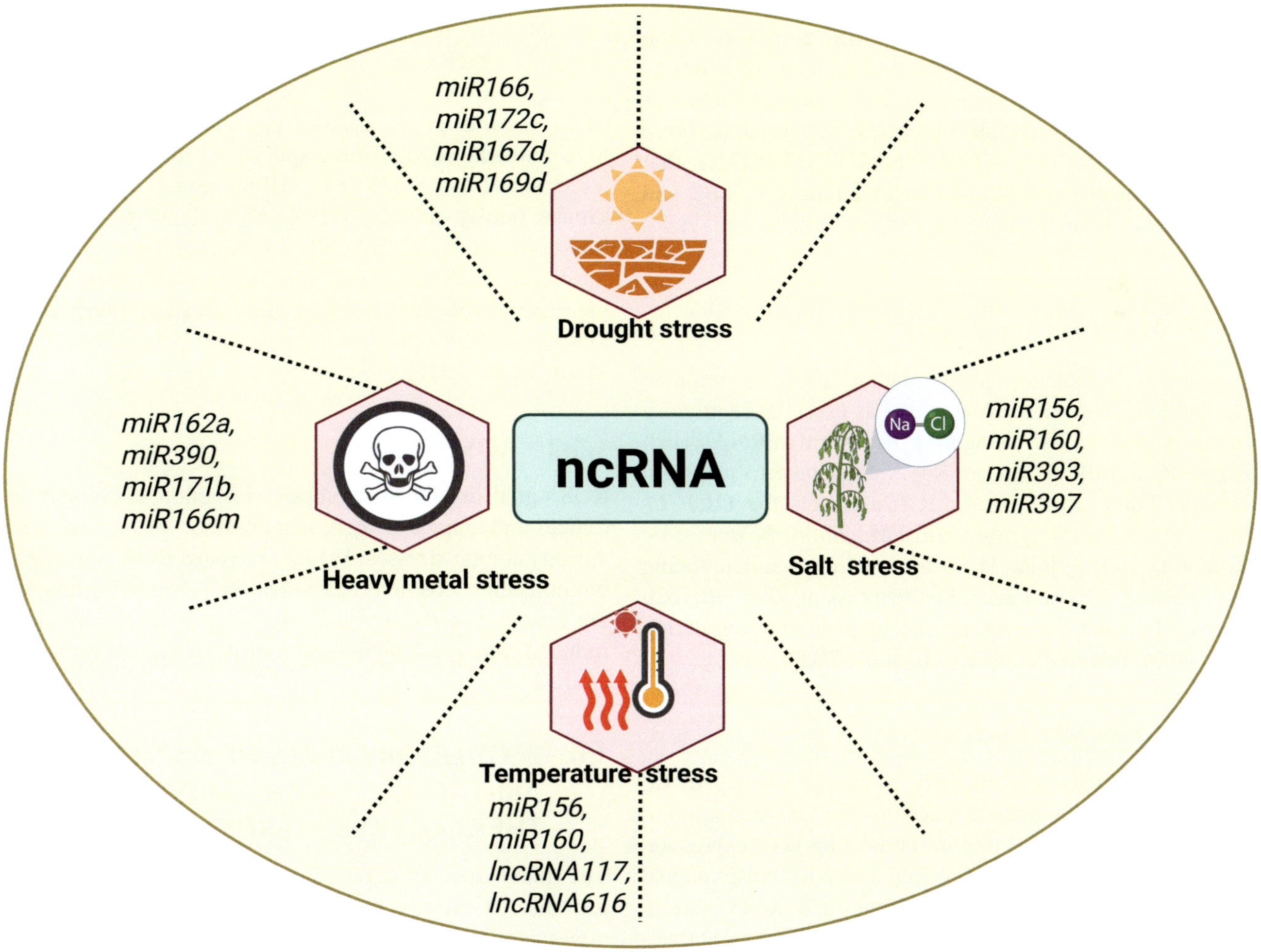

FIGURE 1.3 ncRNAs for the climate adaptation.

enhancing plant resistance to osmotic stress (Hou et al., 2016). Elevated soil salinity results in osmotic stress, ion toxicity and oxidative stress in plants (Zhu, 2002). To mitigate these effects, plants employ various regulatory mechanisms, including stomatal control, osmolyte synthesis, excess ion compartmentalisation and ROS scavenging. The osmotic stress signal triggered by salt stress enhances plant tolerance to high salt and drought conditions by limiting stomatal opening, thus reducing water loss. Under normal conditions, plants effectively utilise Na+ and Cl^- in vesicles and organic solutes in vivo for osmoregulation, avoiding ion toxicity. However, in high-salinity environments, plants require modified water–salt movement pathways to achieve energy-efficient osmoregulation (Zhang et al., 2022).

Arabidopsis may adapt to saline conditions through MiR393 biogenesis, which regulates RACK1A, a negative modulator of ABA-mediated stress pathways. In Arabidopsis, natsiRNAs are involved in salt stress response, as evidenced by a 24-nt-long nat-siRNA associated with SRO5 under saline conditions, targeting the P5CDH gene. Salt stress triggers SRO5 protein production, which decreases P5CDH activity, leading to heightened proline and ROS accumulation. This underscores the significance of the SRO5-P5CDH nat-siRNA in plants' salt stress response (Kumar et al., 2018). Similarly, Jiang et al. (2024) identified molecular mechanisms of ceRNAs in rice's saline stress response, offering genetic resources for breeding salt-resistant rice. Researchers discovered two lncRNAs, *MSTRG.51634.2* and *MSTRG.48576.1*, that interact with osa-miR172d-5p to modulate *OsMYB2* and *OsMADS63* genes, influencing salt tolerance and yield, respectively.

1.4.5 Pathogens and Insects

To combat pathogens, plants deploy intricate immune mechanisms, with ncRNAs serving as essential regulatory factors (Cui et al., 2017) (Figure 1.3). Several miRNAs, such as miR162a, miR535 and miR482, contribute to pathogen defence. For instance, osa-miR162a targets the *OsDCL1* gene, enhancing immunity against *Magnaporthe oryzae* by promoting intracellular H_2O_2 accumulation and inducing cell death at infection sites. Overexpressing *Osa-miR162a* increased resistance to *M. oryzae* by positively regulating various defence-related genes (Li et al., 2020).

1.4.6 Heavy Metal Stress

Heavy metal toxicity critically threatens plant growth, prompting complex defence mechanisms. The regulation of ncRNAs plays a crucial role in modulating gene expression during stress at both transcriptional and post-transcriptional levels (Figure 1.3). MiR398, central to oxidative stress regulation, is suppressed under heavy metal stress, activating CSD1 and CSD2 (Cu/Zn superoxide dismutase genes) to neutralise ROS induced by metals like Cu and Cd. During arsenic (As) stress, miR528 regulates L-ascorbate oxidase, adjusting antioxidant defences (Liu et al., 2015). Additionally, siRNAs target transcripts involved in phytochelatin synthesis (Çelik et al., 2020), crucial for heavy metal chelation and detoxification, emphasising siRNAs' role in regulating detoxification-related genes. Bhattacharya and Roychoudhury (2023) underscore ncRNAs' essential functions in mitigating both salt and heavy metal stress in plants.

1.5 MOLECULAR SIGNALLING OF PLANTS AGAINST ABIOTIC STRESS

1.5.1 Ca^{2+} Signalling

Plants typically respond to cold, drought and high salinity stress with transient increases in intracellular Ca^{2+} concentrations, a universal second messenger in primary stress signals that varies in intensity, frequency and subcellular location (Dong et al., 2020; Zhu, 2016) (Figure 1.3). The cytoplasm's low Ca^{2+} concentration compared to internal or external cellular spaces allows for rapid changes (Martí et al., 2013). Numerous proteins can detect and interpret these Ca^{2+} changes, with specific Ca^{2+} osmotic channels or transporters precisely controlling these fluctuations. Calcium signals are sensed by calcium-binding proteins, which either interact with an associated protein kinase or a directly fused kinase, known as CDPKs or CPKs (Brandt et al., 2015). In *Arabidopsis thaliana*, the SOS pathway recognises a specific cytosolic Ca^{2+} signal under salt stress. SOS3 or SCaBP8 interacts with a SnRK3 kinase family protein (CIPK) to activate SOS2. Plants possess various SCaBP/CBL-CIPK complexes, similar to the SOS3–SOS2 module, which play vital roles in Ca^{2+}-mediated responses to abiotic stresses, particularly in regulating ion transport protein activity (Hrabak et al., 2003; Zhang et al., 2022).

1.5.2 ROS Signalling

ROS signalling, involving superoxide anion, H_2O_2, hydroxyl radical and singlet oxygen, is a common plant response to various abiotic stresses. While excessive ROS can damage biomolecules if cellular detoxification is overwhelmed, they are essential in stress signalling, such as high light stress-induced retrograde signalling and ABA signalling (Zhang et al., 2022).

1.6 REGULATORY ROLES OF ncRNAs TO ABIOTIC AND BIOTIC STRESS WITH THE CONSERVED TARGET SEQUENCE

The plant species show diverse miRNA expression in response to various stressors. However, specific miRNA-target modules exhibit distinct expression patterns when confronted with particular stresses, influencing the regulation of target genes (Figure 1.4). These expression patterns

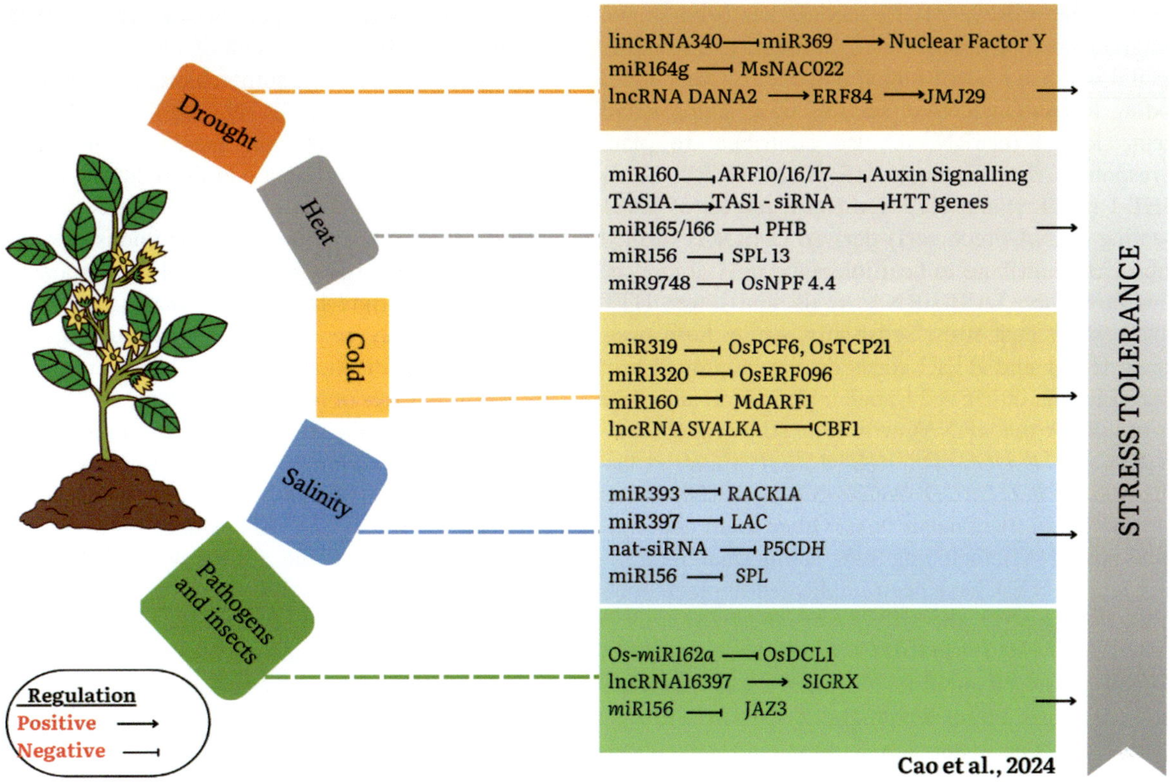

FIGURE 1.4 Regulatory roles of ncRNAs against abiotic and biotic stress mechanism and with their targets involved during stress. (The illustration is made through Canva.)

can be conserved across different plant species (Song et al., 2019). The upregulation of miRNAs under stress conditions may suppress their target genes, which are often negative regulators of stress, such as repressors of stress-responsive genes. On the other hand, stress-induced downregulation of miRNAs can lead to the accumulation of their target mRNAs, potentially enhancing stress tolerance (Sunkar et al., 2007).

Conserved miRNA target modules are vital for both plant growth and development and enhancing stress tolerance by interacting with metabolic pathways. Established miRNA-target modules, including miR159-MYB, miR156-SPL, miR319-TCP *(TEOSINTE BRANCHED/CYCLOIDEA/PROLIFERATING CELL FACTORS)*, miR164-NAC *(NAM/ATAF/CUC)*, miR167-ARF, miR160-ARF, miR169-NFY *(NUCLEAR TRANSCRIPTION FACTOR-Y)*, miR396-GRF, miR398-CSD and miR394-LCR, exert significant regulatory influence under various stress conditions to counteract harmful effects.

The miR167-ARF signalling module is crucial in regulating auxin pathways during drought stress by targeting ARF6 and ARF8, which negatively impact auxin-signalling through *GRETCHEN HAGEN 3 (GH3)*. Drought conditions increase miR167 expression in Arabidopsis, wheat and maize, while decreasing it in rice (Song et al., 2019). The miR169-NFY module also plays a significant role in plant water-deficit response, with reduced miR169 expression resulting in increased NF-Y expression in Arabidopsis, tomato and Medicago. Higher NF-Y levels in guard cells help improve drought resistance by regulating stomatal closure. In Arabidopsis, increased miR398 expression boosts heat tolerance by suppressing the expression of *CSD1*, *CSD2* and *COPPER CHAPERONE OF CSD (CCD)* (Guan et al., 2013; Lu et al., 2013). This allows for the accumulation of HSPs and heat shock transcription factors (HSFs). Other modules, including miR156-SPL and miR172-AP2, contribute to plant heat stress response (Song et al., 2019).

The highly conserved miR394-LCR module is involved in plant cold stress response. Arabidopsis plants overexpressing miR394a exhibit cold resistance by negatively regulating the LCR gene. Additionally, the enhanced expression of genes encoding *CBFs or DREB1* in plants with overexpressed miR394 and lcr mutations shows cold stress tolerance, suggesting that miR394 positively regulates cold stress tolerance through a CBF-dependent pathway (Song et al., 2016). Other studied modules include miR399-PHO2 during phosphate scarcity, miR827-NLA and miR395-SULTR2 in sulphur uptake and transport, miR169-NF-Y in nitrogen deficiency (Song et al., 2019). In Arabidopsis, the miR393-TIR1, AFB2 and AFB3 pathway was the first discovered as a key regulatory module in defending against pathogen *Pseudomonas syringae* pv. tomato DC3000. Upon exposure to the bacterial PAMP flagellin (flg22), elevated expression of miR393 effectively suppresses TIR1, AFB2 and AFB3, boosting resistance to the bacteria.

Similarly, the miR773-MET2 module has been shown to strengthen pathogen-associated immunity, providing enhanced resistance against fungal pathogens. Other small non-coding RNAs (sncRNAs), such as trans-acting small interfering RNAs (tasiRNAs), also contribute to plant stress responses. For instance, HEAT-INDUCED TAS1 TARGET 1 (HTT 1) and HTT 2 mRNA, targets of TAS1 (trans-acting siRNA-precursorl)-derived tasiRNAs formed from miR173, contribute to heat tolerance in Arabidopsis. Plants with higher TAS1-siRNA levels and lower HTT gene levels show heat stress sensitivity, while those overexpressing HTT1 and HTT2 genes display improved heat tolerance (Li et al., 2016). A 24-nucleotide natural antisense transcript siRNA (natsiRNA), in which *D1-PYRROLINE-5-CARBOXYLATE DEHYDROGENASE (P5CDH)* is targeted by *SIMILAR-TO-RCD-ONE 5 (SRO5)* mRNA, leads to the formation of 21 nt natsiRNAs. Other small non-coding RNAs (sncRNAs), including trans-acting small interfering RNAs (tasiRNAs), have been implicated in plant stress responses. For example, the mRNAs of *HEAT INDUCED TAS1 TARGET 1* (*HTT1*) and *HTT 2*, which are regulated by tasiRNAs derived from miR173, contribute to heat tolerance in Arabidopsis. Plants with higher TAS1-siRNA levels and lower HTT gene expression show greater susceptibility to heat stress, whereas those overexpressing HTT1 and HTT2 genes demonstrate increased heat resistance (Li et al., 2016). Furthermore, 24-nt natural antisense transcript siRNAs (natsiRNAs) derived from *SIMILAR-TO-RCD-ONE 5 (SRO5)* mRNA target *D1-PYRROLINE-5-CARBOXYLATE DEHYDROGENASE (P5CDH)*, resulting in 21-nt natsiRNA formation. These natsiRNAs are involved in P5CDH mRNA cleavage. Under salt stress, SRO5 protein induction causes a decrease in P5CDH activity, leading to proline and ROS accumulation. This details how SRO5 natsiRNAs, along with their proteins, contribute to osmo-protection and the mitigation of oxidative stress in salt stress situations (Borsani et al., 2005; Khraiwesh et al., 2012). The plant-endogenous siRNA, nat-siRNAATGB2, was the first discovered to regulate R-gene-mediated effector-triggered immunity (ETI) during bacterial infection by *Pseudomonas syringae* (Ps) (Navarro et al., 2006). PhasiRNAs derived from PPR genes in *Arabidopsis* contribute to the plant's defence mechanism against *Phytophthora capsica* infection.

Various plant species have been found to contain lncRNAs that respond to different environmental stresses, both abiotic and biotic. For example, researchers have identified lncRNAs that react to drought conditions in several plants, including Arabidopsis, populus, maize and rice (Chung et al., 2016; Qin et al., 2017). Extensive research has been conducted on the model plant *A. thaliana* examines the effects of various stressors on lncRNAs. Arabidopsis and other *Brassicaceae* plants harbour some of the most well-studied stress-responsive lncRNAs, including COOLAIR, COLDAIR and COLDWRAP, which collectively suppress FLC expression to promote flowering during cold periods. Under phosphorus-deficient conditions in Arabidopsis, the lncRNA IPS1 was shown to imitate miRNA399, preventing its interaction with its target *PHOSPHATE2(PHO2)*, thus reducing miR399's suppression of PHO_2 (Wu et al., 2013). In Arabidopsis, DRIR, another lncRNA, responds to salt and drought stress by positively regulating the expression of multiple genes. Plants with increased DRIR expression exhibited enhanced tolerance to drought and salinity (Qin et al., 2017).

Under high temperature stress conditions in Arabidopsis, a substantial proportion (40–50%) of light-responsive sense/antisense transcripts linked to H3K9ac and/or H3K27ac were significantly regulated, underscoring the vital role of histone modification and acetylation in light-induced expression changes of NATs (Wang et al., 2014). The RdDM pathway mediates the response of certain lncRNAs to environmental stress. In tomatoes, downregulating SlAGO4A, a pivotal element of this pathway, significantly strengthened resistance to drought and salt stress, outperforming both wild-type and transgenic plants with *SlAGO4A* overexpression (Huang et al., 2016). High-throughput sequencing technologies have identified the expression of stress-responsive circRNAs. Studies in rice have shown that circRNAs play a crucial role in the response to the fungal pathogen *M. oryzae*. The resistant genotype (IR25) exhibits considerable circRNA diversity during *M. oryzae* infection, attributed to increased 3′ and 5′ alternative back splicing and splice site complexity. Moreover, circR5g05160 has been shown to significantly boost immunity against *M. oryzae* (Fan et al., 2020).

1.7 ABIOTIC AND BIOTIC STRESS RESILIENCE USING RNAi AS A TOOL

1.7.1 Biotic Stress Improvement

1.7.1.1 RNAi for Insect Resistance

The RNAi technique has been utilised to reduce crop damage inflicted by insect pests (Kanakala & Ghanim, 2016). Recent research has emphasised host-targeted gene silencing, with numerous RNAi-based transgenes developed and tested as potential insect pest control methods, given RNAi's significant promise in this field (List et al., 2022; Upadhyay et al., 2011). The siRNA pathway could facilitate the development of novel insect pest control products (Niu et al., 2018; Upadhyay et al., 2011). One approach involves using double-stranded RNA (dsRNA) in a BioClay spray to target the sap-sucking whitefly pest, disrupting various stages of its life cycle directly within the plant. This method combines clay-mediated dsRNA delivery with enhanced foliar absorption through adjuvants (Jain et al., 2022). Various novel strategies for dsRNA delivery have been crafted by researchers, encompassing symbiont-mediated techniques, cationic liposome-assisted, nanoparticle-enabled, peptide nanomaterial and plant-mediated (Avila et al., 2018; Zhu & Palli, 2020). Nevertheless, widespread application of RNAi in pest management faces challenges such as varying RNAi effectiveness among insects, limited reliable dsRNA off-target and non-target effects and potential insect population resistance. Whitefly protection was achieved by genetically modifying tobacco to produce siRNA targeting the *Bemisia tabaci* V-ATPaseA gene, establishing insect resistance in the host plant. These transgenic tobacco lines generated dsRNA precursors to form siRNA and silence the V-ATPaseA gene in *B. tabaci* (Tsuji et al., 2006). Reduced V-ATPaseA transcript levels were observed in *B. tabaci* feeding on these plants. These transgenic lines, expressing V-ATPaseA dsRNA, provided siRNA to induce a potent silencing response in whiteflies, leading to their mortality (Nidhi Thakur et al., 2014). *B. tabaci* feeding on these plants showed reduced V-ATPaseA transcript levels. These transgenic lines, expressing V-ATPaseA dsRNA, provided siRNA to induce a strong silencing response in whiteflies, resulting in their mortality. Using a similar method, transgenic corn plants were developed to express dsRNAs targeting tubulin or V-ATPase genes, offering protection against western corn rootworm (WCR) (Baum et al., 2007).

1.7.1.2 RNAi for Phytopathogenic Resistance

Research has shown that host gene silencing through hpRNAi is a more effective method for gene silencing in plants, potentially enhancing resistance to bacterial and fungal diseases by altering pathogenic gene expression in the host plant through genetic manipulation (Senthil-Kumar & Mysore, 2010). Research has shown that using RNAi to suppress *OsSSI2*, a gene responsible for fatty-acid-desaturase activity, has been associated with increased resistance against the leaf blight-causing bacteria (*Xanthomonas oryzae* pv. *Oryzae*) and blast (*Magnaporthe grisea*) pathogens (Jiang et al., 2009). In wheat and barley, host-induced gene silencing has been shown to limit fungal disease attachment by leveraging transient gene regulation, overcoming RNAi resistance caused by silent point alterations. This RNAi-based defence mechanism counter to the disease is a result of RNA transfer from the host plant to the fungus *Blumeria graminis* (Escobar et al., 2001).

1.7.1.3 RNAi for Virus Resistance

RNA interference (RNAi) strategies are employed to exhibit virus-resistant transgenic plants. This procedure contains the creation of double-stranded RNA through an RNAi vector containing viral target genes, which inhibits virus replication within plant cells. The antiviral defence mechanism in plants based on RNAi is orchestrated by multiple Dicer-like enzymes that generate small interfering RNAs (siRNAs) from dsRNA during the replication and/or transcription of DNA and RNA viruses. By associating with Argonaute enzymes, these siRNAs target viral DNA, leading to gene silencing at both the transcriptional and post-transcriptional expression-levels (Pooggin, 2017). Targeting coat protein (CP) genes via RNAi has proven to be an effective strategy for engineering virus resistance in a variety of plants. As an example, papaya genetically modified to carry the CP gene of papaya ringspot-virus typeW (PRSV-W) exhibited resistance to PRSV-W due to the expressed CP (Krubphachaya et al., 2007). This strategy was later utilised to develop resistance to additional viruses, such as banana bract mosaic virus (BBrMV) (Rodoni et al., 1999).

1.7.2 Abiotic Stress Improvement

Plant growth is significantly impacted by abiotic stresses, including extreme temperature, drought, salinity, flood, metal and nutrient imbalances, with temperature, drought and salinity being particularly impactful (Zhang et al., 2022).

1.7.2.1 RNAi for Drought Stress Resistance

Gene silencing through RNAi serves as a protective mechanism counter to non-living environment stresses in plants. miRNAs and siRNAs play pivotal roles in plant responses to abiotic stress, challenging agricultural productivity. The RNA interference (RNAi) process mediates various abiotic stress responses, such as osmo-protectant production, ABA signalling, antioxidant defence and auxin regulation, through targeted gene suppression. Several conserved miRNAs, including miR396, miR169, miR164, miR171, miR2118, miR398, miR399 and miR408, play crucial roles in drought stress adaptation, showing varied expression patterns across different plant species (Banerjee, 2020).

1.7.2.2 RNAi for ABA and Salt Stress Resistance

Targeted RNAi-mediated suppression of FTA in *Brassica napus* significantly boosted drought resistance. MicroRNAs are essential in how plants respond to non-living environmental stresses (Frazier et al., 2011). Genes encoding proteins such as APX and ADH are regulated, influencing the plant's responses to drought and salt stress. According to Liu et al. (2022), a minor rise in internal ABA concentrations, triggered by RNAi-mediated inhibition of OsABA8ox1 (key gene), improved rice plants' lenience to salt–alkali conditions. Boosting internal ABA levels through the inhibition of OsABA8ox1 presents a promising molecular strategy to enhance rice yields in saline-alkali fields, as suggested by these results.

1.7.2.3 RNAi for Heat Stress Resistance

Heat-sensitive transgenic plants expressing miR398-resistant forms of csd1, csd2 or ccs demonstrate reduced expression of genes encoding HSPs and heat stress transcription factors (HSF). However, these genes exhibit enhanced expression in Arabidopsis plants with heat-tolerant csd1, csd2 and ccs (Guan et al., 2013). The suppression of SlMPK1 in tomato plants exhibited enhanced high-temperature tolerance in transgenic varieties, while its overexpression exhibited reduced tolerance (Ding et al., 2018).

1.8 APPLICATION OF CRISPR IN PLANT STRESS RESILIENCE USING ncRNA

CRISPR (Clustered Regularly Interspaced Short Palindromic Repeats) systems are prevalent in prokaryotic organisms, occurring in archaea and bacteria (Jansen et al., 2002). These systems comprise identical repetitive DNA sequences linked with Cas proteins is separated by highly variable spacer sequences (Horvath & Barrangou, 2010; Jansen et al., 2002). CRISPR and Cas proteins form an adaptive immune mechanism in prokaryotes that hinders the entry of the pathogen (Horvath & Barrangou, 2010). This dogmatic finding could make CRISPR/Cas system as a potential tool for DNA and RNA modification across various organisms. CRISPR/Cas systems are categorised into two classes and classes further subdivided into types: class 1 encompasses types I, III and IV, featuring a multi-protein effector complex, and class 2 encompasses types II, V and VI, utilising a single protein for target editing (Koonin et al., 2017; Makarova et al., 2015). A CRISPR/Cas tool consists of a guide RNA (gRNA or CRISPR RNA (crRNA)) that complements the target gene site and a Cas effector protein functioning as an endonuclease to cleave the target (Jinek et al., 2012). Due to their simpler organisation, class 2 Cas systems are more commonly employed as genome editing tools. Within this class, type II and V systems utilise Cas9 and Cas12 enzymes, for DNA editing. Notably, the class 2 CRISPR/Cas9 protein of type II was one of the first Cas proteins to be extensively studied, leading to its widespread application in DNA editing across bacteria, plants and animals.

The events followed in creating the CRISPR-Cas9 genome edited plant viz, the selection of the target gene, designing the single-guide RNA and assessing its expression, assembly of sgRNA to Cas9 forming a complex, target delivery into plants through agrobacterium mediated/protoplast transformation/PEG mediated delivery, target DNA cleavage by sgRNA and Cas9 complex, regeneration and screening of the edited plants for the target events.

1.9 CRISPR-ca-BASED EPIGENETIC REGULATION IN PLANTS

The heritable alterations, caused by activation or suppression of genes without alteration in the DNA sequences, are known as epigenetic changes. Epigenetic changes cause DNA modifications or structural changes in chromatin rather than DNA sequence changes. This happens when plants are exposed to various environmental stresses (Shanker et al., 2021). The plant senses the environmental signals through and changes associated with the regulation of gene through plethora of mechanisms that resisting against the multiple stress conditions (Probst & Scheid, 2015) (Table 1.5; Figure 1.5). In advent of these stresses, the plants undergo several modifications in plants at the level of physiological, morphological and biochemical that changes the activity of the plants in photosynthetic rate, decreased plant growth. Genetic modification in plants is demonstrated either for the overexpression/downregulation of the desired genes in crops (Crisp et al., 2016; Hirayama & Shinozaki, 2010; Lippmann et al., 2019; Shanker et al., 2021). Integration of gene into the host would deter the transcriptional, translational expression levels and that leads to

TABLE 1.5
Application of CRISPR in Plant Stress Resilience Using ncRNA

Plant	Target gene	Stress Response	Reference
Rice	*OsNAC14*	Drought tolerant	Shim et al. (2018)
Rice	*OsDST*	Drought tolerant	Santosh Kumar et al. (2020)
Tomato	*SlARF4*	Drought tolerant	Chen et al. (2021)
Wheat	*TaHAG1*	Salt tolerant	Zheng et al. (2021)
Soybean	*GmAITR*	Salt tolerant	Wang et al. (2021)
Tomato	*SlCPK28*	Heat tolerant	Hu et al. (2021)
Rice	*OsHSA1*	Heat tolerant	Qiu et al. (2018)
Thale cress	*AtCBF1, AtCBF2*	Cold tolerant	Chen et al. (2019)
Rice	*OsAnn5, OsAnn3*	Cold tolerant	Shen et al. (2017)
Tomato	*SlCBF1*	Cold tolerant	Li et al. (2018)
Rice	*OsARM1*	Heavy metal stress tolerant	Wang et al. (2017)
Rice	*OsATX1*	Heavy metal stress tolerant	Zhang et al. (2021)
Cassava	*MeEPSPS*	Herbicide tolerant (glyphosate)	Hummel et al. (2018)
Rice	*OsALS*	Herbicide tolerant	Wang et al. (2021)
Tomato	*WAT1*	Resistant against *Fusarium oxysporum, Verticillium dahliae*	Hanika et al. (2021)
Maize	*ZmFER1*	Resistant against fusarium ear rot disease	Liu et al. (2022)
Chilli	*CaERF28*	Resistant to anthracnose disease	Mishra et al. (2021)
Rice	*OsERF922*	Resistant to *Magnaporthe oryzae*	Wang et al. (2016)
Wheat	*TaNFXL1*	Resistant to *Fusarium graminearum*	Brauer et al. (2020)
Rice	*OsSWEET11, OsSWEET13, OsSWEET14*	Resistant against *Xanthomonas oryzae*	Xu et al. (2019)
Potato	*Nib, Cl, CP, P3 conserved viral regions*	Resistant to potato virus Y	Zhan et al. (2019)
Tomato	*eIF4E1*	Resistant against pepper mottle virus	Zhan et al. (2019)
Benthi	*RNAse3*	Resistant against cucumber mosaic virus	Yu et al. (2022)

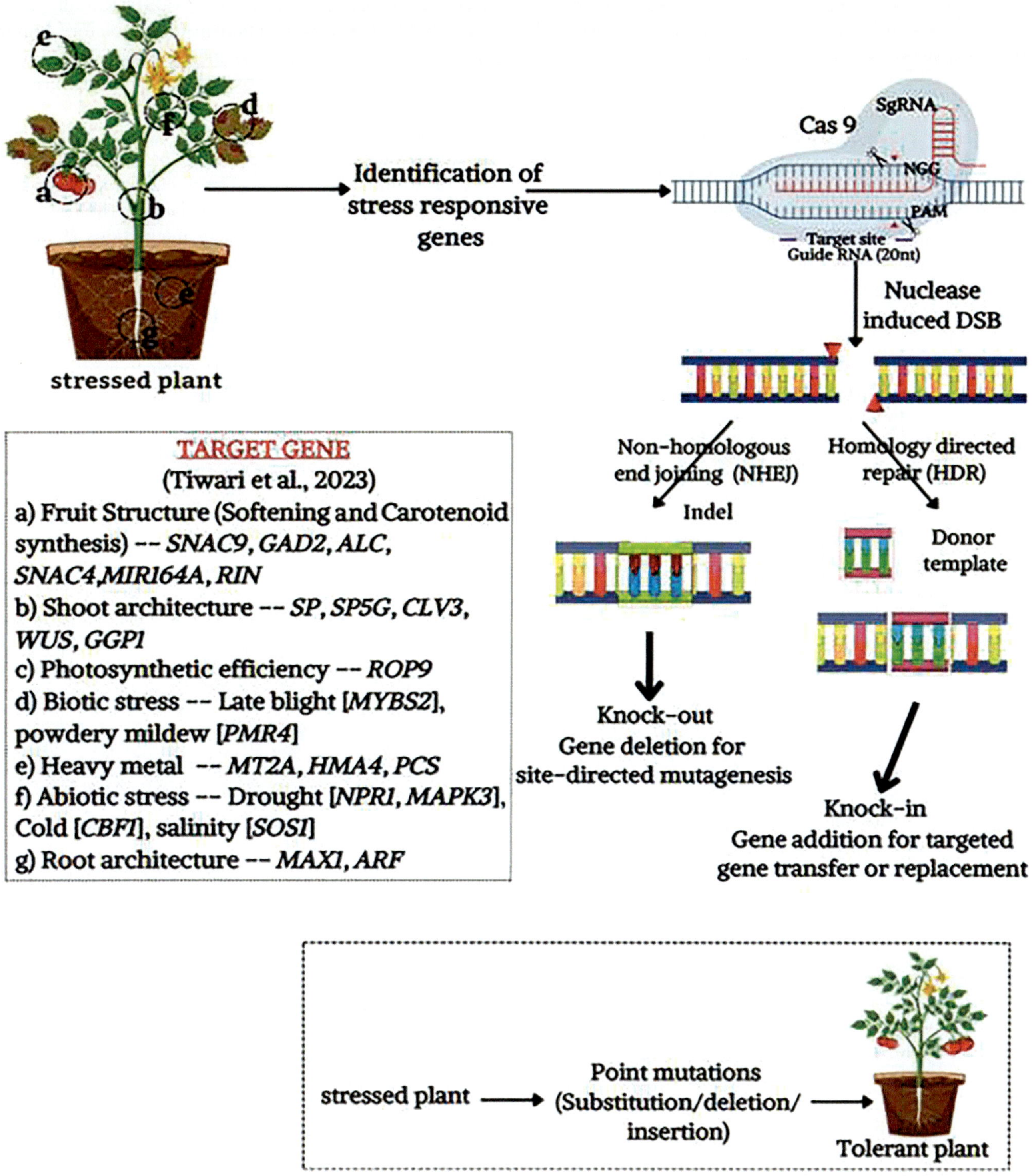

FIGURE 1.5 The CRISPR Cas9 mechanism and the target genes involved in the plant stress (biotic and abiotic). (The illustration is made through Canva.)

the distinct protein modifications. However, the expression of the gene is regulated by the epigenetic modifications. The plants can be epigenetically engineered through different tools such as ZFN, TALEN and CRISPR/Cas9 (Pikaard & Scheid, 2014). These tools modify the plant gene regulation via modifying the sequences in the promoter and transcription region without altering the DNA sequences. The plant epigenetic regulation includes histone-modifying enzymes, non-coding RNAs, and DNA methylation that affect the expressions of various genes. The plant developmental stages are in cohort with epigenetic regulation that may promote gene modifications, chromosomal changes, gene amplification, gene duplication and translocation. The epigenetic modifications are considered vital as most of the stress-responsive genes are controlled by them (Pikaard & Scheid, 2014).

1.10 EPIGENETIC REGULATION IN PLANTS: TYPES AND THE ACTION DISPLAYED

The epigenetic modifications involved in many biological mechanisms as they silence the transposons and transgenes, modifying the chromatin structure, ncRNA, DNA methylation that were lying upfront of the genetic structure that in turn stabilises the nuclear genome. Accessibility of DNA sequences influences chromatin conformation that leads to coding and regulating DNA sequences, intervening transcription and finally regulating gene expression in response to various stresses and development (Shanker et al., 2021; Štorchová, 2017; Struhl & Segal, 2013).

1.10.1 Chromatin Remodelling

The plants can sense the environment stimuli altering the expression of gene under adverse climatic conditions. The dynamics of chromatin influences the gene expression in adverse conditions. Plants holds the transcriptional memory of the stress they encountered (Fan et al., 2005). This kind of counter-strategy is known as priming, associated with chromatin modification, and maybe sustained independently from transcription. Chromatin modifications are heritable epigenetic changes involved in genetic expression without altering the nucleotide sequence. It is involved in several biological functions including biotic and abiotic stress (Kang et al., 2020). Chromatin modelling factors changes the chromatin's nucleosome position and interchanges the histone modifications, DNA methylation, RNA processing at different post-transcriptional levels. For example, when Arabidopsis exposed to different stress like drought, salt that results in no change in the transcriptional level of the chromatin. Histone acetyl transferase and histone deacetyl transferase are exhibited more in the plants during the cold stress. In Arabidopsis, under cold stress, HDA6 gets upregulated and in turn it activated the DREB1 as a stress-responsive mechanism (Luo et al., 2017; Yu et al., 2018).

1.10.2 Histone Modification and DNA Methylation

DNA methylation preserves the epigenetic marks in regulating the gene expression, gene imprinting, chromatin structure and genome integrity as a response against the biotic and abiotic stress. Abnormalities in DNA methylation led to develop CHG, CG and CHH, as it takes place on both cytosine (5mC) and adenine (6mA) residue, and it is highly present on TEs where DNA methylation plays an important role in silencing the expression at transcription levels, so called as transcriptional gene silencing (TGS) (Karanthamalai et al., 2020). DNA methylation in the regulatory regions had the transcriptional gene silencing higher. It also influences on the mRNA splicing mechanisms through repeats and intronic TEs (Zhang et al., 2018).

Histone modifications undergo various post-translational modifications. Histone tails are basic due to the presence of arginine and lysine. The nucleosome is wrapped with histone proteins that provides the accessibility for the DNA–protein and protein–protein interactions (Lauria & Rossi, 2011; Temel et al., 2017). The covalent modifications of histone tails are reversible. The addition and removal of histone marks are catalysed by various enzymes called writers and erasers, respectively. The proteins, recognising these modifications (marks), are called readers (Iwasaki & Paszkowski, 2014). The histone tails' PTMs change the chromatin structure and facilitate binding sites of the regulatory proteins. The alteration in the amino acid structure disrupts the stability and function of the amino acid (Ueda & Seki, 2020).

1.10.3 Epigenetic Regulation in Plants against Abiotic and Biotic Stress

Recent research studies focused on the epigenetic regulation of plants as a tolerance mechanism to tackle against different stress factors like drought, heat, cold, heavy metal, saline condition that includes chromatin remodelling, histone modifications and RNA modifications (Ding & Wang, 2015; Ma et al., 2018). Mostly abiotic and plant developmental stages are controlled/influenced partially by DNA methylation in plants. Among all the epigenetic regulation methods, DNA methylation is considered to be the most preferred and highly studied mechanism against the abiotic stress as it plays a crucial role in the hybrid vigour and flowering (Wang et al., 2015; Zemach et al., 2010). Plants become optimal, if an epigenetically regulated plant relieved from stress (Ueda & Seki, 2020).

1.11 REGULATORY ROLES OF EFFECTOR MOLECULES AND NCRNAS IN ESTABLISHING IMMUNITY

In addition to host-produced ncRNAs that regulate the synthesis and signalling of ncRNAs involved in plant defence, pathogens release pathogenic effectors that exploit the RNA interference pathway to compromise plant immunity and successfully infect the host. These effectors, referred to as RNA silencing antagonists (suppressors), are generated by pathogenic fungi, viruses, oomycetes and bacteria (de Vries et al., 2019; Spanu, 2015).

i. siRNAs for plant-virus interactions: Virus-associated siRNAs straight affect a plant gene responsible for chlorophyll production. Viruses actively generate potent proteins, like P19, which suppress RNA silencing by obstructing DCL activity, sequestering virus-derived-siRNAs, and halting RISC assembly (Silhavy, 2002).

ii. In plant-bacteria interactions, bacterial effectors enter host plant cells and disrupt plant siRNA regulation. The AvrPtoB effector inhibits miR393 accumulation at the transcriptional level, while AvrPto interferes with miR393 processing. HopT1-1 disrupts AGO1 function, thereby inhibiting RNA silencing (Navarro et al., 2008).

iii. In plant–fungi interactions, fungal siRNAs called Bc-siRNAs invade host cells and hijack plant RNA-silencing pathways to suppress key immune genes, including MAPKs and WAKs. The fungal effector PgtSR1 influences specific miRNAs that control plant defence mechanisms. In response, plants transfer distinct small RNAs to fungi, impeding their growth and virulence (Wang et al., 2017; Weiberg et al., 2013).

iv. For plant–oomycete interactions, oomycete effectors PSR1 and PSR2 infiltrate plant cells, disrupt the host's RNA silencing system and weaken plant immunity. Oomycetes also transmit small RNAs to silence host genes like *AtWNK2 [WITH NO LYSINE (K) KINASE2] and AtAED3 [APOPLASTIC, ENHANCED DISEASE SUSCEPTIBILITY1-DEPENDENT3].* In retaliation, host plants deploy numerous small RNAs into oomycetes, undermining pathogen virulence (Dunker et al., 2020).

1.12 REGULATORY ROLES OF NCRNAS FOR CROP ADVANCEMENT

Harnessing the knowledge of ncRNAs and their regulatory mechanisms could prove beneficial in addressing food and nutritional security concerns. Several seminal studies have demonstrated the impact of manipulating individual miRNAs on various traits, including enhanced crop yield, biomass and stress resilience (Zhang et al., 2022). For instance, studies have explicated that the overexpression of the highly conserved miR156 in tomato plants is strongly linked to increased fruit size (Zhang et al., 2016). Additionally, SPL genes, which are targeted by miR156, have been positively linked to rice yield.

Artificial miRNA (*amiRNA/amiR*) has emerged as an important and innovative tool for crop enhancement, enabling precise modulation of target gene expression through the strategic design of amiRNAs (Zhang et al., 2018). An illustration of engineering amiR159b to strategically suppress three key genes governing seed oil metabolism – fatty acid elongase 1 (FAE1), fatty acyl-ACP thioesterase B (FATB) and fatty acid Δ12 desaturase 1 (FAD2) – resulted in enhanced oleic acid levels in Arabidopsis. The versatility of amiR in modulating various traits represents a significant advantage of this approach (Belide et al., 2012). A crucial aspect of crop improvement as potential biomarkers is the quantity of single-nucleotide polymorphisms (SNPs) in ncRNA regions.

Furthermore, research has shown that genome editing using site-specific nucleases, particularly the type II CRISPR (Clustered Regularly Interspaced Short Palindromic Repeats)/CRISPR-associated protein 9 (Cas9) system, emerges as a highly effective method for functional analyses in plants (Zhu & Palli, 2020). The CRISPR/Cas9 system has demonstrated its ability to overcome RNA interference (RNAi) limitations by enabling precise gene knockouts and reducing off-target effects. The nuclear localisation of specific ncRNAs poses a significant hurdle for RNAi, as its action is restricted to the cytoplasm, where the RISC operates. In plants, leveraging genome editing to enhance crops by targeting ncRNAs remains in its early developmental phase, with ongoing efforts to optimise its use (Zhu & Palli, 2020). Altogether, these cutting-edge precision tools hold promise for harnessing ncRNA insights to advance crop improvement by creating cultivars with enhanced traits.

1.13 CONCLUSION

Unpredictable climate shifts impact all forms of life, including vegetation. The increasing population faces food scarcity, while crop yields are affected by both living and non-living stressors. Various techniques are needed to enhance agricultural productivity. Understanding the biomolecular mechanisms in plants, such as yield, nutrient accumulation, pest/pathogen defence and environmental response, is essential for creating improved plants. Scientists have become increasingly interested in studying and molecularly characterising ncRNAs to understand their role in suppressing mRNA translation, which regulates various biological functions, including growth and stress responses. While advancements in high-throughput sequencing technologies and computational tools have aided in predicting new ncRNAs, this progress is restricted to certain plant species. Enhancing single-cell small RNA sequencing techniques and expanding the availability of reference genomes could improve detection capabilities. Furthermore, plants have evolved numerous miRNA–siRNA-lncRNA-target gene networks to manage developmental flexibility during abiotic stress, with some of these mechanisms being evolutionarily preserved to adapt to environmental changes. Addressing food security issues has led to the experimental validation of crucial ncRNAs and their variants through siRNA/miRNA overexpression or knockout studies. The CRISPR-Cas system, an innovative technique, can be specifically employed to mitigate potential off-target mutations. Currently, plant-based ncRNA research primarily focuses on miRNAs, examining their roles in development and stress response regulation. To harness the potential of ncRNAs for advancing crop improvement, a profound grasp of their regulatory and functional dynamics within diverse gene regulatory networks is crucial. Additionally, establishing a trait-specific ncRNA repository is pivotal, providing precise targets for engineering innovative crop varieties. While significant challenges remain unresolved, the strategic application of ncRNA-driven methodologies in molecular-breeding will significantly amplify efforts to address the escalating global food crisis.

REFERENCES

Allen, E., Xie, Z., Gustafson, A. M., Sung, G.-H., Spatafora, J. W., & Carrington, J. C. (2004). Evolution of microRNA genes by inverted duplication of target gene sequences in *Arabidopsis thaliana*. *Nature Genetics*, *36*(12), 1282–1290.

Ariel, F., Romero-Barrios, N., Jégu, T., Benhamed, M., & Crespi, M. (2015). Battles and hijacks: Noncoding transcription in plants. *Trends in Plant Science*, *20*(6), 362–371.

Avila, L. A., Chandrasekar, R., Wilkinson, K. E., Balthazor, J., Heerman, M., Bechard, J., Brown, S., Park, Y., Dhar, S., & Reeck, G. R. (2018). Delivery of lethal dsRNAs in insect diets by branched amphiphilic peptide capsules. *Journal of Controlled Release*, *273*, 139–146.

Axtell, M. J., & Meyers, B. C. (2018). Revisiting criteria for plant microRNA annotation in the era of big data. *The Plant Cell*, *30*(2), 272–284.

Banerjee, P. (2020). Plant abiotic stress responses and microRNAs. *Advanced Agriculture*, 1, 119–118.

Baum, J. A., Bogaert, T., Clinton, W., Heck, G. R., Feldmann, P., Ilagan, O., Johnson, S., Plaetinck, G., Munyikwa, T., & Pleau, M. (2007). Control of coleopteran insect pests through RNA interference. *Nature Biotechnology*, *25*(11), 1322–1326.

Belide, S., Petrie, J. R., Shrestha, P., & Singh, S. P. (2012). Modification of seed oil composition in Arabidopsis by artificial microRNA-mediated gene silencing. *Frontiers in Plant Science*, *3*(JUL).

Bernardi, Y., Ponso, M. A., Belén, F., Vegetti, A. C., & Dotto, M. C. (2022). MicroRNA miR394 regulates flowering time in *Arabidopsis thaliana*. *Plant Cell Reports*, *41*(6), 1375–1388. https://doi.org/10.1007/s00299-022-02863-0

Bhattacharya, A., & Roychoudhury, A. (2023). Plant non-coding RNAs and their regulation in salinity and heavy metal stress. In *Non-coding RNAs* (pp. 128–147). CRC Press.

Borges, F., & Martienssen, R. A. (2015). The expanding world of small RNAs in plants. *Nature Reviews Molecular Cell Biology*, *16*(12), 727–741.

Borsani, O., Zhu, J., Verslues, P. E., Sunkar, R., & Zhu, J.-K. (2005). Endogenous siRNAs derived from a pair of natural cis-antisense transcripts regulate salt tolerance in Arabidopsis. *Cell*, *123*(7), 1279–1291.

Brandt, B., Munemasa, S., Wang, C., Nguyen, D., Yong, T., Yang, P. G., Poretsky, E., Belknap, T. F., Waadt, R., & Alemán, F. (2015). Calcium specificity signaling mechanisms in abscisic acid signal transduction in Arabidopsis guard cells. *Elife*, *4*, e03599.

Brauer, E. K., Balcerzak, M., Rocheleau, H., Leung, W., Schernthaner, J., Subramaniam, R., & Ouellet, T. (2020). Genome editing of a deoxynivalenol-induced transcription factor confers resistance to *Fusarium graminearum* in wheat. *Molecular Plant-Microbe Interactions®*, *33*(3), 553–560. https://doi.org/10.1094/MPMI-11-19-0332-R

Brosnan, C. A., & Voinnet, O. (2009). The long and the short of noncoding RNAs. *Current Opinion in Cell Biology*, *21*(3), 416–425.

Campalans, A., Kondorosi, A., & Crespi, M. (2004). Enod40, a short open reading frame-containing mRNA, induces cytoplasmic localization of a nuclear RNA binding protein in *Medicago truncatula*. *The Plant Cell*, *16*(4), 1047–1059.

Çelik, Ö, Ayan, A., Meriç, S., & Atak, Ç. (2020). Heavy metal stress-responsive phyto-miRNAs. In M. Faisal, Q. Saquib, A. A. Alatar, & A. A. Al-Khedhairy (Eds.), *Cellular and molecular phytotoxicity of heavy metals* (pp. 137–155). Springer International Publishing. https://doi.org/10.1007/978-3-030-45975-8_9

Chekanova, J. A. (2015). Long non-coding RNAs and their functions in plants. *Current Opinion in Plant Biology*, *27*, 207–216.

Chen, G., Hu, J., Dong, L., Zeng, D., Guo, L., Zhang, G., Zhu, L., & Qian, Q. (2019). The tolerance of salinity in rice requires the presence of a functional copy of FLN2. *Biomolecules*, *10*(1), 17.

Chen, H. M., Chen, L. T., Patel, K., Li, Y. H., Baulcombe, D. C., & Wu, S. H. (2010). 22-Nucleotide RNAs trigger secondary siRNA biogenesis in plants. *Proceedings of the National Academy of Sciences of the United States of America*, *107*(34), 15269–15274. https://doi.org/10.1073/PNAS.1001738107

Chen, M., & Penfield, S. (2018). Feedback regulation of COOLAIR expression controls seed dormancy and flowering time. *Science*, *360*(6392), 1014–1017. https://doi.org/10.1126/SCIENCE.AAR7361

Chen, S., Zhang, N., Zhou, G., Hussain, S., Ahmed, S., Tian, H., & Wang, S. (2021). Knockout of the entire family of AITR genes in Arabidopsis leads to enhanced drought and salinity tolerance without fitness costs. *BMC Plant Biology*, *21*(1), 137. https://doi.org/10.1186/s12870-021-02907-9

Chung, P. J., Jung, H., Jeong, D. H., Ha, S. H., Choi, Y. D., & Kim, J. K. (2016). Transcriptome profiling of drought responsive noncoding RNAs and their target genes in rice. *BMC Genomics*, *17*(1). https://doi.org/10.1186/S12864-016-2997-3

Crisp, P. A., Ganguly, D., Eichten, S. R., Borevitz, J. O., & Pogson, B. J. (2016). Reconsidering plant memory: Intersections between stress recovery, RNA turnover, and epigenetics. *Science Advances*, *2*(2), e1501340. https://doi.org/10.1126/sciadv.1501340

Cui, J., Jiang, N., Meng, J., Yang, G., Liu, W., Zhou, X., Ma, N., Hou, X., & Luan, Y. (2019). LncRNA33732-respiratory burst oxidase module associated with WRKY1 in tomato-*Phytophthora infestans* interactions. *The Plant Journal*, *97*(5), 933–946. https://doi.org/10.1111/tpj.14173

Cui, J., Luan, Y., Jiang, N., Bao, H., & Meng, J. (2017). Comparative transcriptome analysis between resistant and susceptible tomato allows the identification of lncRNA16397 conferring resistance to *Phytophthora infestans* by co-expressing glutaredoxin. *Plant Journal*, *89*(3), 577–589. https://doi.org/10.1111/TPJ.13408

D'Ario, M., Griffiths-Jones, S., & Kim, M. (2017). Small RNAs: Big impact on plant development. *Trends in Plant Science*, *22*(12), 1056–1068.

Das, S. S., Karmakar, P., Nandi, A. K., & Sanan-Mishra, N. (2015). Small RNA mediated regulation of seed germination. *Frontiers in Plant Science*, *6*, 828. https://www.frontiersin.org/articles/10.3389/fpls.2015.00828/full

Denver, J. B., & Ullah, H. (2019). miR393s regulate salt stress response pathway in *Arabidopsis thaliana* through scaffold protein RACK1A mediated ABA signaling pathways. *Plant Signaling & Behavior*, *14*(6), 1600394. https://doi.org/10.1080/15592324.2019.1600394

de Vries, S., de Vries, J., & Rose, L. E. (2019). The elaboration of miRNA regulation and gene regulatory networks in plant–microbe interactions. *Genes*, *10*(4), 310.

Ding, B., & Wang, G.-L. (2015). Chromatin versus pathogens: The function of epigenetics in plant immunity. *Frontiers in Plant Science*, *6*, 675.

Ding, H., He, J., Wu, Y., Wu, X., Ge, C., Wang, Y., Zhong, S., Peiter, E., Liang, J., & Xu, W. (2018). The tomato mitogen-activated protein kinase SlMPK1 is as a negative regulator of the high-temperature stress response. *Plant Physiology*, *177*(2), 633–651.

Ding, Y., Chen, Z., & Zhu, C. (2011). Microarray-based analysis of cadmium-responsive microRNAs in rice (Oryza sativa). *Journal of Experimental Botany*, *62*(10), 3563–3573.

Dong, L., Liu, S., Kyaing, M. S., Xu, P., Tharreau, D., Deng, W., Li, X., Bi, Y., Zeng, L., Li, J., Zhou, J., Tao, D., & Yang, Q. (2020). Identification and fine mapping of Pi69(t), a new gene conferring broad-spectrum resistance against magnaporthe oryzae from oryza glaberrima steud. *Frontiers in Plant Science*, *11*, 1190. https://doi.org/10.3389/fpls.2020.01190

Dunker, F., Trutzenberg, A., Rothenpieler, J. S., Kuhn, S., Pröls, R., Schreiber, T., Tissier, A., Kemen, A., Kemen, E., & Hückelhoven, R. (2020). Oomycete small RNAs bind to the plant RNA-induced silencing complex for virulence. *Elife*, *9*, e56096.

Escobar, M. A., Civerolo, E. L., Summerfelt, K. R., & Dandekar, A. M. (2001). RNAi-mediated oncogene silencing confers resistance to crown gall tumorigenesis. *Proceedings of the National Academy of Sciences*, *98*(23), 13437–13442. https://doi.org/10.1073/pnas.241276898

Fan, J., Quan, W., Li, G.-B., Hu, X.-H., Wang, Q., Wang, H., Li, X.-P., Luo, X., Feng, Q., Hu, Z.-J., Feng, H., Pu, M., Zhao, J.-Q., Huang, Y.-Y., Li, Y., Zhang, Y., & Wang, W.-M. (2020). circRNAs are involved in the rice-*Magnaporthe oryzae* interaction. *Plant Physiology*, *182*(1), 272–286. https://doi.org/10.1104/pp.19.00716

Fan, Y., Nikitina, T., Zhao, J., Fleury, T. J., Bhattacharyya, R., Bouhassira, E. E., Stein, A., Woodcock, C. L., & Skoultchi, A. I. (2005). Histone H1 depletion in mammals alters global chromatin structure but causes specific changes in gene regulation. *Cell*, *123*(7), 1199–1212.

Ferdous, J., Sanchez-Ferrero, J. C., Langridge, P., Milne, L., Chowdhury, J., Brien, C., & Tricker, P. J. (2017). Differential expression of microRNAs and potential targets under drought stress in barley. *Plant Cell and Environment*, *40*(1), 11–24. https://doi.org/10.1111/PCE.12764

Frazier, T. P., Sun, G., Burklew, C. E., & Zhang, B. (2011). Salt and drought stresses induce the aberrant expression of microRNA genes in tobacco. *Molecular Biotechnology*, *49*(2), 159–165. https://doi.org/10.1007/s12033-011-9387-5

Gasciolli, V., Mallory, A. C., Bartel, D. P., & Vaucheret, H. (2005). Partially redundant functions of Arabidopsis DICER-like enzymes and a role for DCL4 in producing trans-acting siRNAs. *Current Biology*, *15*(16), 1494–1500.

Gu, S., Jin, L., Huang, Y., Zhang, F., & Kay, M. A. (2012). Slicing-independent RISC activation requires the argonaute PAZ domain. *Current Biology*, *22*(16), 1536–1542.

Guan, Q., Lu, X., Zeng, H., Zhang, Y., & Zhu, J. (2013). Heat stress induction of miR398 triggers a regulatory loop that is critical for thermotolerance in Arabidopsis. *Plant Journal*, *74*(5), 840–851. https://doi.org/10.1111/TPJ.12169

Hanika, K., Schipper, D., Chinnappa, S., Oortwijn, M., Schouten, H. J., Thomma, B. P., & Bai, Y. (2021). Impairment of tomato WAT1 enhances resistance to vascular wilt fungi despite severe growth defects. *Frontiers in Plant Science*, *12*, 721674.

He, J., Jiang, Z., Gao, L., You, C., Ma, X., Wang, X., Xu, X., Mo, B., Chen, X., & Liu, L. (2019). Genome-wide transcript and small RNA profiling reveals transcriptomic responses to heat stress. *Plant Physiology*, *181*(2), 609–629. https://doi.org/10.1104/PP.19.00403

Hirayama, T., & Shinozaki, K. (2010). Research on plant abiotic stress responses in the post-genome era: Past, present and future. *The Plant Journal*, *61*(6), 1041–1052. https://doi.org/10.1111/j.1365-313X.2010.04124.x

Horvath, P., & Barrangou, R. (2010). CRISPR/Cas, the immune system of bacteria and archaea. *Science*, *327*(5962), 167–170. https://doi.org/10.1126/science.1179555

Hou, Q., Ufer, G., & Bartels, D. (2016). Lipid signalling in plant responses to abiotic stress. *Plant, Cell & Environment*, *39*(5), 1029–1048. https://doi.org/10.1111/pce.12666

Hrabak, E. M., Chan, C. W., Gribskov, M., Harper, J. F., Choi, J. H., Halford, N., Kudla, J., Luan, S., Nimmo, H. G., & Sussman, M. R. (2003). The Arabidopsis CDPK-SnRK superfamily of protein kinases. *Plant Physiology*, *132*(2), 666–680.

Hu, Z., Li, J., Ding, S., Cheng, F., Li, X., Jiang, Y., Yu, J., Foyer, C. H., & Shi, K. (2021). The protein kinase CPK28 phosphorylates ascorbate peroxidase and enhances thermotolerance in tomato. *Plant Physiology*, *186*(2), 1302–1317.

Hua, Y., Zhang, C., Shi, W., & Chen, H. (2019). High-throughput sequencing reveals microRNAs and their targets in response to drought stress in wheat (*Triticum aestivum* L.). *Biotechnology and Biotechnological Equipment*, *33*(1), 465–471. https://doi.org/10.1080/13102818.2019.1586586

Huang, J., Yang, M., & Zhang, X. (2016). The function of small RNAs in plant biotic stress response. *Journal of Integrative Plant Biology*, *58*(4), 312–327. https://doi.org/10.1111/jipb.12463

Hummel, A. W., Chauhan, R. D., Cermak, T., Mutka, A. M., Vijayaraghavan, A., Boyher, A., Starker, C. G., Bart, R., Voytas, D. F., & Taylor, N. J. (2018). Allele exchange at the EPSPS locus confers glyphosate tolerance in cassava. *Plant Biotechnology Journal*, *16*(7), 1275–1282. https://doi.org/10.1111/pbi.12868

Hussain, K., Mungikar, K., Kulkarni, A., & Kamble, A. (2018). Identification, characterization and expression analysis of pigeonpea miRNAs in response to Fusarium wilt. *Gene*, *653*, 57–64.

Ingram, J., & Bartels, D. (1996). The molecular basis of dehydration tolerance in plants. *Annual Review of Plant Physiology and Plant Molecular Biology*, *47*(1), 377–403. https://doi.org/10.1146/ANNUREV.ARPLANT.47.1.377

Iwasaki, M., & Paszkowski, J. (2014). Epigenetic memory in plants. *The EMBO Journal*, *33*(18), 1987–1998. https://doi.org/10.15252/embj.201488883

Jain, R., Fletcher, S., & Mitter, N. (2022). Effective RNAi-mediated control of the crop pest whitefly. *Nat. Plant*, *8*(5), 461–462.

Jansen, R., Embden, J. D. A. V., Gaastra, W., & Schouls, L. M. (2002). Identification of genes that are associated with DNA repeats in prokaryotes. *Molecular Microbiology*, *43*(6), 1565–1575. https://doi.org/10.1046/j.1365-2958.2002.02839.x

Jiang, C., Wang, Y., He, Y., Peng, Y., Xie, L., Li, Y., Sun, W., Zhou, J., Zheng, C., & Xie, X. (2024). Identification and characterization of miRNAs and lncRNAs associated with salinity stress in rice panicles. *International Journal of Molecular Sciences*, *25*(15), 8247.

Jiang, C. J., Shimono, M., Maeda, S., Inoue, H., Mori, M., Hasegawa, M., Sugano, S., & Takatsuji, H. (2009). *Suppression of the rice fatty-acid desaturase gene OsSSI2 enhances resistance to blast and leaf blight diseases in rice Molecular Plant-Microbe Interactions : MPMI*, *22*(7), 820–829. https://doi.org/10.1094/MPMI-22-7-0820.

Jinek, M., Chylinski, K., Fonfara, I., Hauer, M., Doudna, J. A., & Charpentier, E. (2012). A programmable dual-RNA–Guided DNA endonuclease in adaptive bacterial immunity. *Science*, *337*(6096), 816–821. https://doi.org/10.1126/science.1225829

Kanakala, S., & Ghanim, M. (2016). RNA interference in insect vectors for plant viruses. *Viruses*, *8*(12), 329.

Kang, H., Wu, D., Fan, T., & Zhu, Y. (2020). Activities of chromatin remodeling factors and histone chaperones and their effects in root apical meristem development. *International Journal of Molecular Sciences*, *21*(3), 771.

Karanthamalai, J., Chodon, A., Chauhan, S., & Pandi, G. (2020). DNA N6-methyladenine modification in plant genomes—A glimpse into emerging epigenetic code. *Plants*, *9*(2), 247.

Khraiwesh, B., Zhu, J.-K., & Zhu, J. (2012). Role of miRNAs and siRNAs in biotic and abiotic stress responses of plants. *Biochimica et Biophysica Acta (BBA)-Gene Regulatory Mechanisms*, *1819*(2), 137–148.

Kindgren, P., Ard, R., Ivanov, M., & Marquardt, S. (2018). Transcriptional read-through of the long non-coding RNA SVALKA governs plant cold acclimation. *Nature Communications*, *9*(1), 4561.

Koonin, E. V., Makarova, K. S., & Zhang, F. (2017). Diversity, classification and evolution of CRISPR-Cas systems. *Current Opinion in Microbiology*, *37*, 67–78.

Krubphachaya, P., Juricek, M., & Kertbundit, S. (2007). Induction of RNA-mediated resistance to papaya ringspot virus type W. *BMB Reports*, *40*(3), 404–411.

Kumar, V., Khare, T., Shriram, V., & Wani, S. H. (2018). Plant small RNAs: The essential epigenetic regulators of gene expression for salt-stress responses and tolerance. *Plant Cell Reports*, *37*(1), 61–75. https://doi.org/10.1007/s00299-017-2210-4

Kuo, Y.-W., Lin, J.-S., Li, Y.-C., Jhu, M.-Y., King, Y.-C., & Jeng, S.-T. (2019). MicroR408 regulates defense response upon wounding in sweet potato. *Journal of Experimental Botany*, *70*(2), 469–483.

Lauria, M., & Rossi, V. (2011). Epigenetic control of gene regulation in plants. *Biochimica et Biophysica Acta (BBA)-Gene Regulatory Mechanisms*, *1809*(8), 369–378.

Li, H., Dong, Y., Yin, H., Wang, N., Yang, J., Liu, X., Wang, Y., Wu, J., & Li, X. (2011). Characterization of the stress associated microRNAs in Glycine max by deep sequencing. *BMC Plant Biology*, *11*(1), 170. https://doi.org/10.1186/1471-2229-11-170

Li, J., Guo, G., Guo, W., Guo, G., Tong, D., Ni, Z., Sun, Q., & Yao, Y. (2012). miRNA164-directed cleavage of ZmNAC1 confers lateral root development in maize (*Zea mays* L.). *BMC Plant Biology*, *12*(1), 220. https://doi.org/10.1186/1471-2229-12-220

Li, J., Song, Q., Zuo, Z.-F., & Liu, L. (2022). MicroRNA398: A master regulator of plant development and stress responses. *International Journal of Molecular Sciences*, *23*(18), 10803.

Li, J., Yue, L., Shen, Y., Sheng, Y., Zhan, X., Xu, G., & Xing, B. (2017). Phenanthrene-responsive microRNAs and their targets in wheat roots. *Chemosphere*, *186*, 588–598.

Li, R., Zhang, L., Wang, L., Chen, L., Zhao, R., Sheng, J., & Shen, L. (2018). Reduction of tomato-plant chilling tolerance by CRISPR–Cas9-mediated *SlCBF1* mutagenesis. *Journal of Agricultural and Food Chemistry*, *66*(34), 9042–9051. https://doi.org/10.1021/acs.jafc.8b02177

Li, W., Wang, T., Zhang, Y., & Li, Y. (2016). Overexpression of soybean miR172c confers tolerance to water deficit and salt stress, but increases ABA sensitivity in transgenic *Arabidopsis thaliana*. *Journal of Experimental Botany*, *67*(1), 175–194.

Li, X.-P., Ma, X.-C., Wang, H., Zhu, Y., Liu, X.-X., Li, T.-T., Zheng, Y.-P., Zhao, J.-Q., Zhang, J.-W., Huang, Y.-Y., Pu, M., Feng, H., Fan, J., Li, Y., & Wang, W.-M. (2020). Osa-miR162a fine-tunes rice resistance to *Magnaporthe oryzae* and yield. *Rice*, *13*(1), 38. https://doi.org/10.1186/s12284-020-00396-2

Lin, B. L., Wang, J. S., Liu, H. C., Chen, R. W., Meyer, Y., Barakat, A., & Delseny, M. (2001). Genomic analysis of the Hsp70 superfamily in Arabidopsis thaliana. *Cell Stress & Chaperones*, *6*(3), 201–208. https://doi.org/10.1379/1466-1268(2001)006<0201:gaoths>2.0.co;2

Lippmann, R., Babben, S., Menger, A., Delker, C., & Quint, M. (2019). Development of wild and cultivated plants under global warming conditions. *Current Biology*, *29*(24), R1326–R1338.

List, F., Tarone, A. M., Zhu-Salzman, K., & Vargo, E. L. (2022). RNA meets toxicology: Efficacy indicators from the experimental design of RNAi studies for insect pest management. *Pest Management Science*, *78*(8), 3215–3225. https://doi.org/10.1002/ps.6884

Liu, C., Kong, M., Zhu, J., Qi, X., Duan, C., & Xie, C. (2022). Engineering null mutants in ZmFER1 confers resistance to ear rot caused by *Fusarium verticillioides* in maize. *Plant Biotechnology Journal*, *20*(11), 2045.

Liu, Q., Hu, H., Zhu, L., Li, R., Feng, Y., Zhang, L., Yang, Y., Liu, X., & Zhang, H. (2015). Involvement of miR528 in the regulation of arsenite tolerance in rice (*Oryza sativa* L.). *Journal of Agricultural and Food Chemistry*, *63*(40), 8849–8861. https://doi.org/10.1021/acs.jafc.5b04191

Liu, X., Xie, X., Zheng, C., Wei, L., Li, X., Jin, Y., Zhang, G., Jiang, C.-J., & Liang, Z. (2022). RNAi-mediated suppression of the abscisic acid catabolism gene OsABA8ox1 increases abscisic acid content and tolerance to saline–alkaline stress in rice (*Oryza sativa* L.). *The Crop Journal*, *10*(2), 354–367.

Lu, Q., Guo, F., Xu, Q., & Cang, J. (2020). LncRNA improves cold resistance of winter wheat by interacting with miR398. *Functional Plant Biology*, *47*(6), 544–557.

Lu, X., Guan, Q., & Zhu, J. (2013). Downregulation of CSD2 by a heat-inducible miR398 is required for thermotolerance in Arabidopsis. *Plant Signaling and Behavior*, *8*(8). https://doi.org/10.4161/PSB.24952

Luo, M., Cheng, K., Xu, Y., Yang, S., & Wu, K. (2017). Plant responses to abiotic stress regulated by histone deacetylases. *Frontiers in Plant Science*, *8*, 2147.

Ma, Y., Min, L., Wang, M., Wang, C., Zhao, Y., Li, Y., Fang, Q., Wu, Y., Xie, S., & Ding, Y. (2018). Disrupted genome methylation in response to high temperature has distinct affects on microspore abortion and anther indehiscence. *The Plant Cell*, *30*(7), 1387–1403.

Makarova, K. S., Wolf, Y. I., Alkhnbashi, O. S., Costa, F., Shah, S. A., Saunders, S. J., Barrangou, R., Brouns, S. J., Charpentier, E., & Haft, D. H. (2015). An updated evolutionary classification of CRISPR–Cas systems. *Nature Reviews Microbiology*, *13*(11), 722–736.

Mallory, A. C., & Vaucheret, H. (2006). Functions of microRNAs and related small RNAs in plants. *Nature Genetics*, *38*(Suppl 6), S31–S36.

Mandal, S., Khairnar, D., & Hazra, S. (2023). Plant Functional Long Non-Coding RNAs: Yesterday's Dark Matter, Today's Regulatory Component *for Abiotic Stress Response. Non-Coding RNAs: Molecular Tools for Crop Improvement*, 172–196. https://doi.org/10.1201/9781003369288-10

Mao, Y.-B., Liu, Y.-Q., Chen, D.-Y., Chen, F.-Y., Fang, X., Hong, G.-J., Wang, L.-J., Wang, J.-W., & Chen, X.-Y. (2017). Jasmonate response decay and defense metabolite accumulation contributes to age-regulated dynamics of plant insect resistance. *Nature Communications*, *8*(1), 1–13.

Martí, M. C., Stancombe, M. A., & Webb, A. A. (2013). Cell- and stimulus type-specific intracellular free Ca2+ signals in Arabidopsis. *Plant Physiology*, *163*(2), 625–634.

Martin, R. C., Asahina, M., Liu, P.-P., Kristof, J. R., Coppersmith, J. L., Pluskota, W. E., Bassel, G. W., Goloviznina, N. A., Nguyen, T. T., & Martínez-Andújar, C. (2010). The microRNA156 and microRNA172 gene regulation cascades at post-germinative stages in Arabidopsis. *Seed Science Research*, *20*(2), 79–87.

Mishra, R., Mohanty, J. N., Mahanty, B., & Joshi, R. K. (2021). A single transcript CRISPR/Cas9 mediated mutagenesis of CaERF28 confers anthracnose resistance in chilli pepper (*Capsicum annuum* L.). *Planta*, *254*(1), 5. https://doi.org/10.1007/s00425-021-03660-x

Munns, R., Passioura, J. B., Colmer, T. D., & Byrt, C. S. (2020). Osmotic adjustment and energy limitations to plant growth in saline soil. *New Phytologist*, *225*(3), 1091–1096. https://doi.org/10.1111/nph.15862

Navarro, L., Dunoyer, P., Jay, F., Arnold, B., Dharmasiri, N., Estelle, M., Voinnet, O., & Jones, J. D. G. (2006). A plant miRNA contributes to antibacterial resistance by repressing auxin signaling. *Science*, *312*(5772), 436–439. https://doi.org/10.1126/SCIENCE.1126088

Navarro, L., Jay, F., Nomura, K., He, S. Y., & Voinnet, O. (2008). Suppression of the MicroRNA pathway by bacterial effector proteins. *Science*, *321*(5891), 964–967. https://doi.org/10.1126/science.1159505

Nguyen, D. Q., Brown, C. W., Pegler, J. L., Eamens, A. L., & Grof, C. P. (2020). Molecular manipulation of microRNA397 abundance influences the development and salt stress response of *Arabidopsis thaliana*. *International Journal of Molecular Sciences*, *21*(21), 7879.

Thakur, N., Upadhyay, S. K., Verma, P. C., Chandrashekar, K., Tuli, R., & Singh, P. K. (2014). Enhanced whitefly resistance in transgenic tobacco plants expressing double stranded RNA of v-ATPase A gene. *PloS One*, *9*(3), e87235. https://doi.org/10.1371/journal.pone.0087235

Niu, J., Taning, C. N. T., Christiaens, O., Smagghe, G., & Wang, J.-J. (2018). Rethink RNAi in insect pest control: Challenges and perspectives. *Advances in Insect Physiology*, *55*, 1–17.

Nozawa, M., Miura, S., & Nei, M. (2010). Origins and evolution of microRNA genes in *Drosophila* species. *Genome Biology and Evolution*, *2*, 180–189.

Origin, V. O. (2009). Biogenesis, and activity of plant microRNAs. *Cell*, *136*(4), 669–687.

Pandey, R., Joshi, G., Bhardwaj, A. R., Agarwal, M., & Katiyar-Agarwal, S. (2014). A comprehensive genome-wide study on tissue-specific and abiotic stress-specific miRNAs in *Triticum aestivum*. *PLoS ONE*, *9*(4), e95800.

Pareek, M., Yogindran, S., Mukherjee, S. K., & Rajam, M. V. (2015). Plant microRNAs: Biogenesis, functions, and applications. In *Plant biology and biotechnology: Volume II: Plant genomics and biotechnology* (pp. 639–661). https://doi.org/10.1007/978-81-322-2283-5_32

Pauli, A., Rinn, J. L., & Schier, A. F. (2011). Non-coding RNAs as regulators of embryogenesis. *Nature Reviews Genetics*, *12*(2), 136–149.

Peragine, A., Yoshikawa, M., Wu, G., Albrecht, H. L., & Poethig, R. S. (2004). SGS3 and SGS2/SDE1/RDR6 are required for juvenile development and the production of trans-acting siRNAs in Arabidopsis. *Genes & Development*, *18*(19), 2368–2379.

Pikaard, C. S., & Scheid, O. M. (2014). Epigenetic regulation in plants. *Cold Spring Harbor Perspectives in Biology*, *6*(12), a019315.

Plaza, S., Menschaert, G., & Payre, F. (2017). In search of lost small peptides. *Annual Review of Cell and Developmental Biology*, *33*, 391–416. https://doi.org/10.1146/ANNUREV-CELLBIO-100616-060516

Pooggin, M. M. (2017). RNAi-mediated resistance to viruses: A critical assessment of methodologies. *Current Opinion in Virology*, *26*, 28–35.

Praveen, A., Dubey, S., Singh, S., & Sharma, V. K. (2023). Abiotic stress tolerance in plants: A fascinating action of defense mechanisms. *3 Biotech*, *13*(3), 102. https://doi.org/10.1007/s13205-023-03519-w

Probst, A. V., & Scheid, O. M. (2015). Stress-induced structural changes in plant chromatin. *Current Opinion in Plant Biology*, *27*, 8–16.

Qin, T., Zhao, H., Cui, P., Albesher, N., & Xiong, L. (2017). A nucleus-localized long non-coding RNA enhances drought and salt stress tolerance. *Plant Physiology*, *175*(3), 1321–1336.

Qiu, Z., Kang, S., He, L., Zhao, J., Zhang, S., Hu, J., Zeng, D., Zhang, G., Dong, G., & Gao, Z. (2018). The newly identified heat-stress sensitive albino 1 gene affects chloroplast development in rice. *Plant Science*, *267*, 168–179.

Rai, M. I., Alam, M., Lightfoot, D. A., Gurha, P., & Afzal, A. J. (2019). Classification and experimental identification of plant long non-coding RNAs. *Genomics*, *111*(5), 997–1005. https://doi.org/10.1016/J.YGENO.2018.04.014

Rajagopalan, R., Vaucheret, H., Trejo, J., & Bartel, D. P. (2006). A diverse and evolutionarily fluid set of microRNAs in *Arabidopsis thaliana*. *Genes & Development*, *20*(24), 3407–3425.

Ren, Y., Li, M., Wang, W., Lan, W., Schenke, D., Cai, D., & Miao, Y. (2022). MicroRNA840 (*MIR840*) accelerates leaf senescence by targeting the overlapping 3′UTRs of *PPR* and *WHIRLY3* in *Arabidopsis thaliana*. *The Plant Journal*, *109*(1), 126–143. https://doi.org/10.1111/tpj.15559

Rodoni, B. C., Dale, J. L., & Harding, R. M. (1999). Characterization and expression of the coat protein-coding region of banana bract mosaic potyvirus, development of diagnostic assays and detection of the virus in banana plants from five countries in southeast Asia. *Archives of Virology*, *144*(9), 1725–1737. https://doi.org/10.1007/s007050050700

Rogers, K., & Chen, X. (2012). microRNA biogenesis and turnover in plants. *Cold Spring Harbor Symposia on Quantitative Biology*, *77*, 183–194.

Sailaja, B., Anjum, N., Prasanth, V. V., Sarla, N., Subrahmanyam, D., Voleti, S. R., Viraktamath, B. C., & Mangrauthia, S. K. (2014). Comparative study of susceptible and tolerant genotype reveals efficient recovery and root system contributes to heat stress tolerance in rice. *Plant Molecular Biology Reporter*, *32*(6), 1228–1240. https://doi.org/10.1007/S11105-014-0728-Y

Salvador-Guirao, R., Hsing, Y., & San Segundo, B. (2018). The polycistronic miR166k-166h positively regulates rice immunity via post-transcriptional control of EIN2. *Frontiers in Plant Science*, *9*, 337.

Santosh Kumar, V. V., Verma, R. K., Yadav, S. K., Yadav, P., Watts, A., Rao, M. V., & Chinnusamy, V. (2020). CRISPR-Cas9 mediated genome editing of drought and salt tolerance (OsDST) gene in indica mega rice cultivar MTU1010. *Physiology and Molecular Biology of Plants*, *26*(6), 1099–1110. https://doi.org/10.1007/s12298-020-00819-w

Scharf, K.-D., Berberich, T., Ebersberger, I., & Nover, L. (2012). The plant heat stress transcription factor (Hsf) family: Structure, function and evolution. *Biochimica et Biophysica Acta (BBA)-Gene Regulatory Mechanisms*, *1819*(2), 104–119.

Senthil-Kumar, M., & Mysore, K. S. (2010). RNAi in plants: Recent developments and applications in agriculture. In *Gene silencing: Theory, techniques and applications* (pp. 183–199). Nova Science Publishers.

Shanker, A. K., Shanker, C., Anand, A., & Maheswari, M. (2021). *Climate change and crop stress: Molecules to ecosystems*. Academic Press.

Shen, C., Que, Z., Xia, Y., Tang, N., Li, D., He, R., & Cao, M. (2017). Knock out of the annexin gene OsAnn3 via CRISPR/Cas9-mediated genome editing decreased cold tolerance in rice. *Journal of Plant Biology*, *60*(6), 539–547. https://doi.org/10.1007/s12374-016-0400-1

Shim, J. S., Oh, N., Chung, P. J., Kim, Y. S., Choi, Y. D., & Kim, J.-K. (2018). Overexpression of OsNAC14 improves drought tolerance in rice. *Frontiers in Plant Science*, *9*, 310.

Silhavy, D. (2002). A viral protein suppresses RNA silencing and binds silencing-generated, 21- to 25-nucleotide double-stranded RNAs. *The EMBO Journal*, *21*(12), 3070–3080. https://doi.org/10.1093/emboj/cdf312

Silva, G. F. F., Silva, E. M., Correa, J. P. O., Vicente, M. H., Jiang, N., Notini, M. M., Junior, A. C., Jesus, F. A. D., Castilho, P., Carrera, E., López-Díaz, I., Grotewold, E., Peres, L. E. P., & Nogueira, F. T. S. (2019). Tomato floral induction and flower development are orchestrated by the interplay between gibberellin and two unrelated microRNA-controlled modules. *New Phytologist*, *221*(3), 1328–1344. https://doi.org/10.1111/NPH.15492

Song, X., Li, Y., Cao, X., & Qi, Y. (2019). MicroRNAs and their regulatory roles in plant-environment interactions. *Annual Review of Plant Biology*, *70*, 489–525. https://doi.org/10.1146/ANNUREV-ARPLANT-050718-100334

Song, X., Liu, G., Huang, Z., Duan, W., Tan, H., Li, Y., & Hou, X. (2016). Temperature expression patterns of genes and their coexpression with LncRNAs revealed by RNA-Seq in non-heading Chinese cabbage. *BMC Genomics*, *17*(1), 297. https://doi.org/10.1186/s12864-016-2625-2

Spanu, P. D. (2015). RNA–protein interactions in plant disease: Hackers at the dinner table. *New Phytologist*, *207*(4), 991–995. https://doi.org/10.1111/nph.13495

Štorchová, H. (2017). The role of non-coding RNAs in cytoplasmic male sterility in flowering plants. *International Journal of Molecular Sciences*, *18*(11), 2429.

Struhl, K., & Segal, E. (2013). Determinants of nucleosome positioning. *Nature Structural & Molecular Biology*, *20*(3), 267–273.

Sun, M., Shen, Y., Chen, Y., Wang, Y., Cai, X., Yang, J., Jia, B., Dong, W., Chen, X., & Sun, X. (2022). Osa-miR1320 targets the ERF transcription factor OsERF096 to regulate cold tolerance via JA-mediated signaling. *Plant Physiology*, *189*(4), 2500–2516.

Sunkar, R., Chinnusamy, V., Zhu, J., & Zhu, J.-K. (2007). Small RNAs as big players in plant abiotic stress responses and nutrient deprivation. *Trends in Plant Science*, *12*(7), 301–309.

Sunkar, R., & Zhu, J.-K. (2004). Novel and stress-regulated microRNAs and other small RNAs from Arabidopsis. *The Plant Cell*, *16*(8), 2001–2019.

Swarup, R., & Denyer, T. (2019). miRNAs in plant development. *Annual Plant Reviews Online*, *2*(3), 689–712. https://doi.org/10.1002/9781119312994.APR0649

Temel, A., Janack, B., & Humbeck, K. (2017). Drought stress-related physiological changes and histone modifications in barley primary leaves at HSP17 gene. *Agronomy*, *7*(2), 43.

Tsuji, H., Aya, K., Ueguchi-Tanaka, M., Shimada, Y., Nakazono, M., Watanabe, R., Nishizawa, N. K., Gomi, K., Shimada, A., Kitano, H., Ashikari, M., & Matsuoka, M. (2006). GAMYB controls different sets of genes and is differentially regulated by microRNA in aleurone cells and anthers. *The Plant Journal*, *47*(3), 427–444. https://doi.org/10.1111/j.1365-313X.2006.02795.x

Ueda, M., & Seki, M. (2020). Histone modifications form epigenetic regulatory networks to regulate abiotic stress response. *Plant Physiology*, *182*(1), 15–26.

Upadhyay, S. K., Chandrashekar, K., Thakur, N., Verma, P. C., Borgio, J. F., Singh, P. K., & Tuli, R. (2011). RNA interference for the control of whiteflies (*Bemisia tabaci*) by oral route. *Journal of Biosciences*, *36*(1), 153–161. https://doi.org/10.1007/s12038-011-9009-1

Verma, S. S., Rahman, M. H., Deyholos, M. K., Basu, U., & Kav, N. N. (2014). Differential expression of miRNAs in *Brassica napus* root following infection with Plasmodiophora brassicae. *PLoS ONE*, *9*(1), e86648.

Wagh, N., Amane, P., Kshirsagar, R., Mandal, S., Kavi Kishor, P., & Suprasanna, P. (2025). Exploring the critical role of non-coding RNAs in plant development and stress adaptation-current status and insights. *Progress in Molecular Biology and Translational Science*. *214*, 19–64. https://doi.org/10.1016/bs.pmbts.2025.01.003

Wang, D., Qu, Z., Yang, L., Zhang, Q., Liu, Z. H., Do, T., Adelson, D. L., Wang, Z. Y., Searle, I., & Zhu, J. K. (2017). Transposable elements (TEs) contribute to stress-related long intergenic noncoding RNAs in plants. *Plant Journal*, *90*(1), 133–146. https://doi.org/10.1111/TPJ.13481

Wang, F., Wang, C., Liu, P., Lei, C., Hao, W., Gao, Y., Liu, Y.-G., & Zhao, K. (2016). Enhanced rice blast resistance by CRISPR/Cas9-targeted mutagenesis of the ERF transcription factor gene OsERF922. *PLoS ONE*, *11*(4), e0154027.

Wang, F., Xu, Y., Li, W., Chen, Z., Wang, J., Fan, F., Tao, Y., Jiang, Y., Zhu, Q.-H., & Yang, J. (2021). Creating a novel herbicide-tolerance OsALS allele using CRISPR/Cas9-mediated gene editing. *The Crop Journal*, *9*(2), 305–312.

Wang, F.-Z., Chen, M.-X., Yu, L.-J., Xie, L.-J., Yuan, L.-B., Qi, H., Xiao, M., Guo, W., Chen, Z., & Yi, K. (2017). OsARM1, an R2R3 MYB transcription factor, is involved in regulation of the response to arsenic stress in rice. *Frontiers in Plant Science*, *8*, 1868.

Wang, H., Beyene, G., Zhai, J., Feng, S., Fahlgren, N., Taylor, N. J., Bart, R., Carrington, J. C., Jacobsen, S. E., & Ausin, I. (2015). CG gene body DNA methylation changes and evolution of duplicated genes in cassava. *Proceedings of the National Academy of Sciences*, *112*(44), 13729–13734. https://doi.org/10.1073/pnas.1519067112

Wang, S., Sun, X., Hoshino, Y., Yu, Y., Jia, B., Sun, Z., Sun, M., Duan, X., & Zhu, Y. (2014). MicroRNA319 positively regulates cold tolerance by targeting OsPCF6 and OsTCP21 in rice (Oryza sativa L.). *PLoS ONE*, *9*(3), e91357.

Wang, T., Xun, H., Wang, W., Ding, X., Tian, H., Hussain, S., Dong, Q., Li, Y., Cheng, Y., & Wang, C. (2021). Mutation of GmAITR genes by CRISPR/Cas9 genome editing results in enhanced salinity stress tolerance in soybean. *Frontiers in Plant Science*, *12*, 779598.

Wang, Y., Fan, X., Lin, F., He, G., Terzaghi, W., Zhu, D., & Deng, X. W. (2014). Arabidopsis noncoding RNA mediates control of photomorphogenesis by red light. *Proceedings of the National Academy of Sciences of the United States of America*, *111*(28), 10359–10364. https://doi.org/10.1073/PNAS.1409457111

Wang, Y., Luo, X., Sun, F., Hu, J., Zha, X., Su, W., & Yang, J. (2018). Overexpressing lncRNA LAIR increases grain yield and regulates neighbouring gene cluster expression in rice. *Nature Communications*, *9*(1). https://doi.org/10.1038/S41467-018-05829-7

Weiberg, A., Wang, M., Lin, F.-M., Zhao, H., Zhang, Z., Kaloshian, I., Huang, H.-D., & Jin, H. (2013). Fungal Small RNAs Suppress Plant Immunity by Hijacking Host RNA Interference Pathways. *Science*, *342*(6154), 118–123. https://doi.org/10.1126/science.1239705

Wierzbicki, A. T., Ream, T. S., Haag, J. R., & Pikaard, C. S. (2009). RNA polymerase v transcription guides ARGONAUTE4 to chromatin. *Nature Genetics*, *41*(5), 630–634. https://doi.org/10.1038/NG.365

Wu, H.-J., Wang, Z.-M., Wang, M., & Wang, X.-J. (2013). Widespread long noncoding RNAs as endogenous target mimics for microRNAs in plants. *Plant Physiology*, *161*(4), 1875–1884.

Xu, W.-B., Guo, Q.-H., Liu, P., Dai, S., Wu, C.-A., Yang, G.-D., Huang, J.-G., Zhang, S.-Z., Song, J.-M., Zheng, C.-C., & Yan, K. (2024). A long non-coding RNA functions as a competitive endogenous RNA to modulate TaNAC018 by acting as a decoy for tae-miR6206. *Plant Molecular Biology*, *114*(3), 36. https://doi.org/10.1007/s11103-024-01448-7

Xu, Z., Xu, X., Gong, Q., Li, Z., Li, Y., Wang, S., Yang, Y., Ma, W., Liu, L., & Zhu, B. (2019). Engineering broad-spectrum bacterial blight resistance by simultaneously disrupting variable TALE-binding elements of multiple susceptibility genes in rice. *Molecular Plant*, *12*(11), 1434–1446.

Yu, C., Chen, Y., Cao, Y., Chen, H., Wang, J., Bi, Y.-M., Tian, F., Yang, F., Rothstein, S. J., & Zhou, X. (2018). Overexpression of miR169o, an overlapping microRNA in response to both nitrogen limitation and bacterial infection, promotes nitrogen use efficiency and susceptibility to bacterial blight in rice. *Plant and Cell Physiology*, *59*(6), 1234–1247.

Yu, X., Meng, X., Liu, Y., Li, N., Zhang, A., Wang, T.-J., Jiang, L., Pang, J., Zhao, X., Qi, X., Zhang, M., Wang, S., Liu, B., & Xu, Z.-Y. (2018). The chromatin remodeler ZmCHB101 impacts expression of osmotic stress-responsive genes in maize. *Plant Molecular Biology*, *97*(4–5), 451–465. https://doi.org/10.1007/s11103-018-0751-8

Yu, Y., Jia, T., & Chen, X. (2017). The 'how' and 'where' of plant microRNAs. *New Phytologist*, *216*(4), 1002–1017. https://doi.org/10.1111/NPH.14834

Yu, Y., Pan, Z., Wang, X., Bian, X., Wang, W., Liang, Q., Kou, M., Ji, H., Li, Y., Ma, D., Li, Z., & Sun, J. (2022). Targeting of SPCSV- *RNase3* via CRISPR-Cas13 confers resistance against sweet potato virus disease. *Molecular Plant Pathology*, *23*(1), 104–117. https://doi.org/10.1111/mpp.13146

Yu, Y., Zhang, Y., Chen, X., & Chen, Y. (2019). Plant noncoding RNAs: Hidden players in development and stress responses. *Annual Review of Cell and Developmental Biology*, *35*, 407–431. https://doi.org/10.1146/ANNUREV-CELLBIO-100818-125218

Zemach, A., Kim, M. Y., Silva, P., Rodrigues, J. A., Dotson, B., Brooks, M. D., & Zilberman, D. (2010). Local DNA hypomethylation activates genes in rice endosperm. *Proceedings of the National Academy of Sciences*, *107*(43), 18729–18734. https://doi.org/10.1073/pnas.1009695107

Zhai, J., Zhang, H., Arikit, S., Huang, K., Nan, G. L., Walbot, V., & Meyers, B. C. (2015). Spatiotemporally dynamic, cell-type-dependent premeiotic and meiotic phasiRNAs in maize anthers. *Proceedings of the National Academy of Sciences of the United States of America*, *112*(10), 3146–3151. https://doi.org/10.1073/PNAS.1418918112

Zhan, X., Zhang, F., Zhong, Z., Chen, R., Wang, Y., Chang, L., Bock, R., Nie, B., & Zhang, J. (2019). Generation of virus-resistant potato plants by RNA genome targeting. *Plant Biotechnology Journal*, *17*(9), 1814–1822. https://doi.org/10.1111/pbi.13102

Zhang, F., Luo, X., Zhou, Y., & Xie, J. (2016). Genome-wide identification of conserved microRNA and their response to drought stress in Dongxiang wild rice (*Oryza rufipogon* Griff.). *Biotechnology Letters*, *38*(4), 711–721. https://doi.org/10.1007/S10529-015-2012-0

Zhang, F., Yang, J., Zhang, N., Wu, J., & Si, H. (2022). Roles of microRNAs in abiotic stress response and characteristics regulation of plant. *Frontiers in Plant Science*, *13*, 919243.

Zhang, H., Lang, Z., & Zhu, J.-K. (2018). Dynamics and function of DNA methylation in plants. *Nature Reviews Molecular Cell Biology*, *19*(8), 489–506.

Zhang, H., Zhu, J., Gong, Z., & Zhu, J.-K. (2022). Abiotic stress responses in plants. *Nature Reviews Genetics*, *23*(2), 104–119.

Zhang, L., Nakagomi, Y., Endo, T., Teranishi, M., Hidema, J., Sato, S., & Higashitani, A. (2018). Divergent evolution of rice blast resistance Pi54 locus in the genus Oryza. *Rice*, *11*(1), 63. https://doi.org/10.1186/s12284-018-0256-8

Zhang, Y., Iaffaldano, B., & Qi, Y. (2021). CRISPR ribonucleoprotein-mediated genetic engineering in plants. *Plant Communications*, *2*(2).

Zheng, M., Lin, J., Liu, X., Chu, W., Li, J., Gao, Y., An, K., Song, W., Xin, M., & Yao, Y. (2021). Histone acetyltransferase TaHAG1 acts as a crucial regulator to strengthen salt tolerance of hexaploid wheat. *Plant Physiology*, *186*(4), 1951–1969.

Zheng, Z., Yang, J., Wang, X., Zhang, N., & Si, H. (2023). Potato Stu-miR398b-3p Negatively Regulates Cu/Zn-SOD Response to Drought Tolerance. *International Journal of Molecular Sciences*, *24*(3), Article 3. https://doi.org/10.3390/ijms24032525

Zhou, M., Li, D., Li, Z., Hu, Q., Yang, C., Zhu, L., & Luo, H. (2013). Constitutive expression of a miR319 gene alters plant development and enhances salt and drought tolerance in transgenic creeping bentgrass. *Plant Physiology*, *161*(3), 1375–1391.

Zhu, J.-K. (2002). Salt and drought stress signal transduction in plants. *Annual Review of Plant Biology*, *53*(1), 247–273. https://doi.org/10.1146/annurev.arplant.53.091401.143329

Zhu, J.-K. (2016). Abiotic stress signaling and responses in plants. *Cell*, *167*(2), 313–324.

Zhu, K. Y., & Palli, S. R. (2020). Mechanisms, Applications, and Challenges of Insect RNA Interference. *Annual Review of Entomology*, *65*(1), 293–311. https://doi.org/10.1146/annurev-ento-011019-025224

Zhu, L.-J., Liu, Q.-Q., Wilson, J. D., Gu, M. H., & Shi, Y. C. (2011). *Digestibility and physicochemical properties of rice (Oryza sativa L.) flours and starches differing in amylose content.*

Zilberman, D., Cao, X., Johansen, L. K., Xie, Z., Carrington, J. C., & Jacobsen, S. E. (2004). Role of Arabidopsis ARGONAUTE4 in RNA-directed DNA methylation triggered by inverted repeats. *Current Biology*, *14*(13), 1214–1220.

Zuo, D.-D., Ahammed, G. J., & Guo, D.-L. (2023). Plant transcriptional memory and associated mechanism of abiotic stress tolerance. *Plant Physiology and Biochemistry*, 201, 107917. https://doi.org/10.1016/j.plaphy.2023.107917

2 Molecular Insights

ncRNAs and Their Impact on Abiotic Stress Tolerance in Plants

Hirdayesh Anuragi, Arun Kumar Handa, Udit Nandan Mishra, Rajesh Kumar Singhal and Ayyanadar Arunachalam

2.1 INTRODUCTION

Climate change and subsequent weather extremes promoted the abiotic stresses like prolonged drought, temperature extremes like heat waves and cold spells, soil salinisation, and biotic stresses such as pest and disease attacks, which cause significant plant growth reduction at different stages and finally the crop yield. The plants' defences are weakened by these forces, leaving them more vulnerable to future insect pest attacks. Enhancing plant resilience to these challenges is crucial for optimising agricultural production to meet the growing food security demand worldwide. Non-coding RNAs (ncRNAs) are the RNA molecules found in the genomes of complex organisms; though they do not encode any proteins, they are still functional and are potentially involved in numerous molecular mechanisms. Through transcriptional and post-transcriptional regulation of gene expression, they play a crucial role in controlling the growth and developmental processes of plants and also significantly contribute to a variety of biotic and abiotic stressors (Jin et al., 2024; Traubenik et al., 2024). Interestingly, the recent advancements in molecular tools and techniques have made it easier to identify the multifaceted roles of ncRNAs in controlling the diverse molecular mechanisms that facilitate plant adaptation and mitigation to these stresses. Various plant growth regulators, like auxin, cytokinin, abscisic acid, gibberellins, ethylene and jasmonic acid, involved in regulating overall growth and stress mechanisms in plants, are also influenced by ncRNA interactions through hormonal signalling pathways and impart the resilience towards environmental abnormalities at various growth stages. They also impart the epigenetic effects on other target genes through influencing chromatin structures, DNA methylation and modification of histones and modify the plant growth and stress tolerance responses. ncRNAs have been classified into constitutive ncRNAs and regulatory ncRNAs (Waititu et al., 2020) on the basis of their functions. Interestingly, contemporary high-throughput NGS and omics technologies have allowed us to identify various classes of constitutive ncRNAs, viz., ribosomal RNA (rRNA), small nuclear RNA (snRNA), small nucleolar RNA (snoRNA), and circular RNAs (circRNAs) and transfer RNA (tRNA) and regulatory ncRNAs viz., small interfering RNAs (siRNAs), Piwi-interacting RNA (piRNA), micro-RNA (miRNA), circular RNA (circRNA), and long non-coding RNA (lncRNA) which are found potentially involved in modulating array of stress-responsive genes, signalling and physiological and mechanisms, as well as intricate gene expression networks, allowing plants to adapt to adverse climatic circumstances (Bordoloi and Agarwala, 2021; Jin et al., 2024; Yadav et al., 2024). The miRNA-based research is quite better documented, while lncRNA and transposon-derived RNA-based studies are required. Additionally, it has been discovered that plants retain stress-induced molecular information as proteins, hormones or metabolites, which aids in creating plant stress memories for future (Kashyap et al., 2024). The non-coding RNA biogenesis with their generalised mode of action at epigenetic, transcription and post-transcription levels and role and different underlying mechanisms of ncRNA for abiotic stress tolerance are highlighted in Figure 2.1. An in-depth understanding of roles of ncRNAs in different regulatory networks will open up a new era for developing climate-resilient crop cultivars. The advancement in high-throughput NGS sequencing and bioinformatics science, and the discovery and quantification of diverse range of ncRNAs along with their roles have become significantly easier in the way like never before.

2.2 MOLECULAR INSIGHTS ON THE ROLE OF ncRNAS ON MITIGATING THE MAJOR ABIOTIC STRESSES

2.2.1 Drought Stress

Drought stress tolerance represents one of the major widespread abiotic stresses and is mostly governed by complex quantitative traits in plants. Molecular mechanisms governing the tolerance to drought stress may vary depending upon gene expression patterns, water transport and osmotic balance, oxidative homoeostasis and repair mechanisms. The game-changing role of ncRNAs in modulating the targeted gene expressions and molecular pathways has been identified in determining the overall response of plants amid varieties of drought stresses. Several drought-responsive ncRNAs have been identified and characterised in different crop plants (Chen et al., 2020; Deng et al.,

DOI: 10.1201/9781003583257-2

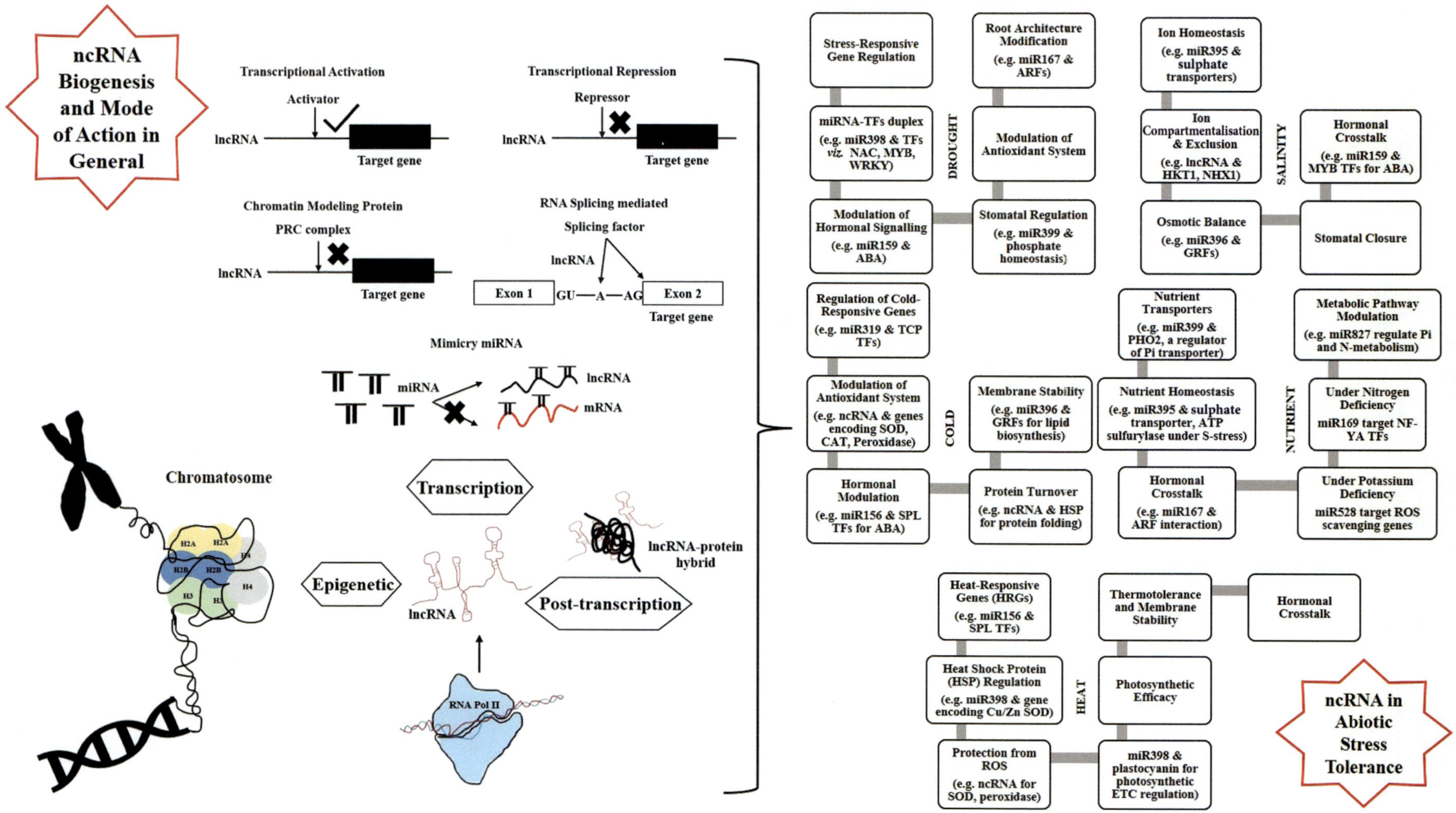

FIGURE 2.1 Non-coding RNA biogenesis with their generalized mode of action at epigenetic, transcription and post-transcription levels and role and different underlying mechanisms of ncRNA for abiotic stress tolerance.

2018). Studies on various plants, such as rice and cotton, have shown that lncRNAs are essential to the resistance of plants to drought stress (Yu et al., 2019). lncRNAs like DRIL1, DRIL2, DRIL3, DRIL4, DRIL5 and DRIL6 targeted the ABA biosynthesis, and MIKKI targeted the root architecture by mimicking the miR171-SCR module, and miRNAs such as miR171f altered the root development and growth during drought stress in *Oryza sativa* (Chung et al., 2016; Oh et al., 2022). The finding has important ramifications for comprehending the intricate relationships between lncRNAs and microRNAs in modulating root architecture, ultimately influencing plant growth and productivity. The role of photosynthesis-related gene expressions modulates the carbon fixation and chlorophyll metabolism to activate plants to improve overall physiological efficiency during drought stress. In *Arabidopsis thaliana*, the physiological mechanisms like photosynthesis, chlorophyll synthesis, fatty acid synthesis and ABA-signal transducers were influenced by ASCO, APOLO, DRIR AT1G09627, AT3G24518, AT2G35945 and MSTRG.35793 lncRNAs to adopt the moisture stress (Chen et al., 2021; Qin et al., 2017). In the same plant, miRNAs, like miR156, miR168, miR158, miR16, miR165, miR159, miR167, miR169, miR171, miR393, miR319, miR394 and miR396, played a major role in controlling the stem cell differentiation and root expansion and regulating phytohormone cross talk (Dong et al., 2022a; Liu et al., 2008). In *Zea mays*, miR164 regulated the NAC1 expression which affected the auxin transmission and lateral root growth during drought (Pandita, 2022). Similarly, in *Solanum lycopersicum*, lncRNA_tomato_467 and lncRNA_tomato_025 governed the control of the stomatal K^+ chloroplast and energy balance transportation (Eom et al., 2019). Secondary metabolites are also extremely important for tolerating abiotic stresses, including drought stress. lncRNAs and mRNAs like TCONS_00055787, TCONS_00038334, DElncRNAs and DEmRNAs were reported to influence the flavonoid biosynthesis to mitigate the drought condition in *Beta vulgaris* (Zou et al., 2023). lncRNAs like lncRNA340 were reported in cassava which regulated the proline and MDA accumulation and mimicked miR169-NF-Y regulatory interaction during reduced water availability (Dong et al., 2022b).

The evolving roles of circRNAs and siRNAs in drought stress tolerance have also been reported in few crops. Drought-responsive circRNAs have recently been identified in plant species, including 33 in *Pyrus betulifolia* Bunge (Wang et al., 2018), maize and *Arabidopsis* (Zhang et al., 2019b). CircGORK was identified to be overexpressed to enhance moisture stress tolerance (Zhang et al., 2019b) in *Z. mays*. Wang et al. (2018) reported the crucial role of circRNA2421 in regulating the drought responses by targeting miR319 and miR156 in commercial crops like *Nicotiana tabacum*, while Shuai et al. (2016) reported the small interfering RNA, SIR57 and SIR 60 to be associated with expressing miR6445-PHA18 for drought stress tolerance in *Populus trichocarpa*.

2.2.2 Salinity Stress

Salinity stress usually arises from the excessive buildup of water-soluble salts, such as NaCl, leading to osmotic stress, ionic stress and oxidative stress in the plants. This condition has numerous damaging consequences on the growth and development of plants. High concentrations of Na^+ and Cl^- ions in the soil significantly impact the overall growth and productivity of glycophytes, particularly crop plants, by causing osmotic stress, ionic toxicity and oxidative stress (Saharan et al., 2022). lncRNAs are found to have a crucial role in helping plants tolerate salinity stress and influence the RNA modification mechanisms under saline conditions. Moreover, lncRNAs may function as endogenous target mimics (eTMs) for miRNAs that respond to salinity stress besides serving as precursors for small non-coding RNAs that are responsive to salinity stress. Salinity stress tolerance in crops has been found to be significantly influenced by a variety of ncRNA types. Altered expressions of *GhPHO2* gene by lncRNA973 and lncRNA354 have been reported in *Gossypium hirsutum* under salinity stress (Deng et al., 2018). Likewise, AK372815 expression responsible for pyridoxal biosynthesis pathway in *Hordeum vulgare* was affected by AK363462, AK370506 and AK370814 (Karlik et al., 2018). During salt stress, lnc_388 targeted PCGs *via* cis-regulation in *G. hirsutum* (Zhang et al., 2019a), and lncRNA.2-FL upregulated the FL478 to induce salinity stress tolerance in *O. sativa* (Mirdar Mansuri et al., 2022). Regulation of auxin-responsive protein and ethylene-responsive TF ABR1 under salinity stress was influenced by MSTRG.139242.1, MSTRG.49718.1 and MSTRG.54711.1 in *Camellia sinensis* (Wan et al., 2020). TCONS_00024229, TCONS_00045028, TCONS_00018576 and TCONS_00037548 influenced chloroplast lipocalin, redox homoeostasis, ABA signalling and ROS metabolism in *Lemna minor* to tolerate the salinity stress (Fu et al., 2020). lncRNAs like Vvlnc_14.529, Vvlnc_16.1373.1, VIT_202s0025n00280, VIT_219s0014n00100, VIT_210s0003n00350 and Vvlnc_un.4287 influenced the gene expression conferring salinity stress tolerance in *Vitis vinifera* (Jin et al., 2020). Similarly, in *Lycopersicon esculentum*, gene expression conferring salt stress tolerance is influenced by sly-miR164a-5p, sly-miR164b-5p and sly-miR319d (Li et al., 2022a). A complex of miRNA-eTM lncRNA interacted with genes, *viz.*, car-miRNA015, Cat-miR159g-3p and CatmiR172c.2 under saline conditions in *Cicer arietinum* (Kumar et al., 2021). In *O. sativa*, lncRNA–miRNA–mRNA network complex influenced the nature of expression of phenylpropanoid biosynthesis pathway (Jain et al., 2021).

2.2.3 Heat Stress

Heat stress or high temperature influences cellular homoeostasis that leads to severe reduction in growth and development, increased risk of disease and even death of plants. Heat stress response mechanisms in plants have received increasing interest in the last decade. ncRNAs play crucial roles by targeting molecular mechanisms and pathways to impart tolerance to heat stress in plants. Genome sequencing technologies and genome-scale approaches have unveiled the roles of ncRNAs and how they influence the responsive genes and mechanisms involved in heat stress mitigation. In recent years, a rapid increase in the number of ncRNAs like miRNAs, siRNAs and lncRNAs has been observed that influence transcriptional and post-transcriptional mechanisms of gene expression, modulation of RNA stability and translation function under heat stress exposures in agricultural and non-agricultural crops. Among miRNAs, miR398 is the most studied one that plays a significant role in such heat stress responses in plants. In *A. thaliana*, *Brassica rapa* and *Populus tomentosa*, miR398 targeted *CSD1*, *CSD2*, *Cox5b-1* and *CCS1* genes that control ROS scavenging mechanisms (Guan et al., 2013; Kotak et al. 2007; Yu et al., 2012; Zhu et al., 2011), while miR156 and miR172 influenced the expression of *SPL*, *AP2*, *FT* and *FUL* for flowering traits in the plants (Kim et al., 2012; Stief et al., 2014). Heat stress downregulated the miR172 and its targeted genes in *A. thaliana*, *Triticum aestivum* and *Helianthus annuus* (Li et al., 2014). In *Arabidopsis*, miR173-cleaved ta-siRNA (*TAS1*) induces thermotolerance by targeting *HTT1* and *HTT2* genes (Li et al., 2014). A major type of ncRNA includes lncRNAs which were reported to be overexpressed in *T. aestivum* (TalnRNA27, TalnRNA5) and *Prunus simonii* (*PsiLncRNA00268512*) and play a crucial role in heat tolerance (Song et al., 2016). A complex of lncRNA–miRNA–mRNA has been reported to be involved in pollen development under heat stress conditions in *T. aestivum* (Babaei et al., 2024).

2.2.4 Cold Stress

Though cold stress is a minor abiotic stress, it can significantly impact a plant's growth, development, yield and geographical distribution. Cold temperatures can affect critical cellular functions like photosynthesis, respiration and membrane permeability. So, it is crucial to comprehend the processes of tolerance to protect crops against chilling stress, especially in temperate regions. Several cold-responsive ncRNAs have been reported in range of crops. An overexpression of *miR156* in *N. tabacum* leaves altered the ERD gene family, *ERD10B*, *ERD10C*, *ERD10D* and *SPL9* which affected the anthocyanin levels, H_2O_2 accumulation and subsequent ROS scavenging during cold stress (Yang et al., 2017). *SPL9*, targeted by miR156, inhibited the genes for anthocyanin synthesis through interfering with MYB–BHLH–WD40 complex, resulting in anthocyanin accumulation in *A. thaliana* (Gou et al., 2011). Similarly, miR319 regulated the cold stress tolerance by targeting *OsPCF6* and *OsTCP21* in *O. sativa* L. (Wang et al., 2014) and AtmiR319 by targeting *AtTCP3* by *A. thaliana* (Li & Zachgo, 2013). Likewise, miR397 in *Arabidopsis* improved the tolerance to cold stress by targeting *LACs* and *CKB3* of cold signalling pathway (Dong & Pei, 2014). In sugarcane, miR319 upregulated the expression of *PCF5* and *PCF6* during cold stress (Thiebaut et al., 2012). The miR397 mimicked FRILAIR lncRNA and modulated the anthocyanin biosynthetic mechanism through *LAC11a* in the process of fruit ripening of *Fragaria ananassa* (Tang et al., 2021).

2.2.5 Mineral Stress

Mineral stress is another abiotic stress often encountered by the plant species. Both macronutrients (N, P, K, Ca, Mg, S) and micronutrients (Fe, Zn) are vital for normal growth and development of the plants. However, mineral excess or deficiency significantly triggers a number of plant responses at phenotypic and molecular levels during its growth cycle. In the case of nitrogen, NRT1, NPF, NRT2 and CLC protein families are majorly involved in uptake, transportation and assimilation in the plants. The miR169 mitigated the N starvation by targeting *AtNFYA* both in roots and shoots under limiting N in *A. thaliana* (Zhao et al., 2011). Osa-miR166 could target a transcription factor gene, OsRDD1, while overexpressing OsRDD1 induced the gene expression of controlling the uptake process of $NH4^+$, Na^+, SO_4^{2-}, Cl^- and PO_4^{3-}, thus enhancing N responsiveness and grain yield of *O. sativa* (Iwamoto & Tagiri, 2016). Genome-wide studies identified the N-responsive intergenic and intronic lncRNAs in *Z. mays* (Lv et al., 2016). miRNAs overexpressed MdATG18a and MdATG19 which improved the resistance to N starvation in *Musca domestica* (Huo et al., 2020). While in the case of phosphorous, in *A. thaliana*, At5G01591.1 and *AtFer1* (ferritin) transcripts along with their promoter containing *P1BS* motif were overexpressed under P_i starvation and indicating that both are also regulated by *PHR1* for P_i and iron homoeostasis (Yuan et al., 2016). In *Medicago truncatula* grown under P stress, three lncRNAs *PDILs*, viz., *PDIL1*, *PDIL2* and *PDIL3*, are involved in suppression of degradation of *MtPHO2* to regulate P_i transport. miR827 was reported to target AT4G23030 and AT1G61890, governing organic acid transport efflux during P deficiency (Vengavasi et al., 2016). Novel lncRNAs have been reported to express under P stress in *M. truncatula* (Wang et al., 2017). In *Phaeodactylum tricornutum*, two lncRNAs – pti-MIR5472 and pti-MIR5471, which are the precursors of annotated miRNAs – were markedly elevated in response to P deficiency, suggesting that such lncRNAs shared characteristics with N and P deficient stresses (Cruz de Carvalho et al., 2016). The ncRNAs are found to regulate potassium (K) stress also, which have been reported in few crops; however, there is still some unclarity. Osa-miR444a, one of the few miRNAs known to regulate the accumulation of both N and P minerals, was downregulated, and its target gene, OsMADS-23, was markedly increased in response to K deprivation (Yan et al., 2014). Low K conditions induced Hvu-miR319 and Hvu-miR169h which suppressed its targeted genes, HvTCP4 and HvGRF, for mitigating K stress in barley (Ye et al., 2021; Zeng et al., 2019). The different ncRNA molecules involved in abiotic stress tolerance mechanisms are highlighted in Table 2.1.

TABLE 2.1
Different ncRNA Molecules Involved in Abiotic Stress Tolerance Mechanisms

Plant Species	Types of ncRNAs	Pathways	Specific Functions	References
		Drought Stress		
Oryza sativa	lncRNAs	*DRIL1, DRIL2, DRIL3, DRIL4, DRIL5, DRIL6,*	ABA biosynthesis	Oh et al. (2022)
O. sativa	miRNAs	*miR171f*	Root development and growth during stress	Chung et al. (2016)
O. sativa	lncRNAs	*MIKKI*	Shapes root architecture by mimicking the miR171-SCR module	Jungnam Cho and Jerzy Paszkowski (2017)
Cassava	lncRNAs	*DIR*	Proline and MDA accumulation	Dong et al. (2022b)
Arabidopsis thaliana	lncRNAs	ASCO, APOLO	Chlorophyll synthesis, photosynthetic pathway, biosynthesis and degradation of fatty acids	Qin et al. (2017)
A. thaliana	lncRNAs	*DRIR*	Regulates ABA-signal transducers	Qin et al. (2017)
A. thaliana	miRNAs	*miR156, miR168, miR158, miR16, miR165, miR159, miR167, miR169, miR171, miR393, miR319, miR394, miR396*	Stem cell differentiation, root expansion, regulation of phytohormone cross talk	Liu et al. (2008); Dong et al. (2022a)
A. thaliana	lncRNAs	*AT1G09627, AT3G24518, AT2G35945, MSTRG.35793*	Chlorophyll synthesis, photosynthetic pathway, biosynthesis and degradation of fatty acids	Chen et al. (2021)
Beta vulgaris	lncRNAs, mRNAs	*TCONS_00055787, TCONS_00038334, DElncRNAs, DEmRNAs*	Flavonoid biosynthesis	Zou et al. (2023)
Zea mays	miRNA	*miR164*	Regulates NAC1 expression, affecting auxin transmission and lateral root growth	Pandita (2022)
Z. mays	*circRNA*	*circGORK*	Improved drought tolerance	Zhang et al. (2019b)
Tobacco	*circRNA*	*circRNA2421*	Regulate drought responses by targeting miR319 and miR156	Wang et al. (2018)
Populus trichocarpa	*siRNA*	*siR57, siR60*	Target expression of miR6445-PHA18 drought adaptation	Shuai et al. (2016)
Gossypium hirsutum	lncRNAs	*XLOC_063105, XLOC_115463*	Regulating plant hormone pathways during moisture stress	Lu et al. (2016)
Solanum lycopersicum	lncRNAs	*lncRNA_tomato_467, lncRNA_tomato_025*	Stomatal K^+ chloroplast and energy balance transportation	Eom et al. (2019)
		Salinity Stress		
O. sativa	lncRNA–miRNA–mRNA network	DEmRNAs, miRNAs, lncRNAs	Integral part of phenylpropanoid biosynthesis pathway (PPP)	Jain et al. (2021)
O. sativa	lncRNAs	lncRNA.2-FL	Regulate the FL478 for tolerance to salinity stress	Mirdar Mansuri et al. (2022)
G. hirsutum	lncRNAs	lnc_388	Target for PCGs via cis-regulation under salinity stress	Zhang et al. (2019a)
G. hirsutum	lncRNAs	lncRNA973, lncRNA354	Influence gene *GhPHO2* expressions	Deng et al. (2018)
Hordeum vulgare	lncRNAs	AK363462, AK370506, AK370814	Affects AK372815 expression for pyridoxal biosynthesis pathway	Karlik et al. (2018)
Camellia sinensis	lncRNAs	MSTRG.139242.1, MSTRG.49718.1, MSTRG.54711.1	Regulation of auxin-responsive protein and ethylene-responsive TF ABR1	Wan et al. (2020)
Lemna minor	lncRNAs	TCONS_00024229, TCONS_00045028	Influence ABA signalling and ROS metabolism	Fu et al. (2020)
L. minor	lncRNAs	TCONS_00018576, TCONS_00037548	Influence chloroplast lipocalin, redox homoeostasis and ROS metabolism	Fu et al. (2020)
Vitis vinifera	lncRNAs	Vvlnc_14.529, Vvlnc_16.1373.1, VIT_202s0025n00280, VIT_219s00, 14n00100, VIT_210s0003n00350, Vvlnc_un.4287	Alter gene expression for conferring salinity tolerance	Jin et al. (2020)

TABLE 2.1 (*Continued*)
Different ncRNA Molecules Involved in Abiotic Stress Tolerance Mechanisms

Plant Species	Types of ncRNAs	Pathways	Specific Functions	References
Lycopersicon esculentum	lncRNAs	sly-miR164a-5p, sly-miR164b-5p, sly-miR319d	Influence gene expression for conferring salinity tolerance	Li et al. (2022)
Cicer arietinum	miRNAs	miRNA-eTM lncRNA	Interact with genes, *viz*., car-miRNA015, Cat-miR159g-3p and CatmiR172c.2	Kumar et al. (2021)
Morus alba	lncRNAs	MuLnc1	Enhances siRNA production	Gai et al. (2018)
		Heat Stress		
Deinococcus radiodurans	ncRNA	*dsr11*	Influence *trmE* and *dr_0651* mRNA	Xue et al. (2019)
Triticum aestivum	lncRNA–miRNA–mRNA	TapmlnRNA8	Involved in pollen development under heat stress	Babaei et al. (2024)
A. thaliana, Brassica rapa, Populus tomentosa	miRNAs	miRNA398	*CSD1, CSD2, Cox5b-1, CCS1*	Zhu et al. (2011); Kotak et al. (2007); Yu et al. (2012)
B. rapa	miRNAs	miR156	Targets *BracSPL2*	Yu et al. (2012)
Helianthus annuus	miRNAs	miR172	Targets *AP2-like* genes	Li S. X. et al. (2014)
Prunus simonii	lncRNAs	*PsiLncRNA00268512*	Influence target genes for heat stress tolerance	Song et al. (2016)
		Cold Stress		
Wheat	miRNAs	*tae-miR398*	Regulate *CSD1* expression	Lu et al. (2020)
Wheat	lncRNAs	*lncR9A, lncR117, lncR616*	Regulate *CSD1* expression	Lu et al. (2020)
Medicago truncatula	lncRNAs	*MtCIR1, MtCBF*s	Transcription regulatory factors of cold-responsive genes	
A. thaliana		*AtCBF2* and *AtCBF3*	Pol II transcription of *CBF*s	
O. sativa	miRNAs	*miR319*	Influences *TCPs, GAMYB*	Yang et al. (2013)
Panicum virgatum	miRNAs	*miR393*	Influences expression of *Osa-miR393a*	Liu et al. (2017)
		Mineral Deficiency Stress		
Brassica napus	miRNAs	miR395, miR398 and miR399	Sulphate, copper or phosphate deficiency homoeostasis	Li et al. (2021)
A. thaliana	miRNAs	miR399, miR827, miR2112, miR2112a	Phosphate homoeostasis	Fan et al. (2022)

2.3 CONCLUSION

ncRNAs play diverse and essential roles in fostering plant resilience to environmental change and agricultural sustainability. Agriculture production and sustainability can be improved indirectly through utilising these ncRNAs for developing stress-tolerant crop varieties. However, type of ncRNA, plant species and stress decides the mechanism of action of tolerance to abiotic stressors in different cases. According to studies, some ncRNAs are either elevated or downregulated under particular stimuli, such as heat, cold, salinity or drought. The ncRNAs from various plant species have been reported to have either up- or downregulation under stress, demonstrating their crucial function in stress management. Breeding efforts can create high-yielding, stress-tolerant agricultural types by incorporating ncRNA expertise. Targeted ncRNA manipulation can be achieved through genome editing and marker-assisted selection. Improved production and quality could result from targeted ncRNA alteration that increases crop tolerance to such environmental stressors. It will be easier to identify, characterise and manipulate ncRNAs for agricultural applications as long as high-throughput NGS sequencing, genomics, bioinformatics and advanced genome editing tools continue to progress. Future studies that concentrate on comprehending the mechanisms, functions and complex interactions between ncRNA and other regulatory molecules along with how these interactions translate into agricultural applications will help ensure food security through sustainable crop production considering the climate change.

REFERENCES

Babaei, S., Bhalla, P. L., & Singh, M. B. (2024). Identifying long non-coding RNAs involved in heat stress response during wheat pollen development. *Frontiers in Plant Science*, *15*, 1344928.

Bordoloi, K. S., & Agarwala, N. (2021). MicroRNAs in plant insect interaction and insect pest control. *Plant Gene*, *26*, 100271.

Chen, K., Huang, Y., Liu, C., Liang, Y., & Li, M. (2021). Transcriptome profile analysis of *Arabidopsis* reveals the drought stress-induced long non-coding RNAs associated with photosynthesis, chlorophyll synthesis, fatty acid synthesis and degradation. *Frontiers in Plant Science*, *12*, 643182.

Chen, L., Zhu, Q. H., & Kaufmann, K. (2020). Long non-coding RNAs in plants: emerging modulators of gene activity in development and stress responses. *Planta*, *252*(5), 92.

Cho, J., & Paszkowski, J. (2017). Regulation of rice root development by a retrotransposon acting as a microRNA sponge. *elife*, *6*, e30038.

Chung, P. J., Jung, H., Jeong, D. H., Ha, S. H., Choi, Y. D., & Kim, J. K. (2016). Transcriptome profiling of drought responsive noncoding RNAs and their target genes in rice. *BMC Genomics*, *17*, 1–12.

Cruz de Carvalho, M. H., Sun, H. X., Bowler, C., & Chua, N. H. (2016). Noncoding and coding transcriptome responses of a marine diatom to phosphate fluctuations. *New Phytologist*, *210*(2), 497–510.

Deng, F., Zhang, X., Wang, W., Yuan, R., & Shen, F. (2018). Identification of *Gossypium hirsutum* long non-coding RNAs (lncRNAs) under salt stress. *BMC Plant Biology*, *18*, 1–14.

Dong, C. H., & Pei, H. (2014). Over-expression of miR397 improves plant tolerance to cold stress in *Arabidopsis thaliana*. *Journal of Plant Biology*, *57*, 209–217.

Dong, Q., Hu, B., & Zhang, C. (2022a). microRNAs and their roles in plant development. *Frontiers in Plant Science*, *13*, 824240.

Dong, S. M., Liang, X. I. A. O., Li, Z. B., Jie, S. H. E. N., Yan, H. B., Li, S. X., … & Ming, P. E. N. G. (2022b). A novel long non-coding RNA, DIR, increases drought tolerance in cassava by modifying stress-related gene expression. *Journal of Integrative Agriculture*, *21*(9), 2588–2602.

Eom, S. H., Lee, H. J., Lee, J. H., Wi, S. H., Kim, S. K., & Hyun, T. K. (2019). Identification and functional prediction of drought-responsive long non-coding RNA in tomato. *Agronomy*, *9*(10), 629.

Fan, F., Shang, X., Ding, G., Zhou, Z., & Tan, J. (2022). Integrated mRNA and miRNA expression analyses of *Pinus massoniana* roots and shoots in long-term response to phosphate deficiency. *Journal of Plant Growth Regulation*, *41*(7), 2949–2966.

Fu, L., Ding, Z., Tan, D., Han, B., Sun, X., & Zhang, J. (2020). Genome-wide discovery and functional prediction of salt-responsive lncRNAs in duckweed. *BMC Genomics*, *21*, 1–14.

Gai, Y. P., Yuan, S. S., Zhao, Y. N., Zhao, H. N., Zhang, H. L., & Ji, X.L. (2018). A novel LncRNA, MuLncl, associated with environmental stress in mulberry (*Morus multicaulis*). *Frontiers in Plant Science*, *9*, 669.

Gou, J. Y., Felippes, F. F., Liu, C. J., Weigel, D., & Wang, J. W. (2011). Negative regulation of anthocyanin biosynthesis in Arabidopsis by a miR156-targeted SPL transcription factor. *The Plant Cell*, *23*(4), 1512–1522.

Guan, Q., Lu, X., Zeng, H., Zhang, Y., & Zhu, J. (2013). Heat stress induction of mi R 398 triggers A regulatory loop that is critical for thermotolerance in *Arabidopsis*. *The Plant Journal*, *74*(5), 840–851.

Huo, L., Guo, Z., Zhang, Z., Jia, X., Sun, Y., Sun, X., … & Ma, F. (2020). The apple autophagy-related gene MdATG9 confers tolerance to low nitrogen in transgenic apple callus. *Frontiers in Plant Science*, *11*, 423.

Iwamoto, M., & Tagiri, A. (2016). Micro RNA-targeted transcription factor gene RDD 1 promotes nutrient ion uptake and accumulation in rice. *The Plant Journal*, *85*(4), 466–477.

Jain, P., Hussian, S., Nishad, J., Dubey, H., Bisht, D. S., Sharma, T. R., & Mondal, T. K. (2021). Identification and functional prediction of long non-coding RNAs of rice (*Oryza sativa L.*) at reproductive stage under salinity stress. *Molecular Biology Reports*, *48*(3), 2261–2271.

Jin, X., Wang, Z., Li, X., Ai, Q., Wong, D. C. J., Zhang, F., Yang, J., Zhang, N., & Si, H. (2024). Current perspectives of lncRNAs in abiotic and biotic stress tolerance in plants. *Frontiers in Plant Science*, *14*, 1334620.

Jin, Z., Gao, S., Ma, W., Lyu, X., Cao, X., & Yao, Y. (2020). Identification and functional prediction of salt stress-related long noncoding RNAs in grapevine roots. *Environmental and Experimental Botany*, *179*, 104215.

Karlik, E., Marakli, S., & Gozukirmizi, N. (2018). Two lncRNAs expression profiles in salt stressed barley (*Hordeum vulgare L.*) roots. *Cytologia*, *83*(1), 37–43.

Kashyap, S., Agarwala, N., & Sunkar, R. (2024). Understanding plant stress memory traits can provide a way for sustainable agriculture. *Plant Science*, *340*, 111954.

Kim, J. J., Lee, J. H., Kim, W., Jung, H. S., Huijser, P., & Ahn, J. H. (2012). The microRNA156-SQUAMOSA PROMOTER BINDING PROTEIN-LIKE3 module regulates ambient temperature-responsive flowering via FLOWERING LOCUS T in Arabidopsis. *Plant Physiology*, *159*(1), 461–478.

Kotak, S., Larkindale, J., Lee, U., von Koskull-Döring, P., Vierling, E., & Scharf, K. D. (2007). Complexity of the heat stress response in plants. *Current Opinion in Plant Biology*, *10*(3), 310–316.

Kumar, N., Bharadwaj, C., Sahu, S., Shiv, A., Shrivastava, A. K., Reddy, S. P. P., … & Varshney, R. K. (2021). Genome-wide identification and functional prediction of salt-stress related long non-coding RNAs (lncRNAs) in chickpea (*Cicer arietinum* L. *Physiology and Molecular Biology of Plants*, *27*, 2605–2619.

Li, C., Nong, W., Zhao, S., Lin, X., Xie, Y., Cheung, M. Y., … & Lam, H. M. (2022). Differential microRNA expression, microRNA arm switching, and microRNA: long noncoding RNA interaction in response to salinity stress in soybean. *BMC Genomics*, *23*(1), 65.

Li, S., Liu, J., Liu, Z., Li, X., Wu, F., & He, Y. (2014). Heat-induced tas1 target1 mediates thermotolerance via heat stress transcription factor A1a–directed pathways in Arabidopsis. *The Plant Cell*, *26*(4), 1764–1780.

Li, S., & Zachgo, S. (2013). TCP 3 interacts with R2R3-MYB proteins, promotes flavonoid biosynthesis and negatively regulates the auxin response in *Arabidopsis thaliana*. *The Plant Journal*, *76*(6), 901–913.

Li, Z., Tian, P., Huang, T., & Huang, J. (2021). Noncoding-RNA-mediated regulation in response to macronutrient stress in plants. *International Journal of Molecular Sciences*, *22*(20), 11205.

Liu, H. H., Tian, X., Li, Y. J., Wu, C. A., & Zheng, C. C. (2008). Microarray-based analysis of stress-regulated microRNAs in *Arabidopsis thaliana*. *RNA*, *14*(5), 836–843.

Liu, Y., Wang, K., Li, D., Yan, J., & Zhang, W. (2017). Enhanced cold tolerance and tillering in switchgrass (*Panicum virgatum* L.) by heterologous expression of Osa-miR393a. *Plant and Cell Physiology*, *58*(12), 2226–2240.

Lu, Q., Guo, F., Xu, Q., & Cang, J. (2020). LncRNA improves cold resistance of winter wheat by interacting with miR398. *Functional Plant Biology*, *47*(6), 544–557.

Lu, X., Chen, X., Mu, M., Wang, J., Wang, X., Wang, D., … & Ye, W. (2016). Genome-wide analysis of long noncoding RNAs and their responses to drought stress in cotton (*Gossypium hirsutum* L.). *PloS one*, *11*(6), e0156723.

Lv, Y., Liang, Z., Ge, M., Qi, W., Zhang, T., Lin, F., … & Zhao, H. (2016). Genome-wide identification and functional prediction of nitrogen-responsive intergenic and intronic long non-coding RNAs in maize (*Zea mays* L.). *BMC Genomics*, *17*, 1–15.

Mirdar Mansuri, R., Azizi, A. H., Sadri, A. H., & Shobbar, Z. S. (2022). Long non-coding RNAs as the regulatory hubs in rice response to salt stress. *Scientific Reports*, *12*(1), 21696.

Oh, N., Seo, J. S., Chung, P. J., Lee, J., Seo, J. K., Kang, J. H., … & Jung, C. (2022). Identification and characterization of drought-induced long noncoding RNAs (DRILs) in rice. *Applied Biological Chemistry*, *65*(1), 79.

Pandita, D., 2022. MicroRNAs shape the tolerance mechanisms against abiotic stress in maize. In *Omics approach to manage abiotic stress in cereals* (pp. 479–493). Springer Nature Singapore.

Qin, T., Zhao, H., Cui, P., Albesher, N., & Xiong, L. (2017). A nucleus-localized long non-coding RNA enhances drought and salt stress tolerance. *Plant Physiology*, *175*(3), 1321–1336.

Saharan, B. S., Brar, B., Duhan, J. S., Kumar, R., Marwaha, S., Rajput, V. D., & Minkina, T. (2022). Molecular and physiological mechanisms to mitigate abiotic stress conditions in plants. *Life*, *12*(10), 1634.

Shuai, P., Su, Y., Liang, D., Zhang, Z., Xia, X., & Yin, W. (2016). Identification of phasiRNAs and their drought-responsiveness in *Populus trichocarpa*. *FEBS letters*, *590*(20), 3616–3627.

Song, Y., Ci, D., Tian, M., & Zhang, D. (2016). Stable methylation of a non-coding RNA gene regulates gene expression in response to abiotic stress in Populus simonii. *Journal of Experimental Botany*, *67*(5), 1477–1492.

Stief, A., Altmann, S., Hoffmann, K., Pant, B. D., Scheible, W. R., & Bäurle, I. (2014). Arabidopsis miR156 regulates tolerance to recurring environmental stress through SPL transcription factors. *The Plant Cell*, *26*(4), 1792–1807.

Tang, Y., Qu, Z., Lei, J., He, R., Adelson, D. L., Zhu, Y., … & Wang, D. (2021). The long noncoding RNA FRILAIR regulates strawberry fruit ripening by functioning as a noncanonical target mimic. *PLoS Genetics*, *17*(3), e1009461.

Thiebaut, F., Rojas, C. A., Almeida, K. L., Grativol, C., Domiciano, G. C., Lamb, C. R. C., … & Ferreira, P. C. (2012). Regulation of miR319 during cold stress in sugarcane. *Plant, Cell & Environment*, *35*(3), 502–512.

Traubenik, S., Charon, C., & Blein, T. (2024). From environmental responses to adaptation: the roles of plant lncRNAs. *Plant Physiology*, *195*(1), 232–244.

Vengavasi, K., Kumar, A., & Pandey, R. (2016). Transcript abundance, enzyme activity and metabolite concentration regulates differential carboxylate efflux in soybean under low phosphorus stress. *Indian Journal of Plant Physiology*, *21*, 179–188.

Waititu, J. K., Zhang, C., Liu, J., & Wang, H. (2020). Plant non-coding RNAs: origin, biogenesis, mode of action and their roles in abiotic stress. *International Journal of Molecular Sciences*, *21*(21), 8401.

Wan, S., Zhang, Y., Duan, M., Huang, L., Wang, W., Xu, Q., … & Yu, Y. (2020). Integrated analysis of long non-coding RNAs (lncRNAs) and mRNAs reveals the regulatory role of lncRNAs associated with salt resistance in *Camellia sinensis*. *Frontiers in Plant Science*, *11*, 218.

Wang, C., Zhao, Y., Gu, P., Zou, F., Meng, L., Song, W., … & Zhang, Y. (2018). Auxin is involved in lateral root formation induced by drought stress in tobacco seedlings. *Journal of Plant Growth Regulation*, *37*, 539–549.

Wang, S. T., Sun, X. L., Hoshino, Y., Yu, Y., Jia, B., Sun, Z. W., … & Zhu, Y. M. (2014). MicroRNA319 positively regulates cold tolerance by targeting OsPCF6 and OsTCP21 in rice (*Oryza sativa* L.). *PloS one*, *9*(3), e91357.

Wang, T., Zhao, M., Zhang, X., Liu, M., Yang, C., Chen, Y., Chen, R., Wen, J., Mysore, K. S., & Zhang, W. H. (2017). Novel phosphate deficiency-responsive long non-coding RNAs in the legume model plant *Medicago truncatula*. *Journal of Experimental Botany*, *68*(21–22), 5937–5948.

Xue, D., Chen, Y., Li, J., Han, J., Zhou, Z., Zhang, W., … & Wang, J. (2019). A novel noncoding RNA dsr11 involved in heat stress tolerance in *Deinococcus radiodurans*. *Biomolecules*, *10*(1), 22.

Yadav, A., Mathan, J., Dubey, A. K., & Singh, A. (2024). The emerging role of non-coding RNAs (ncRNAs) in plant growth, development, and stress response signaling. *Non-coding RNA*, *10*(1), 13.

Yan, Y., Wang, H., Hamera, S., Chen, X., & Fang, R. (2014). miR444a has multiple functions in the rice nitrate-signaling pathway. *The Plant Journal*, *78*(1), 44–55.

Yang, C., Li, D., Mao, D., Liu, X. U. E., Ji, C., Li, X., … & Zhu, L. (2013). Overexpression of micro RNA 319 impacts leaf morphogenesis and leads to enhanced cold tolerance in rice (*Oryza sativa* L.). *Plant, Cell & Environment*, *36*(12), 2207–2218.

Yang, Y., Zhang, X., Su, Y., Zou, J., Wang, Z., Xu, L., & Que, Y. (2017). miRNA alteration is an important mechanism in sugarcane response to low-temperature environment. *BMC Genomics*, *18*, 1–18.

Ye, Z., Zeng, J., Long, L., Ye, L., & Zhang, G. (2021). Identification of microRNAs in response to low potassium stress in the shoots of Tibetan wild barley and cultivated. *Current Plant Biology*, *25*, 100193.

Yu, X., Wang, H., Lu, Y., de Ruiter, M., Cariaso, M., Prins, M., … & He, Y. (2012). Identification of conserved and novel microRNAs that are responsive to heat stress in *Brassica rapa*. *Journal of Experimental Botany*, *63*(2), 1025–1038.

Yu, Y., Zhang, Y., Chen, X., & Chen, Y. (2019). Plant noncoding RNAs: Hidden players in development and stress responses. *Annual review of cell and developmental biology*, *35*(1), 407–431.

Yuan, J., Zhang, Y., Dong, J., Sun, Y., Lim, B. L., Liu, D., & Lu, Z. J. (2016). Systematic characterization of novel lncRNAs responding to phosphate starvation in *Arabidopsis thaliana*. *BMC Genomics*, *17*, 1–16.

Zeng, J., Ye, Z., He, X., & Zhang, G. (2019). Identification of microRNAs and their targets responding to low-potassium stress in two barley genotypes differing in low-K tolerance. *Journal of Plant Physiology*, *234*, 44–53.

Zhang, X., Dong, J., Deng, F., Wang, W., Cheng, Y., Song, L., Hu, M., Shen, J., Xu, Q., & Shen, F. (2019a). The long non-coding RNA lncRNA973 is involved in cotton response to salt stress. *BMC Plant Biology*, *19*(1), 459.

Zhang, P., Fan, Y., Sun, X., Chen, L., Terzaghi, W., Bucher, E., Li, L., & Dai, M. (2019b). A large-scale circular RNA profiling reveals universal molecular mechanisms responsive to drought stress in maize and Arabidopsis. *The Plant Journal*, *98*(4), 697–713.

Zhao, M., Ding, H., Zhu, J. K., Zhang, F., & Li, W. X. (2011). Involvement of miR169 in the nitrogen-starvation responses in Arabidopsis. *New Phytologist*, *190*(4), 906–915.

Zhu, C., Ding, Y., & Liu, H. (2011). MiR398 and plant stress responses. *Physiologia Plantarum*, *143*(1), 1–9.

Zou, C., Guo, Z., Zhao, S., Chen, J., Zhang, C., & Han, H. (2023). Genome-wide analysis of long non-coding RNAs in sugar beet (*Beta vulgaris* L.) under drought stress. *Frontiers in Plant Science*, *14*, 1118011.

3 ncRNAs and Plant Stress Responses
A Molecular Perspective

Pragati, Chetan Awasthi and Sayanti Mandal

3.1 INTRODUCTION

Plants are constantly exposed to biotic (e.g., pathogens, pests) and abiotic (e.g., drought, salinity, temperature extremes) stresses that threaten their growth and productivity. To adapt to these environmental challenges, plants deploy intricate regulatory networks, including non-coding RNAs (ncRNAs), which act as pivotal modulators of stress responses (Gill et al., 2024; Suprasanna et al., 2023; Yadav et al., 2024). Non-coding RNAs (ncRNAs) are functional RNA molecules that do not encode proteins but play critical roles in regulating gene expression, epigenetics and other cellular processes. Their discovery has significantly altered our understanding of molecular biology, shifting the focus from protein-centric models to the broader functional landscape of RNA (Hina et al., 2024; Panchal et al., 2023). The origin of ncRNA research dates back to the 1950s when George Palade and Paul Zamecnik discovered ribosomal RNA (rRNA) and transfer RNA (tRNA), establishing the structural significance of ncRNAs in translational machinery. These findings paved the way for Francis Crick's 'adapter molecule' theory, which postulated a model of RNA translation into protein via a theoretical/assumed 'adapter molecule' now known as t-RNA, cementing RNA's pivotal role in the central dogma of molecular biology. Additionally, the discovery of ribozyme in the 1980s proved that RNA had a catalytic function (Abelson, 2017; Cech, 2002; Cech et al., 1981; Kruger et al., 1982; North, 1989; Wagh et al., 2025). According to Mizuno et al. (1984), the regulatory function of non-coding RNA was first documented in *Escherichia coli* as microRNA F(micF) base pairs outer membrane protein ompF at the complementary 5′ end region of the ompF messenger RNAs (mRNA). It was suggested that this function might inhibit the translation of the ompF mRNA (Delihas, 2015; Mizuno et al., 1984). Lin-4, the first miRNA identified in *Caenorhabditis elegans*, controls the time of development by post-transcriptional silencing (Lee et al., 1993). Small RNAs (sRNAs) such as microRNAs (miRNAs) and small interfering RNAs (siRNAs) were identified as important regulators of gene expression by the discovery of RNA Interference (RNAi) processes, which had an impact on advancements in RNA-based therapies (*Chen* & Kim, 2024; Fire et al., 1998; Timmons & Fire, 1998; Wolfien et al., 2019).

Further housekeeping ncRNAs such as Small nuclear RNAs (snRNAs) and small nucleolar RNAs (snoRNAs) were identified as central to RNA processing, including splicing and modification. Circular RNAs (circRNAs) observed in the 1970s via electron micrography, circRNAs were initially considered byproducts of RNA processing, but recent research reveals their involvement in transcription regulation, chromatin modification and cellular stress responses. It has been established that miRNAs and siRNAs are essential for development, stress tolerance and epigenetic regulation (Reinhart et al., 2002; Xie et al., 2004). Milestones in ncRNA research have been tabulated in Table 3.1 and in Figure 3.1.

Many non-coding sequences were dismissed as biological byproducts or 'junk DNA' for years. Early scholars, however, conjectured that their apparent lack of function did not preclude utility. Since the 1970s, there has been a persistent interest in ncRNAs, demonstrating their vital roles in many biological processes. Transcripts derived from complementary gene sequences in bacteria, for instance, were formerly considered insignificant but are now recognised as crucial regulatory elements (Chen & Kim, 2024).

Our knowledge of molecular biology has evolved due to the extensive study of ncRNAs, which shows that RNA is much more than just a messenger. It is a dynamic regulatory molecule integral to numerous cellular processes, from transcription and translation to stress responses and epigenetic control. ncRNAs are now considered fundamental elements of biology, with applications spanning epigenetics, biotechnology, medicine and plant science.

3.2 ncRNA AND ITS CLASSIFICATION

Non-coding RNAs (ncRNAs) are transcripts that escape translation and play a crucial role in the form of RNA but are crucial in modulating gene expression at both the transcriptional and post-transcriptional levels. ncRNAs involve diverse biological processes, including development, reproduction and stress responses (Gill et al., 2024; Panchal et al., 2023; Suprasanna et al., 2023; Yadav et al., 2024). Through their binding or association with mRNAs, DNA, proteins or other RNAs, they can either promote or repress translation, mRNA stability or chromatin structure, so modifying gene expression. These molecules are integral to numerous biological processes, including transcriptional regulation, RNA splicing and chromatin remodelling (Figure 3.2; Table 3.2).

Over recent years, the significance and diversity of ncRNAs have been increasingly acknowledged. Bioinformatics studies

DOI: 10.1201/9781003583257-3

TABLE 3.1
Milestones in Non-Coding RNA (ncRNA) Research

Year/Period	Milestone	Key Scientists	References
1950s	Discovery of **ribosomal RNA (rRNA)** and **transfer RNA (tRNA),** which are crucial components of translational machinery	George Palade, Paul Zamecnik	Erdmann and Barciszewski, (2011), Hoagland et al. (1958) and Palade (1955)
1957	Concept of '**adapter hypothesis**' that propose theoretical model for existence of encoded information in RNA that determines amino acids sequence in proteins	Francis Crick	Brenner (1995) and Moore (2004)
1960–1961	Discovery of **messenger RNA (mRNA)**	François Jacob, Sydney Brenner, Matthew Meselson	Brenner et al. (1961), lnorkin2013 (2016) and Weinberg (2001)
1960s–1970s	Identification of housekeeping ncRNAs like **snRNAs** and **snoRNAs** and their roles in RNA processing	Reddy R., Busch H., Gary Zieve, Sheldon Penman, Robert A. Weinberg, Sheldon Penman etc.	Krämer (1987), Maxwell and Fournier (1995); Reddy and Singh (1991), Weinberg and Penman (1968) and Zieve and Penman (1976)
1979	Visualisation of **circular RNAs (circRNAs)** in the cytoplasm through electron microscopy	Ming-Ta Hsu and Miguel Coca-Prados	Hsu and Coca-Prados (1979)
1981	Identification of transcripts from complementary sequences in bacteria, termed '**cellular junk'**	R. M. Lacatena and G. Cesareni	Lacatena and Cesareni (1981)
1982	Discovery of catalytic RNA (**ribozymes**) demonstrating enzymatic RNA activity	Thomas Cech, Sidney Altman	Cech et al. (1981), Cech (2002), Kruger et al. (1982) and North (1989)
1990	RNA silencing phenomenon observed in transgenic *Petunia* plants	Rich Jorgensen	Napoli et al. (1990)
1993	Identification of the first microRNA (**lin-4**) in *Caenorhabditis elegans*	Victor Ambros, Rosalind Lee	Lee et al. (1993)
1998	Discovery of **RNA interference (RNAi)** as a gene-silencing mechanism	Andrew Fire, Craig Mello	Fire et al. (1998) and Timmons and Fire (1998)
1999	Identification of small interfering RNAs (**siRNAs**) mediating RNA silencing in plants	David Baulcombe	Hamilton and Baulcombe (1999)
2001	Identification of piRNA from Suppressor of Stellate locus in the testes of *Drosophila melanogaster*	Alexei A. Aravin, Natalia M. Naumova	Aravin et al. (2001)
2001	Genome-wide identification of conserved microRNAs (**miRNAs**) across animals	David Bartel, Victor Ambros	Lagos-Quintana et al. (2001)
2002	Discovery of **plant microRNAs (miRNAs)** in *Arabidopsis*	David Bartel, Detlef Weigel	Reinhart et al. (2002)
2007	Functional roles of lncRNAs in chromatin modification and gene regulation (e.g., **HOTAIR**)	Jeannie T. Lee, Howard Y. Chang	Rinn et al. (2007)
2013	**Circular RNAs (circRNAs)** proved as regulators rather than byproducts in animals	Nikolaus Rajewsky	Kosik (2013) and Memczak et al. (2013)
2015	Discovery of **circular RNAs (circRNAs)** in rice (*Oryza sativa*)	Lu et al.	Lu et al. (2015)
2018	Phased siRNAs (**phasiRNAs**) role in plant development and immunity	Blake Meyers	Fei et al. (2018)
	In Animals		
Year	**Milestone**	**Key Scientists**	**References**
1961	Discovery of regulatory roles of RNA in protein synthesis	François Jacob, Jacques Monod	Jacob and Monod (1961)
1977	Identification of introns and splicing mechanisms in eukaryotic genes	Richard J. Roberts and Phillip A. Sharp	Carr (1993) (Nobel Prize in Physiology in 1993)
1981–1982	Discovery of catalytic RNA (ribozymes)	Thomas Cech, Sidney Altman	Abelson (2017), Cech (2002), Cech et al. (1981), Kruger et al. (1982) and North, (1989)
1993	Identification of the first microRNA (lin-4) in *C. elegans*	Victor Ambros, Rosalind Lee	Lee et al. (1993)
1998	Discovery of RNA interference (RNAi)	Andrew Fire, Craig Mello	(Nobel Prize in 2006) Fire et al. (1998) and Timmons and Fire (1998)

(Continued)

TABLE 3.1 (*Continued*)
Milestones in Non-Coding RNA (ncRNA) Research

Year/Period	Milestone	Key Scientists	References
2001	Genome-wide identification of conserved microRNAs across animals	David Bartel, Victor Ambros	Lagos-Quintana et al. (2001)
2007	Functional roles of lncRNAs in chromatin modification and gene regulation (e.g., HOTAIR)	Jeannie T. Lee, Howard Y. Chang	Rinn et al. (2007)
2013	Emerging evidence of circular RNAs as gene regulators in animals	Nikolaus Rajewsky	Kosik (2013) and Memczak et al. (2013)
	In Plants		
Year	**Milestone**	**Key Scientists**	**References**
1990	Discovery of RNA silencing in transgenic *Petunia*	Rich Jorgensen	Napoli et al. (1990)
1999	Identification of small interfering RNAs (siRNAs) mediating RNA silencing in plants	David Baulcombe	Hamilton and Baulcombe (1999)
2002	Discovery of plant microRNAs (miRNAs) in *Arabidopsis*	David Bartel, Detlef Weigel	Reinhart et al. (2002)
2005	Role of RNA-directed DNA methylation (RdDM) pathways in epigenetic silencing	Craig Pikaard	Xie et al. (2004
2010	Genome-wide identification of long non-coding RNAs (lncRNAs) in plants	Liu et al.	Liu et al. (2015)
2015	Discovery of circular RNAs (circRNAs) in *O. sativa*	Lu et al.	Lu et al. (2015)
2018	Functional roles of phased siRNAs (phasiRNAs) in plant development and immunity	Blake Meyers	Fei et al. (2018)

estimate their numbers to be in the thousands for eukaryotic genomes and in the hundreds for bacterial genomes. ncRNAs are essential across all domains of life – Eukarya, Bacteria and Archaea – and participate in crucial cellular processes. These include DNA replication, chromosome maintenance, transcription regulation, RNA processing (such as cleavage, relegation, modification and editing), translation, mRNA stability and the regulation of protein stability and translocation. Despite their critical functions, many ncRNAs have been discovered incidentally, implying that the known ncRNAs represent only a small fraction of the total population (Huttenhofer, 2006; Palazzo & Lee, 2015).

Most ncRNAs are small, typically less than 500 nucleotides, much shorter than most mRNAs. However, eukaryotes also produce large ncRNAs, such as Xist and Air, which can be several thousand nucleotides long. The specificity of ncRNA functions often depends on their ability to bind selectively to certain proteins and recognise specific RNA

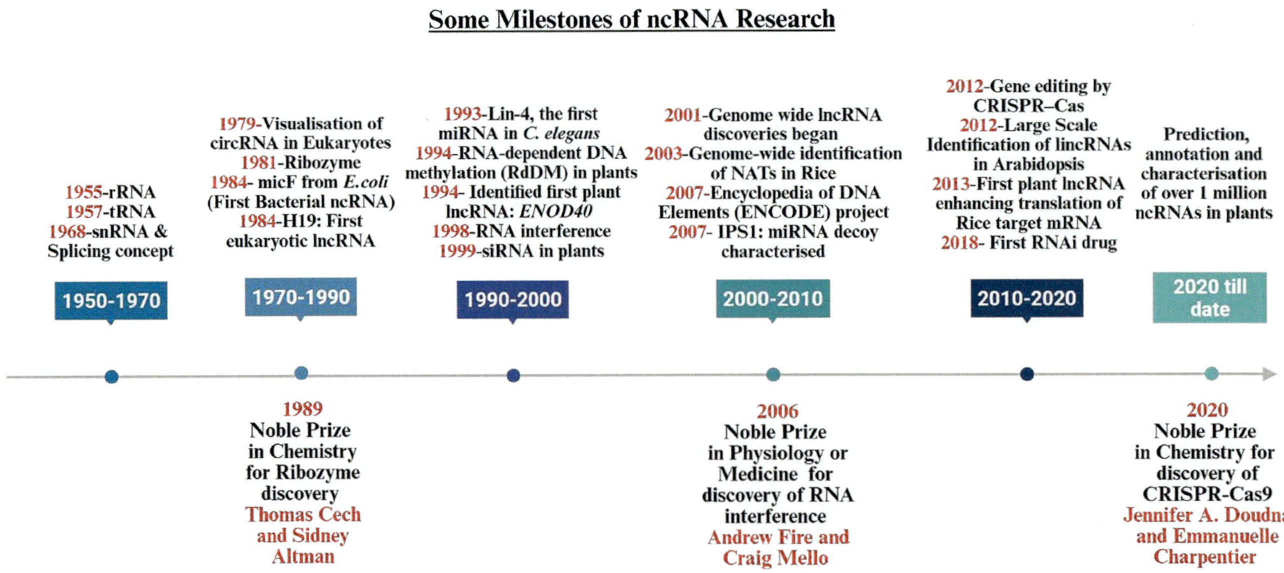

FIGURE 3.1 Timeline for significant milestones achieved in the discovery and the application of ncRNA research. (Image created using BioRender.)

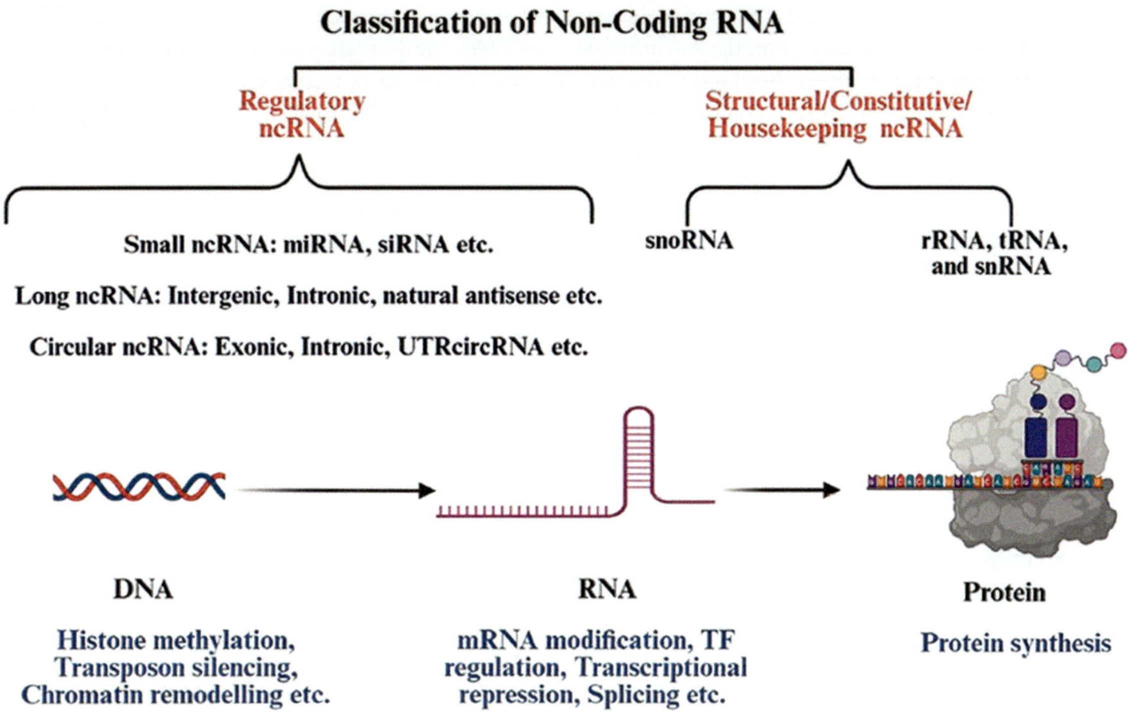

FIGURE 3.2 Classification of non-coding RNAs (ncRNAs). The biological functions of ncRNAs allow for their general classification into two groups: housekeeping RNAs and regulatory RNAs. (Image created using BioRender.)

targets through sequence complementarity (Chen & Kim, 2004). These properties underscore their specialised roles in regulating diverse molecular processes (Huttenhofer, 2006; Wolfien et al., 2019). Based on their size, function and biogenesis, ncRNA can be classified as follows (Amin et al., 2019; Bhogireddy et al., 2021; Liu et al., 2017; Yadav et al., 2024).

3.3 MicroRNAs (MIRNAs)

miRNAs belong to a class of RNA molecules which are single-stranded, noncoding in nature and endogenous in origin. These short molecules of about 18–25 nucleotide length modulate eukaryotic gene expression post-transcriptionally in a trans-acting sequence-specific manner. These

TABLE 3.2
Classification of Non-Coding RNAs (ncRNAs)

Classification	Type	Description	Examples
By size	**Small ncRNAs (<200 nt)**	Involved in post-transcriptional regulation and RNA processing.	miRNAs, siRNAs, piRNAs, snRNAs
	Long ncRNAs (>200 nt)	Function in transcriptional regulation, chromatin remodelling and scaffolding of protein complexes.	Xist, HOTAIR
By function	**Housekeeping ncRNAs**	Essential for basic cellular functions.	rRNAs, tRNAs, snRNAs, snoRNAs
	Regulatory ncRNAs	Regulate gene expression at transcriptional or post-transcriptional levels.	miRNAs, siRNAs, lncRNAs, circRNAs
By biogenesis	**Canonical ncRNAs**	Generated through standard transcription pathways.	miRNAs, lncRNAs, snRNAs
	Non-canonical ncRNAs	Arise from unconventional pathways such as introns or antisense transcripts.	Antisense RNAs, circRNAs
By cellular function	**Cis-acting ncRNAs**	Regulate nearby genes at their locus of transcription.	Xist, Air
	Trans-acting ncRNAs	Act on distant loci to modulate gene expression.	miRNAs
By organism-specific classes	**Prokaryotic ncRNAs**	Play roles in adaptive immunity and stress responses in bacteria and archaea.	CRISPR RNAs, sRNAs
	Plant-specific ncRNAs	Modulate gene expression and stress responses unique to plants.	tasiRNAs, nat-siRNAs

molecules are found across eukaryotic kingdoms, including bryophytes, fungi, higher plants and animals, but no evidence supports their presence in algae. In plants, miRNAs primarily suppress gene expression by targeting mRNAs for degradation or translational repression through near-perfect base pairing with their target sequences (Mandal et al., 2023; Riolo et al., 2020; Zhao et al., 2021). Among mRNA targets, transcription factors form the largest group, followed by genes involved in metabolic processes, stress responses, defence mechanisms and hormonal regulation (Raza et al., 2023; Reinhart et al., 2002; Waititu et al., 2020; Wang et al., 2019).

miRNAs were first identified in *C. elegans*, where *lin-4* and *let-7* regulated developmental timing through partial base pairing with target mRNAs (Cao et al., 2024). To date, approximately 35,000 miRNA sequences have been identified across 271 organisms, revealing evolutionary conservation and recent emergence across species (Riolo et al., 2020). While plant and animal miRNAs share some similarities, their mechanisms differ significantly. Plant miRNA precursors are much longer (up to 300 nucleotides) than those in animals (~70 nucleotides). Plants require near-perfect complementarity for miRNA–mRNA interactions, resulting in mRNA degradation alongside translational repression. In contrast, animal miRNAs often bind with partial complementarity, targeting the 3′-UTR of mRNAs for translational repression alone (Wang et al., 2019).

miRNAs play critical roles in regulating spatiotemporal gene expression during development, including tissue patterning and organ formation. They influence diverse physiological and biochemical processes, such as cell division, differentiation, stress responses, host–pathogen interactions and signalling pathways. miRNAs also contribute to agricultural traits, including yield and stress resilience (Ma et al., 2022; Mondal et al., 2023). miRNAs significantly impact reproductive processes and fruit development. They regulate fruit ripening by modulating genes involved in softening, sugar accumulation, pigment production and secondary metabolite synthesis. These regulatory interactions highlight miRNAs as essential mediators in plant developmental pathways (Wang et al., 2019). Overall, despite their small size, miRNAs exert a vast influence over plant development, stress responses and metabolic regulation, establishing them as central players in plant biology.

3.4 SMALL INTERFERING RNAs (siRNAs)

siRNAs, are small sized approximately 21–24-nucleotides-long transcripts known for their key role in RNA silencing. Generated from double-stranded RNA (dsRNA) by Dicer-like (DCL) proteins, siRNAs mediate post-transcriptional gene silencing by incorporating into the RNA-induced silencing complex (RISC) with ARGONAUTE (AGO) proteins. siRNAs are classified into various types based on their origin, such as heterochromatic siRNAs, phased siRNAs and transposon-derived siRNAs. They are involved in stress responses, genome stability and regulating processes like thermotolerance, pathogen defence and abiotic stress adaptation, suggesting their essential role in plant stress responses and development (Bhogireddy et al., 2021; Chen & Kim, 2024; Panchal et al., 2023; Patil et al., 2014; Yadav et al., 2024).

For instance, hc-siRNAs, the most abundant, are essential for genome integrity and epigenetic modifications, while cis-NATs regulate plant development and environmental responses. PhasiRNAs, generated through miRNA-triggered mechanisms, regulate transcription factors, hormone-responsive genes and nucleotide-binding leucine-rich repeat (NLR) genes, often acting as negative regulators in biological processes (Rajewsky et al., 2017; Song et al., 2021). In *Arabidopsis*, tasiRNAs derived from TAS loci are crucial for responses to hypoxia, where elevated levels of TAS1, TAS2 and TAS3 correspond to stress adaptation. Heat-induced tasiRNAs, such as TAS1 targets HTT1 and HTT2, regulate thermotolerance by modulating heat shock protein (HSP) interactions and transcription factors like HSFA1a. Similarly, TAS4-siR81(–) influences anthocyanin biosynthesis under phosphate or nitrogen deficiency. Transposon-derived rasiRNAs, such as those in maize hybrids, contribute to heterosis by maintaining genome stability, while lsiRNAs mediate pathogen response, as seen in *Arabidopsis* with AtlsiRNA-1 silencing RAP-domain proteins during bacterial infection. Stress-specific examples include heat-responsive nat-siRNAs in *Brassica rapa* and the activation of retrotransposon ONSEN in *Arabidopsis* under heat stress, illustrating siRNA-mediated control of stress-induced genomic changes. Additionally, cold stress triggers tasiRNAs, like TAS3-derived molecules in wheat and cassava, which modulate auxin signalling and transcriptional regulators such as NAC domains, highlighting siRNAs' central role in abiotic stress tolerance and adaptation across diverse plant species (Felippes & Weigel, 2009; Li et al., 2014; Waititu et al., 2020).

3.5 LONG NON-CODING RNAs (lncRNAs)

Long non-coding RNAs (lncRNAs) are more than 200-nucleotide-length RNA molecules that lack an evident coding potential, and they are ubiquitous across plants, animals, fungi and prokaryotes (Domínguez-Rosas et al., 2023; Mandal et al., 2023). These molecules are predominantly nuclear, associating with chromatin, though they also function in the cytoplasm. Eukaryotic cells use RNA polymerases I, II and III for lncRNA transcription, with Pol II primarily producing lncRNAs despite its major role in mRNA biogenesis. In plants, unique enzymes like Pol IV and Pol V generate lncRNAs that silence transposable elements (Yadav et al., 2024). LncRNAs operate at pre-transcriptional, transcriptional and post-transcriptional levels through diverse mechanisms, such as modulating chromatin topology, interacting with transcription factors and acting as competitive endogenous RNAs (ceRNAs) to regulate miRNA activity. Examples include the regulation of auxin-responsive genes and stress adaptation through lncRNAs like APOLO and COOLAIR, demonstrating their importance in plant development and environmental responses. In *Arabidopsis*, the lncRNA APOLO forms R-loops to alter chromatin structure, coordinating the expression of auxin-responsive genes, while the lncRNA asDOG1 suppresses DOG1 expression to promote seed germination. ENOD40, isolated from a fabaceous plant, *Medicago sativa*, was the first plant lncRNA which influences the localisation of the RNA-binding protein (RBP) MtRBP1 at the subcellular level. Subsequently, lncRNAs like COOLAIR, COLDAIR and CIL1 have been implicated in cold stress response by silencing genes involved in flowering through epigenetic changes. The lncRNA IPS1 in *Arabidopsis* serve as a target mimic for miRNA399 under phosphate-deficient conditions, preventing miRNA-mediated repression of PHO2, which enhances phosphate acquisition. Similarly, in *Populus*, drought-responsive lncRNAs act as target mimics for miRNAs, modulating stress tolerance. These molecules also act as precursors of miRNAs and siRNAs; for instance, lncRNAs in *Populus* and *Brassica rapa* have also been found to be precursors for miRNAs under cold, heat and nitrogen-deficiency stresses. In tomatoes, silencing the RdDM pathway factor SlAGO4A enhanced salt and drought tolerance, showcasing lncRNA involvement in the RdDM pathway during stress adaptation (Ariel et al., 2020; Borah et al., 2018; Gonzales et al., 2024; Huang, Xian, et al., 2016; Moison et al., 2021; Statello et al., 2021; Traubenik et al., 2024; Yang et al., 2023, 2024). These findings highlight the versatility and crucial roles of lncRNAs in regulating gene expression and stress responses in plants.

3.6 CIRCULAR RNAs (circRNAs)

circRNAs, a distinct group of single-stranded, endogenously formed non-coding RNAs, are designated by their covalently closed-loop structure formed through a non-canonical process called back-splicing, where the 5′ and 3′ ends are covalently linked (Zhao et al., 2019). This structural feature renders circRNAs highly stable and resistant to degradation by ribonucleases such as RNase R, granting them longer half-lives compared to linear RNAs. First identified in tomato viroids in the 1970s, circRNAs are now recognised as ubiquitously present across all eukaryotic organisms, as well as in prokaryotes and archaea. They can originate from various genomic regions, including exons (exonic circRNAs), introns (circular intronic RNAs or ciRNAs), exon-intron combinations (ElciRNAs) and intergenic regions, reflecting their diverse biogenesis (Cao et al., 2024;Memczak et al., 2013; Yadav et al., 2024; Zhang, Zhan, et al., 2024). Their formation is thought to involve two major models: either through base pairing of inverted repeats in flanking introns or RBP interactions at specific motifs or from intron lariats that escape debranching. The unique stability and diversity of circRNAs make them important regulatory molecules in gene expression, acting as sponges for miRNAs and RBPs or influencing alternative splicing by forming R-loop structures that pause transcription. CircRNAs have been implicated in crucial regulatory roles in plants, particularly in stress responses. For example, in *Arabidopsis thaliana*, a circRNA derived from exon 6 of the SEPALLATA3 (SEP3) gene forms a circRNA:DNA R-loop, which downregulates the transcription of SEP3 and increases the abundance of the exon-skipped splicing variant SEP3.3, highlighting a mechanism where circRNAs influence gene expression through transcriptional pausing. Similarly, the QS-circQS-miR6024 regulatory network in *Nicotiana tabacum* shows circRNAs acting as miRNA sponges; miR6024 is down-regulated in response to topping treatments, leading to increased expression of circQS and its host gene QS (Cao et al., 2024). In another example, circRNAs are involved in stress adaptation, as seen in *Populus*, where drought-responsive circRNAs modulate miRNA activity, demonstrating their role in regulating gene networks under environmental stress (Shuai et al., 2013; Yadav et al., 2024; Zhang, Ma, et al., 2024). These findings illustrate the multifunctional nature of circRNAs in plants, from mediating transcriptional regulation to facilitating stress tolerance and underscore the need for further research to unravel the detailed mechanisms underpinning their formation and functions.

3.7 BIOGENESIS OF DIFFERENT TYPES OF ncRNA

ncRNA biogenesis involves specific pathways for different types, such as Dicer-dependent processing for miRNAs. These molecules modulate plant stress responses by altering gene expression pathways, interacting with transcription factors and influencing hormonal signalling networks (Hombach & Kretz, 2016; Wang et al., 2019) (Table 3.3).

Three key steps in the biogenesis of ncRNAs are.

i. *Transcription*: ncRNAs are transcribed from DNA by RNA polymerases.
ii. *Processing*: Small ncRNAs undergo processing by enzymes such as Dicer, which cleaves precursor molecules into functional forms.
iii. *Modification*: Many lncRNAs undergo post-transcriptional modifications that influence their stability and function.

ncRNA biogenesis is an intricate process that may differ based on the ncRNA class.

TABLE 3.3
Types of ncRNA and Their Biogenesis

ncRNA Type	Biogenesis Pathway	Biogenesis Details	Size/Length	Key Functions	Examples
tRNA-derived fragments (tRFs)	Non-canonical	Cleaved from precursor tRNA during maturation. Fragmented into tRF-5, tRF-3 or tRNA halves.	13–40 nucleotides	Regulate transcription, translation, RNA degradation.	Influence root symbiosis in soybean, pollen grain development in *Arabidopsis*.
Small interfering RNAs (siRNAs)	Canonical	Processed from double-stranded RNA (dsRNA) by dicer-like (DCL) enzymes.	21–24 nucleotides	Post-transcriptional gene silencing (PTGS), gene regulation, maintaining genome integrity.	Regulate transposons, viral RNA, reproduction, *Arabidopsis* phasiRNAs, hcsiRNAs and rasiRNAs.
MicroRNAs (miRNAs)	Canonical	Transcribed by MIR genes using RNA polymerase II. Processed into mature forms by DCL1 in plants.	20–24 nucleotides	Transcript cleavage or translational repression affecting gene expression.	miR156, miR319 in *Arabidopsis*, miPEP165a for root development, miPEP858a for secondary metabolite regulation.
Long non-coding RNAs (lncRNAs)	Non-canonical/ canonical	Transcribed by RNA polymerases (I, II, III, Pol IV and Pol V). Often include alternative splicing.	>200 nucleotides	Regulate gene expression at transcriptional/post-transcriptional levels, scaffold functions, epigenetic regulation.	COLDAIR, COOLAIR for flowering, SVALKA for dormancy, DEVIL1, POLARIS for growth regulation.
Circular RNAs (circRNAs)	Non-canonical	Produced through back-splicing, forming closed-loop structures.	Variable, often short	Act as miRNA/protein sponges, regulate gene expression.	Tissue-specific accumulation, potential biomarkers, regulate transcription and post-transcription activities.
Piwi-interacting RNAs (piRNAs)	Non-canonical	Processed from long single-stranded RNA precursors interacting with PIWI proteins.	21–34 nucleotides	Regulate transposons, protect against viruses, gene silencing in animals.	Not applicable in plants, involved in animal gene regulation and defence.

3.8 MicroRNA BIOGENESIS

miRNA biogenesis begins with the transcription of MIR genes, which can be intronic or intergenic, by RNA polymerase II. This process produces primary miRNAs (pri-miRNAs) that undergo capping, adenylation and co-transcriptional folding into stem-loop structures. Processing occurs in two key stages within the nucleus, involving the DCL endonucleases and accessory proteins such as SERRATE (SE) and HYPONASTIC LEAVES 1 (HYL1) within nuclear processing centres like SmD3/SmB bodies.

1. *Primary Processing*: DCL1 cleaves the pri-miRNAs into precursor miRNAs (pre-miRNAs).
2. *Secondary Processing*: Pre-miRNAs are further cleaved into miRNA/miRNA* duplexes.

The duplex undergoes methylation by HUA ENHANCER 1 (HEN1) to enhance stability, followed by loading onto AGO proteins. The miRNA* passenger strand is degraded, leaving the guide miRNA strand to form the RISC. Although originally believed to occur in the cytoplasm, RISC assembly is now known to happen within the nucleus, assisted by the molecular chaperone HSP90. The mature RISC is transported to the cytoplasm via the HASTY protein, a plant ortholog of animal exportin.

Once in the cytoplasm, the RISC complex, guided by the mature miRNA, base-pairs with target mRNAs to mediate gene silencing through post-transcriptional mechanisms such as mRNA degradation or translational repression (Bhogireddy et al., 2021; Reinhart et al., 2002; Riolo et al., 2020; Wang et al., 2019). This intricate pathway ensures the precise regulation of gene expression, enabling plants to respond effectively to developmental and environmental cues.

3.9 lncRNA BIOGENESIS

Following the same post-transcriptional changes as mRNAs, such as capping, splicing and polyadenylation, lncRNAs are typically generated by RNA polymerase II. They have complicated biogenesis, and their secondary structures play a major role in how they function.

Plant lncRNA transcription is accomplished by RNA polymerases I, II, III, IV and V. Besides mostly transcribed mRNAs, polymerase II also creates a number of lncRNAs.

The production of lncRNAs by plant polymerases IV and V is very unique since it contributes to transposable element silencing and RNA-directed DNA methylation (RdDM), two essential adaptive processes in plants. Similar structural features of mRNAs and lncRNAs include polyadenylated 3′ ends, 5′ caps, promoter regions and splicing signals.

The transcription of plant lncRNAs is carried out by RNA polymerases I, II, III, IV and V. Polymerase II produces several lncRNAs in addition to its primary function of transcription of mRNAs. Plant polymerases IV and V are particularly distinctive because they produce lncRNAs that are involved in RdDM, a critical adaptive process in plants and transposable element silencing. Splicing signals, promoter regions, 5′ caps and polyadenylated 3′ ends are structural characteristics that lncRNAs and mRNAs have in common. However, other plant lncRNAs, particularly those produced by Pol V, may have capped or tri-phosphorylated 5′ ends and are not polyadenylated (Bhogireddy et al., 2021; Borah et al., 2018; Chen & Kim, 2024; Hombach & Kretz, 2016).

3.10 siRNA BIOGENESIS

dsRNA precursors, which are frequently generated by RNA-dependent RNA polymerase (RDR), are the source of siRNAs. DCL proteins convert these dsRNAs into siRNA duplexes. Target mRNA cleavage and degradation are facilitated by the loading of one strand of the duplex into RISC (Bhogireddy et al., 2021; Li et al., 2024; Zhang et al., 2023). ncRNA biogenesis can also be classified as Canonical or non-Canonical based on the conserved processing steps, as depicted in the diagram (Table 3.3).

3.11 MOLECULAR MECHANISMS OF ncRNA ACTION IN RESPONSE TO PLANT STRESS

Non-coding RNAs (ncRNAs), such as sRNAs, miRNAs, siRNAs, and lncRNAs, are pivotal regulators of gene expression at both transcriptional and post-transcriptional levels. ncRNAs play versatile multifaceted roles in plant stress responses by modulating gene expression through post-transcriptional, transcriptional and hormonal regulation (Figure 3.3). Plant ncRNAs act through the following ways to modulate stress response mechanisms (Bhogireddy et al., 2021; Statello et al., 2021; Waititu et al., 2020; Wang & Chekanova, 2017; Wolfien et al., 2019):

i. ***Transcriptional Regulation***: ncRNAs play a vital role in regulating gene expression at the transcriptional level in plants by influencing chromatin structure and transcription factor activity. ncRNAs modulate transcription by binding to DNA, interacting with transcription factors or altering chromatin structure. They can recruit chromatin-modifying enzymes to specific genomic regions, resulting in modifications such as DNA methylation or histone modifications, which regulate gene accessibility and activity (Statello et al., 2021; Waititu et al., 2020).
   a. *Chromatin Modification*: ncRNAs guide chromatin-modifying complexes to specific genomic loci, inducing histone modifications or DNA methylation that repress or activate gene expression. In *Arabidopsis*, siRNAs facilitate RdDM, which silences transposons and stress-responsive genes, maintaining genomic stability during abiotic stress (Xie & Yu, 2015).
   b. *Transcriptional Interference*: Certain ncRNAs interfere with transcription by preventing transcription factor binding to gene promoters. The lncRNA COOLAIR in *Arabidopsis* prevents transcription of FLC by inducing repressive chromatin modifications during vernalisation (Numan et al., 2024).
   c. *Regulation of Transcription Factors*: ncRNAs can interact with transcription factors, modulating their activity. For example, HIDDEN TREASURE 1 (HID1), a conserved lncRNA in land plants, modulates photomorphogenesis by binding to the promoter of the transcription factor PIF3 and inhibiting its transcription (Wang & Chekanova, 2017).

ii. *Post-Transcriptional Regulation*: At the post-transcriptional level, ncRNAs influence mRNA stability, splicing and translation in plants, fine-tuning gene expression in response to environmental cues.
   a. *mRNA Cleavage and Stability*: miRNAs and siRNAs mediate the cleavage or degradation of target mRNAs. miR398, for example, regulates oxidative stress in *Arabidopsis* by targeting Cu/Zn superoxide dismutase (CSD) mRNAs, reducing their stability and regulating stress responses (Sun et al., 2020).
      - *Translation Repression*: ncRNAs can also inhibit translation by preventing ribosome binding or movement along the mRNA. In *Arabidopsis*, miR156 controls developmental transitions by repressing the translation of SPL (SQUAMOSA promoter binding protein-like) genes under stress conditions. siRNAs, in association with AGO1 or AGO10, bind imperfectly to target mRNAs, interfere with ribosome recruitment or movement, resulting in translational repression. The RISC-AGO1 complex can bind to the 5′ untranslated region (UTR) or open reading frame (ORF) of target genes. Proteins such as

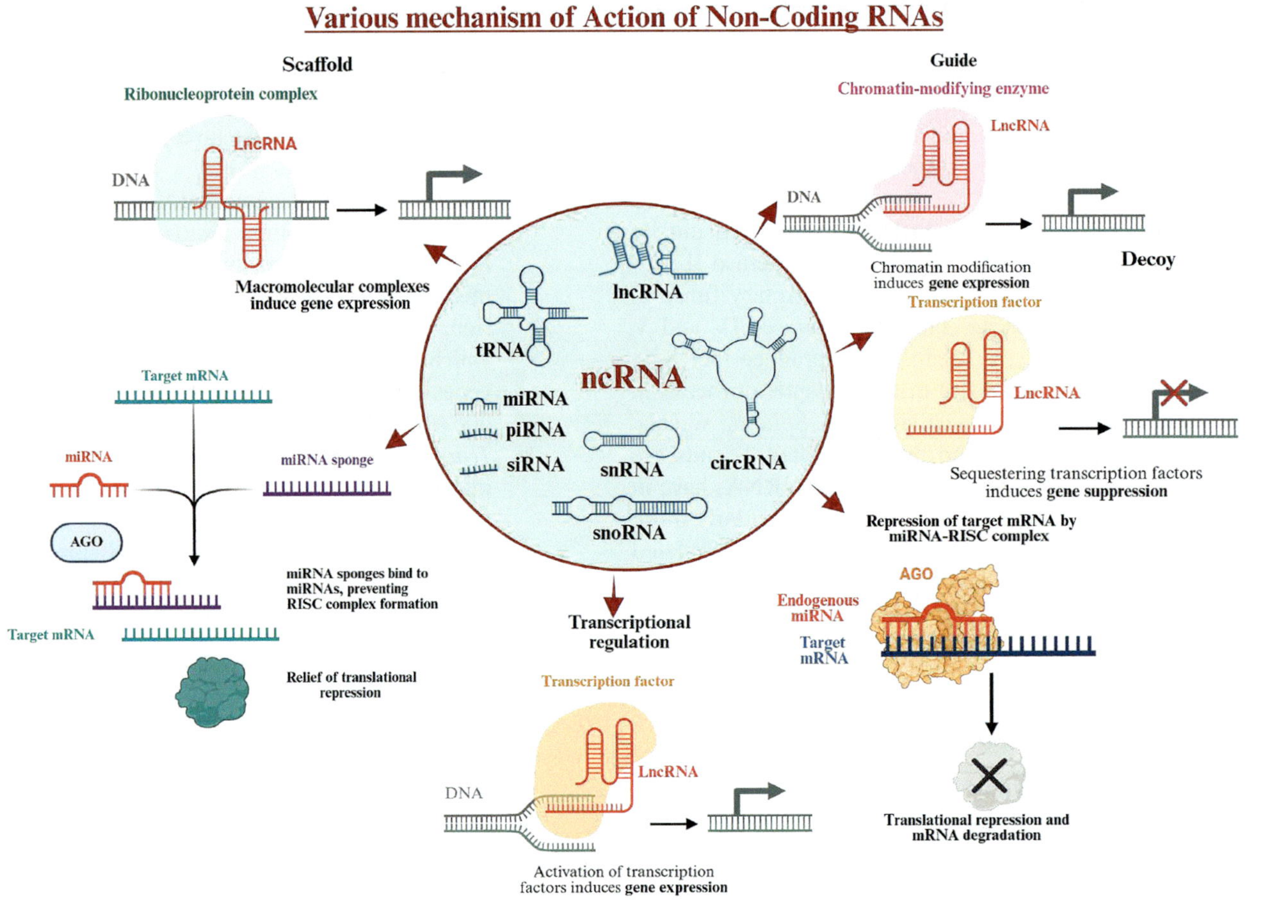

FIGURE 3.3 Molecular mechanisms underlying ncRNA Action. (Image created using BioRender.)

ALTERED MERISTEM PROGRAM 1 (AMP1), VARICOSE.

(VCS), GW-repeat proteins and KATANIN (KTN1) influence this inhibition (Bhogireddy et al., 2021; Waititu et al., 2020; Wang & Chekanova, 2017; Yadav et al., 2024).

b. *Alternative Splicing Regulation*: lncRNAs influence alternative splicing events by interacting with the spliceosome machinery. The lncRNA SVALKA in *Arabidopsis* modulates splicing under salt stress, generating different isoforms of stress-responsive proteins (Bhogireddy et al., 2021; Waititu et al., 2020; Wang & Chekanova, 2017; Yadav et al., 2024; Zhang et al., 2019).

iii. *Gene Silencing*: miRNAs and siRNAs are critical regulators of gene silencing, guiding the RISC to target mRNAs for degradation or translational repression. miRNAs bind to complementary sequences in target mRNAs, leading to their cleavage or inhibition of translation. miRNAs and siRNAs bind to complementary sequences on target mRNAs, leading to mRNA cleavage. miRNAs are processed from MIR genes by the DCL1-HYL1-SE complex into miRNA–miRNA* duplexes. The mature miRNA is transported by HASTY to the cytoplasm, where it integrates into the RISC containing AGO1. siRNAs also integrate with RISC-AGO complexes, which cleave target transcripts at specific sites. The cleaved fragments are degraded by the EXONUCLEASE 4 (XRN4) enzyme. AGO1, AGO2, AGO4 and AGO10 contribute to the complexity of sRNA-induced silencing. In *Arabidopsis*, miR398 targets CSD mRNAs to regulate oxidative stress responses, while siRNAs participate in the silencing of viral RNAs during pathogen defence (Bhogireddy et al., 2021; Cao et al., 2024; Waititu et al., 2020; Yadav et al., 2024; Zhang et al., 2023).

iv. *Epigenetic Regulation by ncRNAs*: ncRNAs can induce epigenetic modifications, including DNA methylation and histone modifications, establishing long-term gene silencing in plants.

a. *RdDM*: siRNAs and lncRNAs guide chromatin-modifying complexes to specific loci, leading to DNA methylation and transcriptional silencing. In *Arabidopsis*, RdDM plays a key role in silencing transposons and stress-responsive genes during abiotic stress such as drought and salinity. DCL1 generates 21-nt miRNAs from partially paired dsRNA precursors.

DCL2/DCL4 produce 20–22-nt siRNAs, and DCL3 generates 24-nt heterochromatic siRNAs (hc-siRNAs) involved in RdDM. hc-siRNAs direct DNA and histone modifications, such as H3K9 methylation, to silence genes via AGO4-RISC complexes. This process integrates enzymes like RNA-dependent RNA polymerase (RDR), AGO proteins and Pol IV/V (Hombach & Kretz, 2016; Statello et al., 2021; Waititu et al., 2020; Wang & Chang, 2011; Xie & Yu, 2015; Yadav et al., 2024; Zhang et al., 2023).

v. Molecular Scaffolds, Decoys, Guides and Sponges:

a. Non-coding RNAs (ncRNAs) serve as molecular scaffolds in plants, facilitating interactions between proteins, RNA molecules and chromatin to regulate gene expression. ncRNAs guide the assembly of molecular complexes involved in transcriptional regulation or chromatin remodelling. ncRNAs act as structural frameworks that mediate interactions between proteins and RNA molecules. By base-pairing with specific sequences, they guide molecular complexes to their targets and facilitate their assembly and function. For example, COOLAIR, an lncRNA in *Arabidopsis*, recruits the polycomb repressive complex 2 (PRC2) to the FLC gene during vernalisation, mediating its repression in response to temperature changes. lncRNAs function as scaffolds, facilitating the assembly of ribonucleoprotein complexes to regulate gene expression. In rice, the lncRNA LAIR (LRK Antisense Intergenic RNA), derived from the antisense strand of the *LRK* gene cluster, enhances grain yield by upregulating *LRK* gene expression. LAIR interacts with chromatin at the *LRK1* locus and recruits histone modification proteins OsMOF and OsWDR5, increasing H3K4me3 and H4K16ac levels to activate gene transcription. In *A. thaliana*, the lncRNA ELENA1 contributes to immune responses by promoting the expression of *pathogenesis-related gene 1* (*PR1*). ELENA1 directly binds to the Mediator subunit MED19a and the repressor protein FIBRILLARIN2 (FIB2). By inhibiting FIB2's repressive effects, ELENA1 enhances MED19a enrichment at the *PR1* promoter, activating its transcription and strengthening the plant's defence mechanism (Hombach & Kretz, 2016; Statello et al., 2021; Waititu et al., 2020; Wang & Chang, 2011; Yadav et al., 2024; Zhang et al., 2023).

b. ncRNAs can act as decoys that sequester regulatory proteins, preventing them from interacting with their intended targets. lncRNAs regulate gene expression by acting as decoys that sequester proteins and disrupt their interactions. In *Populus simonii*, the lncRNA PROMPT_1281 responds to osmotic stress by binding to MYB transcription factors, preventing their DNA interactions and subsequently enhancing the expression of target genes both locally (cis) and distantly (trans) (Song et al., 2019). In *Medicago truncatula*, the lncRNA MtENOD40, containing a short ORF (sORF), plays a crucial role in root nodule development. It interacts with MtRBP1 (also known as MtNSR1), causing its relocation from the nucleus to the cytoplasm. This relocation mechanism is conserved, as shown by similar behaviour in *Arabidopsis* orthologs AtNSRa and AtNSRb, which regulate pre-mRNA alternative splicing. In *A. thaliana*, the auxin-induced lncRNA ASCO (Lnc351) competitively binds to AtNSRs, thereby influencing alternative splicing and diversifying the transcriptome during lateral root formation. ASCO also associates with splicing factors like PRP8a and SmD1b, adjusting splicing patterns in response to bacterial flagellin, further underscoring its role in dynamic transcriptome regulation (Bhogireddy et al., 2021; Statello et al., 2021; Waititu et al., 2020; Wang & Chekanova, 2017; Wolfien et al., 2019).

c. Some ncRNAs function as sponges, sequestering miRNAs or proteins and preventing them from interacting with their targets. In *Arabidopsis*, the lncRNA IPS1 sequesters miR399, a key regulator of phosphate homeostasis, ensuring proper phosphate signalling under nutrient stress (Borah et al., 2018).

d. lncRNAs regulate gene expression by recruiting protein complexes to specific chromatin regions, acting as guides. For example, the lncRNA APOLO forms R-loops by binding to complementary DNA, guiding the polycomb repressive complex 1 (PRC1) component like heterochromatic protein 1 (LHP1) to deposit repressive H3K27me3 histone marks and suppress auxin-responsive gene transcription (Ariel et al., 2020; Chen et al., 2021; Gonzales et al., 2024).

vi. *Hormonal Regulation*: ncRNAs also regulate plant hormone signalling pathways, modulating gene expression related to hormone biosynthesis, perception and signalling in response to environmental stimuli. miRNAs regulate key genes in hormone signalling pathways. For example, miR160 targets ARF10, ARF16 and ARF17, which are involved in auxin signalling during stress. Similarly, miR393 regulates auxin perception by targeting TIR1/AFB receptors, influencing plant responses to pathogen attack (Dai et al., 2021; Šečić et al., 2021).

In plants, non-coding RNAs are integral to regulating gene expression through various mechanisms, including scaffolding, transcriptional control, post-transcriptional modifications, epigenetic regulation and gene silencing (Gill et al., 2024; Yadav et al., 2024). By modulating key processes such as chromatin remodelling, translation, mRNA stability and alternative splicing, ncRNAs enable plants to adapt to environmental stresses like drought, salinity, nutrient deprivation and pathogen attack. These regulatory networks are essential for maintaining plant survival and optimising growth under fluctuating environmental conditions.

3.12 ROLE OF ncRNA IN ABIOTIC AND BIOTIC STRESS

Non-coding RNAs (ncRNAs) are pivotal in regulating gene expression in plants, particularly in adverse stress conditions. These RNA molecules play crucial roles in orchestrating plant responses to biotic stressors like pathogen infection and abiotic stressors like drought and salinity (Table 3.4).

3.12.1 Non-coding RNAs (ncRNAs) in Abiotic Stress

To coordinate a response to abiotic stressors, ncRNAs engage with a number of cellular processes, including transcription factor networks, ROS scavenging and hormone signalling. For instance, auxin signalling pathways are influenced by miRNAs such as miR393, which affects root architecture in salinity and drought. Likewise, miR319 controls JA signalling, which impacts the expression of genes that respond to stress. By connecting ncRNA networks with stress-responsive transcription factors, lncRNAs can serve as hubs, forming multi-layered feedback loops that regulate gene expression in response to stress. lncRNAs such as COLDAIR influence epigenetic modifications that improve resistance to cold. In addition to protecting against temperature extremes, these adaptive reactions aid in preserving ion homeostasis and water balance. Multiple

TABLE 3.4
List of ncRNA and Its Functional Role Involved in Abiotic and Biotic Stress

		Abiotic Stress		
ncRNA	**Target**	**Plant Species**	**Function**	**References**
miR396	Growth-regulating factor (GRF)	Creeping bentgrass (*Agrostis stolonifera*)	Enhances water retention, chlorophyll amount, plasma membrane stability and Na^+ removal under salt stress	Yuan et al. (2019)
miR156	SQUAMOSA promoter-binding protein (SBP)	Rice (*Oryza sativa*)	Significance in antioxidant system involving ROS scavenging, anthocyanin and proline metabolism	Lan et al. (2019) and Zhau et al. (2020)
miR393	*TIR1*	Maize (*Zea mays*)	Regulates degradation of *Aux/IAA* protein through ubiquitination in drought conditions	Seeve et al. (2019)
miR399 and miR398	CCS (copper chaperone for SOD) and phosphate transporter	Alfalfa (*Medicago sativa* L.)	Affects phosphate uptake and miRNA-mediated regulation of root-shoot growth.	Kosik (2013) and Memczak et al. (2013)
nat-siRNA	*P5CDH*	*Arabidopsis thaliana*	Drought responsive Proline and ROS accumulation	Kumar et al. (2018)
IPS1 (induced by phosphate starvation 1)	miR399	*A. thaliana*	Phosphate homeostasis: forms a non-cleavable lncRNA–miRNA duplex with miR399, enhancing PHO2 transcription and reducing shoot phosphate levels.	Franco-Zorrilla et al. (2007)
PDIL1 (phosphate deficiency-induced lncRNA1)	miR399	*Medicago truncatula*	Prevents miR399-mediated degradation of MtPHO2, ensuring phosphate homeostasis under deficiency.	Wang et al. (2017)
PILNCR1 (phosphate-deficiency-induced lncRNA1)	miR399	*Zea mays*	Blocks miR399-directed cleavage of ZmPHO2, ensuring phosphate regulation under deficiency.	Du et al. (2018)
UPAT (UHRF1-protein-associated transcript)	VIM1 and LHP1	Human (studied in *Arabidopsis*)	Structural convergence between plant and mammalian lncRNAs allows them to interact with conserved plant epigenetic regulators, disrupting the *Arabidopsis* transcriptome and interfering with YUCCA2 promoter regulation.	Fonouni-Farde et al., 2022
APOLO	WRKY42	*A. thaliana*	Epigenetic activation of RHD6, promoting root hair elongation via RSL2, RSL4 and EXT3 under cold stress, in association with WRKY42 as a ribonucleoprotein complex.	Moison et al. (2021) and Pacheco et al. (2021)
	VIM1and LHP1		Coordinates VIM1 and LHP1 to regulate DNA methylation and H3K27me3 on YUCCA2, enabling auxin biosynthesis and thermomorphogenesis in response to warmth.	Fonouni-Farde et al. (2022)

TABLE 3.4 (*Continued*)
List of ncRNA and Its Functional Role Involved in Abiotic and Biotic Stress

			Biotic Stress		
ncRNA	**Target**	**Pest/Pathogen**	**Plant species**	**Function**	**References**
Sl-miR482a	*Sl-NBS-LRR*	*Phytophthora infestans*	Tomato (*Solanum pimpinellifolium*)	Regulate the homeostasis of nucleotide-binding site–leucine-rich repeat (NBS-LRR) protein in tomato for resistance against *P. infestans.*	Jiang et al. (2020)
miRNA167, miRNA160, miRNA390, miRNA393 and miRNA846	Transport inhibitor response 1 and auxin-signalling F-box (TIR1/AFB) transcripts	*Pseudomonas syringae* and *Xanthomonas axonopodis*	Tomato (*Solanum lycopersicum* L.)	Inhibit the expression of auxin response factor 8 (ARF8) for inducing immunity	Jodder et al. (2017)
miR164	*Pseudomonas syringae*	*A. thaliana*	Regulates the immune response by targeting *NAC* transcription factors to enhance defence.	*NAC transcription factors*	Lee et al. (2017)
miR156	*SPL3, JAZ3*	*Helicoverpa armigera* (*cotton bollworm*)	*Arabidopsis*	MiR156-regulated SPL9 proteins stabilise JAZs, leading to an increase in SPL protein levels during ageing, which further accumulates JAZ proteins, thereby impairing the JA response.	Mao et al. (2017)
miR482/2118b	*NLRs*	Bacterial and oomycete pathogens	Solanum spp.	Suppression of the generation of secondary siRNAs from NLRs and TAS5.	Canto-Pastor et al. (2019)
Osa-miR535	*OsSPL4, OsSPL14*	*Magnaporthe oryzae*	Rice (*O. sativa*)	Targets OsSPL14, OsSPL4 that affect WRKY45 and GH3.2 expression respectively, impacting both immunity and IAA homeostasis	Zhang et al. (2022)
lncRNA16397	*SlGRX*	*P. infestans*	Tomato (*S. pimpinellifolium*)	Increases SlGRX expression, resulting in reduced ROS accumulation and attenuating plasma membrane damage	Cui et al. (2017)
lncRNA33732	*RBOH*	*P. infestans*	Tomato (*S. pimpinellifolium*)	Induces expression of RBOH resulting in accumulation of hydrogen peroxide (H_2O_2)	Cui et al. (2019)
lncRNACHI	*SlCHTB3, SlCHI*	*Botrytis cinerea*	Tomato (*S. pimpinellifolium*)	lncRNACHI-VIGS impairs chitinase activity	Chen et al. (2024)
lncRNAMMP	*SlMMP2, SlMMP3*	*Botrytis cinerea*	Tomato (*S. pimpinellifolium*)	Silencing of lncRNAMMP and lncRNASBT1.9 inhibit	Chen et al. (2024)
lncRNASBT1.9	*SlSBT1.9*	*Botrytis cinerea*	Tomato (*S. pimpinellifolium*)	expression of Jasmonic acid (JA)-related genes	Chen et al. (2024)
lncRNA MSTRG18363	miR1918	*Botrytis cinerea*	Tomato (*Solanum lycopersicum*)	Induced by *Bacillus subtilis* SL18r, sequesters miR1918 to increase accumulation of target mRNAs, enhancing resistance against the pathogen *Botrytis cinerea*	Zhou et al. (2020)
ELENA1	FIB2, MED19a	*Pseudomonas syringae*	*A. thaliana*	Regulate expression pathogenesis-related 1(PR1) via recruitment of MED19a and removal of FIB2 from complex at promoter of PR1	Seo et al. (2017, 2019)

Source: Modified from Cao et al. (2024), Chorostecki et al. (2023), Raza et al. (2023) and Samarfard et al. (2022).

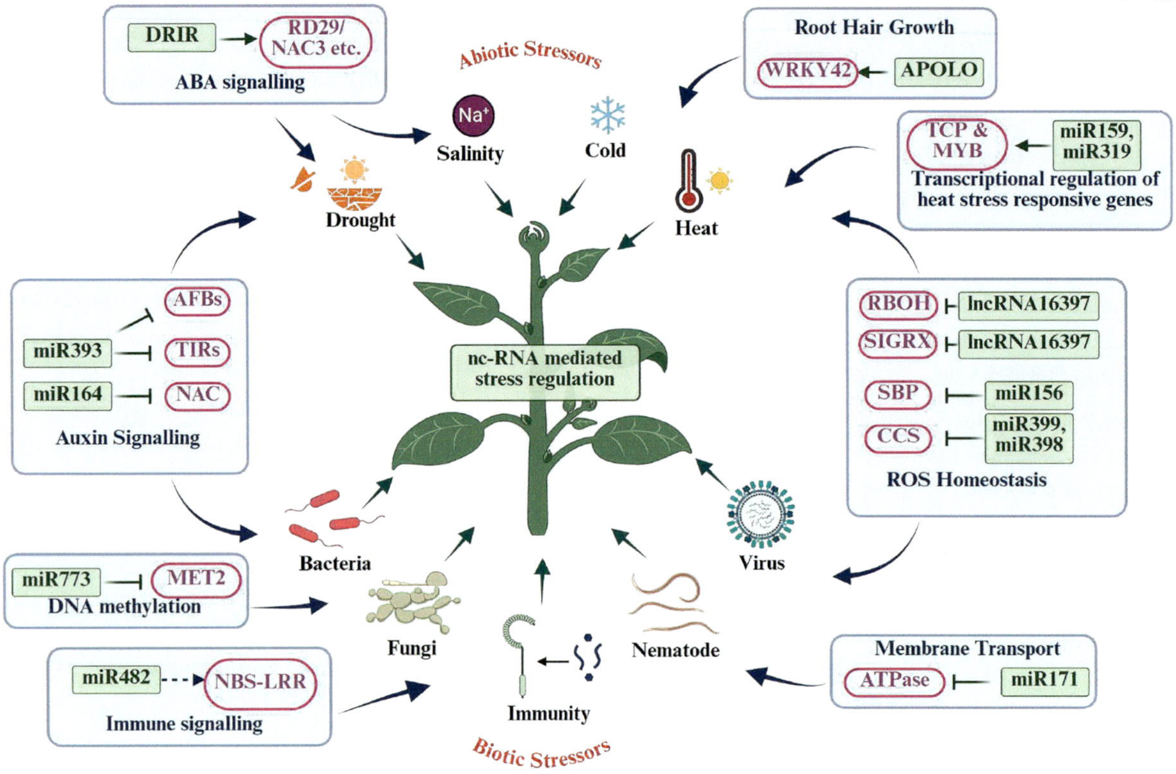

FIGURE 3.4 Role of different ncRNAs involved in regulating Abiotic and Biotic stress. (Image created using BioRender.)

abiotic stresses, such as drought and high temperatures, are frequently experienced by plants at the same time. Studies have shown that ncRNAs with dual functions, such as miR169 and miR393, target general stress-responsive genes in response to various stress signals. Although more research is required to completely understand their functions, lncRNAs and circRNAs may also modify expression under combined stressors (Liu et al., 2015; Mondal et al., 2023; Panchal et al., 2023; Yadav et al., 2024). Abiotic stresses, including heat, drought and salinity, significantly impact plant growth, development and agricultural productivity. Among the regulatory mechanisms plants employ, sRNAs, particularly miRNAs and lncRNAs, play a pivotal role in mediating adaptive responses (Figure 3.4).

3.12.1.1 Heat Stress

Heat stress poses a major threat to global crop productivity, with climate change exacerbating its effects. For instance, rising temperatures are projected to reduce wheat and maize yields by 5.5% and 3.8%, respectively. To combat heat stress, plants utilise a combination of HSPs, heat shock transcription factors (HSFs), reactive oxygen species (ROS)-scavenging enzymes and sRNAs to maintain thermotolerance. sRNAs, together with HSPs, HSFs, hormones and alternative splicing mechanisms, form a sophisticated regulatory network that ensures thermotolerance. In *Arabidopsis*, heat stress triggers the expression of miR160, which represses the expression of ARF10, ARF16 and ARF17 to regulate the heat tolerance mechanisms of plants, including the high expression of HSP genes. Similarly, miR393 is involved in auxin signalling and increases heat tolerance in creeping bentgrass by repressing AsAFB2 and AsTIR1, leading to higher HSP expression. Additionally, miR159 and miR319, which target TCP and MYB genes, regulate transcription levels of heat-stress-responsive genes, conferring heat stress tolerance in *Solanum lycopersicum*. The miR156-SPL module is also involved in sustaining the expression of HSFA2 and HSPs during recovery from heat stress in *Arabidopsis*, potentially integrating stress responses with development. miR398 regulates the accumulation of ROS by downregulating copper/zinc superoxide dismutase (CSD1/2) and copper chaperone for superoxide dismutase (CCS), which activates HSP and HSF levels, thus forming a regulatory loop. miR396, in sunflower, targets HaWRKY6, which is involved in early responses to high temperatures, and miR396-resistant HaWRKY6 plants exhibit sensitivity to heat damage (Bhogireddy et al., 2021; Cao et al., 2024; Li et al., 2014; Liu et al., 2014; Mondal et al., 2023; Panchal et al., 2023; Yadav et al., 2024).

In rice, miR166 and miR169 target SGT1 (G2 allele of skp1) and regulate thermal stress during flowering by interacting with HSP90 and HSP70.

miR156 regulates the *SQUAMOSA promoter-binding protein-like* (SPL) transcription factors, which are critical for developmental phase transitions (e.g., juvenile-to-adult and vegetative-to-reproductive stages). Under heat stress, miR156 levels increase, promoting the expression of heat-responsive genes and contributing to 'heat stress memory'. This memory allows plants to retain thermotolerance even after returning to favourable conditions, a process mediated by the HSFA2 cascade, *heat-stress-associated 32* (HSA32) and ROF1. miR172 targets the APETALA2 (AP2) transcription factor family, regulating thermosensory flowering under normal conditions and enhancing heat stress responses. The miR172-AP2 module is critical in balancing growth and stress adaptation. In species like sweet sorghum, miR172 contributes to heat stress tolerance by targeting genes involved in stress response and protein stability. miR159 and miR319 are involved in regulating male fertility and leaf development. In crops like cucumber, under heat stress, miR159 downregulation leads to increased expression of GIBBERELLIC ACID MYB (GAMYB) genes, which support stress adaptation by enhancing reproductive processes. Overexpression of miR160 in cotton (*Gossypium hirsutum*) under extreme heat stress suppresses ARF10 and ARF17, resulting in reduced anther dehiscence and lower fertility, which negatively affects crop productivity. miR166 regulates xylem development and plays a role in heat tolerance by modulating transpiration rates and promoting leaf rolling, which reduces water loss. miR396 targets *GRF* (growth-regulating factors) transcription factors, which play a key role in regulating plant growth under heat stress. miR169 and miR393 have been shown to modulate ROS production and regulate the expression of HSPs, contributing to plant resilience under high temperatures. In *Arabidopsis*, miR398 is upregulated in response to heat stress, suppressing targets such as superoxide dismutase (SOD) genes (*CSD1*, *CSD2* and *Copper Chaperone for SOD*). This modulation helps manage oxidative stress by fine-tuning ROS-scavenging enzyme activity. Osa-miR5144 enhances heat tolerance in rice by regulating *OsPDIL1;1*, a gene involved in protein disulphide bond formation, which is essential for protein stability under stress. miR400 acts as a negative regulator during heat stress. Alternative splicing of its primary transcript reduces mature miR400 levels, enabling better stress adaptation in heat-resistant plants (Bhogireddy et al., 2021; Cao et al., 2024; Guan et al., 2013; Li et al., 2014, 2017; Liu et al., 2014; Mehtab-Singh et al., 2024; Mondal et al., 2023; Panchal et al., 2023; Xu et al., 2016; Yadav et al., 2024).

3.12.1.2 Cold Stress

Cold stress disturbs the physiological and metabolic balance in plants, impairing cellular homeostasis. A key pathway for cold tolerance is the abscisic acid (ABA)-independent route, which involves the *Inducer of CBF Expression* (ICE). ICE activates *CBF/DREB1* transcription factors, which bind to C-repeat (CRT) or Low Temperature-Response Elements (LTRE), driving the expression of cold-responsive (COR) genes. In rice, cold stress induces the upregulation of Osa-miR319b, which represses *Ospcf6* and *OsTCP21* while activating *Bdrebl/CBF*, *DREB2A* and *TPP1/2*, thereby enhancing cold stress tolerance (Wang et al., 2014). OsmiR156 targets *OsSPL3*, which in turn positively regulates *OsWRKY71*. This interaction inhibits *OsMYB2* and *OsMYB3R-2*, leading to the activation of stress-responsive genes such as *OsLEA3*, *OsDREB24* and *OsCTP1* (Yang et al., 2012; Zhou & Tang, 2019).

tasiRNA-ARF and miR167 in wheat regulate auxin signalling to support the cold stress response. OsSPL3, regulated by miR156, interacts with OsWRKY71, which in turn regulates cold stress tolerance through the expression of *OsLEA3*, *OsDREB2A* and *OsCTP1*. Furthermore, miR394 in *Arabidopsis* downregulates LCR (Leaf Curling Responsiveness) and activates CBF1, CBF2 and CBF3, enhancing cold tolerance. The miR169/NF-YA module in *Arabidopsis* plays a crucial role in the Aux/IAA14-mediated cold stress response, while miR167 and tasiRNA-ARF in wheat are involved in auxin signalling and developmental responses to cold stress. miR408 and miR397 also contribute to low-temperature stress responses by modulating ROS homeostasis and the lignin biosynthesis pathway. miR408 targets the phytocyanin family and laccases, enhancing tolerance to low temperatures, while miR397 targets laccases (LAC2, LAC17, LAC4) and casein kinase β subunit 3, promoting freezing tolerance (Aslam et al., 2020; Ma et al., 2022b; Nowak et al., 2024; Singh et al., 2020; Song et al., 2013, 2016; Tang et al., 2012; Yang et al., 2012).

3.12.1.3 Drought Stress

Water availability is a key factor regulating plant growth and development. Plants face drought conditions, which are defined as a decrease in water availability or, more specifically, a decrease in the water potential (Ψw) of soil (Kramer & Boyer, 1995). It is one of the most significant challenges to plant survival. Changes in sRNAs profiles are observed in response to drought, indicating their involvement in regulating adaptive mechanisms to dehydration. In *Arabidopsis*, drought stress leads to the upregulation of miR393, miR397b and miR402, with a downregulation of miR319c and miR389a. In rice, 16 miRNAs are upregulated and 17 downregulated in response to drought. Similarly, maize shows a distinct pattern of miRNA regulation in leaves and roots under drought conditions. In apple, miR160 promotes drought adaptation by regulating *MdARF17*, forming a positive feedback loop that enhances root formation and stress resilience. miR165/166 regulates HD-ZIP III transcription factors, which are crucial for ABA signalling and drought tolerance. They act in concert with ARFs, providing a cooperative mechanism for drought adaptation. Both miR159 and miR169 are involved in ABA-dependent responses to drought. miR159 regulates *MYB33* and *MYB101*, which modulate ABA signalling, while miR169 targets NFYA transcription factors, enhancing drought tolerance. MiR169 regulates plant responses to water stress by modulating stomatal closure and maintaining cellular water balance. Overexpression of NFYA5 in *Arabidopsis* improves drought resilience, a trend also observed in *Brassica napus* and soybean. miR156 is upregulated during drought stress and targets *SPL* (SQUAMOSA promoter-binding-like) genes, which influence plant growth and stress adaptation. Overexpression of miR156 enhances drought tolerance by modulating the expression of stress-responsive genes. miRNAs like miR156, miR166 and miR393 regulate key genes involved in stress response pathways, enhancing drought resilience. miR398 and miR399: These miRNAs regulate antioxidant defence mechanisms and phosphate metabolism under drought conditions, protecting cells from oxidative damage. DRIR (drought and salt-inducible RNA) is a key lncRNA that is highly expressed under drought stress. It modulates ABA signalling and enhances stress tolerance by regulating genes involved in osmotic stress response and stomatal closure, critical for water conservation. ZmNAC48, a cis-NAT in *maize*, regulates NAC transcription factors involved in root development under drought conditions. These lncRNAs help the plant cope with dehydration by regulating genes that affect root growth and water absorption. Under drought conditions, siRNAs help regulate the expression of genes involved in water retention and osmotic stress tolerance. In *Arabidopsis*, siRNAs are involved in regulating RD (Responsive to Drought) genes and signalling components like ABA, a hormone critical for drought tolerance. miR390-mediated production of tasi-ARFs influences auxin signalling under drought stress. These tasiRNAs modulate root development and gene expression related to water uptake, enhancing drought tolerance by optimising the plant's root architecture and stress response. In *poplar*, phasiRNAs generated from miR482 and miR828 help regulate drought-responsive NAC transcription factors, involved in water stress and stomatal regulation. These phasiRNAs play a role in improving plant tolerance by controlling osmotic stress pathways and ROS detoxification (Liu et al., 2019; Ma et al., 2022; Mondal et al., 2023; Panchal et al., 2023; Sharma et al., 2023; Singroha et al., 2021; Yadav et al., 2024).

3.12.1.4 Salinity Stress

Salinity stress hampers plant growth and reproductive processes by disrupting ion balance and inducing oxidative stress. sRNAs help mitigate the adverse effects of salinity by regulating key genes involved in stress tolerance. miRNAs are integral in regulating genes that control ion transport, osmotic balance and stress tolerance during salt exposure. miR160 expression is induced by salt stress in peanuts, and overexpression of miR160 protects seedling development by regulating the ARF18 pathway, which protects against ROS damage (Ma et al., 2022; Panchal et al., 2023; Yadav et al., 2024). In apple, overexpression of miR156a weakens salt resistance, while its target gene, MdSPL13, strengthens salt tolerance. Osa-miR393, overexpressed in rice and *Arabidopsis*, reduces tolerance to salt, while expressing a miR393-resistant form in *Arabidopsis* enhances salt tolerance. In species such as rice and creeping bentgrass, miR393 regulates the expression of auxin response factors (ARFs), which are involved in salt stress tolerance by modulating auxin signalling and root architecture. Overexpression of miR393a in creeping bentgrass enhances salt stress tolerance by improving potassium uptake. In rice and *Arabidopsis*, the overexpression of miR393 reduces salt tolerance, whereas a mutant form (*MTIR1*) enhances salt resistance. This demonstrates the complex role of miR393 in salt stress responses. Moreover, miR396c, which targets GRF, reduces salt tolerance in rice and *Arabidopsis*, whereas it enhances tolerance in creeping bentgrass through the AsSOS1 pathway. Similarly, miR414c and its target GhFSD1 play roles in salt tolerance in cotton, with miR414c overexpression leading to sensitivity to salt stress by regulating ROS metabolism. Other miRNAs, such as PeNAC070 and miR164, which target GmNFYA3, also affect salt tolerance. Overexpressing miR164 in *Arabidopsis* increases salt stress sensitivity. miRNVL5, targeting GhCHR, represses key genes involved in stress tolerance, leading to hypersensitivity to salt stress. In rice, the salt-tolerant cultivar *Pokkali* has been used to identify key miRNAs that regulate salt stress responses. MiRNAs like miR156, miR166 and miR393 target genes involved in ion transport and stress signalling, enhancing the plant's ability to cope with high salinity. Additionally, miRNAs such as miR6225-5p in sweet sorghum regulate calcium ion uptake, influencing salt tolerance. miR169 acts as a key regulator in salt stress tolerance by targeting *NF-YA* genes, which control salt-responsive gene expression. miR172: Serves as a positive regulator of salt stress tolerance in monocots. Functional studies have shown that specific miR172 variants contribute to salt stress tolerance by activating salt-responsive genes. In cotton, miR414c targets *GhFSD1* (Iron Superoxide Dismutase 1), modulating ROS metabolism and influencing salt tolerance. Overexpression of miR414c makes cotton plants more susceptible to salt stress. In *Arabidopsis*, *cis*-NATs like AtNAC5 regulate NAC transcription factors, which control salt tolerance mechanisms such as ion transport and osmotic regulation. These lncRNAs ensure the plant adapts effectively to salt stress. miR390-mediated tasiRNA production targets ARFs, which play a key role in auxin signalling and root growth under salt stress. These tasiRNAs help modulate the plant's response to salt by regulating ion transport and enhancing stress adaptation. PhasiRNAs generated by miR482 and miR828 regulate NAC and AP2/ERF transcription factors under salt stress. These sRNAs control the expression of genes related to salt tolerance, such as those involved in ion transport and oxidative stress management. Salt stress activates the production of siRNAs that target genes involved in Na^+/K^+ homeostasis, such as SOS1, a salt-responsive gene that helps maintain ionic balance. These siRNAs ensure proper ion transport and protect the plant from excessive salt buildup (Chen et al., 2015; Ma et al., 2020, 2022; Mondal et al., 2023; Panchal et al., 2023; Xu et al., 2021; Yadav et al., 2024).

In *Arabidopsis*, the lncRNA DRIR positively regulates ABA-mediated drought and salt-stress responses by acting at or upstream of gene transcription stages within the ABA signalling pathway. Overexpression of DRIR in plants results in enhanced tolerance to both drought and salt stress. In cotton, lncRNA973 modulates miR399 and its target PHO2, which are involved in the response to salt stress. Additionally, lncRNA973 influences the expression of ROS-scavenging genes, transcription factors and other genes related to salt stress tolerance. lncRNA354 in cotton functions as a competing endogenous RNA (ceRNA) for miR160b, regulating GhARF17/18 genes and modulating auxin signalling in response to salt stress. In cold stress, XH123 is a cotton lncRNA whose silencing increases sensitivity to cold, causing chloroplast damage and elevated ROS levels. Moreover, the transcriptional read-through of SVALKA in *Arabidopsis* results in the expression of an antisense lncRNA asCBF1, which suppresses the expression of CBF1, thereby reducing freezing tolerance. Another lncRNA, APOLO, interacts with the transcription factor WRKY42 to activate the expression of RHD6 (ROOT HAIR DEFECTIVE 6) by forming a chromatin loop around its promoter, promoting root hair growth in response to low temperatures (Baek et al., 2016; Cao et al., 2021; Kindgren et al., 2018; Ma et al., 2022; Pegler et al., 2020; Qin et al., 2017; Yadav et al., 2024; Zhang et al., 2019).

sRNAs are key regulators of plant responses to abiotic stresses such as heat, cold, drought and salinity. They orchestrate the expression of stress-responsive genes and pathways, making them critical for stress tolerance. These molecular tools not only protect plants from stress but also offer promising avenues for breeding and engineering stress-resilient crops, addressing the challenges of climate change and ensuring sustainable agricultural productivity. Understanding the roles of specific miRNAs and their target genes can offer new avenues for improving crop resilience through genetic manipulation, ultimately enhancing agricultural productivity under adverse environmental conditions.

3.12.2 Non-coding RNAs (ncRNAs) in Biotic Stress

Plants are subject to biotic stress from pests, herbivores and pathogens. ncRNAs have a crucial role in the immunological response of plants for modulating plant defence. To improve immune responses, for instance, miR393 is elevated after pathogen invasion. Auxin receptors are suppressed, and pathogen-associated molecular pattern (PAMP)-triggered immunity (PTI) is enhanced when miR393 is increased during pathogen invasion (Figure 3.4; Table 3.4). In herbivory, siRNAs play a role in defence mechanisms by inhibiting genes that are altered by insect effectors while they are feeding. Defensive proteins can be recruited by lncRNAs via scaffolds or molecular decoys (Bhogireddy et al., 2021; Cao et al., 2024; Raza et al., 2023; Song et al., 2021).

3.12.2.1 ncRNAs in Response to Bacterial or Fungal Pathogens

miR164 targets the NAC transcription factor (*NAC4*) in *A. thaliana* during infection by *Pseudomonas syringae*. This promotes programmed cell death, helping to restrict bacterial spread. It downregulates *NAC4* to activate hypersensitive response (Šečić et al., 2021; Song et al., 2021; Yuan et al., 2019). In tomato (*Solanum lycopersicum*), miR482 targets nucleotide-binding site leucine-rich repeat (NBS-LRR) resistance genes, leading to mRNA degradation and the generation of secondary siRNAs; therefore, it suppresses immunity when overexpressed (Liao et al., 2022; Shivaprasad et al., 2012). miR393 regulates auxin receptor genes (*TIR1*, *AFB2*, *AFB3*) during *Pseudomonas syringae* and *Xanthomonas axonopodis* infections in tomato. osa-miR169 targets HSPs, ABC transporters and NBS-LRR genes in *Ralstonia solanacearum*-infected rice and activates defence-related gene expression. miR168a regulates AGO1 (AGO protein) in *Malus hupehensis* (tea crab apple) infected by *Botryosphaeria dothidea* to enhance pathogen resistance by fine-tuning AGO1-mediated defence. miR773 targets *MET2* in *Arabidopsis* during *Plectosphaerrella cucumerina* infection and activates PAMP-triggered immunity. miR319a regulates *BEE1* and *PERK2* genes in cucumber during *Fusarium oxysporum* infection, controlling immune-related transcriptional reprogramming. miR156ab suppresses WRKY transcription factors in apple during *Alternaria alternaria* infection by reducing WRKY-mediated PR protein expression. miR168a and miR164a activate garlic (*Allium sativum*) defence genes against *Fusarium oxysporum*. miR393 targets *TIR1*, *AFB2* and *AFB3* to enhance resistance against *Pseudomonas syringae* by downregulating auxin signalling in response to PAMPs. miR773-MET2 Module: enhances fungal pathogen resistance by regulating DNA methylation. miR528-AO module in rice enhances antiviral defence by accumulating ROS during rice stripe virus infection. tasiRNAs in *Arabidopsis* derived from TAS1 and TAS2 loci, tasiRNAs reduce virulence against *Botrytis cinerea*. In tomato, lncRNA16397 enhances resistance to *Phytophthora infestans* by inducing the *GLUTAREDOXIN 22* gene in cis. circRNAs are identified in crops like wheat, rice and potato, and circRNAs are implicated in responses to pathogens such as *Magnaporthe oryzae* in rice and bacterial pathogens in kiwi and potato (Jiang et al., 2022; Raza et al., 2023; Song et al., 2021; Yuan et al., 2019).

3.12.2.2 ncRNAs in Viral Infection, Nematode and Insect Herbivore Attack

miR168 regulates AGO1, a key antiviral protein, in *Arabidopsis* during *Tombus virus* infection. Viral pathogens exploit this regulation to evade immunity. miR482: Enhances resistance by regulating NBS-LRR genes during viral infections in common bean (*Vigna mungo*), modulating host immune responses. miR17 targets GRAS transcription factors in tomato during *Bemisia tabaci* infestation, regulating defence-related transcription factors. miR396d-3p regulates calcium-dependent protein kinases (CDPKs) and mitogen-activated protein kinases (MAPKs) in *Camellia sinensis* during herbivory by *Ectropis oblique* to activate defence signalling cascades. miR398b modulates MAPK cascade genes in rice during *Nilaparvata lugens* infestation, controlling herbivory-related defence pathways. miR159 and miR319 regulate auxin-related genes (*TAS3*, *MYB33*) during root-knot nematode (*Meloidogyne javanica*) infection in *Arabidopsis*, which modulate auxin signalling to control gall formation. miR827 targets nitrogen limitation adaptation genes, promoting susceptibility to cyst nematodes (*Heterodera schachtii*), fine-tuning host nutrient pathways (Raza et al., 2023; Reinhart et al., 2002; Song et al., 2021).

3.12.3 ncRNAs in Plant Immunity

R proteins play a pivotal role in recognising pathogen-derived effector proteins, thereby triggering effector-triggered immunity (ETI) to protect plants from infections (Figure 3.4; Table 3.4). The expression of R genes is tightly regulated by various miRNAs, ensuring a balanced immune response in plant species such as *Arabidopsis*, legumes and members of the *Solanaceae* family (Basak & Nithin, 2015; Bhogireddy et al., 2021; Hong et al., 2020; Raza et al., 2023; Song et al., 2021; Yuan et al., 2019). Many miRNAs, typically 22 nucleotides in length, initiate the production of phased siRNAs (phasiRNAs) from NLR gene targets to modulate immune responses. For instance, in *Arabidopsis*, miR472 – a member of the miR482 family – targets the NLR gene RPS5, promoting phasiRNA production to suppress ETI-mediated resistance. Similarly, in *M. truncatula*, miR2118a, miR1507 and miR2109 regulate distinct NLR genes, triggering extensive phasiRNA generation. In tobacco, miR6019 and miR6020 mediate the cleavage of transcripts from the **N** gene, a TIR-NB-LRR receptor gene that confers resistance to the tobacco mosaic virus. In barley, the miR9863 family targets specific alleles of the NLR gene Mla, conferring resistance to powdery mildew in a race-specific manner (Fei et al., 2015; Huang, Yang, et al., 2016; Jyothsna & Alagu, 2022; Liao et al., 2022; Liu et al., 2014; Song et al., 2021; Sós-Hegedűs et al., 2020).

The miR482/2118 superfamily is particularly notable for targeting conserved NLR gene motifs, such as the P-loop, TIR1 and kinase-2 domains, to induce phasiRNA-mediated suppression. This highlights the conserved nature and redundancy of miRNA-mediated NLR regulation across species. However, some miRNAs bypass phasiRNA production and directly cleave NLR mRNA to reduce gene expression. For instance, miR1510 in soybean targets the NLR gene 16G135500, leading to decreased resistance to *Phytophthora sojae*. Similarly, Md-miRLn11 in apple targets Md-NBS, an NLR gene, to compromise the plant's defence against apple leaf spot disease. nat-siRNAATGB2 in *Arabidopsis* regulates ETI against *Pseudomonas syringae* by suppressing a negative regulator (*PPRL*) of RPS2-mediated defence. PhasiRNAs from miR482/2118 in Tomato target NLR (nucleotide-binding leucine-rich repeat) genes, crucial for resistance against bacterial and oomycete pathogens. These diverse regulatory mechanisms underscore the conserved miRNA-mediated fine-tuning of plant immune responses (Canto-Pastor et al., 2019; Cui et al., 2017; Hou et al., 2009; Liao et al., 2022; López-Márquez et al., 2023; Ruiz-Ferrer & Voinnet, 2009; Song et al., 2021).

Beyond NLR regulation, miRNAs and siRNAs are critical modulators of immune signalling pathways. In *Arabidopsis*, miR172b enhances the expression of FLS2, a receptor crucial for immune perception, by repressing the transcriptional repressors TOE1 and TOE2 that bind the FLS2 promoter. Additionally, miR863-3p in *Arabidopsis* targets ARLPK1 and ARLPK2, two receptor-like pseudokinase genes that negatively regulate immunity during early stages of *Pseudomonas syringae* pv. *tomato* infection. Furthermore, a natural antisense transcript-derived siRNA, natsiRNAATGB2, silences a gene encoding a pentatricopeptide repeat protein, a negative regulator of RPS2-mediated ETI during infections involving the *avrRpt2* effector. In tomato, circRNAs such as circRNA45 and circRNA47 have been implicated in positively regulating immunity against *P. infestans* by interacting with miR477-3p and SpRLK, respectively (Hong et al., 2020; Niu et al., 2016; Song et al., 2021; Zou et al., 2018).

miRNAs also contribute to transcriptional reprogramming during pathogen infection by targeting transcription factors associated with immune responses. For example, during *Plectosphaerella cucumerina* infection in *Arabidopsis*, the downregulation of miR396 allows its targets – GRF transcription factors – to activate immune gene expression. In rice, miR169 targets NF-YA transcription factors, negatively regulating immunity against *M. oryzae*, while suppression of miR164a during infection promotes the activity of OsNAC60, enhancing ROS accumulation and callose deposition. miR156, also in rice, suppresses SPL transcription factors, reducing resistance to bacterial blight. Additionally, a transposable element-derived siRNA, TE-siR815, modulates WRKY45-mediated immune responses by repressing the leucine-rich repeat receptor kinase protein ST1, thereby fine-tuning immunity against *Xanthomonas oryzae pv. Oryzae* (Li et al., 2017, 2019; Song et al., 2021; Yuan et al., 2019).

Collectively, these findings illustrate the extensive role of miRNAs and siRNAs in orchestrating plant immunity through the regulation of NLR genes, signalling pathways and transcription factors. These mechanisms represent an evolutionarily conserved strategy to balance robust immune responses with the metabolic costs associated with pathogen defence. The diverse roles of ncRNAs in plant defence demonstrate their immense potential for developing biotic stress-tolerant crops. By fine-tuning immune signalling, modulating stress-responsive transcription factors and controlling host–pathogen interactions, miRNAs provide a flexible and efficient regulatory mechanism. This understanding can guide genetic engineering and crop improvement strategies to enhance resistance against various biotic stresses.

3.13 APPROACHES FOR ncRNA DISCOVERY AND FUNCTIONAL CHARACTERISATION

Non-protein-coding RNAs (ncRNAs) encompass a diverse group of cellular transcripts which don't code for proteins. Even though these molecules have been known for a while, their importance has recently come into focus as a result of genome-wide studies that found thousands of ncRNAs across various model organism (Hazra et al., 2023). These findings, which allowed for both individual and global identification of ncRNAs, were made possible by direct biochemical isolation of ncRNAs from cells, tissues or species as well as computational predictions of ncRNA candidates in entire genomes (Huttenhofer, 2006; Wolfien et al., 2019) (Table 3.5).

Uncovering new ncRNAs has been made possible in a substantial manner by experimental techniques referred to as 'Experimental RNomics' (Huttenhofer, 2006).

These consist of the following:

i. RNA sequencing (enzymatic or chemical)
ii. Shotgun cloning of ncRNAs
iii. Microarray analysis
iv. Genomic SELEX

Broadly, several methods employed to discover and functionally characterise ncRNAs in plants can be categorised as follows:

- *High-throughput sequencing*: RNA-seq and sRNA sequencing are frequently employed to identify novel ncRNAs and their expression analysis in various conditions.
- *Computational or Bioinformatics tools*: Target prediction algorithms and gene ontology (GO) analyses are examples of computational or bioinformatics methods that aid in the prediction of ncRNA functions. All-inclusive search Using computational tools (such as ncRNA databases like NONCODE, miRBase and Ensembl) and algorithms (like Rfam, CMfinder, RNAz and Infernal), putative ncRNAs across the genome can be identified based on sequence patterns and secondary structure conservation.
- *Functional studies*: Techniques such as gene knockdown or overexpression studies help elucidate the role of specific ncRNAs. CRISPR/Cas9 and other gene editing techniques are employed to functionally validate ncRNAs and their roles in stress response.

Strategic analysis of ncRNA requires integration of the above-mentioned approaches to design a robust workflow

TABLE 3.5
Approaches to Study, Analyses and Characteristics of ncRNA

Category	Approach	Purpose	Pros	Cons
Experimental approaches for ncRNA discovery	Northern blotting	Detect and size-specific ncRNAs in tissue or cell samples.	High specificity, visualises RNA size.	Labour-intensive, low throughput, requires prior knowledge.
	Quantitative PCR (qPCR)	Measure ncRNA abundance in tissues or under different conditions.	Sensitive and quantitative.	Requires prior knowledge of ncRNA sequence.
	Ribonuclease protection assay (RPA)	Detect and quantify specific ncRNAs using protected probes.	High specificity, quantitative accuracy.	Complex setup, limited throughput.
	RNA interference (RNAi) screening	Knockdown ncRNAs to examine their roles in biological processes.	High-throughput screening.	Off-target effects, incomplete knockdown.
	Small RNA sequencing (small RNA-Seq)	Discover small ncRNAs like miRNAs and piRNAs.	Unbiased discovery, deep profiling.	Requires small RNA enrichment.
	Chromatin immunoprecipitation (ChIP)	Identify ncRNAs involved in chromatin modifications.	Reveals RNA–chromatin interaction.	Indirect evidence requires further validation.
Functional characterisation of ncRNAs	Gene knockout or knockdown	Knock out or knockdown ncRNA genes to study biological functions.	High specificity (CRISPR).	Potential compensatory mechanisms.
	Overexpression systems	Introduce synthetic ncRNAs to observe phenotypic changes.	Direct functional insights.	May not replicate natural conditions.
	Crosslinking immunoprecipitation (CLIP)	Identify ncRNA–protein interactions (HITS-CLIP, iCLIP, eCLIP).	High-resolution RNA–protein interaction mapping.	Technically demanding, requires expertise.
	RNA–protein interaction profiling (RIP-Seq)	Identify proteins interacting with ncRNAs.	Efficient for identifying regulatory complexes.	Cannot distinguish direct vs. indirect interactions.
	RNA-Seq upon ncRNA depletion or overexpression	Examine transcriptomic changes after ncRNA depletion or overexpression.	Comprehensive insights into regulatory networks.	Challenging to distinguish direct vs. indirect effects.
Functional pathways and network analysis	Gene ontology (GO) and pathway enrichment analysis	Identify biological processes and pathways associated with ncRNAs.	Contextualises ncRNA functions.	Requires high-confidence target datasets.
	Target prediction	Predict potential targets and interactions of ncRNAs.	Aids in hypothesis generation.	Requires experimental validation.
	Network analysis	Visualise interactions between ncRNAs and their regulatory pathways.	Simplifies complex data interpretation.	Dependent on input data quality.
Single-cell approaches	Single-cell RNA-Seq	Study ncRNA expression at the single-cell level.	High-resolution cellular insights.	High cost, computationally intensive.
	Spatial transcriptomics	Map ncRNA expression localisation in tissues.	Links expression to spatial context.	Limited resolution compared to single-cell RNA-Seq.

which includes Integrating discovery (e.g., RNA-Seq, ChIP), functional characterisation (e.g., CRISPR, RIP-Seq) and computational analyses (e.g., pathway enrichment, network analysis) to unveil their regulatory mechanisms and biological significance in various cellular processes, including development, stress responses and defence mechanisms (Basak & Nithin, 2015; Huttenhofer, 2006; Liu et al., 2017; Wolfien et al., 2019). Table 3.5 briefly mentions various approaches, highlighting their pros and cons.

3.14 ncRNAs AS TARGETS FOR CROP IMPROVEMENT

Through a variety of hybridisation, mutation and transgenic breeding techniques, horticultural plants – such as fruits, vegetables, ornamental trees and flowers, herbs and tea trees – have been tamed to meet human dietary and aesthetic needs. In the early stages of transgenic breeding, protein-coding genes associated with particular target agricultural traits were selected as primary targets. Non-coding RNAs (ncRNAs) have been shown to have important functions in transcriptional or post-transcriptional regulation of plant growth, development and response to environmental stressors. Consequently, ncRNAs are becoming a focus material to hasten the domestication of horticulture crops (Huttenhofer, 2006; Liu et al., 2017; Wolfien et al., 2019). Certain ncRNA pathways can be genetically modified to increase resistance to environmental stressors, hence increasing crop yields and stability in adverse environments, enhancing stress resilience in crops. ncRNA profiles can be used to serve as biomarkers for the selection of stress-resistant crop varieties.

sRNAs, particularly miRNAs, are key regulators of plant stress responses and have significant potential for the development of climate-resilient crops. As global agricultural systems face increasing challenges due to climate change—such as drought, temperature extremes, soil contamination and variable precipitation—breeding crops that can endure these stresses is essential for ensuring food security. miRNAs are central to plant adaptation to abiotic stresses by modulating the expression of target genes involved in stress tolerance. Advances in high-throughput sequencing and bioinformatics have made it easier to identify miRNAs with their targets, offering a new perspective on their roles in stress response mechanisms. For instance, in *A. thaliana* and *Cicer arientinum,* overexpression of miRNAs-miR408 has enhanced resistance to drought, with transgenic plants exhibiting reduced chlorosis and wilting. Similarly, miR393a from *Oryza sativa* has been shown to confer multi-stress tolerance in *Agrostis stolonifera*, improving resilience to both salinity and heat stress. Other miRNAs, such as miR172 and miR319b, have also demonstrated effectiveness in enhancing tolerance to salt, water deficit and cold stress across various species. In *O. sativa*, miR319b overexpression improved recovery from cold stress, underscoring its potential for developing cold-tolerant crops. Furthermore, the knock-out approach using short tandem target mimic (STTM) technology has gained attention for its ability to silence negative regulatory miRNAs, thus enhancing stress resilience. For example, knocking out miR166 in *O. sativa* resulted in improved drought tolerance by reducing stomatal conductance and increasing photosynthetic efficiency (Mondal et al., 2023; Raza et al., 2023). These findings highlight the promise of miRNA-based strategies, including both overexpression and knockdown approaches, in engineering climate-resilient crops.

3.15 CONCLUSION AND FUTURE PROSPECTS

Plant stress adaptation relies substantially on ncRNAs, which offer sophisticated regulatory networks. Future investigations would seek to clarify and better understand their intricate relationships and use this understanding to promote sustainable agriculture and increased food security. New opportunities for enhancing agricultural resilience arise as we gain molecular understanding of the role of ncRNAs in plant stress responses. To fully comprehend the complex networks involving ncRNAs, future research should integrate proteomic, transcriptomic, and genomic approaches. Our ability to modify these molecules specifically to improve crop performance under stress will increase as our understanding of the subject grows. With the use of reference sequences, a better knowledge of ncRNAs will promote the development of improved cultivars to feed the world's population.

The use of non-coding RNAs (ncRNAs) to enhance crops has great promise. It is possible to target stress-resilient crops using ncRNA-mediated genetic engineering in order to improve traits of interest that influence stress regulation. On the other hand, there are downsides to ncRNA alteration, including fitness trade-offs, off-target effects and unforeseen regulatory consequences. Future research is necessary to guarantee stability and efficacy under varying environmental circumstances, even if RNA interference (RNAi) and CRISPR-based gene editing provide precise tools for ncRNA-based crop engineering.

REFERENCES

Abelson, J. (2017). The discovery of catalytic RNA. *Nature Reviews Molecular Cell Biology*, *18*(11), 653–653. https://doi.org/10.1038/nrm.2017.105

Amin, N., McGrath, A., & Chen, Y.-P. P. (2019). Evaluation of deep learning in non-coding RNA classification. *Nature Machine Intelligence*, *1*(5), 246–256. https://doi.org/10.1038/s42256-019-0051-2

Aravin, A. A., Naumova, N. M., Tulin, A. V., Vagin, V. V., Rozovsky, Y. M., & Gvozdev, V. A. (2001). Double-stranded RNA-mediated silencing of genomic tandem repeats and transposable elements in the *D. melanogaster* germline. *Current Biology*, *11*(13), 1017–1027. https://doi.org/10.1016/S0960-9822(01)00299-8

Ariel, F., Lucero, L., Christ, A., Mammarella, M. F., Jegu, T., Veluchamy, A., Mariappan, K., Latrasse, D., Blein, T., Liu, C., Benhamed, M., & Crespi, M. (2020). R-loop mediated trans action of the APOLO long noncoding RNA. *Molecular Cell*, *77*(5), 1055–1065.e4. https://doi.org/10.1016/j.molcel.2019.12.015

Aslam, M., Sugita, K., Qin, Y., & Rahman, A. (2020). Aux/IAA14 regulates microRNA-mediated cold stress response in *Arabidopsis* roots. *International Journal of Molecular Sciences*, *21*. https://doi.org/10.3390/ijms21228441

Baek, D., Chun, H. J., Kang, S., Shin, G., Park, S. J., Hong, H., Kim, C., Kim, D. H., Lee, S. Y., Kim, M. C., & Yun, D.-J. (2016). A role for *Arabidopsis* miR399f in salt, drought, and ABA signaling. *Molecules and Cells*, *39*(2), 111–118. https://doi.org/10.14348/molcells.2016.2188

Basak, J., & Nithin, C. (2015). Targeting non-coding RNAs in plants with the CRISPR-Cas technology is a challenge yet worth accepting. *Frontiers in Plant Science*, *6*. https://doi.org/10.3389/fpls.2015.01001

Bhogireddy, S., Mangrauthia, S. K., Kumar, R., Pandey, A. K., Singh, S., Jain, A., Budak, H., Varshney, R. K., & Kudapa, H. (2021). Regulatory non-coding RNAs: A new frontier in regulation of plant biology. *Functional & Integrative Genomics*, *21*(3–4), 313–330. https://doi.org/10.1007/s10142-021-00787-8

Borah, P., Das, A., Milner, M. J., Ali, A., Bentley, A. R., & Pandey, R. (2018). Long non-coding RNAs as endogenous target mimics and exploration of their role in low nutrient stress tolerance in plants. *Genes*, *9*(9), 459. https://doi.org/10.3390/genes9090459

Brenner, S. (1995). Molecular biology by numbers …. three. *Current Biology*, *5*(12), 1464. https://doi.org/10.1016/S0960-9822(95)00283-1

Brenner, S., Jacob, F., & Meselson, M. (1961). An unstable intermediate carrying information from genes to ribosomes for protein synthesis. *Nature*, *190*(4776), 576–581. https://doi.org/10.1038/190576a0

Canto-Pastor, A., Santos, B., Valli, A., Summers, W., Schornack, S., & Baulcombe, D. (2019). Enhanced resistance to bacterial and oomycete pathogens by short tandem target mimic RNAs in tomato. *Proceedings of the National Academy of Sciences*, *116*, 201814380. https://doi.org/10.1073/pnas.1814380116

Cao, W., Yang, L., Zhuang, M., Lv, H., Wang, Y., Zhang, Y., & Ji, J. (2024). Plant non-coding RNAs: The new frontier for the regulation of plant development and adaptation to stress. *Plant Physiology and Biochemistry*, *208*, 108435. https://doi.org/10.1016/j.plaphy.2024.108435

Cao, Z., Ting, Z., Wang, L., Han, J., Chen, J., Hao, Y., & Guan, X. (2021). The lincRNA XH123 is involved in cotton cold-stress regulation. *Plant Molecular Biology*, *106*. https://doi.org/10.1007/s11103-021-01169-1

Carr, K. (1993). Nobel goes to discoverers of "split genes." *Nature*, *365*.

Cech, T. R. (2002). Ribozymes, the first 20 years. *Biochemical Society Transactions*, *30*(Pt 6), 1162–1166. https://doi.org/10.1042/bst0301162

Cech, T. R., Zaug, A. J., & Grabowski, P. J. (1981). *In vitro* splicing of the ribosomal RNA precursor of Tetrahymena: Involvement of a guanosine nucleotide in the excision of the intervening sequence. *Cell, 27*(3, Part 2), 487–496. https://doi.org/10.1016/0092-8674(81)90390-1

Chen X. (2004). A microRNA as a translational repressor of APETALA2 in Arabidopsis flower development. *Science, 303* (5666), 2022–2025. https://doi.org/10.1126/science.1088060

Chen, L.-L., & Kim, V. N. (2024). Small and long non-coding RNAs: Past, present, and future. *Cell, 187*(23), 6451–6485. https://doi.org/10.1016/j.cell.2024.10.024

Chen, Q., Liu, K., Yu, R., Zhou, B., Huang, P., Cao, Z., Zhou, Y., & Wang, J. (2021). From "Dark Matter" to "Star": Insight into the regulation mechanisms of plant functional long non-coding RNAs. *Frontiers in Plant Science, 12*. https://doi.org/10.3389/fpls.2021.650926

Chen, Z., Hu, L., Han, N., Hu, J., Yang, Y., Xiang, T., Zhang, X., & Wang, L. (2015). Overexpression of a miR393-resistant form of transport inhibitor response protein 1 (mTIR1) enhances salt tolerance by increased osmoregulation and Na^+ exclusion in *Arabidopsis thaliana*. *Plant and Cell Physiology, 56*(1), 73–83. https://doi.org/10.1093/pcp/pcu149

Chorostecki, U., Bologna, N. G., & Ariel, F. (2023). The plant noncoding transcriptome: A versatile environmental sensor. *The EMBO Journal*. https://doi.org/10.15252/embj.2023114400

Cui, J., Jiang, N., Meng, J., Yang, G., Liu, W., Zhou, X., Ma, N., Hou, X., & Luan, Y. (2019). LncRNA33732-respiratory burst oxidase module associated with WRKY1 in tomato- *Phytophthora infestans* interactions. *The Plant Journal: For Cell and Molecular Biology, 97*(5).

Cui, X., Yan, Q., Gan, S., Xue, D., Dou, D., Guo, N., & Xing, H. (2017). Overexpression of gma-miR1510a/b suppresses the expression of a NB-LRR domain gene and reduces resistance to *Phytophthora sojae*. *Gene, 621*, 32–39. https://doi.org/10.1016/j.gene.2017.04.015

Dai, X., Lu, Q., Wang, J., Wang, L., Xiang, F., & Liu, Z. (2021). MiR160 and its target genes ARF10, ARF16 and ARF17 modulate hypocotyl elongation in a light, BRZ, or PAC-dependent manner in *Arabidopsis*: miR160 promotes hypocotyl elongation. *Plant Science: An International Journal of Experimental Plant Biology, 303*, 110686. https://doi.org/10.1016/j.plantsci.2020.110686

Delihas, N. (2015). Discovery and characterization of the first non-coding RNA that regulates gene expression, *micF* RNA: A historical perspective. *World Journal of Biological Chemistry, 6*(4), 272. https://doi.org/10.4331/wjbc.v6.i4.272

Domínguez-Rosas, E., Hernández-Oñate, M. Á., Fernandez-Valverde, S. L., & Tiznado-Hernández, M. E. (2023). Plant long non-coding RNAs: identification and analysis to unveil their physiological functions. *Frontiers in Plant Science, 14*, 1275399. https://doi.org/10.3389/fpls.2023.1275399

Du, Q., Wang, K., Zou, C., Xu, C., & Li, W. X. (2018). The *PILNCR1*-miR399 regulatory module is important for low phosphate tolerance in maize. *Plant Physiology, 177*(4), 1743–1753. https://doi.org/10.1104/pp.18.00034

Erdmann, V. A., & Barciszewski, J. (2011). 2011: 50th anniversary of the discovery of the genetic code. *Angewandte Chemie International Edition, 50*(41), 9546–9552. https://doi.org/10.1002/anie.201103895

Fei, Q., Li, P., Teng, C., & Meyers, B. C. (2015). Secondary siRNAs from Medicago-s modulated via miRNA–target interactions and their abundances. *The Plant Journal, 83*(3), 451–465. https://doi.org/10.1111/tpj.12900

Fei, Q., Yu, Y., Liu, L., Zhang, Y., Baldrich, P., Dai, Q., Chen, X., & Meyers, B. C. (2018). Biogenesis of a 22-nt microRNA in Phaseoleae species by precursor-programmed uridylation. *Proceedings of the National Academy of Sciences, 115*(31), 8037–8042. https://doi.org/10.1073/pnas.1807403115

Felippes, F. F., & Weigel, D. (2009). Triggering the formation of tasiRNAs in *Arabidopsis thaliana*: The role of microRNA miR173. *EMBO Reports, 10*(3), 264–270. https://doi.org/10.1038/embor.2008.247

Fire, A., Xu, S., Montgomery, M. K., Kostas, S. A., Driver, S. E., & Mello, C. C. (1998). Potent and specific genetic interference by double-stranded RNA in *Caenorhabditis elegans*. *Nature, 391*(6669), 806–811. https://doi.org/10.1038/35888

Fonouni-Farde, C., Christ, A., Blein, T. et al. (2022). The Arabidopsis APOLO and human UPAT sequence-unrelated long noncoding RNAs can modulate DNA and histone methylation machineries in plants. *Genome Biology, 23*, 181. https://doi.org/10.1186/s13059-022-02750-7

Franco-Zorrilla, J. M., Valli, A., Todesco, M., Mateos, I., Puga, M. I., Rubio-Somoza, I., Leyva, A., Weigel, D., García, J. A., & Paz-Ares, J. (2007). Target mimicry provides a new mechanism for regulation of microRNA activity. *Nature Genetics, 39*(8), 1033–1037. https://doi.org/10.1038/ng2079

Gill, S. S., Khan, N. A., Agarwala, N., Singh, K., Sunkar, R., & Tuteja, N. (2024). NcRNAs in plant development and stress responses. *Plant Physiology and Biochemistry, 214*, 108950. https://doi.org/10.1016/j.plaphy.2024.108950

Gonzales, L. R., Blom, S., Henriques, R., Bachem, C. W. B., & Immink, R. G. H. (2024). LncRNAs: The art of being influential without protein. *Trends in Plant Science, 29*(7), 770–785. https://doi.org/10.1016/j.tplants.2024.01.006

Guan, Q., Lu, X., Zeng, H., Zhang, Y., & Zhu, J. (2013). Heat stress induction of *miR398* triggers a regulatory loop that is critical for thermotolerance in *Arabidopsis*. *The Plant Journal: For Cell and Molecular Biology, 74*(5), 840–851. https://doi.org/10.1111/tpj.12169

Hamilton, A. J., & Baulcombe, D. C. (1999). A species of small antisense RNA in posttranscriptional gene silencing in plants. *Science (New York, N.Y.), 286*(5441), 950–952. https://doi.org/10.1126/science.286.5441.950

Hazra, S., Moulick, D., Mukherjee, A., Sahib, S., Chowardhara, B., Majumdar, A., Upadhyay, M. K., Yadav, P., Roy, P., Santra, S. C., Mandal, S., Nandy, S., & Dey, A. (2023). Evaluation of efficacy of non-coding RNA in abiotic stress management of field crops: Current status and future prospective. *Plant Physiology and Biochemistry, 203*, 107940. https://doi.org/10.1016/j.plaphy.2023.107940

Hina, A., Abbasi, A., Hassan, A. U., Arshad, T., Sanaullah, T., Kausar, F., Chaudhry, A., & Abdelsalam, N. R. (2024). The dynamic landscape of regulatory ncRNAs in plants under environmental constraints. In C. S. Prakash, M. Waseem, & S. Fiaz (Eds.), *ncRNAs: Mediated regulation*. Springer. https://doi.org/10.1007/978-3-031-69354-0_8

Hoagland, M. B., Stephenson, M. L., Scott, J. F., Hecht, L. I., & Zamecnik, P. C. (1958). A Soluble Ribonucleic Acid Intermediate in Protein Synthesis. *Journal of Biological Chemistry, 231*(1), 241–257. https://doi.org/10.1016/S0021-9258(19)77302-5

Hombach, S., & Kretz, M. (2016). Non-coding RNAs: Classification, biology and functioning. In O. Slaby & G. A. Calin (Eds.), *Non-coding RNAs in colorectal cancer* (pp. 3–17). Springer International Publishing. https://doi.org/10.1007/978-3-319-42059-2_1

Hong, Y.-H., Meng, J., Zhang, M., & Luan, Y.-S. (2020). Identification of tomato circular RNAs responsive to *Phytophthora infestans*. *Gene*, *746*, 144652. https://doi.org/10.1016/j.gene.2020.144652

Hou, S., Yang, Y., & Zhou, J.-M. (2009). The multilevel and dynamic interplay between plant and pathogen. *Plant Signaling & Behavior*, 4:4, 283–293, DOI: 10.4161/ psb.4.4.8155

Hsu, M.-T., & Coca-Prados, M. (1979). Electron microscopic evidence for the circular form of RNA in the cytoplasm of eukaryotic cells. *Nature*, *280*(5720), 339–340. https://doi.org/10.1038/280339a0

Huang, J., Yang, M., & Zhang, X. (2016). The function of small RNAs in plant biotic stress response. *Journal of Integrative Plant Biology*, *58*(4), 312–327. https://doi.org/10.1111/jipb.12463

Huang, W., Xian, Z., Hu, G., & Li, Z. (2016). SlAGO4A, a core factor of RNA-directed DNA methylation (RdDM) pathway, plays an important role under salt and drought stress in tomato. *Molecular Breeding*, *36*(3), 28. https://doi.org/10.1007/s11032-016-0439-1

Huttenhofer, A. (2006). Experimental approaches to identify non-coding RNAs. *Nucleic Acids Research*, *34*(2), 635–646. https://doi.org/10.1093/nar/gkj469

Jacob, F., & Monod, J. (1961). Genetic regulatory mechanisms in the synthesis of proteins. *Journal of Molecular Biology*, *3*, 318–356. https://doi.org/10.1016/s0022-2836(61)80072-7

Jiang, N., Cui, J., Hou, X., Yang, G., Xiao, Y., Han, L., Meng, J., & Luan, Y. (2020). Sl-lncRNA15492 interacts with Sl-miR482a and affects Solanum lycopersicum immunity against Phytophthora infestans. *The Plant Journal: for Cell and Molecular Biology*, *103*(4), 1561–1574. https://doi.org/10.1111/tpj.14847

Jiang, J., Zhu, H., Li, N., Batley, J., & Wang, Y. (2022). The miR393-target module regulates plant development and responses to biotic and abiotic stresses. *International Journal of Molecular Sciences*, *23*(16), Article 16. https://doi.org/10.3390/ijms23169477

Jodder, J., Basak, S., Das, R., & Kundu, P. (2017). Coherent regulation of miR167a biogenesis and expression of auxin signaling pathway genes during bacterial stress in tomato. *Physiological and Molecular Plant Pathology*, *100*, 97–105.

Jyothsna, S., & Alagu, M. (2022). Role of phasiRNAs in plant-pathogen interactions: Molecular perspectives and bioinformatics tools. *Physiology and Molecular Biology of Plants*, *28*(5), 947–961. https://doi.org/10.1007/s12298-022-01189-1

Kindgren, P., Ard, R., Ivanov, M., & Marquardt, S. (2018). Transcriptional read-through of the long non-coding RNA SVALKA governs plant cold acclimation. *Nature Communications*, *9*(1), 4561. https://doi.org/10.1038/s41467-018-07010-6

Kosik, K. S. (2013). Circles reshape the RNA world. *Nature*, *495*(7441), 322–324. https://doi.org/10.1038/nature11956

Krämer, A. (1987). Fractionation of HeLa cell nuclear extracts reveals minor small nuclear ribonucleoprotein particles. *Proceedings of the National Academy of Sciences*, *84*(23), 8408–8412. https://doi.org/10.1073/pnas.84.23.8408

Kramer, P., & Boyer, J. (1995). Water Relations of Plants and Soils. Academic Press, London, UK. https://doi.org/10.1016/B978-012425060-4/50003-6

Kruger, K., Grabowski, P. J., Zaug, A. J., Sands, J., Gottschling, D. E., & Cech, T. R. (1982). Self-splicing RNA: Autoexcision and autocyclization of the ribosomal RNA intervening sequence of Tetrahymena. *Cell*, *31*(1), 147–157. https://www.cell.com/cell/pdf/0092-8674(82)90414-7.pdf

Kumar, V., Khare, T., Shriram, V., & Wani, S. H. (2018). Plant small RNAs: The essential epigenetic regulators of gene expression for salt-stress responses and tolerance. *Plant Cell Reports*, *37*(1), 61–75. https://doi.org/10.1007/s00299-017-2210-4

Lacatena, R. M., & Cesareni, G. (1981). Base pairing of RNA I with its complementary sequence in the primer precursor inhibits ColE1 replication. *Nature*, *294*(5842), 623–626. https://doi.org/10.1038/294623a0

Lagos-Quintana, M., Rauhut, R., Lendeckel, W., & Tuschl, T. (2001). Identification of novel genes coding for small expressed RNAs. *Science*, *294*(5543), 853–858. https://doi.org/10.1126/science.1064921

Lan, T., Zheng, Y., Su, Z., Yu, S., Song, H., Zheng, X., Lin, G., & Wu, W. (2019). *OsSPL10*, a SBP-Box gene, plays a dual role in salt tolerance and trichome formation in rice (*Oryza sativa* L.). *G3* (*Bethesda, Md.*), *9*(12), 4107–4114. https://doi.org/10.1534/g3.119.400700

Lee, M. H., Jeon, H. S., Kim, H. G., & Park, O. K. (2017). An Arabidopsis NAC transcription factor NAC4 promotes pathogen-induced cell death under negative regulation by microRNA164. *The New Phytologist*, *214*(1), 343–360. https://doi.org/10.1111/nph.14371

Lee, R. C., Feinbaum, R. L., & Ambros, V. (1993). The *C. elegans* heterochronic gene lin-4 encodes small RNAs with antisense complementarity to lin-14. *Cell*, *75*(5), 843–854. https://doi.org/10.1016/0092-8674(93)90529-y

Li, S., Liu, J., Liu, Z., Li, X., Wu, F., & He, Y. (2014). *HEAT-INDUCED TAS1 TARGET1* Mediates Thermotolerance via HEAT STRESS TRANSCRIPTION FACTOR A1a– Directed Pathways in *Arabidopsis*. *The Plant Cell*, *26*(4), 1764–1780. https://doi.org/10.1105/tpc.114.124883

Li, Q., Wang, Y., Sun, Z., Li, H., & Liu, H. (2024). The biosynthesis process of small RNA and its pivotal roles in plant development. *International Journal of Molecular Sciences*, *25*(14), 7680. https://doi.org/10.3390/ijms25147680

Li, Y., Jeyakumar, J. M. J., Feng, Q., Zhao, Z.-X., Fan, J., Khaskheli, M. I., & Wang, W.-M. (2019). The roles of rice microRNAs in rice-*Magnaporthe oryzae* interaction. *Phytopathology Research*, *1*(1), 33. https://doi.org/10.1186/s42483-019-0040-8

Li, Y., Zhao, S.-L., Li, J.-L., Hu, X.-H., Wang, H., Cao, X.-L., Xu, Y.-J., Zhao, Z.-X., Xiao, Z.-Y., Yang, N., Fan, J., Huang, F., & Wang, W.-M. (2017). Osa-miR169 negatively regulates rice immunity against the blast fungus *Magnaporthe oryzae*. *Frontiers in Plant Science*, *8*. https://doi.org/10.3389/fpls.2017.00002

Liao, L., Xie, B., Guan, P., Jiang, N., & Cui, J. (2022). New insight into the molecular mechanism of miR482/2118 during plant resistance to pathogens. *Frontiers in Plant Science*, *13*, 1026762. https://doi.org/10.3389/fpls.2022.1026762

Liu, D., Mewalal, R., Hu, R., Tuskan, G. A., & Yang, X. (2017). New technologies accelerate the exploration of non-coding RNAs in horticultural plants. *Horticulture Research*, *4*(1), 1–8. https://doi.org/10.1038/hortres.2017.31

Liu, J., Cheng, X., Liu, D., Xu, W., Wise, R., & Shen, Q.-H. (2014). The miR9863 family regulates distinct *Mla* alleles in barley to attenuate NLR receptor-triggered disease resistance and cell-death signaling. *PLoS Genetics*, *10*(12), e1004755. https://doi.org/10.1371/journal.pgen.1004755

Liu, J., Wang, H., & Chua, N.-H. (2015). Long noncoding RNA transcriptome of plants. *Plant Biotechnology Journal*, *13*(3), 319–328. https://doi.org/10.1111/pbi.12336

Liu, X., Zhang, X., Sun, B., Hao, L., Liu, C., Zhang, D., Tang, H., Li, C., Li, Y., Shi, Y., Xie, X., Song, Y., Wang, T., & Li, Y. (2019). Genome-wide identification and comparative analysis of drought-related microRNAs in two maize inbred lines with contrasting drought tolerance by deep sequencing. *PLoS One*, *14*(7), e0219176. https://doi.org/10.1371/journal.pone.0219176

lnorkin2013. (2016, October 6). A most "Elegant" experiment: Sydney Brenner, Francois Jacob, Mathew Meselson, and the discovery of messenger RNA. *Leonard Norkin Virology Site*.

https://norkinvirology.wordpress.com/2016/10/06/a-most-elegant-experiment-sydney-brenner-frjacob-mathew-meselson-and-the-discovery-of-messenger-rna/

López-Márquez, D., Del-Espino, Á., Ruiz-Albert, J., Bejarano, E. R., Brodersen, P., & Beuzón, C. R. (2023). Regulation of plant immunity via small RNA-mediated control of NLR expression. *Journal of Experimental Botany*, *74*(19), 6052–6068. https://doi.org/10.1093/jxb/erad268

Lu, T., Cui, L., Zhou, Y., Zhu, C., Fan, D., Gong, H., Zhao, Q., Zhou, C., Zhao, Y., Lu, D., Luo, J., Wang, Y., Tian, Q., Feng, Q., Huang, T., & Han, B. (2015). Transcriptome-wide investigation of circular RNAs in rice. *RNA*, *21*(12), 2076–2087. https://doi.org/10.1261/rna.052282.115

Ma, X., Zhao, F., & Zhou, B. (2022). The characters of non-coding RNAs and their biological roles in plant development and abiotic stress response. *International Journal of Molecular Sciences*, *23*(8), 4124. https://doi.org/10.3390/ijms23084124

Ma, Y., Xue, H., Zhang, F., Jiang, Q., Yang, S., Yue, P., Wang, F., Zhang, Y., Li, L., He, P., & Zhang, Z. (2020). The miR156/SPL module regulates apple salt stress tolerance by activating MdWRKY100 expression. *Plant Biotechnology Journal*, *19*. https://doi.org/10.1111/pbi.13464

Mandal, S., Khairnar, D., & Hazra, S. (2023). Plant functional long non-coding RNAs: Yesterday's dark matter, today's regulatory component for abiotic stress response. *Non-coding RNAs: Molecular Tools for Crop Improvement*, 172–196. https://doi.org/10.1201/9781003369288-10

Mao, Y. B., Liu, Y. Q., Chen, D. Y., Chen, F. Y., Fang, X., Hong, G. J., Wang, L. J., Wang, J. W., & Chen, X. Y. (2017). Jasmonate response decay and defense metabolite accumulation contributes to age-regulated dynamics of plant insect resistance. *Nature Communications*, *8*, 13925. https://doi.org/10.1038/ncomms13925

Maxwell, E. S., & Fournier, M. J. (1995). The small nucleolar RNAs. *Annual Review of Biochemistry*, *64*(64), 897–934. https://doi.org/10.1146/annurev.bi.64.070195.004341

Mehtab-Singh, Tripathi, R. K., Bekele, W. A., Tinker, N. A., & Singh, J. (2024). Differential expression and global analysis of miR156/SQUAMOSA promoter binding-like proteins (SPL) module in oat. *Scientific Reports*, *14*(1), 9928. https://doi.org/10.1038/s41598-024-60739-7

Memczak, S., Jens, M., Elefsinioti, A., Torti, F., Krueger, J., Rybak, A., Maier, L., Mackowiak, S. D., Gregersen, L. H., Munschauer, M., Loewer, A., Ziebold, U., Landthaler, M., Kocks, C., Le Noble, F., & Rajewsky, N. (2013). Circular RNAs are a large class of animal RNAs with regulatory potency. *Nature*, *495*(7441), 333–338. https://doi.org/10.1038/nature11928

Mizuno, T., Chou, M. Y., & Inouye, M. (1984). A unique mechanism regulating gene expression: Translational inhibition by a complementary RNA transcript (micRNA). *Proceedings of the National Academy of Sciences*, *81*(7), 1966–1970. https://doi.org/10.1073/pnas.81.7.1966

Moison, M., Pacheco, J. M., Lucero, L., Fonouni-Farde, C., Rodríguez-Melo, J., Mansilla, N., Christ, A., Bazin, J., Benhamed, M., Ibañez, F., Crespi, M., Estevez, J. M., & Ariel, F. (2021). The lncRNA *APOLO* interacts with the transcription factor WRKY42 to trigger root hair cell expansion in response to cold. *Molecular Plant*, *14*(6), 937–948. https://doi.org/10.1016/j.molp.2021.03.008

Mondal, R., Das, A., & Bandyopadhyay, A. (2023). Implications of small RNAs in plant development, abiotic stress response and crop improvement in changing climate. *The Nucleus*, *66*(3), 321–339. https://doi.org/10.1007/s13237-023-00447-1

Moore, P. (2004). Francis Crick dies. *Genome Biology*, *5*(1), spotlight-20040730-01. https://doi.org/10.1186/gb-spotlight-20040730-01

Napoli, C., Lemieux, C., & Jorgensen, R. (1990). Introduction of a chimeric chalcone synthase gene into petunia results in reversible co-suppression of homologous genes in trans. *The Plant Cell*, 279–289. https://doi.org/10.1105/tpc.2.4.279

Niu, D., Lii, Y., Chellappan, P., Lei, L., Peralta, K., Jiang, C.-H., Guo, J.-H., Coaker, G., & Jin, H. (2016). MiRNA863-3p sequentially targets negative immune regulator ARLPKs and positive regulator SERRATE upon bacterial infection. *Nature Communications*, *7*, 11324. https://doi.org/10.1038/ncomms11324

North, G. (1989). RNA's catalytic role. *Nature*, *341*(6243), 556–556. https://doi.org/10.1038/341556a0

Nowak, K., Wójcik, A. M., Konopka, K., Jarosz, A., Dombert, K., & Gaj, M. D. (2024). miR156-SPL and miR169-NF-YA modules regulate the induction of somatic embryogenesis in *Arabidopsis* via LEC-and auxin-related pathways. *International Journal of Molecular Sciences*, *25*(17), Article 17. https://doi.org/10.3390/ijms25179217

Numan, M., Sun, Y., & Li, G. (2024). Exploring the emerging role of long non-coding RNAs (lncRNAs) in plant biology: Functions, mechanisms of action, and future directions. *Plant Physiology and Biochemistry*, *212*, 108797. https://doi.org/10.1016/j.plaphy.2024.108797

Pacheco, J. M., Mansilla, N., Moison, M., Lucero, L., Gabarain, V. B., Ariel, F., & Estevez, J. M. (2021). The lncRNA *APOLO* and the transcription factor WRKY42 target common cell wall EXTENSIN encoding genes to trigger root hair cell elongation. *Plant Signaling & Behavior*, *16*(8), 1920191. https://doi.org/10.1080/15592324.2021.1920191

Palade, G. E. (1955). A small particulate component of the cytoplasm. *The Journal of Biophysical and Biochemical Cytology*, *1*(1), 59–68. https://doi.org/10.1083/jcb.1.1.59

Palazzo, A. F., & Lee, E. S. (2015). Non-coding RNA: What is functional and what is junk? *Frontiers in Genetics*, *6*, 2. https://doi.org/10.3389/fgene.2015.00002

Panchal, A., Maurya, J., Seni, S., Singh, R. K., & Prasad, M. (2023). An insight into the roles of regulatory ncRNAs in plants: An abiotic stress and developmental perspective. *Plant Physiology and Biochemistry*, *201*, 107823. https://doi.org/10.1016/j.plaphy.2023.107823

Patil, V. S., Zhou, R., & Rana, T. M. (2014). Gene regulation by noncoding RNAs. *Critical Reviews in Biochemistry and Molecular Biology*, *49*(1), 16–32. https://doi.org/10.3109/10409238.2013.844092

Pegler, J. L., Oultram, J. M. J., Grof, C. P. L., & Eamens, A. L. (2020). Molecular Manipulation of the miR399/*PHO2* Expression Module Alters the Salt Stress Response of *Arabidopsis thaliana*. *Plants*, *10*(1), 73. https://doi.org/10.3390/plants10010073

Qin, T., Zhao, H., Cui, P., Albesher, N., & Xiong, L. (2017). A nucleus-localized long non-coding RNA enhances drought and salt stress tolerance. *Plant Physiology*, *175*, 00574–2017. https://doi.org/10.1104/pp.17.00574

Rajewsky, N., Jurga, S., & Barciszewski, J. (Eds.). (2017). *Plant epigenetics*. Springer International Publishing. https://doi.org/10.1007/978-3-319-55520-1

Raza, A., Charagh, S., Karikari, B., Sharif, R., Yadav, V., Mubarik, M. S., Habib, M., Zhuang, Y., Zhang, C., Chen, H., Varshney, R. K., & Zhuang, W. (2023). miRNAs for crop improvement. *Plant Physiology and Biochemistry*, *201*, 107857. https://doi.org/10.1016/j.plaphy.2023.107857

Reddy, R., & Singh, R. (1991). Synthesis of small nuclear RNAs. In P. Jeanteur, Y. Kuchino, W. E. G. Müller, & P. L. Paine (Eds.), *Progress in molecular and subcellular biology* (pp. 1–36). Springer. https://doi.org/10.1007/978-3-642-76553-7_1

Reinhart, B. J., Weinstein, E. G., Rhoades, M. W., Bartel, B., & Bartel, D. P. (2002). MicroRNAs in plants. *Genes & Development*, *16*(13), 1616–1626. https://doi.org/10.1101/gad.1004402

Rinn, J. L., Kertesz, M., Wang, J. K., Squazzo, S. L., Xu, X., Brugmann, S. A., Goodnough, L. H., Helms, J. A., Farnham, P. J., Segal, E., & Chang, H. Y. (2007). Functional demarcation of active and silent chromatin domains in human *HOX* loci by noncoding RNAs. *Cell*, *129*(7), 1311–1323. https://doi.org/10.1016/j.cell.2007.05.022

Riolo, G., Cantara, S., Marzocchi, C., & Ricci, C. (2020). miRNA Targets: From prediction tools to experimental validation. *Methods and Protocols*, *4*(1), 1. https://doi.org/10.3390/mps4010001

Ruiz-Ferrer, V., & Voinnet, O. (2009). Roles of plant small RNAs in biotic stress responses. *Annual Review of Plant Biology*, *60*, 485–510. https://doi.org/10.1146/annurev.arplant.043008.092111

Samarfard, S., Ghorbani, A., Karbanowicz, T. P., Lim, Z. X., Saedi, M., Fariborzi, N., McTaggart, A. R., & Izadpanah, K. (2022). Regulatory non-coding RNA: The core defense mechanism against plant pathogens. *Journal of Biotechnology*, *359*, 82–94. https://doi.org/10.1016/j.jbiotec.2022.09.014

Šečić, E., Kogel, K.-H., & Ladera-Carmona, M. J. (2021). Biotic stress-associated microRNA families in plants. *Journal of Plant Physiology*, *263*, 153451. https://doi.org/10.1016/j.jplph.2021.153451

Seeve, C.M., Sunkar, R., Zheng, Y. et al. (2019). Water-deficit responsive microRNAs in the primary root growth zone of maize. *BMC Plant Biology, 19*, 447. https://doi.org/10.1186/s12870-019-2037-y

Sharma, B., Prall, W., Bhatia, G., & Gregory, B. D. (2023). The diversity and functions of plant RNA modifications: What we know and where we go from here. *Annual Review of Plant Biology*, *74*(74), 53–85. https://doi.org/10.1146/annurev-arplant-071122-085813

Shivaprasad, P. V., Chen, H.-M., Patel, K., Bond, D. M., Santos, B. A. C. M., & Baulcombe, D. C. (2012). A microRNA superfamily regulates nucleotide binding site-leucine-rich repeats and other mRNAs. *The Plant Cell*, *24*(3), 859–874. https://doi.org/10.1105/tpc.111.095380

Shuai, P., Liang, D., Zhang, Z., Yin, W., & Xia, X. (2013). Identification of drought-responsive and novel *Populus trichocarpa* microRNAs by high-throughput sequencing andtheir targets using degradome analysis. *BMC Genomics*, *14*(1), 233. https://doi.org/10.1186/1471-2164-14-233

Singh, A., Gandhi, N., Mishra, V., Yadav, S., Rai, V., & Sarkar, A. (2020). Role of abiotic stress responsive miRNAs in *Arabidopsis* root development. *Journal of Plant Biochemistry and Biotechnology*, *29*. https://doi.org/10.1007/s13562-020-00626-0

Singroha, G., Sharma, P., & Sunkur, R. (2021). Current status of microRNA-mediated regulation of drought stress responses in cereals. *Physiologia Plantarum*, *172*(3), 1808–1821. https://doi.org/10.1111/ppl.13451

Song, J. B., Gao, S., Sun, D., Li, H., Shu, X. X., & Yang, Z. M. (2013). miR394 and LCR are involved in *Arabidopsis* salt and drought stress responses in an abscisic acid-dependent manner. *BMC Plant Biology*, *13*, 210. https://doi.org/10.1186/1471-2229-13-210

Song, J. B., Gao, S., Wang, Y., Li, B. W., Zhang, Y. L., & Yang, Z. M. (2016). miR394 and its target gene LCR are involved in cold stress response in *Arabidopsis*. *Plant Gene*, *5*, 56–64. https://doi.org/10.1016/j.plgene.2015.12.001

Song, L., Fang, Y., Chen, L., Wang, J., & Chen, X. (2021). Role of non-coding RNAs in plant immunity. *Plant Communications*, *2*(3). https://doi.org/10.1016/j.xplc.2021.100180

Song, X., Li, Y., Cao, X., & Qi, Y. (2019). MicroRNAs and Their Regulatory Roles in Plant-Environment Interactions. *Annual Review of Plant Biology*, *70*, 489–525. https://doi.org/10.1146/annurev-arplant-050718-100334

Sós-Hegedűs, A., Domonkos, Á, Tóth, T., Gyula, P., Kaló, P., & Szittya, G. (2020). Suppression of NB-LRR genes by miRNAs promotes nitrogen-fixing nodule development in *Medicago truncatula. Plant, Cell & Environment*, *43*(5), 1117–1129. https://doi.org/10.1111/pce.13698

Statello, L., Guo, C.-J., Chen, L.-L., & Huarte, M. (2021). Gene regulation by long non-coding RNAs and its biological functions. *Nature Reviews Molecular Cell Biology*, *22*(2), 96–118. https://doi.org/10.1038/s41580-020-00315-9

Sun, Z., Shu, L., Zhang, W., & Wang, Z. (2020). Cca-miR398 increases copper sulfate stress sensitivity via the regulation of CSD mRNA transcription levels in transgenic *Arabidopsis thaliana*. *PeerJ*, *8*, e9105. https://doi.org/10.7717/peerj.9105

Suprasanna, P., Kumar, V., & Srivastava, A. K. (Eds.). (2023). *Non-coding RNAs: Molecular tools for crop improvement* (1st ed.). CRC Press. https://doi.org/10.1201/9781003369288

Tang, Z., Zhang, L., Xu, C., Yuan, S., Zhang, F., Zheng, Y., & Zhao, C. (2012). Uncovering small RNA-mediated responses to cold stress in a wheat thermosensitive genic male-sterile line by deep sequencing. *Plant Physiology*, *159*(2), 721–738. https://doi.org/10.1104/pp.112.196048

Timmons, L., & Fire, A. (1998). Specific interference by ingested dsRNA. *Nature*, *395*(6705), 854–854. https://doi.org/10.1038/27579

Traubenik, S., Charon, C., & Blein, T. (2024). From environmental responses to adaptation: The roles of plant lncRNAs. *Plant Physiology*, *195*(1), 232–244. https://doi.org/10.1093/plphys/kiae034

Wagh, N., Amane, P., Kshirsagar, R., Mandal, S., Kavi Kishor, P., & Suprasanna, P. (2025). Exploring the critical role of non-coding RNAs in plant development and stress adaptation-current status and insights. *Progress in Molecular Biology and Translational Science*.

Waititu, J. K., Zhang, C., Liu, J., & Wang, H. (2020). Plant non-coding RNAs: Origin, biogenesis, mode of action and their roles in abiotic stress. *International Journal of Molecular Sciences*, *21*(21), 8401. https://doi.org/10.3390/ijms21218401

Wang, H.-L. V., & Chekanova, J. A. (2017). Long noncoding RNAs in plants. *Advances in experimental medicine and biology* (Vol. 1008, pp. 133–154). https://doi.org/10.1007/978-981-10-5203-3_5

Wang, J., Mei, J., & Ren, G. (2019). Plant microRNAs: Biogenesis, homeostasis, and degradation. *Frontiers in Plant Science*, *10*, 360. https://doi.org/10.3389/fpls.2019.00360

Wang, K. C., & Chang, H. Y. (2011). Molecular mechanisms of long noncoding RNAs. *Molecular Cell*, *43*(6), 904–914. https://doi.org/10.1016/j.molcel.2011.08.018

Wang, S., Sun, X., Hoshino, Y., Yu, Y., Jia, B., Sun, Z., Sun, M., Duan, X., & Zhu, Y. (2014). MicroRNA319 positively regulates cold tolerance by targeting OsPCF6 and OsTCP21 in rice (*Oryza sativa* L.). *PLoS One*, *9*(3), e91357. https://doi.org/10.1371/journal.pone.0091357

Weinberg, A. M. (2001). Messenger RNA: Origins of a discovery. *Nature*, *414*(6863), 485. https://doi.org/10.1038/35107234

Weinberg, R. A., & Penman, S. (1968). Small molecular weight monodisperse nuclear RNA. *Journal of Molecular Biology*, *38*(3), 289–304. https://doi.org/10.1016/0022-2836(68)90387-2

Wolfien, M., Brauer, D. L., Bagnacani, A., & Wolkenhauer, O. (2019). Workflow development for the functional characterization of ncRNAs. In X. Lai, S. K. Gupta, & J. Vera (Eds.), *Computational biology of non-coding RNA* (Vol. 1912, pp. 111–132). Springer. https://doi.org/10.1007/978-1-4939-8982-9_5

Xie, M., & Yu, B. (2015). siRNA-directed DNA methylation in plants. *Current Genomics*, *16*(1), 23–31. https://doi.org/10.2174/1389202915666141128002211

Xie, Z., Johansen, L. K., Gustafson, A. M., Kasschau, K. D., Lellis, A. D., Zilberman, D., Jacobsen, S. E., & Carrington, J. C. (2004). Genetic and functional diversification of small RNA pathways in plants. *PLOS Biology*, *2*(5), e104. https://doi.org/10.1371/journal.pbio.0020104

Xu, M., Hu, T., Zhao, J., Park, M.-Y., Earley, K. W., Wu, G., Yang, L., & Poethig, R. S. (2016). Developmental functions of miR156-regulated SQUAMOSA PROMOTER BINDING PROTEIN-LIKE (SPL) genes in *Arabidopsis thaliana*. *PLoS Genetics*, *12*(8), e1006263. https://doi.org/10.1371/journal.pgen.1006263

Xu, T., Zhang, L., Yang, Z., Wei, Y., & Dong, T. (2021). Identification and functional characterization of plant MiRNA under salt stress shed light on salinity resistance improvement through MiRNA manipulation in crops. *Frontiers in Plant Science*, *12*, 665439. https://doi.org/10.3389/fpls.2021.665439

Yadav, A., Mathan, J., Dubey, A. K., & Singh, A. (2024). The emerging role of non-coding RNAs (ncRNAs) in plant growth, development, and stress response signaling. *Non-Coding RNA*, *10*(1), 13. https://doi.org/10.3390/ncrna10010013

Yang, A., Dai, X., & Zhang, W.-H. (2012). A R2R3-type MYB gene, OsMYB2, is involved in salt, cold, and dehydration tolerance in rice. *Journal of Experimental Botany*, *63*(7), 2541–2556. https://doi.org/10.1093/jxb/err431

Yang, W., Bai, Q., Li, Y., Chen, J., & Liu, C. (2023). Epigenetic modifications: Allusive clues of lncRNA functions in plants. *Computational and Structural Biotechnology Journal*, *21*, 1989–1994. https://doi.org/10.1016/j.csbj.2023.03.008

Yang, Y., Liang, Y., Wang, C., & Wang, Y. (2024). MicroRNAs as potent regulators in nitrogen and phosphorus signaling transduction and their applications. *Stress Biology*, *4*(1), 38. https://doi.org/10.1007/s44154-024-00181-x

Yuan, X., Wang, H., Jiating, C., Li, D., & Song, F. (2019). NAC transcription factors in plant immunity. *Phytopathology Research*, *1*, 3. https://doi.org/10.1186/s42483-018-0008-0

Zhang, D., Ma, Y., Naz, M., Ahmed, N., Zhang, L., Zhou, J.-J., Yang, D., & Chen, Z. (2024). Advances in CircRNAs in the past decade: Review of CircRNAs biogenesis, regulatory mechanisms, and functions in plants. *Genes*, *15*(7), 958. https://doi.org/10.3390/genes15070958

Zhang, L. L., Huang, Y. Y., Zheng, Y. P., Liu, X. X., Zhou, S. X., Yang, X. M., Liu, S. L., Li, Y., Li, J. L., Zhao, S. L., Wang, H., Ji, Y. P., Zhang, J. W., Pu, M., Zhao, Z. X., Fan, J., & Wang, W. M. (2022). Osa-miR535 targets SQUAMOSA promoter binding protein-like 4 to regulate blast disease resistance in rice. *The Plant Journal: For Cell and Molecular Biology*, *110*(1), 166–178. https://doi.org/10.1111/tpj.15663

Zhang, X., Dong, J., Deng, F., Wang, W., Cheng, Y., Song, L., Hu, M., Shen, J., Xu, Q., & Shen, F. (2019). The long non-coding RNA lncRNA973 is involved in cotton response to salt stress. *BMC Plant Biology*, *19*(1), 459. https://doi.org/10.1186/s12870-019-2088-0

Zhang, X., Du, M., Yang, Z., Wang, Z., & Lim, K.-J. (2023). Biogenesis, mode of action and the interactions of plant non-coding RNAs. *International Journal of Molecular Sciences*, *24*(13), 10664. https://doi.org/10.3390/ijms241310664

Zhang, Y., Zhan, L., Jiang, X., & Tang, X. (2024). Comprehensive review for non coding RNAs: From mechanisms to therapeutic applications. *Biochemical Pharmacology*, *224*, 116218. https://doi.org/10.1016/j.bcp.2024.116218

Zhao, W., Chu, S., & Jiao, Y. (2019). Present Scenario of Circular RNAs (circRNAs) in Plants. Frontiers in Plant Science, 10, 379. https://doi.org/10.3389/fpls.2019.00379

Zhao, M., Wang, H., Chen, J., Xi, Y., Wang, F., Huo, C., Li, W., Chu, Y., Xu, P., Huang, Q., & Bu, S. (2021). Expression of long non-coding RNA H19 in colorectal cancer patients with type 2 diabetes. *Archives of Physiology and Biochemistry*, *127*(3), 228–234.

Zhou, M., & Tang, W. (2019). MicroRNA156 amplifies transcription factor-associated cold stress tolerance in plant cells. *Molecular Genetics and Genomics*, *294*. https://doi.org/10.1007/s00438-018-1516-4

Zhu, T., Liu, Y., Ma, L., Wang, X., Zhang, D., Han, Y., Ding, Q., & Ma, L. (2020). Genome-wide identification, phylogeny and expression analysis of the SPL gene family in wheat. *BMC Plant Biology*, *20*(1), 420. https://doi.org/10.1186/s12870-020-02576-0

Zieve, G., & Penman, S. (1976). Small RNA species of the HeLa cell: Metabolism and subcellular localization. *Cell*, *8*(1), 19–31. https://doi.org/10.1016/0092-8674(76)90181-1

Zou, Y., Wang, S., Zhou, Y., Bai, J., Huang, G., Liu, X., Zhang, Y., Tang, D., & Lu, D. (2018). Transcriptional Regulation of the immune receptor FLS2 controls the ontogeny of plant innate immunity. *The Plant Cell*, *30*(11), 2779–2794. https://doi.org/10.1105/tpc.18.00297

4 ncRNAs
Key Regulators of Abiotic Stress Responses in Plants

Arslan Asim, Alyan Ashraf, Saboor Muarij Bunny, Muhammad Muneeb Ullah, Junaid Iqbal and Usman Khalid Chaudhry

4.1 INTRODUCTION

Abiotic stresses, including drought, salinity, extreme temperatures, nutrient imbalances and waterlogging, pose significant challenges to plant growth, development and yield, particularly in arid and semi-arid regions. Climate changes add up to these stresses, inducing oxidative damage and disrupting essential physiological processes. For instance, severe drought can limit photosynthesis and cellular growth, reducing yields by up to 82% (Ashokan et al., 2024). Salinity problem occurs especially in irrigated and coastal regions, impairs ion balance, water uptake and nutrient absorption, hampering plant metabolism (Bartas, 2024). Extreme temperatures further disrupt cellular stability, with low temperatures reducing membrane fluidity and enzyme activity, while excessive heat leads to protein denaturation and membrane destabilisation (Kaur et al., 2024). Nutrient imbalances whether deficiencies or toxicities compromise plant health, particularly when key nutrients like potassium, phosphorus and nitrogen are limited, leading to stunted growth (Suvarna et al., 2024). Waterlogging, prevalent in regions with heavy rainfall or poor drainage, restricts root oxygen availability and hinders plant function in anaerobic conditions (Naqvi et al., 2025). Addressing these stresses requires advanced strategies, such as molecular breeding, nutrient management and genomic tools to enhance crop resilience.

Plants utilise a range of physiological, biochemical and molecular mechanisms to adapt to abiotic stresses, supporting resilience in harsh environmental conditions. Osmoregulation is a key response where plants accumulate osmolytes like proline, glycine betaine and sugars, stabilising cell structures to prevent dehydration under drought and salinity stress (Rane & Jangid, 2024; Sharma et al., 2024). Additionally, antioxidative enzymes such as peroxidase, catalase and superoxide dismutase mitigate oxidative damage caused by reactive oxygen species (ROS), preserving cellular integrity (Sharma et al., 2024). Hormonal regulation also plays a critical role, with abscisic acid (ABA) being particularly important during drought stress. ABA induces stomatal closure to reduce water loss and triggers the expression of stress-responsive genes (Li et al., 2024; Sharma et al., 2024). Transcription factors (TFs) like WRKY further enhance adaptation by regulating genes involved in drought and salinity tolerance (Li et al., 2024). Morphological adaptations, including optimised root architecture and altered leaf shapes, improve water uptake and water-use efficiency in water-limited conditions (Singh et al., 2024). Symbiotic relationships with rhizobacteria and mycorrhizal fungi enhance nutrient uptake and support tolerance to waterlogging and nutrient deficiencies.

Non-coding RNAs (ncRNAs), such as microRNAs (miRNAs), long non-coding RNAs (lncRNAs) and circular RNAs (circRNAs), play essential roles in gene expression regulation and stress adaptation. These ncRNAs help plants maintain osmotic balance, antioxidant defences and support physiological processes under various stress conditions. miRNAs are small ncRNAs that regulate gene expression by targeting messenger RNAs (mRNAs) for degradation or translational repression, which aids in managing stress responses and maintaining cellular homeostasis (Deng et al., 2024). This modulation helps maintain osmotic balance and enhances antioxidant defences, crucial for drought and salinity tolerance. lncRNAs (>200 nucleotides) contribute to stress resilience through chemical priming and chromatin remodelling, supporting homeostasis during salinity, drought and temperature stress (Cao et al., 2024a; Imaduwage & Hewadikaram, 2024). They influence gene expression at the chromatin level, acting as scaffolds for chromatin-modifying proteins and affecting RNA stability and alternative splicing, which impact stress response gene networks. circRNAs are stable due to their circular structure and resist degradation, allowing them to interact with miRNAs and mRNAs to enhance stress signalling and regulate stress-responsive gene networks (Cao et al., 2024; Yadav et al., 2024). Together, these ncRNAs form a complex regulatory network that boosts plant resilience, highlighting their potential for use in breeding stress-tolerant crops.

ncRNAs contribute to epigenetic regulation by facilitating chromatin remodelling and gene silencing. Specific lncRNAs guide chromatin-modifying proteins to target genes, inducing DNA methylation and histone modifications that establish long-term gene silencing. This silencing helps maintain genomic stability and cellular identity across generations, enhancing stress resilience (Cao et al., 2024; García-Padilla et al., 2022). miRNAs, in conjunction

DOI: 10.1201/9781003583257-4

with the RNA-induced silencing complex (RISC), target mRNAs for degradation, contributing to transcriptomic stability and allowing precise control of gene expression in stress response pathways (Abaturov & Babych, 2022). These mechanisms are essential for plants to cope with environmental stresses, as they regulate the expression of stress-responsive genes through chromatin and post-transcriptional modifications. Research on ncRNAs has paved the way for advancements in agricultural sustainability. Modifying ncRNAs, such as miRNAs and lncRNAs, offers promising avenues for enhancing crop resistance to abiotic stresses like drought, salinity and temperature extremes (Hazra et al., 2023). Through marker-assisted breeding, genomic selection and transcriptome analysis, plant scientists can leverage ncRNAs to develop resilient crop varieties. Techniques such as chemical priming with compounds like ABA can augment ncRNA-mediated stress tolerance, creating crops better adapted to adverse conditions (Imaduwage & Hewadikaram, 2024). However, ongoing research is crucial to fully understand ncRNA mechanisms across diverse crops, facilitating effective integration into breeding programmes and offering agricultural solutions for climate adaptation.

4.2 microRNAs (miRNAs)

A family of tiny, ncRNAs with a length of 20–24 nucleotides, known as miRNAs, are essential for the post-transcriptional control of plant gene expression. These molecules are made from primary miRNA (pri-miRNA) transcripts, which are converted to precursor miRNA (pre-miRNA) in the nucleus by a complex comprising Dicer-like 1 (DCL1). Following their export to the cytoplasm, mature miRNAs join the RISC and use sequence complementarity to direct the complex to target mRNAs, which results in translational inhibition or degradation of mRNA (Bartel, 2004).

4.2.1 miRNAs Regulating Abiotic Stress Responses

In addition to that, miRNAs react to particular abiotic stressors and the ways in which they modulate these reactions. Plant development and production are seriously threatened by drought stress, which causes oxidative damage as ROS build up. By focusing on transcripts that encode antioxidant enzymes, including cytochrome c oxidase and copper/zinc superoxide dismutase (CSD1 and CSD2), miRNAs like miR398 are crucial in reducing this damage. The overexpression of these antioxidant enzymes under drought stress circumstances is made possible by the downregulation of miR398; this improves the plant's capacity to scavenge ROS and increases drought tolerance (Ferdous et al., 2015). Furthermore, by controlling APETALA2 (AP2) TFs, miR172 has been linked to drought responses; downregulating miR172 raises AP2 levels, which alter stress-responsive pathways and foster drought resistance (Moradi & Khalili, 2018).

Plant cellular homeostasis is affected by salinity stress, which also causes osmotic and ionic imbalances. Because of its function in controlling auxin signalling, miR393 is among the most researched miRNAs in this field. It modifies the plant's growth and stress responses by targeting TIR1/AFB auxin receptors. Reduced root development and auxin sensitivity are adaptive responses to high salt environments caused by overexpression of miR393, which suppresses auxin signalling under salinity stress (Sunkar et al., 2006). Through its targeting of laccase genes, which are involved in the synthesis of lignin, miR397 also helps to withstand salinity. MiR397-regulated reduced lignin content improves salt tolerance and water transport effectiveness (Wang et al., 2014).

Plants undergo a number of reactions in response to heat stress, one of which is the modification of miRNAs to preserve cellular homeostasis. One of the important miRNAs in the control of heat stress is miR156. It specifically targets SPL TFs, which are crucial for regulating stress response genes and developmental processes. SPLs are downregulated as a result of increased expression of miR156 under heat stress, allowing plants to put stress survival ahead of growth (Stief et al., 2014). Under heat stress, another miRNA, miR398, exhibits dynamic regulation. Although miR398 is normally downregulated during oxidative stress to enhance the expression of CSD, its regulation may vary during heat stress, striking a balance between reducing oxidative stress and preventing heat-induced cellular damage (Guan et al., 2013). In plants, cold stress severely impairs cellular and metabolic functions. Targeting TCP TFs, miR319 affects the expression of genes linked to cold tolerance and photosynthesis. Modification of miR319 levels enhances cold tolerance and modulates leaf morphology (Zhou et al., 2013). Through its targeting of pathways linked to auxin signalling and cell cycle regulation, miR393 also contributes to cold stress. Plant development is impacted by miR393 modulation in cold climates, which changes the primary concern from active growth to survival (Figure 4.1) (Islam et al., 2022).

Low phosphate (Pi) and other nutrient shortages cause plants to activate miRNA-based adaptation mechanisms. The Pi homeostasis regulator miR399 is well-known. It targets the E2 ubiquitin-conjugating enzyme encoded by the PHO2 gene. Upregulating miR399 during Pi famine causes PHO2 to be downregulated, increasing the expression of Pi transporters and improving Pi recycling and acquisition (Liu et al., 2012). The complex regulatory network that allows a single miRNA to control several pathways and promote plant tolerance to compound stressors is demonstrated by the overlapping activities of miRNAs like miR393 and miR398. Plants frequently experience several abiotic stressors at once, necessitating a coordinated miRNA response. This is best demonstrated by miR444, which controls stress-responsive genes across many pathways, impacting reactions to nutritional deficits, salt and drought all at once (Zhang et al., 2022).

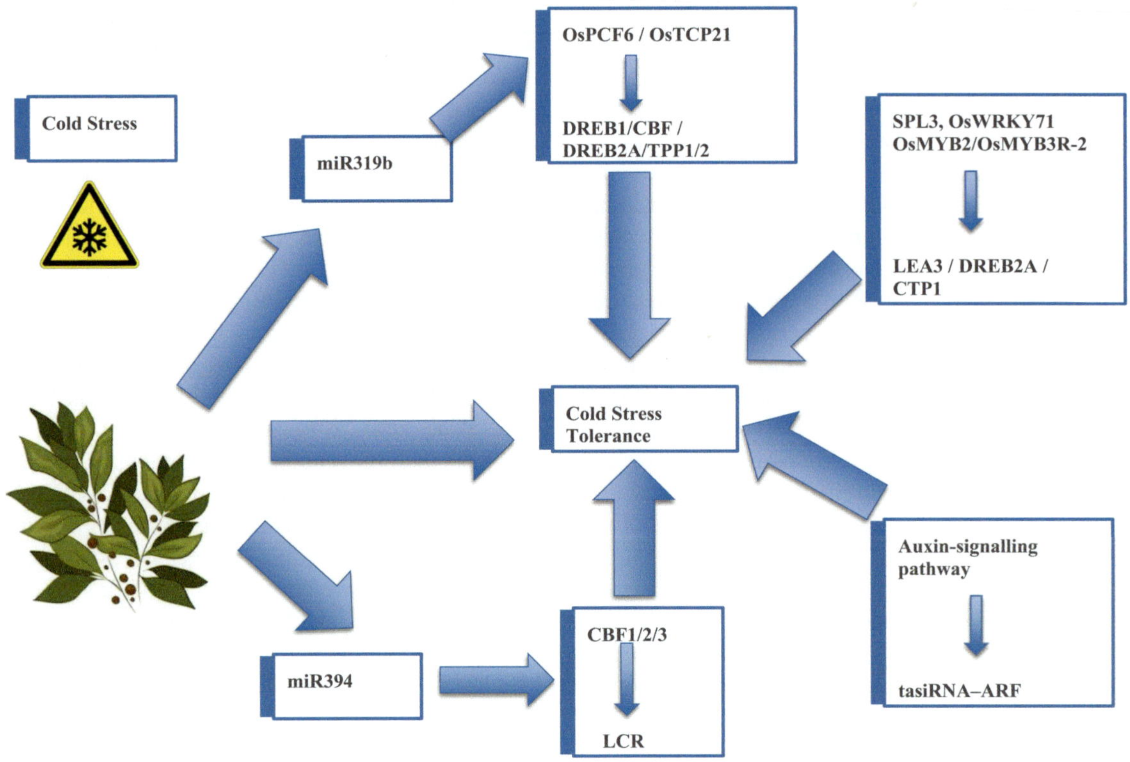

FIGURE 4.1 An example illustrating the mode of action of ncRNA in regulating the cold stress in plants.

4.2.2 Biogenesis and Mechanism of Action

The multi-step process of miRNA synthesis starts in the nucleus and proceeds in the cytoplasm. RNA polymerase II converts miRNA genes into pri-miRNA, which has a shape like a hairpin. Like protein-coding mRNAs, these transcripts are polyadenylated and capped (Bartel, 2004). MiRNAs interact with target mRNAs via sequence complementarity as part of their functional processes. Either translational suppression or mRNA cleavage results from this binding. Depending on the degree of sequence complementarity, the exact mechanism varies: translational inhibition is often caused by poor matches, whereas mRNA degradation is typically caused by perfect or nearly perfect matches (Song et al., 2019).

4.3 BIOGENESIS OF SMALL INTERFERING RNAs (siRNAs)

Double-stranded RNA (dsRNA) molecules known as small interfering RNAs (siRNAs) are crucial for controlling gene expression and protecting plants from viruses. Processing of lengthy dsRNA precursors, which might originate from transposons, viral replication intermediates or endogenous transcripts with similar sequences, is the first step in their synthesis (Vaucheret, 2006). These dsRNA molecules are broken down into siRNA duplexes of 21–24 nucleotides by the DCL proteins, particularly DCL2, DCL3 and DCL4. Following their loading onto ARGONAUTE (AGO) proteins, these duplexes form the RISC (Matzke & Birchler, 2005).

4.3.1 Mechanism of Action of siRNAs

The siRNA-loaded RISC locates and binds to complementary sequences on target mRNAs, directing translational inhibition or cleavage to achieve gene silencing (Hannon, 2002). Specificity is guaranteed by the complementarity between the siRNA and its target sequence, while AGO proteins aid in the cleavage and subsequent destruction of mRNA (Vaucheret, 2006). siRNAs have the ability to stimulate RNA-directed DNA methylation (RdDM) in addition to post-transcriptional silencing. Through the development of heterochromatin, this mechanism directs chromatin alterations and transcriptional gene silencing via 24-nucleotide siRNAs processed by DCL3 (Matzke & Birchler, 2005).

4.4 BIOGENESIS AND FUNCTIONAL MECHANISMS OF LONG NON-CODING RNAs (lncRNAs)

RNA molecules longer than 200 nucleotides that do not code for proteins yet are essential for gene control are known as lncRNAs. They undergo alterations such as 5′ capping, splicing and 3′ polyadenylation during their biogenesis, which is comparable to that of mRNAs (Ponting et al., 2009). They are transcribed by RNA polymerase II. Because of their varied regulatory functions, lncRNAs can come from a variety of genomic locations, such as the intergenic, intronic and antisense strands of protein-coding genes (Rinn & Chang, 2012).

LncRNAs function by acting in a variety of ways. They can behave as decoys that sequester TFs or miRNAs to stop them from binding their targets, scaffolding for putting protein complexes together or guides for attracting chromatin-modifying enzymes to certain genomic locations (Wang & Chang, 2011). For example, lncRNAs change chromatin states or function in cis or trans to control genes that are close or far away (Cech & Steitz, 2014).

4.4.1 lncRNAs Regulating Abiotic Stress Tolerance

By modifying gene expression through epigenetic processes, lncRNAs have gained recognition for their function in improving plant resistance to a variety of abiotic stressors. LNC_973, which has been found in *Zea mays* and is upregulated under drought stress, is a noteworthy example. It controls target genes by changing histone modification patterns to improve drought tolerance (Qin et al., 2017). According to Swiezewski et al. (2009), COOLAIR, an lncRNA in *Arabidopsis thaliana*, also regulates cold tolerance by directing polycomb repressive complex 2 (PRC2) to mute the blooming LOCUS C (FLC) gene, hence modifying flowering time in regions with low temperatures (Swiezewski et al., 2009). When plants are under stress, these epigenetic changes are essential for regulating their growth and developmental responses.

The regulatory potential of various ncRNAs, in addition to miRNAs, siRNAs and lncRNAs, is drawing attention. It has been demonstrated that circRNAs, which are put together covalently closed loop structures, function as miRNA sponges, interfering with miRNAs and preventing them from attaching to their targets. According to Zhang et al. (2019), this activity aids in regulating the expression of genes that respond to stress (Liu et al., 2023). Transfer RNA-derived fragments (tRFs), which are produced from tRNAs and can alter gene expression by attaching to RNA-binding proteins or by competing with miRNAs, are another new type. It has been discovered that tRFs take part in how plants react to oxidative stress and nutritional deficiency (Thompson & Parker, 2009).

4.5 OVERVIEW OF LESSER-KNOWN NCRNAS AND THEIR POTENTIAL ROLES IN STRESS RESPONSES

The covalently closed-loop structures of circRNAs, a subtype of non-coding RNA, are distinguished by the absence of free 5′ or 3′ ends. Compared to linear RNAs, they are much more stable because of their structures. According to recent research, circRNAs act as miRNA sponges, capturing miRNAs and stopping them from attaching to their mRNA targets. This allows them to modify the expression of genes linked to stress. For example, circRNAs have been demonstrated to affect the activity of miRNAs that govern the synthesis of antioxidant enzymes in response to oxidative stress, assisting plants in better controlling ROS levels (Liu et al., 2023).

RNA molecules from the opposing strand of a coding gene are called natural antisense transcripts (NATs), and they produce complementary RNA sequences that have the ability to construct dsRNA structures. Through processes including RNA duplex formation, which can impact mRNA stability, translation efficiency and splicing, these NATs are essential for regulating gene expression. For instance, NATs can function as regulators in drought or nutrient stress situations by attaching to complementary mRNAs and either boosting or inhibiting their degradation, so modifying the plant's reaction to shifting environmental circumstances (Meena et al., 2024).

NATs and circRNAs both add to the intricacy of plant stress responses, emphasising how complicated and multilayered ncRNA regulation is. These lesser-known ncRNAs have important, if still unclear, functions in boosting plant resistance to environmental stresses by altering important pathways and interacting with different parts of the gene expression machinery.

4.5.1 Potential of ncRNAs in Synthetic Biology and Plant Stress Adaptation

miRNAs, tiny interfering RNAs (siRNAs), lncRNAs and circRNAs are examples of ncRNAs that have great potential in synthetic biology to improve plant stress response. Synthetic biology's use of ncRNAs reduces the strain on cellular machinery by precisely regulating gene networks without the need for additional proteins. For instance, by regulating gene silencing or triggering protective pathways, synthetic miRNA constructions may be designed to target certain stress-related genes, enabling more precise responses to drought, salt or temperature changes (Rodriguez et al., 2010).

4.5.2 Technological Advances in ncRNA Research

The thorough profile of ncRNAs made possible by high-throughput sequencing methods, such as RNA-seq, enables scientists to identify and measure a variety of ncRNA species across plant tissues and stressors. New ncRNAs, including miRNAs, siRNAs, lncRNAs and circRNAs, have been discovered in large part due to this technology, which has also shed light on their many functions in stress responses and gene regulation (Wang et al., 2009). Furthermore, ncRNA expression at the cellular level has been studied through the use of single-cell RNA sequencing, which has revealed regulatory networks and ncRNA activities unique to particular cell types under stress (Tang et al., 2009). In agriculture, ncRNAs have revolutionary potential, especially when it comes to improving crop resilience, productivity and nutritional quality. To increase plant resistance to environmental stressors, including drought, salt, heat and nutrient shortages, ncRNAs like miRNAs and siRNAs can be used to target genes that respond to stress. For instance, synthesising miRNAs that alter root structure or oxidative stress pathways might result in crops that are more resilient to saline soils and water scarcity (Sunkar et al., 2012).

4.5.3 Harnessing ncRNA for Sustainable Agriculture and Stress-Resilient Crops

Plant biology's understanding of ncRNAs offers a fascinating path towards developing stress-tolerant crops and sustainable farming methods. Plant responses to environmental stressors, including drought, salt, heat and cold, as well as nutrient availability, are mostly regulated by ncRNAs, which include miRNAs, siRNAs, lncRNAs and developing classes like circRNAs. Researchers can increase crop stress tolerance, lessen reliance on chemical inputs and increase yield stability in a variety of environmental circumstances by utilising these regulatory molecules (Sunkar & Zhu, 2004). Through the use of ncRNAs, including circRNAs, lncRNAs, siRNAs and miRNAs, scientists hope to modify important regulatory networks that affect yield stability, stress tolerance and nutrient use efficiency in a variety of environmental settings. Direct editing or modulation of ncRNAs implicated in stress responses is being investigated using gene-editing tools like CRISPR/Cas9 and CRISPR/Cas13. For example, scientists can enhance drought or salt tolerance by fine-tuning the expression of stress-responsive pathways by modifying the regulatory regions of miRNA genes or their target mRNAs (Dong et al., 2022). In the realm of ncRNA, synthetic biology is also making progress. Synthetic miRNA constructs are being developed to target several stress-responsive genes at once, enabling combinatorial resistance to stressors like nutritional shortages, excessive salinity and drought (Rodriguez et al., 2010).

4.6 MECHANISMS OF NCRNA-MEDIATED ABIOTIC STRESS REGULATION IN CROP PLANTS

ncRNAs have emerged as crucial regulators of gene expression and cellular processes in plants, particularly in response to various abiotic stresses. These molecular players orchestrate complex regulatory networks that enable plants to adapt to environmental challenges. Here, we discuss the diverse mechanisms through which ncRNAs mediate abiotic stress responses in crop plants, with a focus on the major classes of ncRNAs and their specific roles in stress adaptation.

4.6.1 Long Non-Coding RNAs (lncRNAs) in Abiotic Stress Response

lncRNAs employ multiple mechanisms to regulate gene expression at the transcriptional level during abiotic stress responses. One primary mechanism involves the interaction of lncRNAs with chromatin-modifying complexes. For instance, in rice (*Oryza sativa*), the drought-responsive lncRNA drought-induced lncRNAs (DRIR) has been shown to interact with the H3K4me3 methyltransferase complex, leading to enhanced expression of stress-responsive genes (Wang et al., 2018). This interaction results in the establishment of active chromatin marks at target loci, facilitating transcriptional activation of drought-responsive genes. In wheat (*Triticum aestivum*), the lncRNA *TaWDR* has been characterised as a crucial regulator of drought stress response. This lncRNA functions by recruiting chromatin remodelling factors to the promoter regions of stress-responsive genes, modulating their accessibility to TFs (Zhang et al., 2019). The mechanism involves direct interaction with the SWI/SNF chromatin remodelling complex, highlighting the sophisticated nature of lncRNA-mediated transcriptional regulation. The lncRNAs also execute their regulatory functions through various post-transcriptional mechanisms. In maize (*Z. mays*), the cold-responsive lncRNA cold-induced long antisense intergenic RNA (COOLAIR) has been demonstrated to regulate the expression of its target genes through RNA–RNA interactions (Li et al., 2020). This mechanism involves base-pairing with complementary mRNA sequences, affecting their stability or translation efficiency. Another example is the salt-responsive lncRNA salt-induced LOC_Os02g14800 (SINE1) in rice, which functions as a competing endogenous RNA (ceRNA) by sequestering specific miRNAs, thereby modulating the expression of stress-responsive genes (Yang et al., 2019). This mechanism represents a complex regulatory network where lncRNAs serve as molecular sponges, fine-tuning the availability of other regulatory ncRNAs.

4.6.2 MicroRNAs in Stress Response Regulation

miRNAs represent one of the most well-characterised classes of ncRNAs in plant stress responses. These small regulatory RNAs primarily function through sequence-specific targeting of mRNAs, leading to either mRNA degradation or translational repression. In cotton (*Gossypium hirsutum*), miR399 has been identified as a key regulator of Pi homeostasis under drought stress conditions. This miRNA targets PHO2, a ubiquitin-conjugating E2 enzyme, thereby modulating Pi uptake and distribution during water-deficit conditions (Zhou et al., 2020). Similarly, in soybean (*Glycine max*), miR394 has been shown to regulate salt stress tolerance by targeting the F-box gene GmF-BOX. This regulatory mechanism affects protein degradation pathways, ultimately influencing the plant's ability to cope with salt stress (Wang et al., 2019). The specificity and efficiency of miRNA-mediated regulation make these molecules crucial players in stress response pathways.

4.6.3 Regulation of Transcription Factors

One of the most significant mechanisms of miRNA-mediated stress regulation involves the targeting of TFs. In rice, miR319 has been demonstrated to regulate a family of TCP TFs involved in leaf development and stress response. Under cold stress conditions, the expression of miR319 is modulated, leading to altered expression of its target TCPs, which in turn affects the expression of downstream stress-responsive genes (Zhang et al., 2020). miRNAs also play crucial roles in modulating hormone signalling pathways during stress responses. In tomato (*Solanum lycopersicum*), miR160 has been shown to regulate auxin response factors (ARFs) under drought stress conditions. This regulation affects auxin signalling, which is crucial for stress-induced morphological adaptations (Liu et al., 2019). The integration of miRNA-mediated regulation with hormone signalling pathways demonstrates the complex nature of stress response networks in plants.

4.6.4 Small Interfering RNAs (siRNAs) in Stress Response

siRNAs participate in transcriptional gene silencing through DNA methylation and chromatin modification. In rice, stress-induced nat-siRNAs have been shown to direct DNA methylation at specific loci, resulting in transcriptional silencing of stress-responsive genes (Chen et al., 2020). This epigenetic regulation provides an additional layer of control over stress response pathways.

The siRNAs also function through post-transcriptional gene silencing mechanisms. In maize, drought-responsive siRNAs have been identified that target specific stress-related transcripts for degradation through the RISC pathway (Saeed et al., 2022. This mechanism provides precise control over gene expression during stress conditions.

4.6.5 Circular RNAs (circRNAs) in Stress Response

circRNAs have emerged as important players in abiotic stress responses, primarily functioning as miRNA sponges. In rice, the circRNA Os08circ01881 has been shown to act as a molecular sponge for miR528, affecting the expression of ascorbate oxidase genes under drought stress conditions (Wang et al., 2020). This regulatory mechanism adds another layer of complexity to ncRNA-mediated stress responses.

Some circRNAs have been found to interact directly with proteins, modulating their function or localisation. In cotton, specific stress-responsive circRNAs have been shown to interact with RNA-binding proteins, affecting their availability for other cellular processes (Zhang et al., 2021). This protein interaction mechanism represents a novel mode of circRNA-mediated stress regulation.

4.7 INTEGRATION OF ncRNA REGULATORY NETWORKS

The various classes of ncRNAs do not function in isolation but rather form intricate regulatory networks. For example, in wheat, lncRNAs have been shown to compete with miRNAs for binding to target mRNAs, creating complex regulatory circuits that fine-tune stress responses (Li et al., 2021). This cross-talk between different ncRNA classes allows for precise control over stress-responsive gene expression. The ncRNAs integrate with traditional stress response pathways, including hormone signalling and TF networks. In rice, the interaction between miR394 and the ABA signalling pathway has been well-documented, demonstrating how ncRNAs can modulate classical stress response mechanisms (Yang et al., 2020). This integration ensures coordinated cellular responses to abiotic stresses. Under drought conditions, specific ncRNA-mediated mechanisms are activated to regulate water use efficiency and osmolyte production. In maize, the drought-responsive lncRNA drought-induced lncRNAs 1 (DRIL1) has been shown to regulate the expression of genes involved in proline biosynthesis through interaction with chromatin-modifying complexes (Zhang et al., 2019). This regulation helps maintain cellular osmotic balance under water-deficit conditions. Salt stress tolerance involves complex ncRNA-mediated regulatory networks. In cotton, the salt-responsive miR857 has been demonstrated to regulate sodium transporters through direct targeting of their transcripts, affecting ion homeostasis under saline conditions (Liu et al., 2020). This mechanism represents a crucial adaptation to salt stress in crop plants. Both heat and cold stress responses involve specific ncRNA-mediated mechanisms. In rice, the temperature-responsive lncRNA cold-assisted intronic non-coding RNA (CAIN) has been shown to regulate the expression of cold-responsive genes through interaction with splicing factors (Wang et al., 2019). This regulation affects alternative splicing patterns under temperature stress conditions. Oxidative stress, often a secondary consequence of various abiotic stresses, is regulated by specific ncRNA mechanisms. In wheat, miR398 has been shown to target copper/zinc superoxide dismutase genes, affecting the plant's ability to scavenge ROS under stress conditions (Zhang et al., 2020).

4.8 ncRNAs IN SPECIFIC ABIOTIC STRESS RESPONSES

Globally, drought stress events are on the rise due to unpredictable rainfall patterns, significantly impacting plant growth and threatening food security. Plants employ a variety of physiological and biochemical strategies to adapt to drought conditions, utilising complex molecular networks and signalling pathways. Recent advances in molecular biology have identified key genes and quantitative trait loci (QTLs) involved in drought stress responses. ncRNAs play a significant role in drought responses (Ghorbanzadeh et al., 2022). The discovery of drought-responsive miRNAs has illuminated the molecular mechanisms underlying plant adaptation to water scarcity. These miRNAs regulate target genes that are crucial for enhancing drought tolerance. ncRNAs interact with miRNAs, protein-coding genes and TFs to form co-expression networks that help plants cope with water stress. Research has documented numerous drought-responsive ncRNAs across various species, including six in *Arabidopsis*, 504 in *Populus* spp., 98 in rice, 664 in maize, 19 in foxtail millet, 185 in cassava and 1,597 in switchgrass (Jha et al., 2020). In foxtail millet, deep sequencing identified 584 ncRNAs, with specific lincRNAs and NAT lncRNAs showing differential expression under drought stress. In *Populus trichocarpa*, over 2,500 lncRNAs were catalogued, with 504 being responsive to drought (Peng et al., 2020). Functional validation revealed that some of these lncRNAs were upregulated or downregulated during water stress (Gervais et al., 2021).

In cassava, strand-specific RNA sequencing identified 318 lncRNAs responding to both cold and drought stresses. Among these, several exhibited significant changes in expression levels under drought conditions. Notably, some lncRNAs acted as target mimics for specific miRNAs involved in regulating stress-responsive genes (Ding et al., 2019). miRNAs such as miR167 and miR169 have been shown to modulate plant responses to drought by affecting root architecture and regulating key genes associated with drought tolerance. Other miRNAs like miR156 promote drought tolerance by silencing specific targets that affect root growth and water retention. In apple trees, miR160 regulates root development through a feedback loop involving MdARF17 and MdHYL1 (Luo et al., 2022). Overall, the interplay between ncRNAs and their target genes is crucial for plant adaptation to drought stress. Understanding these complex interactions can pave the way for developing crops with enhanced resilience to water scarcity.

4.9 ROLE OF ncRNAs UNDER DIFFERENT ABIOTIC STRESSES

ncRNAs, including miRNAs and lncRNAs, have emerged as critical regulators in plants' responses to salinity stress. Salinity poses a significant challenge to plant growth by disrupting ion homeostasis, leading to osmotic stress and ion toxicity. Recent studies have highlighted the role of specific ncRNAs in enhancing salinity tolerance through various mechanisms (Yadav et al., 2024). For instance, miR156 has been shown to regulate several target genes involved in stress response pathways, promoting plant resilience under saline conditions. This miRNA influences the expression of TFs that modulate the plant's ability to manage osmotic stress and maintain ion balance (Jeyakumar et al., 2020).

In addition to miR156, other ncRNAs such as lncRNA.2-FL have been identified as salt-responsive in rice. This particular lncRNA has been linked to the regulation of genes involved in salt tolerance mechanisms. Through RNA sequencing, researchers have identified differentially expressed lncRNAs (DE-lncRNAs) that interact with both protein-coding genes and miRNAs, forming complex regulatory networks that enhance the plant's ability to cope with salt stress. The study of these lncRNAs reveals their potential as regulatory hubs that can influence gene expression patterns critical for salinity tolerance (Mirdar Mansuri et al., 2022).

The mechanisms by which ncRNAs facilitate ion homeostasis and signalling under salt stress are multifaceted. They regulate the expression of ion transporters, such as sodium-potassium exchangers and high-affinity potassium transporters, which are essential for maintaining cellular ionic balance. By modulating these transporters' expression, ncRNAs help plants exclude excess sodium ions while retaining potassium ions, thereby mitigating the detrimental effects of salinity (Assaha et al., 2017). Furthermore, ncRNAs are involved in signalling pathways that respond to salt stress, including ABA signalling and ROS management. These pathways are crucial for activating stress-responsive genes that enhance plant survival under saline conditions (Ahmad et al., 2024).

Recent advancements in high-throughput sequencing technologies have significantly improved our understanding of the roles of ncRNAs in salinity tolerance. Studies have revealed a wide range of ncRNAs that respond to salt stress across different plant species, highlighting their importance in agricultural resilience against salinisation. As research continues to uncover the intricate networks involving ncRNAs, there is potential for developing crops with enhanced salinity tolerance through genetic engineering and breeding strategies (Gill et al., 2024).

Heat and cold stress are significant environmental challenges that affect plant growth and productivity. Recent studies have highlighted the essential roles of ncRNAs, particularly heat-responsive miRNAs and lncRNAs, in mediating plant responses to these stresses. One of the most studied miRNAs in the context of heat stress is miR398 (Guan et al., 2013). This miRNA is rapidly induced under heat stress conditions and plays a critical role in regulating the expression of its target genes, such as CSD1, CSD2 and CCS, which are involved in oxidative stress responses. The induction of miR398 leads to the downregulation of these genes, thereby enhancing thermotolerance in plants by reducing oxidative damage, particularly in reproductive tissues. Transgenic plants that express miR398-resistant forms of these target genes exhibit increased sensitivity to heat stress, while mutant plants lacking these genes show improved heat tolerance, underscoring the importance of miR398 in thermoregulation and stress adaptation (Jha et al., 2023).

In addition to miR398, various other ncRNAs have been implicated in heat stress responses. For instance, studies have identified numerous heat-responsive miRNAs that regulate a range of target genes associated with antioxidant activity and signalling pathways. These include members of the miR156, miR166 and miR393 families, which are involved in controlling gene expression related to stress tolerance mechanisms (Sihag et al., 2021). The interplay between these ncRNAs and their targets is crucial for managing the physiological responses to heat stress, such as the activation of heat shock proteins (HSPs) and heat stress HSFs. Furthermore, advancements in sequencing technologies have allowed researchers to uncover a diverse array of ncRNAs that respond to heat stress across different plant species, revealing their potential for improving crop resilience through genetic manipulation (Zhao et al., 2020).

On the other hand, cold stress also triggers significant changes in the expression of ncRNAs, which play regulatory roles in plant responses to low temperatures. Cold-responsive ncRNAs help modulate gene expression related to cold acclimation processes. For example, certain lncRNAs have been shown to interact with cold-responsive TFs, enhancing the plant's ability to withstand chilling temperatures. The regulatory functions of these ncRNAs often involve complex networks that integrate various signalling pathways, including those mediated by ABA and other phytohormones. These interactions allow plants to balance growth and stress responses effectively (Hina et al., 2024).

Recent findings suggest that ncRNAs not only contribute to immediate stress responses but also play roles in establishing stress memory, enabling plants to respond more effectively to subsequent stress events. This capability is particularly important as climate change leads to more frequent and severe temperature fluctuations. Understanding the mechanisms by which ncRNAs regulate responses to both heat and cold stress can provide valuable insights for developing crops with enhanced resilience to abiotic stresses.

Heavy metal stress significantly impacts plant health, leading to reduced growth and productivity. ncRNAs, particularly miRNAs like miR398 and miR395, play vital roles in regulating plant responses to heavy metal toxicity (Yang et al., 2022). miR398 is particularly important for managing oxidative stress caused by heavy metals. It targets genes such as CSDl and CSD2, which encode copper/zinc superoxide dismutases that help detoxify ROS generated during metal exposure. Under conditions of heavy metal stress, the expression of miR398 is downregulated, allowing the accumulation of its target proteins, which enhances the plant's ability to scavenge ROS and mitigate oxidative damage. This regulatory mechanism highlights miR398's role in maintaining cellular redox balance and promoting heavy metal tolerance (Ghosh et al., 2022).

Similarly, miR395 is implicated in sulphate metabolism and plays a crucial role in heavy metal detoxification. It regulates the expression of sulphate transporters and ATP sulphurylase genes, facilitating sulphate uptake and assimilation, which are essential for synthesising glutathione – a key antioxidant that aids in detoxifying heavy metals. The upregulation of miR395 has been associated with enhanced tolerance to cadmium stress in various plant species, as it promotes the detoxification processes necessary for coping with metal exposure (Zhang et al., 2013).

In addition to miRNAs, lncRNAs have been identified as important regulators of heavy metal stress responses. These lncRNAs can modulate gene expression through various mechanisms, including acting as molecular scaffolds that interact with miRNAs or TFs. For instance, certain lncRNAs have been shown to regulate the expression of genes involved in metal ion transport and detoxification pathways. By influencing the transcriptional landscape under heavy metal stress, lncRNAs contribute to enhancing plant tolerance to toxic metals (Table 4.1) (Imaduwage & Hewadikaram, 2024).

TABLE 4.1

List of Most Important ncRNAs and Their Roles in Abiotic Stress Regulation in Different Crops

Crop	ncRNA Type	Mode of Action	Abiotic Stress	References
Rice	miR398	Regulates superoxide dismutase to enhance ROS scavenging	Drought, salinity	Cui et al. (2023) and Zhou et al. (2024)
	lncRNA DRO1-AS	Modulates root growth for drought resistance	Drought	Cui et al. (2023)
Maize	miR164	Controls NAC transcription factors to mitigate water stress	Drought	Fang et al. (2023)
	tasiRNAs	Modulates auxin response pathways	Salinity, heat	Fang et al. (2023)
Soybean	miR393	Regulates auxin signalling by targeting TIR1 genes	Drought	Cui et al. (2023)
	lncRNA GmDi19	Enhances dehydration tolerance by regulating ion homeostasis	Salinity	Cui et al. (2023)
Wheat	miR159	Targets MYB transcription factors for improved ABA signalling	Cold, Salinity	Zhou et al. (2024)
	lncRNA TAS3	Modulates stomatal development and stress signalling	Heat	Zhou et al. (2024)
Tomato	miR319	Regulates cell proliferation and photosynthesis under heat stress	Heat	Fang et al. (2023)
	siRNA	Targets-specific mRNAs involved in ROS detoxification	Oxidative stress	Cui et al. (2023)
Arabidopsis	miR169	Downregulates NF-Y transcription factors for drought responses	Drought	Cui et al. (2023) and Zhou et al. (2024)
	lncRNA COLDWRAP	Stabilises cold-responsive proteins	Cold	Zhou et al. (2024)

4.10 MOLECULAR APPROACHES FOR MANIPULATING NCRNAS IN PLANTS

Abiotic stresses, including drought, salinity, extreme temperatures and nutrient deficiency, significantly impact plant productivity and crop yield worldwide. With the growing need for sustainable agriculture, especially under challenging climate conditions, enhancing plants' tolerance to these stresses is a crucial need. Recent advancements in molecular biology have revealed that ncRNAs—particularly miRNAs and lncRNAs—play crucial roles in regulating gene expression associated with stress responses in plants. ncRNAs are transcripts that do not code for proteins but instead regulate various biological processes by interacting with DNA, RNA or proteins. Among ncRNAs, miRNAs and lncRNAs are particularly significant in plants. miRNAs are small (20–24 nucleotides), single-stranded RNAs that modulate gene expression at the post-transcriptional level by targeting mRNA for degradation or translational repression (Singh et al., 2021). lncRNAs, typically more than 200 nucleotides long, interact with miRNAs, mRNAs and proteins to form regulatory complexes that influence gene expression in diverse ways, including chromatin remodelling and transcriptional activation or repression (Zhang et al., 2022).

One of the most promising molecular approaches to enhancing abiotic stress tolerance involves the manipulation of miRNA levels. Many miRNAs are associated with stress responses in plants. For instance, miR156, miR159 and miR398 are known to regulate drought, heat and oxidative stress responses, respectively (Li et al., 2020). By altering the expression levels of these miRNAs, scientists can potentially enhance plants' resilience to such stresses. For example, overexpression of miR156 has been shown to improve drought tolerance in rice by modulating SPL (SQUAMOSA promoter-binding protein-like) genes, which control leaf morphology and other physiological traits associated with water conservation (Tang et al., 2021). Gene editing tools, such as CRISPR-Cas9, have made it possible to specifically target miRNA genes to modulate their expression. In rice, CRISPR-based knockout of miR397, which suppresses laccase genes, has enhanced drought tolerance by promoting lignin biosynthesis and improving plant cell wall integrity (Wang et al., 2023). These findings underscore the potential of miRNA manipulation in promoting stress resilience through precise genetic engineering. lncRNAs have emerged as critical regulators of plant stress responses, primarily through their interactions with miRNAs and mRNAs. Recent studies have identified numerous stress-responsive lncRNAs that act as 'miRNA sponges', sequestering miRNAs and preventing them from targeting mRNAs (Sun et al., 2019). For instance, lncRNA-derived miRNA sponges can regulate drought and salinity responses in various crops by targeting stress-responsive miRNAs (Ma et al., 2022). Such lncRNAs could be manipulated to increase their binding affinity for specific miRNAs, reducing the regulatory impact of those miRNAs on stress-sensitive genes. RNA interference (RNAi) is another approach to manipulate lncRNAs. Through RNAi-mediated knockdown or overexpression, researchers can alter lncRNA levels to influence plant stress responses. For example, overexpression of a drought-responsive lncRNA in *Arabidopsis* was found to confer greater drought resistance by influencing hormone signalling pathways involved in water retention and root growth (Liu et al., 2020).

4.11 ROLE OF CRISPR-Cas9 IN ncRNA MANIPULATION

CRISPR-Cas9 technology is revolutionising plant genetic engineering by allowing precise, targeted manipulation of genes, including ncRNAs. By editing or silencing ncRNAs, CRISPR-Cas9 enables researchers to regulate plant stress responses without introducing foreign DNA. CRISPR-Cas9 has been successfully applied to modify miRNA genes, such as miR396, in *Arabidopsis*, resulting in enhanced tolerance to salinity and drought by altering growth-related genes (Shi et al., 2023). CRISPR interference (CRISPRi) and CRISPR activation (CRISPRa) are also useful for manipulating ncRNA expression. CRISPRi can be used to silence stress-responsive ncRNAs, while CRISPRa can increase the expression of beneficial ncRNAs. This approach has been demonstrated in rice, where CRISPRa-induced overexpression of a heat-responsive lncRNA improved thermotolerance by regulating stress-associated TFs (Wu et al., 2022).

4.12 TRANSCRIPTION FACTORS AND ncRNA NETWORKS IN STRESS RESPONSES

ncRNAs do not function in isolation but rather as part of a complex regulatory network that includes TFs. Many TFs regulate ncRNA expression, while ncRNAs can also modulate TF activity, creating feedback loops that fine-tune stress responses (Jiang et al., 2021). By targeting TFs involved in ncRNA regulation, it may be possible to modify entire regulatory networks that govern plant stress responses. For instance, manipulating the TF DREB1A, which activates stress-responsive lncRNAs, has been shown to improve drought tolerance in various crops (Khan et al., 2023). This approach highlights the possibility of indirect ncRNA manipulation through TFs, providing an additional layer of control over plant stress adaptation mechanisms. While the manipulation of ncRNAs offers immense potential for improving abiotic stress tolerance in plants, challenges remain. One key issue is the off-target effects associated with gene editing technologies, which could inadvertently affect other regulatory processes (Rana et al., 2024). Additionally, the regulatory frameworks governing genetically modified crops may restrict the commercialisation of ncRNA-manipulated plants. Despite these challenges, advancements in bioinformatics and high-throughput sequencing are helping to improve target specificity and identify novel stress-responsive ncRNAs. In the future, combining ncRNA manipulation with traditional breeding methods and other genetic approaches could lead to the development of crops that are both high-yielding and stress-resilient (Zhou et al., 2023).

4.13 CROSSTALK BETWEEN ncRNAs AND OTHER MOLECULAR REGULATORS

ncRNAs are essential for plant stress responses, functioning not only as gene regulators but also by engaging in complex crosstalk with TFs and protein-coding genes. This interplay is crucial for plants to survive and adapt to various abiotic stresses, including drought, salinity and extreme temperatures. By coordinating the expression of stress-responsive genes, ncRNAs form a dynamic regulatory network that enables plants to mount rapid and precise responses to environmental challenges (Liu et al., 2023).

A key function of ncRNAs involves their interaction with TFs, which are central players in stress regulation. Among ncRNAs, miRNAs and lncRNAs have attracted significant attention for their roles in fine-tuning TF activity. MiRNAs typically regulate gene expression by binding to complementary sequences on TF mRNAs, resulting in either mRNA degradation or translational inhibition. For example, during drought stress, miRNAs target specific TFs that promote growth-related genes, thereby suppressing these pathways and allowing the plant to conserve resources for stress tolerance (Zhang et al., 2023). LncRNAs regulate TFs through varied mechanisms. Some lncRNAs act as 'molecular sponges' for miRNAs, sequestering them away from target mRNAs and allowing specific TFs to be translated. This indirect regulation contributes to a balanced network in which lncRNAs modulate miRNA availability, thereby impacting TF activity and downstream gene expression. Certain lncRNAs can also interact directly with TF proteins, modulating their stability, localisation or DNA-binding affinity to influence the transcriptional regulation of stress-responsive genes (Chen et al., 2023).

The interplay between ncRNAs and TFs exemplifies a sophisticated regulatory mechanism that fine-tunes gene expression in response to stress. This flexibility allows plants to adjust their physiological responses to stress severity and duration. For example, in prolonged drought or salinity stress, ncRNAs can modulate multiple TFs in a coordinated manner, resulting in the selective activation or repression of gene networks that optimise water retention, cell wall fortification or antioxidant defences (Wang et al., 2023). miRNAs and lncRNAs often function synergistically with protein-coding genes and other regulatory molecules, creating a multilayered network that amplifies or modulates stress responses as needed. This interaction is particularly valuable under fluctuating environmental conditions, where rapid adjustment of gene expression is critical. MiRNAs, for instance, may target TFs that enhance growth under optimal conditions, but downregulate these TFs during stress to redirect resources towards defence mechanisms. Concurrently, lncRNAs may interact with other TFs or miRNAs in the same pathway, reinforcing adaptive changes at multiple regulatory levels (Kim et al., 2023).

Such crosstalk between ncRNAs and TFs demonstrates an intricate system of checks and balances within plant cells, ensuring tight regulation of stress-responsive genes. This complexity allows plants to handle multiple stressors simultaneously, integrating signals across pathways to maintain cellular homeostasis. As research into ncRNAs advances, it is increasingly evident that these molecules not only supplement TF-based regulation but also introduce unique layers of control, making them indispensable for plant adaptation to stress (Park et al., 2023).

4.14 ncRNA-MEDIATED REGULATION AND INTEGRATIVE STRESS RESPONSE NETWORKS IN PLANTS

ncRNAs, particularly miRNAs and lncRNAs, are emerging as central regulators in plant responses to environmental stresses, such as drought, heat and salinity. Through their interactions with TFs and protein-coding genes, ncRNAs establish a multilayered regulatory network that enables plants to adapt to stressful conditions with precision and efficiency. By modulating the expression of stress-responsive genes at transcriptional, post-transcriptional and translational levels, ncRNAs ensure that plant stress responses are balanced, resource-efficient and suited to the intensity and type of stress.

The dynamic interaction between ncRNAs and TFs plays a critical role in orchestrating stress responses. TFs are key regulators that control the expression of genes involved in stress adaptation, and ncRNAs, particularly miRNAs, modulate TF activity to fine-tune these responses. For instance, miR156 targets members of the SQUAMOSA Promoter Binding Protein-like (SPL) family of TFs during drought stress. By inhibiting SPL TFs, miR156 downregulates developmental pathways, helping the plant conserve resources that would otherwise be expended on growth, thereby promoting stress resilience (Liu et al., 2023). In response to heat stress, miR159 has been shown to regulate MYB TFs, which control the expression of HSPs. By modulating MYB TFs, miR159 reduces the production of HSPs, balancing the plant's thermotolerance with its metabolic demands (Park et al., 2021). Similarly, miR398 regulates copper/zinc superoxide dismutase (CSD) enzymes and associated TFs during oxidative stress, allowing the plant to mitigate oxidative damage without excessive production of defence proteins, thus maintaining cellular homeostasis (Zhao et al., 2022). These miRNA-TF interactions exemplify how ncRNAs enable plants to dynamically adjust TF activity, aligning gene expression with the immediate demands of stress adaptation.

LncRNAs further contribute to the regulatory network by modulating miRNAs that target TFs or by directly interacting with TF proteins. Certain lncRNAs act as molecular 'sponges', sequestering miRNAs and preventing them from binding to TF mRNAs. This indirect regulation enhances TF availability, adding flexibility to the regulatory network (Kim et al., 2023). LncRNAs can also bind TFs directly, altering their stability, localisation or DNA-binding affinity, thereby influencing the transcription of stress-responsive genes (Chen et al., 2023). Such interactions between lncRNAs, miRNAs and TFs exemplify the sophisticated, multilayered regulation that ncRNAs provide under stress. In addition to modulating TFs, ncRNAs directly regulate stress-responsive protein-coding genes, particularly those involved in critical tolerance mechanisms such as ion transport, osmotic adjustment and antioxidant responses. MiRNAs achieve this primarily through mRNA cleavage or translational inhibition. For example, miR393 targets TIR1, a key protein in the auxin signalling pathway, modulating plant response to salt stress by adjusting gene expression to favour tolerance pathways (Li et al., 2022).

LncRNAs also play a role in modulating protein-coding genes. Acting as molecular scaffolds, some lncRNAs bring together RNA-binding proteins and chromatin modifiers, influencing the transcription of stress-related genes. During drought stress, the drought-induced lncRNAs (DRIR) recruits chromatin remodellers to the promoters of genes involved in water retention and osmotic balance, thereby enhancing drought tolerance (Wang et al., 2023). In salinity stress, specific lncRNAs regulate ion transporter proteins that help the plant maintain ionic homeostasis, preventing toxic ion buildup and supporting growth in high-salinity environments (Zhao et al., 2023). The interplay between ncRNAs, TFs and protein-coding genes forms a comprehensive regulatory network that allows plants to adapt to diverse and fluctuating stress conditions. This network operates in a coordinated fashion, where ncRNAs act both as modulators of TF activity and as direct regulators of stress-responsive genes. For instance, miRNAs may repress TFs linked to growth pathways during drought while activating TFs involved in water conservation. Concurrently, lncRNAs could influence the availability of these miRNAs, adding an additional layer of regulation to match stress severity and duration (Zhang et al., 2023).

This coordinated regulation helps plants avoid excessive accumulation of stress-response proteins, which would otherwise deplete energy reserves and disrupt cellular homeostasis. By modulating both upstream regulators like TFs and downstream targets like protein-coding genes, ncRNAs ensure that stress responses are not only activated but are also fine-tuned to fit the demands of specific stressors. This ability to modulate stress response pathways flexibly and in real-time positions ncRNAs as key mediators in plant adaptation, enhancing resilience and maintaining cellular balance during stress.

4.15 FUTURE DIRECTIONS AND RESEARCH PERSPECTIVES

Despite considerable progress in identifying ncRNAs, particularly lncRNAs, much is still unknown about their exact roles in plant stress responses. Key gaps remain regarding specific mechanisms and interactions within complex regulatory networks influencing plant development and stress tolerance (Zhang et al., 2024; Cao et al., 2024). Further research to validate ncRNA targets and pathways essential for stress resilience is crucial. Interactions between lncRNAs, siRNAs and miRNAs remain largely unexplored but are critical to understanding ncRNAs' roles in mediating stress responses (Yadav et al., 2024). Although ncRNAs are implicated in abiotic stress tolerance, functional validation across various plant species is limited, with most studies focused on model plants like *Arabidopsis* (Imaduwage & Hewadikaram, 2024). Environmental factors like temperature, soil composition and drought significantly influence ncRNA expression, but their full impact on ncRNA activity and stability is still unknown. Uncovering these mechanisms could support the development of crops resilient to specific environmental stressors (Cao et al., 2024).

Technical challenges in ncRNA research include understanding the complex regulatory networks and diverse roles of ncRNAs. Current methods may not capture the complete range of ncRNA interactions, especially given their tissue-specific functions (Zhang et al., 2024; Yadav et al., 2024). Standardised, cross-species approaches are necessary to accurately assess ncRNA functions across different plants, as inconsistencies in methodologies limit the generalisability of findings (Imaduwage & Hewadikaram, 2024). Elucidating regulatory mechanisms governing ncRNA synthesis, stability and degradation under stress could help pinpoint key areas for enhancing stress tolerance (Zhang et al., 2024; Yadav et al., 2024). Further research into ncRNAs in stress memory and epigenetic adaptation could reveal how plants maintain resilience across generations (Shi et al., 2024).

Evidence suggests that lncRNAs contribute to gene silencing and chromatin remodelling, processes critical for stress memory. These epigenetic modifications enable plants to 'remember' previous stress exposures, though the underlying mechanisms are not yet fully characterised. Some lncRNAs appear to respond specifically to certain stressors, such as salinity, drought or extreme temperatures, indicating their role in targeted stress responses (Imaduwage & Hewadikaram, 2024). Similarly, miRNAs modulate pathways and TFs essential for both biotic and abiotic stress adaptation, but their roles in multi-stress situations require further exploration. Clarifying how ncRNAs coordinate responses to simultaneous stressors could be transformative for developing multi-stress-tolerant crops.

High-throughput sequencing has led to the discovery of numerous ncRNAs, yet functional characterisation remains challenging. Inconsistent research methods complicate cross-species comparisons, and current ncRNA databases lack comprehensive information for non-model and crop species, hindering agricultural applications (Yadav et al., 2024; Zhang et al., 2024). ncRNAs, especially lncRNAs, influence epigenetic modifications like DNA methylation and histone acetylation, which help regulate gene expression during stress responses and confer 'stress memory' (Zhang et al., 2024; Cao et al., 2024). lncRNAs can also act as ceRNAs, sequestering miRNAs and altering their effects on target genes, adding a layer to stress response regulation (Yadav et al., 2024).

Recent findings indicate that ncRNAs, including lncRNAs and circRNAs, serve as ceRNAs that modulate stress-responsive gene expression by sequestering miRNAs. This network enables plants to adapt gene expression to environmental changes (Cao et al., 2024; Yadav et al., 2024). ncRNAs also interact with hormone-regulated genes in stress pathways, such as those involving ABA and gibberellins, crucial for managing drought tolerance (Imaduwage & Hewadikaram, 2024; Shi et al., 2024). Additionally, some lncRNAs encode small open reading frames (sORFs) that produce stress-responsive micropeptides, contributing to resilience by regulating ion transport and ROS scavenging. Integrating ncRNA regulatory networks into crop breeding could enhance agricultural resilience to environmental stresses. Understanding and modifying these pathways may improve crop yields and sustainability under climate stress. To achieve this, expanding research to diverse crops, standardising methodologies and focusing on functional characterisation across environmental contexts are essential steps. Addressing these gaps could unlock the potential of ncRNA-based strategies for sustainable agriculture and food security in a changing climate.

4.16 CONCLUSION

This chapter highlights the crucial role of ncRNAs, in plant stress resilience by regulating gene expression, modulating chromatin and influencing hormonal signalling. ncRNAs enable plants to adapt to abiotic stresses like drought, salinity and extreme temperatures. Their involvement in stress memory and epigenetic regulation further enhances long-term stress tolerance, offering potential for crop improvement. Despite significant advances, many functional aspects of ncRNAs remain underexplored, especially across diverse plant species beyond model organisms. The complexity of ncRNA interactions, their tissue-specific roles and their responses to varying environmental conditions present challenges for comprehensive understanding. To fully harness the potential of ncRNAs in agriculture, future research must focus on addressing these gaps, standardising methodologies and validating ncRNA functions across a broader range of crops. This will facilitate the development of crops with enhanced resilience, ensuring food security and sustainability in the face of climate change.

REFERENCES

Abaturov, A. E., & Babych, E. A. (2022). Mechanisms of miRNA-mediated gene silencing. *Cellular Biochemistry and Function*, *40*(3), 371–378.

Ahmad, Y. et al. (2024). Crosstalk between ncRNAs and stress signaling pathways unmasking the regulatory networks. In C.S. Prakash, M. Waseem, & S. Fiaz (Eds.), *ncRNAs: Mediated regulation: The unarguable biological function to environmental constrains* (pp. 265–282). Springer Nature Switzerland. Available at: https://doi.org/10.1007/978-3-031-69354-0_10

Ashokan, R., Kumar, A., & Reddy, K. (2024). Impact of drought on agricultural productivity. *Agricultural Sciences Journal*, *11*(3), 234–248.

Assaha, D. V. M. et al. (2017) The role of Na^+ and K^+ transporters in salt stress adaptation in glycophytes. *Frontiers in Physiology*, *8*, 509. Available at: https://doi.org/10.3389/fphys.2017.00509

Bartas, J. (2024). Salinity stress and its effects on crop physiology. *Journal of Plant Nutrition*, *45*(2), 67–80.

Bartel, D. P. (2004). MicroRNAs: Genomics, biogenesis, mechanism, and function. *Cell*, *116*(2), 281–297.

Cao, J., Wei, L., & Zhang, Q. (2024a). ncRNA interactions in plant stress signaling pathways. *Plant Science Innovations*, *45*(3), 88–102.

Cao, Q., Wang, H., & Zhang, Z. (2024b). Roles of long non-coding RNAs in regulating abiotic stress tolerance in plants: A review et al. *Plant Molecular Biology*, *39*(3), 278–290.

Cech, T. R., & Steitz, J. A. (2014). The noncoding RNA revolution—Trashing old rules to forge new ones. *Cell*, *157*(1), 77–94.

Chen, X., Zhang, Y., & Wang, J. (2020). Genome-wide identification and functional analysis of nat-siRNAs in response to salt stress in rice. *Plant Physiology*, *182*(3), 1580–1595.

Chen, Y., Li, Z., Wang, X., Zhang, L., & Wu, Q. (2023). lncRNAs modulating transcription factor binding and activity in plant stress responses. *Plant Cell Reports*, *42*(3), 210–218.

Cui, Y., Feng, Y., Hu, Y., Liu, L., & Duan, L. (2023). Long non-coding RNAs of plants in response to abiotic stresses and their regulating roles in promoting environmental adaptation. *Cells*, *12*(5), 729.

Deng, X., Li, Y., & Zhou, L. (2024). MicroRNA-mediated regulation of stress responses in plants. *Plant Molecular Biology*, *42*(2), 201–215.

Ding, Z. et al. (2019). Strand-specific RNA-seq based identification and functional prediction of lncRNAs in response to melatonin and simulated drought stresses in cassava. *Plant Physiology and Biochemistry*, *140*, 96–104. Available at: https://doi.org/10.1016/j.plaphy.2019.05.008

Dong, Q., Hu, B., & Zhang, C. (2022). MicroRNAs and their roles in plant development. *Frontiers in Plant Science*, *13*, 824240.

Fang, W., Zhang, J., & Liu, T. (2023). *Roles of miRNAs in abiotic stress regulation in plants*. Plant Science Today.

Ferdous, J., Hussain, S. S., & Shi, B.-J. (2015). Role of microRNAs in plant drought tolerance. *Plant Biotechnology Journal*, *13*(3), 293–305.

García-Padilla, C. et al. (2022). Functions of lncRNAs in chromatin remodeling and epigenetic regulation. *Epigenetics & Chromatin*, *15*(1), 78.

Gervais, T. et al. (2021). Potato response to drought stress: Physiological and growth basis. *Frontiers in Plant Science*, *12*, 698060. Available at: https://doi.org/10.3389/fpls.2021.698060

Ghorbanzadeh, Z. et al. (2022). Non-coding RNA: Chief architects of drought-resilient roots. *Rhizosphere*, *23*, 100572. Available at: https://doi.org/10.1016/j.rhisph.2022.100572

Ghosh, S. et al. (2022). Contribution of plant miRNAome studies towards understanding heavy metal stress responses: Current status and future perspectives. *Environmental and Experimental Botany*, *194*, 104705. Available at: https://doi.org/10.1016/j.envexpbot.2021.104705

Gill, S. S. et al. (2024). ncRNAs in plant development and stress responses. *Plant Physiology and Biochemistry*, *214*, 108950. Available at: https://doi.org/10.1016/j.plaphy.2024.108950.

Guan, Q. et al. (2013). Heat stress induction of *miR398* triggers a regulatory loop that is critical for thermotolerance in *Arabidopsis*. *The Plant Journal*, *74*(5), 840–851. Available at: https://doi.org/10.1111/tpj.12169

Hannon, G. J. (2002). RNA interference. *Nature*, *418*(6894), 244–251.

Hazra, S. et al. (2023). Non-coding RNAs in plant stress responses: Emerging mechanisms and applications. *Journal of Plant Biology*, *66*(4), 455–467.

Hina, A. et al. (2024). The dynamic landscape of regulatory ncRNAs in plants under environmental constraints. In C. S. Prakash, M. Waseem, & S. Fiaz (Eds.), *ncRNAs: Mediated regulation: The unarguable biological function to environmental constrains* (pp. 209–234). Springer Nature Switzerland. Available at: https://doi.org/10.1007/978-3-031-69354-0_8.

Imaduwage, D., & Hewadikaram, S. (2024). Mechanisms of ncRNA crosstalk in abiotic stress responses in crops. *Frontiers in Plant Science*, *15*, 1045–1053.

Islam, W., Tauqeer, A., Waheed, A., & Zeng, F. (2022). MicroRNA mediated plant responses to nutrient stress. *International Journal of Molecular Sciences*, *23*(5), 2562.

Jeyakumar, J. M. J. et al. (2020). Characterizing the role of the miR156-SPL network in plant development and stress response. *Plants*, *9*(9), 1206. Available at: https://doi.org/10.3390/plants9091206

Jha, U. C. et al. (2020). Long non-coding RNAs: emerging players regulating plant abiotic stress response and adaptation. *BMC Plant Biology*, *20*(1), 466. Available at: https://doi.org/10.1186/s12870-020-02595-x

Jha, U. C. et al. (2023). Non-coding RNAs (ncRNAs) in plant: Master regulators for adapting to extreme temperature conditions. *Plant Physiology and Biochemistry*, *205*, 108164. Available at: https://doi.org/10.1016/j.plaphy.2023.108164

Jiang, X., Li, Y., Wang, L., & Chen, Q. (2021). Non-coding RNAs and transcription factors in plant responses to abiotic stress. *Plant Molecular Biology Reporter*, *39*(1), 78–89.

Kaur, G., Singh, H., & Dhillon, M. (2024). Temperature stress: Effects on plant growth and survival mechanisms. *Environmental Plant Physiology*, *29*(4), 123–139.

Khan, S., Ali, M., & Hassan, R. (2023). Transcription factors and non-coding RNAs in abiotic stress tolerance: Future directions for plant molecular breeding. *Journal of Plant Research*, *136*(3), 321–333.

Kim, S., Lee, J., Cho, H., Park, Y., & Kim, J. (2023). Molecular sponging and other lncRNA mechanisms in plant stress regulation. *Plant Molecular Biology*, *92*(1), 25–34. https://doi.org/10.1007/s11103-022-01289-4

Li, H., Zhang, L., & Chen, Y. (2024). Hormonal and transcriptional regulation in plant stress responses. *Plant Molecular Biology*, *51*(3), 122–135.

Li, M., Zhang, Q., Liu, H., & Wang, Z. (2022). Salinity tolerance in rice: miRNA regulatory mechanisms. *Plant Science*, *283*, 110098. https://doi.org/10.1016/j.plantsci.2022.110098

Li, Z., Tang, Y., & Zhou, Q. (2020). MicroRNAs in plant drought stress tolerance: Molecular mechanisms and applications. *Frontiers in Plant Science*, *11*, 1568.

Liu, H., Sun, Y., Zhang, S., & Xu, Q. (2020). Long non-coding RNAs as regulators of plant stress responses. *Plant Science*, *291*, 110399.

Liu, H., Yang, Y., & Zhang, L. (2019). MiR160-mediated regulation of auxin response factors modulates drought tolerance in tomato. *Plant Science*, *280*, 397–409.

Liu, T., Gao, Y., Chen, F., Sun, R., & Xiao, J. (2023). MiRNAs and their role in TF regulation under drought stress in plants. *Plant Physiology*, *184*(2), 401–415. https://doi.org/10.1104/pp.23.00217

Liu, T.-Y., Huang, T.-K., Tseng, C.-Y., Lai, Y.-S., Lin, S.-I., Lin, W.-Y., Chen, J.-W., & Chiou, T.-J. (2012). PHO2-dependent degradation of PHO1 modulates phosphate homeostasis in *Arabidopsis*. *The Plant Cell*, *24*(5), 2168–2183.

Luo, P. et al. (2022). MicroRNAs are involved in regulating plant development and stress response through fine-tuning of TIR1/AFB-dependent auxin signaling. *International Journal of Molecular Sciences*, *23*(1), 510. Available at: https://doi.org/10.3390/ijms23010510

Ma, W., Chen, T., Sun, Y., & Li, J. (2022). The role of lncRNAs in plant tolerance to abiotic stresses. *BMC Plant Biology*, *22*, 67.

Matzke, M. A., & Birchler, J. A. (2005). RNAi-mediated pathways in the nucleus. *Nature Reviews. Genetics*, *6*(1), 24–35.

Meena, S.K., Quevedo, M., Nardeli, S.M., Verez, C., Bhat, S.S., Zacharaki, V., & Kindgren, P. (2024). Antisense transcription from stress-responsive transcription factors fine-tunes the cold response in Arabidopsis. *The Plant Cell*, 36(9), 3467–3482.

Mirdar Mansuri, R. et al. (2022). Long non-coding RNAs as the regulatory hubs in rice response to salt stress. *Scientific Reports*, *12*(1), 21696. Available at: https://doi.org/10.1038/s41598-022-26133-x

Moradi, K., & Khalili, F. (2018). Assessment of pattern expression of miR172 and miR169 in response to drought stress in *Echinacea purpurea* L. *Biocatalysis and Agricultural Biotechnology*, *16*, 507–512.

Naqvi, S.L.H, Ayub, F., Nadeem, M., Ansar, J., Ul Ain, Q. & Naqvi, S.H.Z. (2025). Boosting crop resilience: adaptive strategies and exogenous applications for surviving salt, drought, and waterlogging stress. *Journal of Plant Nutrition*, 1–29.

Park, J., Seo, H., Kim, J., & Lee, S. (2021). miR159 and heat shock protein regulation in thermotolerance. *Plant Molecular Biology*, *98*(4–5), 345–359. https://doi.org/10.1007/s11103-021-01134-0

Park, J., Yang, H., Cho, M., & Lee, S. (2023). Functional crosstalk between ncRNAs and transcription factors in abiotic stress tolerance. *Journal of Experimental Botany*, *74*(5), 1345–1355. https://doi.org/10.1093/jxb/erad056

Peng, Y. et al. (2020). Genome-wide identification and expression analysis of auxin response factor (ARF) gene family in longan (*Dimocarpus longan* L.). *Plants*, *9*(2), 221. Available at: https://doi.org/10.3390/plants9020221

Ponting, C. P., Oliver, P. L., & Reik, W. (2009). Evolution and functions of long noncoding RNAs. *Cell*, *136*(4), 629–641.

Qin, T., Zhao, H., Cui, P., Albesher, N., & Xiong, L. (2017). A nucleus-localized long non-coding RNA enhances drought and salt stress tolerance. *Plant Physiology*, *175*(3), 1321–1336.

Rana, R., Kumar, V., & Sharma, P. (2024). CRISPR-Cas tools in engineering plant stress tolerance: Opportunities and challenges. *Biotechnology Advances*, *55*, 107971.

Rane, S., & Jangid, V. (2024). Osmoprotectants in plant stress tolerance. *Journal of Agricultural Biotechnology*, *18*(2), 85–98.

Rinn, J. L., & Chang, H. Y. (2012). Genome regulation by long noncoding RNAs. *Annual Review of Biochemistry*, *81*(1), 145–166.

Rodriguez, R. E., Mecchia, M. A., Debernardi, J. M., Schommer, C., Weigel, D., & Palatnik, J. F. (2010). Control of cell proliferation in *Arabidopsis thaliana* by microRNA miR396. *Development (Cambridge, England)*, *137*(1), 103–112.

Saeed, F., Chaudhry, U. K., Bakhsh, A., Raza, A., Saeed, Y., Bohra, A., & Varshney, R.K., 2022. Moving beyond DNA sequence to improve plant stress responses. *Frontiers in Genetics*, 13, 874648.

Sharma, P., Verma, A., & Singh, R. (2024). Abiotic stress responses in plants: Osmoregulation and antioxidant mechanisms. *Plant Stress Physiology*, *40*(1), 15–27.

Shi, T., Liu, Z., & Wang, L. (2023). CRISPR-Cas-based manipulation of miRNAs in plant abiotic stress tolerance. *Plant Biotechnology Journal*, *21*(5), 987–997.

Sihag, P. et al. (2021). Discovery of miRNAs and development of heat-responsive miRNA-SSR markers for characterization of wheat germplasm for terminal heat tolerance breeding. *Frontiers in Genetics*, *12*, 699420. Available at: https://doi.org/10.3389/fgene.2021.699420

Singh, P., Kumar, M., & Kaur, S. (2024). Root traits and crop resilience in changing climates. *Agronomy Advances*, *32*(2), 112–130.

Singh, R., Jha, P., & Kumar, A. (2021). MicroRNAs in plants under abiotic stress: An overview. *Environmental and Experimental Botany*, *186*, 104463.

Song, X., Li, Y., Cao, X., & Qi, Y. (2019). MicroRNAs and their regulatory roles in plant-environment interactions. *Annual Review of Plant Biology*, *70*(1), 489–525.

Stief, A., Altmann, S., Hoffmann, K., Pant, B. D., Scheible, W.-R., & Bäurle, I. (2014). *Arabidopsis* miR156 regulates tolerance to recurring environmental stress through SPL transcription factors. *The Plant Cell*, *26*(4), 1792–1807.

Sun, Q., Wu, X., & Zhou, Y. (2019). lncRNAs as miRNA sponges in plant stress responses. *Plant Physiology and Biochemistry*, *137*, 73–79.

Sunkar, R., Kapoor, A., & Zhu, J.-K. (2006). Posttranscriptional induction of two Cu/Zn superoxide dismutase genes in *Arabidopsis* is mediated by downregulation of miR398 and important for oxidative stress tolerance. *The Plant Cell*, *18*(8), 2051–2065.

Sunkar, R., Li, Y.-F., & Jagadeeswaran, G. (2012). Functions of microRNAs in plant stress responses. *Trends in Plant Science*, *17*(4), 196–203.

Sunkar, R., & Zhu, J.-K. (2004). Novel and stress-regulated microRNAs and other small RNAs from *Arabidopsis*. *The Plant Cell*, *16*(8), 2001–2019.

Suvarna, P., Ahmed, S., & Gupta, R. (2024). Nutrient imbalances and their role in crop yield reduction. *Crop Science Innovations*, *39*(2), 88–102.

Swiezewski, S., Liu, F., Magusin, A., & Dean, C. (2009). Cold-induced silencing by long antisense transcripts of an *Arabidopsis* Polycomb target. *Nature*, *462*(7274), 799–802.

Tang, F., Barbacioru, C., Wang, Y., Nordman, E., Lee, C., Xu, N., Wang, X., Bodeau, J., Tuch, B. B., Siddiqui, A., Lao, K., & Surani, M. A. (2009). mRNA-seq whole-transcriptome analysis of a single cell. *Nature Methods*, *6*(5), 377–382.

Tang, Z., Yang, Y., & Zhu, H. (2021). Overexpression of miR156 improves drought tolerance in rice. *Rice*, *14*, 32.

Thompson, D. M., & Parker, R. (2009). Stressing out over tRNA cleavage. *Cell*, *138*(2), 215–219.

Vaucheret, H. (2006). Post-transcriptional small RNA pathways in plants: Mechanisms and regulations. *Genes & Development*, *20*(7), 759–771.

Wang, B., Li, X., & Zhang, Y. (2018). Genome-wide identification and characterization of long non-coding RNAs involved in drought stress response in rice. *Journal of Experimental Botany*, *69*(15), 3575–3590.

Wang, C.-Y., Zhang, S., Yu, Y., Luo, Y.-C., Liu, Q., Ju, C., Zhang, Y.-C., Qu, L.-H., Lucas, W. J., Wang, X., & Chen, Y.-Q. (2014). MiR397b regulates both lignin content and seed number in *Arabidopsis* via modulating a laccase involved in lignin biosynthesis. *Plant Biotechnology Journal*, *12*(8), 1132–1142.

Wang, H., Chen, Y., & Wu, S. (2019). The salt-responsive lncRNA SINE1 mediates salt tolerance through interaction with miR399 in rice. *Plant Cell and Environment*, *42*(5), 1071–1086.

Wang, K. C., & Chang, H. Y. (2011). Molecular mechanisms of long noncoding RNAs. *Molecular Cell*, *43*(6), 904–914.

Wang, L., Yu, C., & Chen, Q. (2020). Functional characterization of circular RNAs during ABA and osmotic stress in rice. *Scientific Reports*, *10*(1), 1–12.

Wang, Y., Zhang, W., & Li, S. (2023). Engineering rice for enhanced drought tolerance through CRISPR/Cas9-mediated gene editing of miR397. *Molecular Plant Breeding*, *11*, 22–32.

Wang, Z., Gerstein, M., & Snyder, M. (2009). RNA-seq: A revolutionary tool for transcriptomics. *Nature Reviews. Genetics*, *10*(1), 57–63.

Wu, J., Xu, Y., & Li, Q. (2022). CRISPRa-based overexpression of a heat-responsive lncRNA in rice for improved heat tolerance. *Plant Molecular Biology Reporter*, *40*, 22–34.

Yadav, A., Mathan, J., Dubey, A.K. and Singh, A., 2024. The emerging role of non-coding RNAs (ncRNAs) in plant growth, development, and stress response signaling. *Non-coding RNA*, *10*(1), 13.

Yang, Y. et al. (2022). microRNAs: Key players in plant response to metal toxicity. *International Journal of Molecular Sciences*, *23*(15), 8642. Available at: https://doi.org/10.3390/ijms23158642

Yang, Y., Chen, L., & Li, W. (2020). Cross-regulation of microRNA394 and ABA signaling modulates plant response to stress in rice. *Molecular Plant*, *13*(8), 1178–1195.

Yang, Z., Zhang, T., & Wang, F. (2019). Genome-wide identification of long non-coding RNAs responsive to drought stress in rice. *Genes*, *10*(4), 302.

Zhang, F., Yang, J., Zhang, N., Wu, J., & Si, H. (2022). Roles of microRNAs in abiotic stress response and characteristics regulation of plant. *Frontiers in Plant Science*, *13*, 919243.

Zhang, H., Hu, W., & Hao, J. (2019). Genome-wide identification and functional prediction of novel drought-responsive lncRNAs in *Triticum aestivum*. *BMC Genomics*, *20*(1), 1–15.

Zhang, H., Xu, J., Li, W., & Ma, Y. (2023). Multilayered ncRNA interactions in transcription factor regulation during plant stress. *Plant Biotechnology Journal*, *21*(4), 1224–1236. https://doi.org/10.1111/pbi.13978

Zhang, K., Shi, X., & Zhao, X. (2020). MiR319-regulated TCPs modulate cold stress response through interaction with ICE1 in rice. *Molecular Plant*, *13*(8), 1264–1279.

Zhang, L. et al. (2013). MIR395 is involved in detoxification of cadmium in *Brassica napus*. *Journal of Hazardous Materials*, *250/251C*, 204–211.

Zhao, J. et al. (2020). Plant Responses to Heat Stress: Physiology, Transcription, Noncoding RNAs, and Epigenetics. *International Journal of Molecular Sciences*, *22*(1), 117. Available at: https://doi.org/10.3390/ijms22010117

Zhao, L., Wu, M., Li, C., & Zhang, X. (2023). LncRNAs and ion transport regulation in plant salinity tolerance. *Plant Cell Reports*, *42*(5), 780–791. https://doi.org/10.1007/s00299-023-03067-5

Zhao, X., Wang, L., Gao, Y., & Zhang, H. (2022). miRNA-mediated antioxidant response in oxidative stress. *Journal of Experimental Botany*, *73*(8), 2298–2310.

Zhou, B., Zheng, B., & Wu, W. (2024). The ncRNAs involved in the regulation of abiotic stress-induced anthocyanin biosynthesis in plants. *Antioxidants*, *13*(1), 55.

Zhou, H., Liu, X., & Lin, Y. (2023). Advances in non-coding RNA research for improving plant stress resilience. *Current Opinion in Plant Biology*, *68*, 102182.

Zhou, J., Liu, M., & Jiang, J. (2020). MiR399-mediated regulation of PHO2 plays a critical role in drought stress tolerance in cotton. *Plant Physiology and Biochemistry*, *152*, 51–61.

Zhou, M., Li, D., Li, Z., Hu, Q., Yang, C., Zhu, L., & Luo, H. (2013). Constitutive expression of a miR319 gene alters plant development and enhances salt and drought tolerance in transgenic creeping bentgrass. *Plant Physiology*, *161*(3), 1375–1391.

5 Impact of Non-Coding RNA Networks in Plant Abiotic Stress Tolerance

Nadia Sharif, Ambreen Zari, Ummey Farwa Hassan, Noor Fatima and Zainul Abideen

5.1 INTRODUCTION

Abiotic stressors such as drought, salinity, extreme temperatures and nutrient deficiencies are critical factors that severely limit plant growth and productivity (Abideen et al., 2021; Shoukat et al., 2020). As sessile organisms, plants have evolved complex regulatory networks to cope with these stresses, including complex gene expression programmes. Recent advances in molecular biology and genomics have revealed that non-coding RNAs (ncRNAs), a class of RNA molecules that do not encode proteins, play essential roles in regulating gene expression in response to abiotic stress. These ncRNAs include microRNAs (miRNAs), small interfering RNAs (siRNAs), long non-coding RNAs (lncRNAs) and other small regulatory RNAs, all of which contribute to the post-transcriptional and epigenetic regulation of stress-responsive genes (EL Sabagh et al., 2022). Understanding the role of ncRNAs in these networks is essential to developing strategies to improve plant stress tolerance and increase agricultural productivity in rapidly changing environments.

ncRNAs function as key regulators in many biological processes, influencing developmental plasticity and stress tolerance. In plants, ncRNAs have emerged as key components of abiotic stress responses and facilitate the fine-tuning of stress response pathways. Understanding the role of ncRNAs in these networks is crucial for developing strategies to improve plant stress tolerance and enhance agricultural productivity in rapidly changing environments. The ncRNAs range can be classified into the basics of categories, their basics of size: illegal ncRNAs (including miRNAs, siRNAs, piRNAs) and long ncRNAs (lnRNAs) (Gelaw & Sanan-Mishra, 2021).

These lncRNAs regulate gene expression by controlling RNA degradation, inhibiting translation or altering the chromatin state. Among them, miRNAs and lncRNAs have been the most studied in the context of responses to abiotic stresses (Abdulraheem et al., 2024).

miRNAs are small endogenous ncRNAs, typically 20–24 nucleotides in length, that play a key role in the post-transcriptional regulation of gene expression by directing the cleavage or translational inhibition of target mRNAs (Sarropoulou & Fernández, 2023). In plants, miRNAs are involved in various biological processes, including development, hormone signalling and responses to biotic and abiotic stresses (Kar & Raichaudhuri, 2021). Several miRNAs have been implicated in plant responses to abiotic stress (Zhang et al., 2022a). For example, miR398 is downregulated under oxidative stress conditions such as drought and high salinity, leading to upregulation of its target genes, copper/zinc superoxide dismutase (CSD), which play a critical role in the detoxification of reactive oxygen species (ROS). Another well-characterised miRNA, miR319, modulates plant responses to extreme temperatures. In *Arabidopsis*, miR319 targets TCP transcription factors (TFs) that regulate the expression of jasmonic acid (JA)-responsive genes involved in leaf morphogenesis and stress tolerance (Fang et al., 2021; Sun et al., 2020).

miRNAs also play a role in hormone signalling during stress responses: for example, miR159 regulates the ABA signalling pathway by targeting MYB TFs, which are key components of the ABA response pathway under drought stress. Similarly, miR167 is involved in auxin signalling under abiotic stress conditions and targets auxin response factors (ARFs) that control root architecture and adaptability to stress environments (Singh et al., 2023).

Drought stress, one of the most studied abiotic stress factors, triggers a series of miRNA-mediated responses in plants, and several miRNAs, including miR398, miR169, miR172 and miR393, are differentially expressed under drought conditions. For example, miR169 is downregulated in plants under drought stress, leading to the upregulation of the nuclear factor Y subunit A (NFYA) gene, which plays a role in ABA-mediated drought stress tolerance. Similarly, miR393 regulates auxin signalling by targeting auxin receptor genes, thereby controlling root growth and adaptation to drought stress (Banerjee et al., 2016).

A known lncRNA in plants is COOLAIR, which controls vernalisation, the process that leads to flowering after prolonged exposure to cold. COOLAIR acts by regulating the chromatin state of the FLOWERING LOCUS C (FLC) gene, a key regulator of flowering time, in response to cold stress. Similarly, other ncRNAs are involved in regulating gene expression under drought, salinity and heat stress (Zhao et al., 2022).

In response to salt stress, a number of lncRNAs have been identified that regulate the expression of stress-responsive genes in rice. For example, lncRNA OsDREH1 is induced by salt stress and promotes the regulation of genes involved in ROS homeostasis, thereby improving salt tolerance in rice. Another salt-responsive lncRNA, DRIR, was found to positively regulate drought and salt tolerance in *Arabidopsis* by regulating the expression of stress-responsive genes such as RD29A and RD22 (Jiang et al., 2024).

DOI: 10.1201/9781003583257-5

LncRNAs also play a role in regulating plant responses to heat stress. In *Arabidopsis*, lncRNA heat-induced lncRNA (HILNCR) has been identified as a key player in regulating the expression of heat stress transcription factors (HSFs) and heat shock proteins (HSPs), which are essential for thermotolerance (Heo & Sung, 2011). By regulating the expression of these stress-responsive genes, lncRNAs help plants adapt to extreme temperatures and other abiotic stresses (Zhang et al., 2022a).

siRNAs are another class of RNA involved in the regulation of the expression of genes by means of an interference of RNA (RNAi). In plants, miRNA is crucial for protection against viral infections, regulation of gene expression and the maintenance of the stability of the genome. It also plays a role in abiotic stress tolerance by repressing genes involved in stress signalling pathways (Kryovrysanaki et al., 2022).

One notable example is the involvement of phased siRNAs (phasiRNAs) in controlling plant responses to phosphate deficiency: PhasiRNAs are produced from specific RNA precursors in response to stress and target genes involved in phosphate uptake and metabolism. In *Arabidopsis*, phasiRNA production is controlled by miR399, which is induced by phosphate starvation and promotes phosphate homeostasis by acting on the PHO2 gene, a negative regulator of phosphate uptake (Yang et al., 2024).

NcRNAs do not function in isolation; instead, they form complex networks involving crosstalk between different types of ncRNAs and protein-coding genes. For example, miRNAs and lncRNAs often interact to regulate stress response pathways. In rice, lncRNA Mir168A modulates the expression of MIR168, which, in turn, regulates the expression of Argonaut1 (AGO1), the key component of the RNA-induced silencing complex (RISC), which mediates the miRNA function. This interaction between miRNA-lncRNA illustrates how ncRNAs can configure gene expression networks during stress reactions (Yadav et al., 2024).

Similarly, siRNAs and miRNAs cooperate to regulate stress-responsive genes: during drought, for example, miR172 targets genes encoding AP2-type TFs involved in ABA signalling. This regulation is enhanced by siRNAs that target the same gene, providing an additional level of control over the ABA response pathway (Shi & Hussain, 2020).

5.2 ABIOTIC STRESSES AND TYPES OF ABIOTIC STRESSES

Plants, like all living organisms, face various challenges and stresses throughout their life cycle. One of the most important problems plants face is abiotic stress. Abiotic stress is the adverse effect of non-living factors on plants in a particular environment (Bartas, 2024). Abiotic stress factors that affect plants include excessive heat, extreme cold, salinity, drought, mineral availability, or toxicity. Such abiotic stress factors have a negative impact on the overall crop yield, and there is a need to create resistant plant varieties that can withstand abiotic stress factors (Kumar, 2020). Plants are also adversely affected by such non-biological stressors because toxins are also considered to be the surrounding stressors that are affected by the environment (Shah & Smith, 2020).

Abiotic stress is known to affect the internal metabolism of plant parts, resulting in reduced productivity of the whole plant, which can be a significant negative effect, as it is widely recognised that abiotic stressors generally only cause adverse effects (Zhang et al., 2023). The various abiotic stresses include the following:

5.2.1 Cold and Heat Stress

It is known that each plant can only survive at a certain temperature, and that changes in temperature alter the overall nutritional pattern of an individual plant. If plants are grown at low temperatures, the low temperature can damage the plant tissues, adversely affecting the plant's life cycle (Shubham & Ayan, 2022). Both extremes of temperature – freezing at <0°C and chilling at 0–15°C – have an effect on the growth, development and physiological processes of plants. Many temperature-responsive regulatory genes, including C-repeat/dehydration-responsive element binding factors (CBF/DREB1s) and heat shock factors, bind to the promoters of cold-regulated genes (COR) and heat stress response genes, respectively, to increase plant resistance to cold and heat stress (Yadav et al., 2024).

Numerous siRNAs, circRNAs and miRNA-targets are among the many ncRNAs that have been found to respond to heat and cold stress. The notable examples of miRNAs and their targets associated with heat and cold stress are miR156-SPL, miR159-GAMYB, miR160-ARF, miR164-NAC, miR165/166-homeodomain leucine zipper class-III TF primarily PHBULOSA and ATHB-8, miR167-ARF6, miR168-AGO1, miR171-GRAS gene family, miR172-APETALA2, miR319-TCP, miR393-auxin receptors with F-box domains; TIR1/AFB2, miR396-GRF, miR397-laccase gene; lignin-degrading enzyme, miR402-DEMETER-LIKE PROTEIN3; a 5-methylcytosine DNA glycosidase. Additionally, conserved miRNA (miR166) and novel miRNAs (miR168, miR444 and miR528) in maize leaves during the V3 stage under heat stress have been discovered using genome-wide transcriptome analysis. Certain miRNAs, including IbmiR162, IbmiR164, IbmiR171 and IbmiR397, have been linked to physiological alterations in sweet potato seedlings, namely oxidative stress during chilling and heat stress (Saeed et al., 2024).

Chip-based microarray and strand-specific sequencing techniques identified a number of lncRNAs for both heat and cold stress, with excellent specificity in tissue types and developmental stages. For example, during heat stress, only the target gene – rather than the lncRNA itself – was expressed. Lnc-173 targets the SUCROSE SYNTHASE4 gene. The sensitivity and adaptation of eukaryotes to heat stress are significantly influenced by heat shock transcription factors (HSFs). Heat stress was discovered to produce asHSFB2a (natural antisense transcript of HSFB2a) in *Arabidopsis*, which in turn adversely regulates HSFB2a expression. Pears have recently been found to contain heat-induced long intergenic ncRNA 1 (HILincl). Multiprotein bridging factor lc is upregulated as a result of this lncRNA's target binding with HSF to increase its transcriptional activity during heat response (Zhao et al., 2024).

Plants that experience cold stress go through a process called vernalisation, which stops them from flowering during winter vegetative development and encourages them to do so during the reproductive phase when conditions are right. FLC is in charge of this vernalisation regulation mechanism. A number of lncRNAs, including COLDAIR, COLDWRAP and COOLAIR, use chromatin modification to inhibit FLC expression. Additionally, freezing tolerance is controlled by the transcription of the CBF1 gene by the lncRNAs SVALKA, cryptic antisense CBF1 (asCBF1) and CIR (CBF intergenic lncRNA). While the cis-acting NAT lncRNA_2962 stimulates transcription of cold-responsive genes, the COLD INDUCED lncRNA1 (CIL) contributes to the low-temperature stress response by influencing the reactive oxygen pathway or osmotic adjustment substance. By controlling the expression of several stress-related genes during the developmental stage, these lncRNAs influence how plants react to cold stage (Gelaw & Sanan-Mishra, 2021).

5.2.2 Salinity Stress

Salt concentration is one of the main factors that affect plant growth and development. More salt leads to the reconstructive reality of the gene to minimise the stress on salt concentration and thus optimise surviving plants in such dangerous situations (Atta et al., 2023). Plants developed mechanisms to cope with the stress of salt, by synthesising osmolites, the areas of vacuole salt and adjusting the ion movement to the cell membrane. These mechanisms help plants maintain cellular ion homeostasis and enable plants to survive in saline environments (Atta et al., 2023; Zhao et al., 2020).

Salt stress is a significant impediment to plant growth and development, affecting approximately 33% of cultivated land. It increases ionic and intracellular osmotic pressure in plants, leading to increased toxicity, oxidative damage and osmotic stress. Rice is sensitive to salt stress. The conductivity threshold of most cultivars is 3 dSm^{-1}. There is limited data on ncRNAs related to the regulation of salt stress in plants. Nearly 75 conserved miRNAs were identified from the control and NaCl-treated salt-tolerant rice varieties. Among them, osa-miR12477, which targets L-ascorbate oxidase, showed an accumulation of oxidative stress in the plant upon salt treatment. Similarly, Jain et al. identified 1500 and 2100 putative lncRNAs from root and leaf tissues of salt-tolerant (Horkuch) and salt-sensitive (IR-29) rice genotypes, respectively (Zhao et al., 2021).

5.2.3 Drought Stress

Drought stress poses a major risk to agriculture, and global efforts to enhance drought-tolerant crops are urgently needed. Drought stress has adverse effects on plants, including osmotic imbalance, degradation of membrane systems and reduced respiration and photosynthetic rates. These responses not only hinder plant growth and metabolism throughout its life cycle, but also decrease the yield and the quality of cultures (Wu et al., 2023).

The discovery of the lncRNA sensitive to drought (DRIR) demonstrates its crucial role in the regulation of tolerance to plants with drought and saline stress. Recently, Zhang et al. conducted large-scale profiling and identified 2174 and 1354 high-confidence circRNAs in maize and *Arabidopsis*, respectively. In particular, they discovered two circRNAs associated with genes GRMZM2G032852 and GRMZM5G817173, which encode calcium-dependent protein kinase (CPK) and cytokinin oxidase/dehydrogenase (CKX), respectively. These circRNAs, named circCPK and circCPK, were formed by backsplicing between exons, resulting in allelic insertions and deletions in the maize panel. The survival rate of corn plants with CIRCPK accompanied by an insertion anti-gene and the interpreted anti-containing genes was significantly higher than that of normal and drought conditions. This observation suggests a potential relationship between the expression of siRNA and the resistance to drought in plants (Qiao et al., 2024).

5.2.4 Flooding Stress

Flooding stress is another abiotic stress that affects plant growth and development. It can lead to oxygen deprivation, which can have a significant impact on plant metabolism. Plants have evolved various mechanisms to cope with flood stress, including ethylene synthesis and adventitious root formation. Ethylene helps plants tolerate flood stress by promoting the elongation of stem and root tissues, whereas adventitious roots help plants absorb water and nutrients from waterlogged soil (Eerdekens et al., 2024; Sahu et al., 2022).

Flooding is a temporary inundation of land by water, while waterlogging occurs when the soil is either completely saturated or nearly saturated most of the time. Both processes impede various physiological processes in plants, such as photosynthesis, disrupt cellular energy production and upset ionic balance. These effects are exacerbated when plants experience lower than optimal oxygen content, a phenomenon known as hypoxia, which occurs frequently in plants' natural environments. Limited oxygen availability in the root zone negatively impacts the plant's overall metabolism, stunting growth and development. Several ncRNAs were found in various plant species susceptible to waterlogging or hypoxia, such as *Arabidopsis*, maize, rice, *Populus*, *Medicago*, tomato and cucumber. For example, in rice, about 291 and 58 ncRNAs responsive to deep water stress were found in internodes and nodes, respectively.

Furthermore, degradome analysis predicted nearly 1586 and 744 targets to be cleaved by 230 and 53 miRNAs from internodes and nodes, respectively. Notably, miR156 targeted the SQUAMOSA promoter-binding protein box (SBP) gene (cleaved at 1163 nt), and the TF NF-Y (cleaved at 1729 nt) was a target of miR169. The study further identified miRNA-lncRNA-mRNA regulatory networks such as nov-miR145-TCONS_00011544 and miR159e-TCONS_00059798 targeting ZFD (Os11g36430) and MYB (Os01g36460) genes, respectively. Flooding is also associated with decreased energy metabolism, detected by sucrose fermentation-unrelated protein kinase (SnRK1). Overexpression of miR156 has previously been shown to improve abiotic stress tolerance in various plants through the regulation of TFs such as SQUAMOSA promoter binding protein (SPL) (Božić et al., 2024; Sharma et al., 2022).

5.2.5 Heavy Metal Stress

Heavy metal toxicity is caused by the accumulation of heavy metals in soil and can have significant effects on plant growth and development. Plants have evolved various mechanisms to combat heavy metal toxicity, including the synthesis of metal chelating compounds and the regulation of metal transport across cell membranes (Feng et al., 2023).

Rapid industrialisation, the use of large amounts of chemical fertilisers and the careless contamination of arable land and irrigation water with heavy metals – especially cadmium – have created a major threat to crop production and human health. Plants employ a number of regulatory molecular mechanisms to reduce the amount of heavy metals that enter their systems from the soil; lncRNAs may be particularly important in regulating this process. In *Brassica napus*, RNA-seq profiling revealed 301 cadmium-responsive lncRNAs, 67 of which were eTMs for 36 Cd-responsive miRNAs. Target mimicry for EL628609, TC182597 and TC203372 mRNAs involved in Cd absorption, and detoxification was demonstrated by the functional validation of TCONS_00091906, TCONS-00033487 and TCONS-00097191 lncRNAs under Cd stress by qRT-PCR analysis (Rashid et al., 2023; Wan et al., 2024).

5.3 LOSS OF PLANT PRODUCTIVITY DUE TO ABIOTIC STRESS

Plant growth and yield are greatly affected by cold, drought, salinity and heavy metals. Abiotic stresses affect plants from the morphological to the molecular levels and are found at all stages of plant development. Drought affects three main stages of plant growth: vegetative development, pre-flowering and terminal growth. Physiological responses of plants to stress include wilting, leaf drop, reduced leaf area and reduced water loss through transpiration. In drought conditions, crop development faces the challenge of extreme water use in agriculture, and the reduction in turgor pressure, which is important for cell growth, is one of the most sensitive physiological mechanisms affected. Drought stress disrupts water flow in higher plants from xylem to neighbouring elongating cells, suppressing cell elongation. This results in decreased leaf area, plant height and overall crop development due to impaired cell elongation, mitosis and expansion (Yadav et al., 2024).

Abiotic resistance to stress includes mechanisms of avoidance, evasion and tolerance. Osmotic modifications, which promote the active accumulation of solutes in the cytoplasm, can reduce the adverse effects of stress by maintaining the cellular water balance. Low temperatures have a significant impact on plant survival and geographic distribution: cold stress causes cell and tissue dehydration and crystallisation of cellular water, thereby reducing plant growth and productivity. It also causes a decrease in membrane conductance, an increase in water viscosity, closure of hydraulically active stomata and increased electrolyte leakage. Cold stress slows metabolism, dissipates energy and causes the production of free radicals due to oxidative stress (Sachdev et al., 2021).

5.4 PHYSIOLOGICAL AND BIOCHEMICAL RESPONSE OF PLANTS DUE TO ABIOTIC STRESS

When plants are exposed to abiotic stress conditions, various physiological and biochemical changes occur to cope with these adverse conditions. These responses are essential to maintain cellular homeostasis, promote survival and minimise damage.

One of the main physiological responses of plants to abiotic stress is the closure of stomatal apertures, which helps to reduce water loss through transpiration during drought or high salinity, but also limits CO_2 uptake, reducing the rate of photosynthesis and affecting plant growth (Shoukat et al., 2019). Advancing adjustment is another important reaction when plants accumulate osmotic pressure, such as prin, beta and sugar glycine (Ahluwalia et al., 2021; dos Santos et al., 2022). Additionally, there is a change in root architecture, with plants increasing root growth and increasing root depth to more efficiently access water and nutrients under stress. In saline environments, plants also exhibit ionic homeostasis to avoid toxic levels of ions such as sodium (Na^+) by regulating ion transport mechanisms, including via the salt overly sensitive (SOS) pathway (Shelden & Munns, 2023).

At the biochemical level, plants activate several methods and absorb reactive active oxygen (AFC). These AFCs are harmful and can cause lipids, proteins and nucleic acids. To combat this, plants increase the levels of antioxidant enzymes such as superoxide dismutase (SOD), catalase (CAT) and peroxidase (POD) (Mansoor et al., 2022). Moreover, secondary metabolites such as flavonoids, terpenoids and phenols play important roles in signalling pathways involved in plant defence against oxidative stress and stress tolerance (Qasim et al., 2017). Plants also enhance hormone signalling networks by regulating the levels of hormones such as abscisic acid (ABA), ethylene and salicylic acid (SA), which play central roles in plant responses to stress (Panahirad et al., 2023).

In response to extreme heat stress, plants activate HSPs, which act as molecular chaperones that stabilise and refold denatured proteins. Similarly, cold stress induces the production of cold-regulated proteins (CORs), which help maintain membrane fluidity and protect the photosynthetic apparatus (Sadura & Janeczko, 2024).

5.5 NON-CODING RNAs IN STRESS TOLERANCE

ncRNA is a functional RNA molecule that is not translated into protein. The DNA sequences from which functional ncRNA are transcribed are often referred to as RNA genes. Abundant and functionally important types of ncRNAs include transfer RNAs (tRNAs) and ribosomal RNAs (rRNAs), as well as small RNAs such as miRNAs, siRNAs, piwi-interacting RNAs, snoRNAs, snRNAs, exRNAs, scaRNAs and the long ncRNAs such as Xist and HOTAIR. The first ncRNA to be characterised was an alanine tRNA found in baker's yeast; its structure was published in 1965. In the early 1980s, the discovery of new ncRNAs continued with the help of snoRNA, Xist, CRISPR and others (George et al., 2024).

ncRNA is a complex class of RNA that controls cells without coding for proteins. They are divided into two main types: housekeeping and regulatory (Zhang et al., 2019). As shown in Figure 5.1, housekeeping ncRNAs include rRNA, tRNA, small nuclear RNAs (snRNA) and small nucleolar RNAs (snoRNAs). Regulatory ncRNAs are divided into small ncRNAs (< 200 nt) and long ncRNAs (> 200 nt). Some examples of regulatory ncRNAs include: miRNAs, small interfering RNAs (siRNAs), Piwi-associated RNAs (piRNAs), antisense RNAs (asRNAs), activating RNAs (eRNAs) (Samarfard et al., 2022). ncRNAs are involved in many biological processes, including cell cycle and cancer. For example, lncRNAs have been studied in cancer and are involved in almost all stages of the cell cycle. asRNA can regulate gene expression by base pairing with coding mRNA (Mattick et al., 2023; Zangouei et al., 2023).

5.6 NON-CODING RNA NETWORK IN PLANTS

Plants are secular species and cannot escape stress, but when exposed to constantly changing environmental conditions, they require avoidance and/or tolerance mechanisms for survival. Abiotic stresses such as cold, heat, hypoxia, drought, nutrient deficiency and salinity affect plant growth and productivity, leading to significant yield reductions. In addition, a global population in constant growth, a decrease in arable land due to rapid urbanisation, significant changes in the quality of the soil and global warming, according to forecasts, will be even more aggravated by food security problems (Anwar et al., 2021).

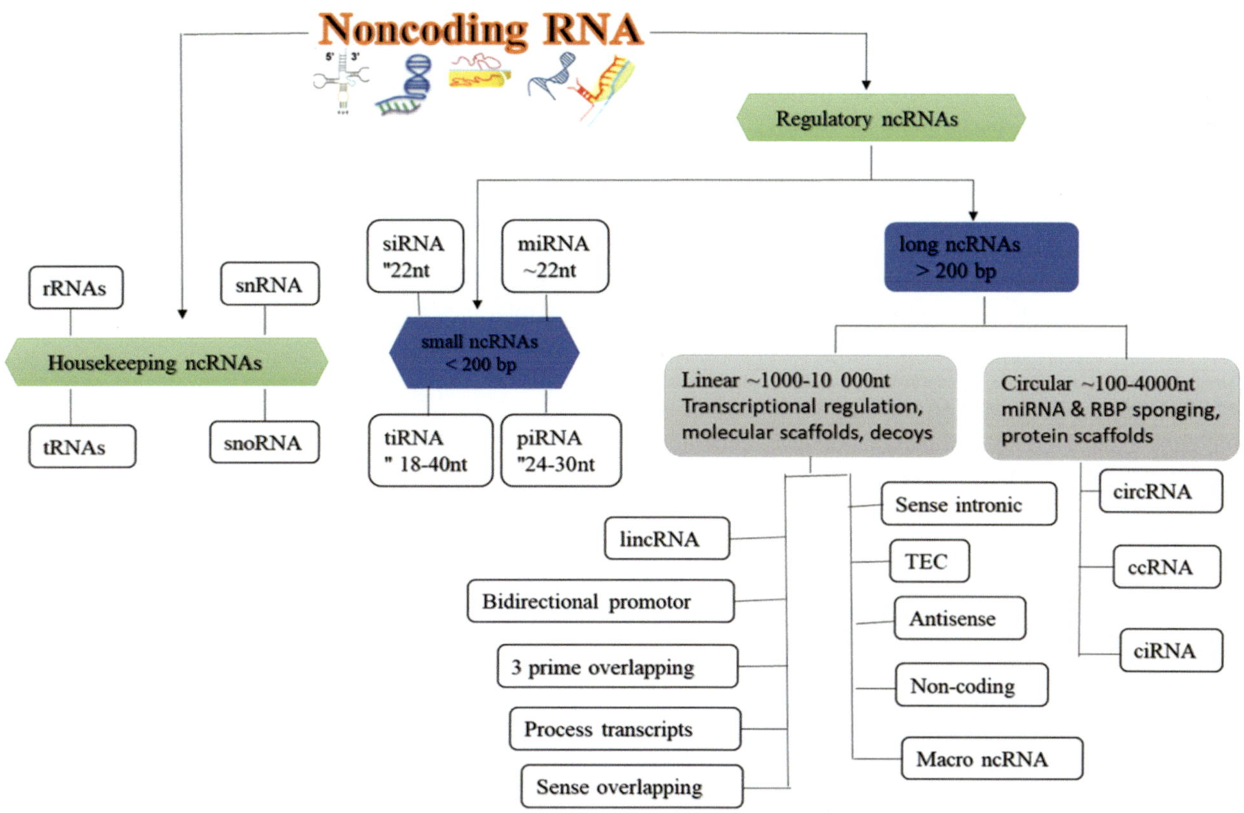

FIGURE 5.1 Classification of non-coding RNA.

In recent years, RNA, which is not HEAD (ARNC), has rapidly increased as a biological molecule that contributes to genome and expression diversity. Up to 90% of the eukaryotic creature genome is transferred by RNA, but only about 2% of tRNA protein products. The remaining transcriptome contains ncRNA transcripts, which were previously considered transcriptional noise due to their lack of protein-coding potential and low sequence conservation (Gámez-Valero et al., 2020). However, high-throughput sequencing analysis and experimental validation have confirmed that ncRNAs are involved in myriad different levels of gene regulation, including epigenetic, transcriptional and post-transcriptional. Therefore, the development of efficient next-generation sequencing (NGS) techniques combined with homology-based and/or experimental approaches is essential for ncRNA discovery in plants. Since their initial discovery, ncRNAs have been classified into several different classes based on their origin, biogenesis and mechanism of action. Housekeeping and regulation ARNC are major ARNC transfer groups (Magar et al., 2023). Housekeeping ncRNAs are primarily associated with cellular and ribosomal functions, and they include tRNAs, snRNAs, rRNAs and snRNAs. On the other hand, regulatory ncRNAs consist of miRNAs, short interfering RNAs (siRNAs), piwi-interacting RNAs (piRNAs) and lncRNAs. These regulatory ncRNAs are transcribed from DNA but cannot be translated into proteins. Nevertheless, they perform a lot of important functions of growth, development and non-biological stress reactions at the level of transfer and transcription.

Over the past few decades, basic development has been achieved during the development of dynamic and complex mechanisms in response to plant environmental stress. Extensive efforts have been made in identifying stress-responsive genes and corresponding protein interaction networks, which have led to the observation that plant responses to stress depend on precise gene expression and regulation, which is achieved through multiple mechanisms at different levels, including transcriptional and post-transcriptional regulation. However, despite significant scientific research focused primarily on the control mechanisms of protein coding, recent discoveries have revealed major functions of ncRNAs in controlling transcriptional and post-transcriptional gene expression levels during plant development and in response to environmental stresses (Waititu et al., 2020).

5.7 MECHANISMS OF ACTION OF ncRNAs IN ABIOTIC STRESS

ARNC interacts with targets and proteins related to miRNA or protein additional arrays, affecting these target functions, stability or diffusion. miRNA is associated with the additional site of the MRNC target, which leads to the decomposition or inhibition of miRNA. This is a highly specific interaction governed by sequence complementarity: for example, under drought conditions, miR169 targets the TF NF-YA, thereby decreasing the expression of genes related to water transport and preventing excessive water loss (Ma et al., 2023; Ratti et al., 2020).

siRNAs are generally derived from double-stranded RNA and also act by sequence-specific binding to target mRNAs, resulting in RNA interference (RNAi) and silencing of target genes. For example, siRNAs are involved in the removal of transposable elements that can become destabilised under stress, compromising genome stability (Neumeier & Meister, 2021). ncRNAs have a wider range of interactions because they can form secondary structures that interact with TFs, chromatin remodellers and even other ncRNAs. For example, the ncRNA SVALKA stabilises cold-responsive gene expression by binding to proteins involved in chromatin modification, thereby protecting important gene regions from downregulation in response to cold (Gonzales et al., 2024). Some circRNAs sequester miRNAs, preventing them from binding to their normal targets. This soaking effect maintains the expression of specific genes by indirectly suppressing the action of miRNAs. During stress, these circRNAs act as molecular buffers to maintain the expression of stress-responsive genes (Bizzarri & Cannistraro, 2021).

ncRNAs facilitate crosstalk between different abiotic stress pathways, with some ncRNAs responding to multiple stresses by providing a single response mechanism. For example, the MIR398 is involved in both the oxidative and drought stress reaction and adjusts the expression of copper/zinc enzymes from superoxide dismutase (CSD1 and CSD2), which protects cells from stressed AFC damage. Some lncRNAs reproduce the remodelling of chromatin in a specific gene, causing genetic changes in the expression of the gene. These epigenetic changes can make stress-responsive genes more readily accessible in future generations, potentially contributing to transgenerational stress tolerance. Hormones like ABA, JA and SA play key roles in stress signalling, and ncRNAs frequently target hormone signalling genes. Under drought stress, ABA accumulation upregulates miRNAs such as miR159, which target MYB TFs to modulate the ABA signalling pathway and enhance stress tolerance (Nguyen et al., 2022).

5.7.1 Transcriptional Regulation

Transcriptional regulation involves controlling gene expression at the DNA level, which is often achieved by ncRNAs that regulate epigenetic modifications. ncRNAs help activate or repress stress-responsive genes by influencing TFs or directly interacting with DNA. miRNAs control transcription by directing RNA-directed DNA methylation (RdDM), which adds methyl groups to specific regions of DNA, often repressing stress-responsive genes, and also regulate the expression of TFs that activate or repress target genes under stress conditions. miR393 targets genes involved in auxin signalling, suppressing growth processes that are non-essential during drought, thereby conserving resources. This action helps the plant adapt by reducing water loss and enhancing root structure (Matzke & Mosher, 2014).

siRNAs play an important role in driving RdDM, which adds methylation marks to the promoter regions of target genes, effectively repressing them. This allows plants to save energy and resources by temporarily silencing certain genes. siRNAs mediate the methylation of promoter regions of genes involved in salt sensitivity, such as genes controlling ion uptake, helping to maintain ion homeostasis and improve salt tolerance in plants (Yang et al., 2022).

lncRNAs act as molecular scaffolds or guides to recruit and direct chromatin-modifying complexes to specific DNA sites to add or remove epigenetic marks that alter gene transcription. The lncRNA COLDAIR is known to direct polycomb repressive complex 2 (PRC2) to the FLC gene. This effect can lead to Histon's methylation, removal of the expression of FLC and adaptation to the long-term effects of cold and delayed flowering (Gonzales et al., 2024).

5.7.2 Post-Transcriptional Regulation

Post-transcriptional regulation includes the control of mRNA processing, stability, translation and degradation. ncRNAs play a major role at this level by directly interacting with target mRNAs, influencing their availability for protein synthesis. miRNAs bind to complementary sequences on target mRNAs, causing their cleavage or inhibiting their translation, as represented in Figure 5.2, which is necessary to rapidly adjust protein levels in response to stress. miR398 targets and suppresses SOD genes under normal conditions. However, under oxidative stress, miR398 is suppressed, allowing the accumulation of SOD proteins, which help to scavenge ROS and reduce oxidative damage (Li et al., 2022).

siRNAs can also guide the degradation of mRNAs by recognising and binding complementary sequences, which is especially useful for quickly downregulating non-essential genes during stress. siRNAs target mRNAs encoding non-essential enzymes during salinity stress, reducing their expression to conserve energy and resources that can be redirected to essential salt-tolerance mechanisms (Agrawal et al., 2003).

RNA functions as a sponge or molecular food by contacting miRNA and isolates miRNA, preventing the interaction with the mRNA target. This action adjusts the availability of miRNA and adds flexibility to the response to stress. During thermal stress, isolation of miRNA lncRNA, which inhibits the spread of thermal impact proteins (HSP), can increase HSP to protect cell structures from thermal injury (Ratti et al., 2020).

RNAS non-decelerated RNA plays an important role in epigenetic adjustments in non-biological stress of plants. This adjustment includes guidance changes such as DNA methylation and changing stress-sensitive gene, which helps plants adapt to environmental conditions where they are difficult. The main types of ncRNAs involved in these processes include miRNAs, small interfering RNAs (siRNAs) and lncRNAs (Abdulraheem et al., 2024).

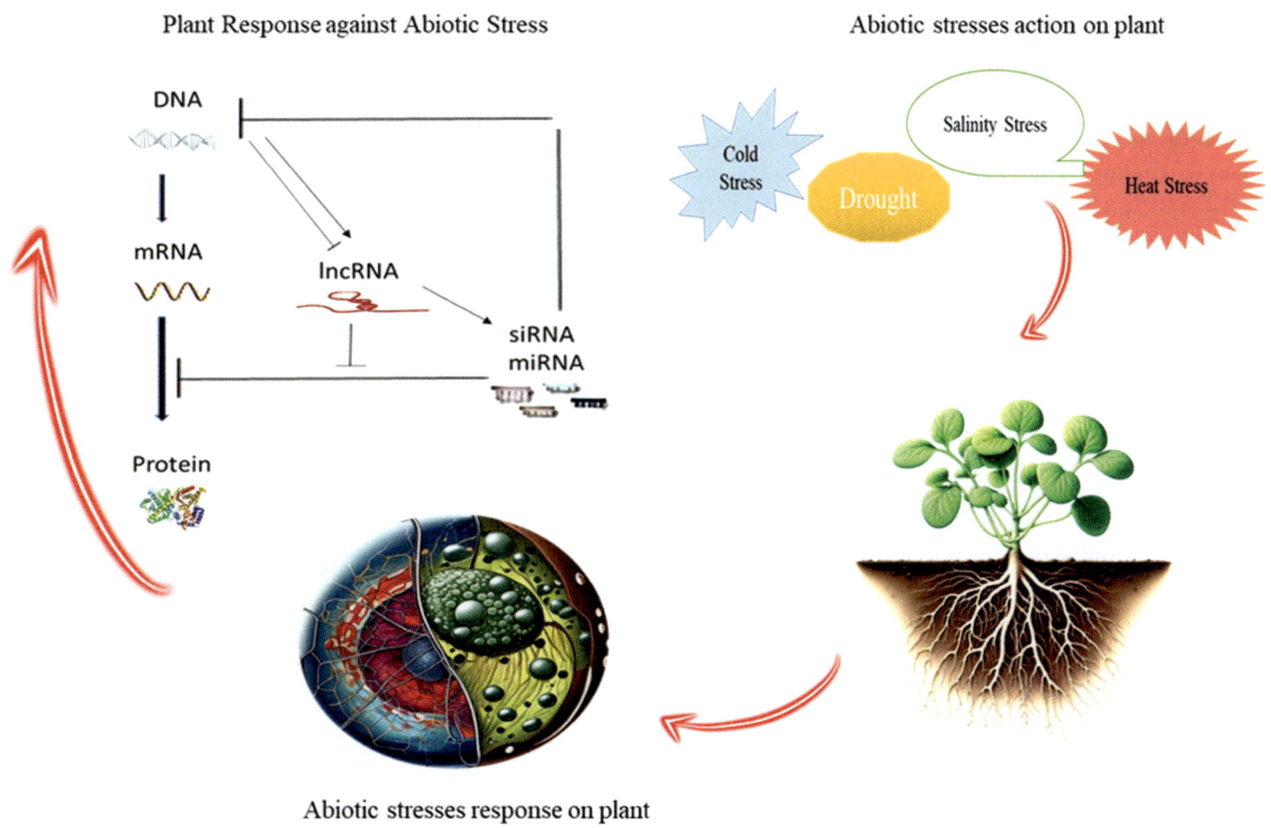

FIGURE 5.2 Plant responses to abiotic stress on DNA and RNA moieties.

5.8 ROLE OF NCRNAS IN EPIGENETIC REGULATION UNDER ABIOTIC STRESS

5.8.1 miRNAs and DNA Methylation

miRNAs can affect gene expression without altering the DNA sequence because they can direct DNA methylation of specific stress-responsive genes. They often function through the RdDM pathway to direct the methylation machinery to target genes, repressing or activating genes as needed for stress resistance. miR156 was found to affect drought-responsive TFs and genes controlling water conservation and root development via RdDM, which helps plants conserve resources and increase drought tolerance (Erdmann & Picard, 2020).

5.8.2 siRNAs in Histone Modifications

siRNAs play a key role in the RdDM pathway, which directs DNA methylation and histone modifications at specific loci. During stress, siRNAs can affect histone methyltransferases or deacetylases, resulting in chromatin modifications that affect the accessibility of stress-responsive genes. siRNAs involved in salt stress have been shown to modulate histone acetylation and methylation at salt tolerance genes, thus altering gene expression and enhancing ion transport and osmotic balance, which are critical for salt stress tolerance (Delarue et al., 2025).

5.8.3 lncRNAs as Epigenetic Modulators

lncRNAs act as scaffolds or molecular guides to deliver epigenetic modifiers, such as methyltransferases and demethylases, to specific gene regions. This targeted epigenetic modification helps to activate or repress stress-responsive genes depending on environmental conditions. The lncRNA COLDAIR is associated with vernalisation (cold response) and recruits PRC2 to the FLC gene, leading to histone methylation to repress FLC expression, thereby preparing plants for seasonal changes (Gonzales et al., 2024).

5.9 CROSSTALK WITH OTHER EPIGENETIC MECHANISMS

Multiple miRNAs and lncRNAs interact to form a complex network that cooperatively regulates gene expression under stress conditions, for example, lncRNAs can act as sponges for miRNAs and sequester miRNAs to relieve repression on target mRNAs, thus enhancing the flexibility of stress responses. During heat stress, some lncRNAs sequester miRNAs that would otherwise inhibit HSPs, allowing increased HSP expression and better cellular protection against heat-induced damage (Han et al., 2022). Epigenetic stress memory, or the ability of plants to 'remember' past stress exposures, involves ncRNAs. Through persistent DNA methylation or histone modification guided by ncRNAs, plants can retain stress tolerance traits across generations (Ramakrishnan et al., 2022).

5.10 TARGET INTERACTIONS OF ncRNA-MEDIATED STRESS TOLERANCE

In order to adapt their growth and development to stressful environmental conditions, plants have evolved a number of processes. Recent studies have demonstrated that under such adverse circumstances, ncRNAs express themselves differently. These ncRNAs either directly respond to stress or regulate gene expression in associated cellular networks. The effects of miRNAs, siRNAs and lncRNAs on plants' responses to abiotic stress were the main topics of this section (Jha et al., 2020).

5.10.1 miRNA-Target Interactions

Because miRNAs have a wide range of functions, they have emerged as potential targets for enhancing plant productivity and abiotic stress tolerance. It seems that miRNAs' functions in the abiotic stress response are what regulate them. Target genes that may be negative regulators of stress tolerance (such as stress-responsive gene repressors) may be downregulated by stress-upregulated miRNAs. Furthermore, the accumulation of their target gene mRNAs due to the downregulation of miRNAs during stress may favourably modulate stress tolerance. The common important functions of miRNAs and the challenges of methodically characterising these miRNAs in response to abiotic stress in plants were the main topics of this section. Different abiotic stressors either increase or decrease miRNA expression. In *Arabidopsis*, there was an up-regulation of miR393, miR397b and miR402. Different abiotic stressors either increase or decrease miRNA expression. Under dehydration stress, *Arabidopsis* showed up-regulation of miR393, miR397b and miR402 and downregulation of miR319c and miR389a (Karimi et al., 2023).

Similarly, 17 and 16 miRNAs were repressed and increased, respectively, in rice seedlings under drought stress. Eight and thirteen miRNAs were upregulated and downregulated in the leaves of maize under drought stress, whereas seven miRNAs were upregulated and seven more were downregulated in the roots. Conversely, drought stress boosted the expression of miR156, miR160 and miR162, miR164 and miR171, which cause their target genes' expression to decline. ARF genes are targeted by miR160. ARF17 functions as a gene regulator that is almost auxin-responsive, whereas ARF16 is said to be important in the development of root cap cells. By suppressing plant development, downregulating ARF16 and ARF17 will make maize more resilient to drought stress. Similarly, nuclear factor Y (NFY), which controls the expression of nitrate transporters AtNRT2.1 and AtNRT1.1, accumulated as a result of downregulating miR169. Furthermore, a single plant's abiotic stress response may cause one miRNA to be simultaneously up- and downregulated. Various publications have reported both elevated and inhibited expression of miR398a/b in the *Medicago truncatula* in response to drought stress (Kaur et al., 2024).

Similarly, in maize, miR156, miR164 and miR171 showed distinct regulatory trends under drought stress, whereas in two separate studies of rice under drought stress, miR156 and miR171 showed opposing regulatory patterns. Furthermore, two distinct plant species have been shown to exhibit differing miRNA expression under the same abiotic stressors. In response to drought stress, miR159 expression was downregulated in maize and *Arabidopsis* but upregulated in rice and wheat, whereas miR399 expression was downregulated in rice but upregulated in maize and wheat. Enhancements were made to miR168, miR397 and miR188, which were inhibited in maize and in *Arabidopsis*. The disparity in the expression of these miRNAs, which may be caused by variations in the intensity of abiotic stress and/or other plant growth circumstances, needs further investigation. Plant species have evolved to conserve some miRNAs for stress adaptation. Several miRNAs were upregulated in *Arabidopsis* and maize in response to nitrogen (N) shortage, including miR169, miR171, miR395, miR397, miR398, miR399, miR408 and miR827. Similarly, in response to drought stress, miR166, miR172, miR397, miR827 and miR1432 were all downregulated in maize, *Arabidopsis*, rice and wheat, whereas miR160, miR162, miR395 and miR827 were all upregulated (Božić et al., 2024; Zhang et al., 2017).

Certain miRNAs exhibit tissue-dependent expression patterns, according to numerous studies. Peach roots showed higher expression of miRNAs than leaves under drought stress, suggesting that roots are more vulnerable to drought than leaf tissue. While six miRNAs (miR159, miR172, miR319, miR399, miR528 and miR4393) were increased in leaves but inhibited in roots under drought stress, miR166 was downregulated in maize leaves and upregulated in roots. Similarly, drought stress produced four miRNAs (hvu-miR156a, hvu-miR166, hvu-miR171 and hvu-miR408) in barley leaves, while hvu-miR166 was repressed in barley roots, while the other three remained unaltered (Eldem et al., 2012; Gómez-Martín et al., 2023).

Variations in the expression patterns of the same miRNAs are also influenced by the treatment strategy used for abiotic stressors in plants. For example, the real field drought and the drought stress simulated by polyethylene glycol (PEG) were very different. PEG treatment increased the expression of miR398 in maize, while natural dry conditions caused it to be downregulated. In maize leaves, PEG treatment increased miR159 and miR166, whereas field-simulated dryness decreased both miRNAs. While 16 and 24 hours of PEG-simulated therapy upregulated and downregulated miR156, respectively, moderate and severe field drought downregulated and upregulated miR156. In both leaves and roots, field-simulated drought stress increased miR399 expression, whereas PEG-simulated treatment decreased miR399 expression. These results imply that the effects of field and PEG-simulated drought on miRNA expression are not entirely clear. Remarkably, certain miRNAs react to different abiotic stressors in a general way. Oxidative stress, salt, ABA and elevated Cu^{2+} all decreased the expression of miR398 in *Arabidopsis*, whereas drought and ABA stress suppressed the expression of miR169a and miR169c. In contrast, salt, drought, cold, ABA and UV-B stressors all caused *Arabidopsis* to express miR393. Additionally, some miRNAs react differently to particular abiotic stressors. Salt, cold and UVB irradiation all increased miR169 in *Arabidopsis*, but dryness, heat and ABA treatment inhibited it (Gómez-Martín et al., 2023; Zhang et al., 2022a). Few studies have been done on non-coding miRNAs that are involved in plant abiotic stress responses, and these are combined in Table 5.1.

TABLE 5.1
Non-Coding miRNAs Involved in Plant Abiotic Stress Responses

Abiotic Stress	Factors	ncRNA Type	Name	Plants	Function	References
Drought/nutrient deficiency	Water scarcity/ nutrient limitation	miRNA	miR156/ miR395	*Arabidopsis*	Regulates drought-responsive gene expression	Shriram et al. (2016)
Drought/oxidative stress	Water scarcity/ ROS accumulation	lncRNA	FPS/OXLR1	Maize	Regulates stomatal closure	Zheng et al. (2019)
Drought	Water scarcity	siRNA	nat-siRNA	*Arabidopsis*	Regulates drought-responsive gene expression	Khraiwesh et al. (2012)
Salt stress	Ion toxicity	miRNA	miR169	Rice	Targets NF-YA transcription factor	Ding et al. (2020)
Salt stress	Ion toxicity	snoRNA	snoR14	*Arabidopsis*	Regulates alternative splicing	Wu et al. (2021)
Salt stress/ oxidative stress	Ion toxicity/ROS accumulation	tRF	tRF5	Wheat	Regulates antioxidant gene expression	Li et al. (2020)
Heat stress/ oxidative stress	Temperature/ROS accumulation	miRNA	miR398	*Arabidopsis*	Regulates antioxidant gene expression	Guan et al. (2013)
Heat stress	Temperature	lncRNA	DR2	Rice	Regulates heat stress-responsive gene expression	Zhang et al. (2018)
Cold stress	Temperature	miRNA	miR172	*Arabidopsis*	Regulates cold-responsive gene expression	Liu et al. (2018)

5.10.2 siRNA-Target Interactions

The involvement of siRNAs in plant responses to abiotic stress has been recently investigated. In *Arabidopsis*, nat-siRNA pairing of SRO5 and pyrroline 5-carboxylate dehydrogenase (P5CDH) regulated proline metabolism, thereby attenuating ROS induced by high salt stress. TASIRNOV TAS1, TAS2 and TAS3 levels are samples of hyperxy naps treated with hypoxia, which significantly increase the functions of stress reactions. It is shown that these changes at the Tajilna level are correlated with the miRNA TAS level (MIR173 and MIR390). Repeated pentipeptide (PPR) is a TAS2 target, and the abolition of these targets is associated with the protection of low-sex mitochondria. Targets targeted by TAS1, which are led by heat, are generally guided by the conditions of Targets targeted by TAS1 thermal response. Overexpression of TAS1 downregulates HTT gene expression, thereby inducing decreased thermotolerance, whereas overexpression of the HTT gene upregulates HSF genes, thereby enhancing thermotolerance. Transcription of the HTT gene was induced in thermotolerant transgenic plants overexpressing the HSFA1a gene. It has been reported that this HSFA1a gene activates the expression of HTTs by binding to their promoters. In *Arabidopsis*, HTT1 interacts with HSP70-14 and HSP40. Taken together, these results suggest that HTT1 is a cofactor of HSP70-14 complexes, mediates thermotolerance pathways targeted by TASla and is triggered by HSFA1a. On the other hand, miR828-regulated ta-siRNA (TAS4) transcription is essential for anthocyanin biosynthesis in response to phosphate deficiency in *Arabidopsis*. Induction of TAS4-siR81(–) led to anthocyanin accumulation under nitrogen-starvation conditions. TAS4-siR81(–) targets MYB transcripts essential for anthocyanin biosynthesis. In a recent study, a copia-like retrotransposon called ONSEN was activated in heat-stressed *Arabidopsis* seedlings, and furthermore, novel ONSEN insertions occurred in stressed plants defective in siRNA biogenesis via retrotransposition. These insertions were inherited by the next generation and conferred thermosensitivity to the adjacent genes, thus indicating that the siRNA-directed pathway plays an important role in controlling transgenerational retrotransposition induced by abiotic stress. Research studies on Chinese cabbage suggested that some chloroplast small RNAs (csRNAs) and nat-siRNAs are sensitive to heat stress. Differential expression analysis in *Brassica rapa* showed that nat-siRNAs derived from 12 cis-NATs were sensitive to heat stress, with transcripts generating heat-responsive nat-siRNAs being upregulated, whereas transcripts from the opposite strand of the same locus were downregulated under heat stress. Similarly, the induced gene induced by slaughtering ABA dehydration has been reported to instruct the synthesis of endogenous miRNA, which acts on the tolerance of dehydration of Cratystigma. Analysis of wheat sowing northern blot has shown that the expression of four miRNA has changed significantly in response to cold, heat, salt and dehydration tension. Heating and NACL tension have been deleted by 002061_0636_3054.1 siRNA and 005047_0654_1904.1 siRNA. These genes targeted for miRNA may be important in adjusting the stress reactions of wheat. In *Arabidopsis*, transgenerational responses are based on hcsiRNA-mediated DNA methylation during abiotic stress. In addition, modifications triggered by histone deacetylases 2C (HD2C) and HDA6 contribute to water deficit and ABA responses. The re-activated transfer RDDM was targeted for HDA6, and HDA6 and HD2C interchanged with methyltransferase Met1 and ATDNMT2, respectively. At the same time, the relationship between histone acetyltransferase and methyltransferase indicates that HDA6 and HD2C are part of the sncRNA regulator path involving ABA deficiency and reactions. Other studies reported that ta-siRNA TAS3a-50D6(+) regulates the auxin signalling pathway by directing the degradation of ARF, thereby playing a key role in the adaptation of wheat to cold stress. Similarly, three TAS3-derived ta-siRNAs were shown to target cassava ARF under cold stress. Thus, the response of ta-siRNA to cold stress correlates with auxin signalling in plants. In cassava, two nat-siRNAs were differentially expressed in response to cold stress. One of them encoded a NAC (no apical meristem) domain transcriptional regulator involved in plant responses to cold.

Table 5.2 represents the different types of siRNA, targets, and roles in plant abiotic stress tolerance.

5.10.3 lncRNA-Target Interactions

Some plant lncRNAs participate in the response to abiotic stress through mimicry. They act as competitive endogenous RNAs (ceRNAs) targeting miRNAs, thus blocking the interactions of miRNAs with their targets. Under the condition of D, the P lncRNA IPS1 miRNA399 defects, and the miRNA399 is blocked from the interaction with the target phosphoric acid 2 (PHO2), which leads to the weakening of the suppression of the mir399. The similar role of RNA with similar imitation is recorded in rice, with a shortage of P. *Populus trichocarpa* reports PTC-MIR482A.1, PTC-MI476A and PTC-MI156C as a sensitive target for drought imitation. It is the anti-era of vegetable interactions with the sensual mRNA during reactions to non-biological stress. Such an interaction leads to the formation of double-stranded RNA double chains and thus affects the expression of the chain on the other side. Five antisense lncRNAs were found in poplar under nitrogen deficiency conditions. Similar findings have also been reported in maize under drought stress and in *Arabidopsis* under light and heat stress conditions. Moreover, some plant ncRNAs respond to abiotic stress through chromatin remodelling. Expression of COOLAIR and COLDAIR represses FLC in cold-exposed *Arabidopsis* through lncRNA (lncR2Epi)-mediated chromatin modifications. COOLAIR promotes the reduction of H3K36me3 or H3K4me2 at the FLC, while COLDAIR combines with the PRC2 to promote the accumulation of H3K27me3 at the FLC. In *Arabidopsis*, the expression of 40–50% of light-responsive sense/antisense transcripts associated with

TABLE 5.2
siRNA Types, Targets, and Roles in Plant Abiotic Stress Tolerance

siRNA Type	Stress	Target Genes	Role in Stress Response	Effect on Plant
nat-siRNA (SRO5)	High salt	*P5CDH* (Chinnusamy et al., 2007)	Regulates proline metabolism to mitigate ROS induced by salt stress	Enhances salt stress tolerance by reducing ROS
TAS1, TAS2, TAS3 ta-siRNAs	Hypoxia	*PPR mRNAs* (Howell et al., 2007)	Targets PPR mRNAs, aiding mitochondrial defence during hypoxic stress	Protects mitochondria under low oxygen conditions
HTT1 and HTT2 (TAS1 targets)	Heat	*HTT1, HTT2, HSP70-14, HSP40* (Liu et al., 2017)	Modulates thermotolerance via interactions with heat shock proteins and HSFA1a	Increases thermotolerance by enhancing the heat shock response
TAS4-siR81(−)	Phosphate deficiency	*MYB* (Liang et al., 2014)	Induces anthocyanin biosynthesis in response to nutrient deficiency	Accumulation of anthocyanin helps plants cope with low nutrient levels
ONSEN siRNA	Heat	Retrotransposon *ONSEN* (Masuta et al., 2017)	Controls retrotransposition activation under heat stress, with effects passed to the next generation	Provides heat responsiveness to nearby genes in progeny
csRNAs and nat-siRNAs	Heat	Various chloroplast-related genes (Liu et al., 2021)	Regulates chloroplast gene expression under heat stress conditions	Enhances heat tolerance by stabilising chloroplast function
nat-siRNAs (e.g., 002061_0636_3054.1)	Cold, heat, salt, dehydration	Genes associated with stress response (Waititu et al., 2020)	Regulates multiple stress responses, including cold and salt, by modifying gene expression	Adaptive responses to multiple abiotic stresses in wheat
hc-siRNAs	Abiotic stress	DNA methylation loci (Gautam et al., 2024)	Mediate DNA methylation in response to stress, affecting transgenerational response	Contributes to stress memory in *Arabidopsis*
TAS3-derived ta-siRNAs	Cold	*ARF* (Dotto et al., 2014)	Targets auxin-responsive factors in cassava under cold stress	Improves cold stress tolerance through auxin regulation
nat-siRNAs (cassava)	Cold	*NAC domain* transcriptional regulator (Li et al., 2017)	Modulates NAC domain regulators involved in cold response	Provides enhanced cold tolerance in cassava
siRNAs regulated by HD2C and HDA6	Water deficit, ABA response	DNA methyltransferases *MET1, AtDNMT2* (Contreras-Cubas et al., 2012)	Interacts with DNA methylation and histone deacetylases, affecting ABA and water deficit responses	Helps plants manage water stress by altering epigenetic markers

H3K9ac and/or H3K27ac was significantly upregulated under high light stress conditions. Thus, histone modifications and acetylation are important for light-responsive NAT expression changes. Some lncRNAs respond to environmental stresses through the RdDM pathway. In tomato, deletion of SlAGO4A, which encodes a key factor in the RdDM pathway, significantly increased tolerance to salt and drought stress compared with wild-type and transgenic plants overexpressing SlAGO4A. In response to heat stress, *Arabidopsis* rdr2 and dcl3 mutants showed reduced viability. In *Arabidopsis*, heat stress induces the expression of the vacuolar protein 4 (Vps4) gene, and increased expression of the same gene was observed in the nrpd2 mutant. The RdDM pathway regulates *Arabidopsis* MYB domain protein 74 (MYB74) under salt stress. Plant lncRNAs function as sRNA precursors during responses to abiotic stress. Nine lincRNAs functioning as precursors of 11 known miRNAs were identified in a *Populus* study under N deficiency, and another five lincRNAs were precursors of 14 novel miRNAs. In a study by Song et al., they reported 9687 novel ncRNAs and 50 ncRNAs functioning as miRNA precursors in *B. rapa* under cold and heat stress. Similar results were reported in *Phaeodactylum tricornutum* under P-deficient conditions, and in *Populus simonii*, two lncRNAs, Psi-lncRNA00020674 and Psi-lnc00201294, were suggested to be precursors of miRNA166a and miRNA27 by overlapping with the genomic regions that generate these miRNAs. In maize under water stress conditions and wheat (*Triticum aestivum*) under heat stress conditions, several lncRNAs were identified as miRNAs, siRNAs and shRNAs.

5.11 REGULATORY NETWORKS INVOLVING ncRNAs IN PLANT STRESS RESPONSES

ncRNAs play pivotal roles in the complex regulatory networks that govern plant responses to various stresses. These networks often involve intricate interactions with

TFs, hormonal pathways and other regulatory molecules. Multifaceted interactions enable plants to adapt to changing environmental conditions such as drought, salinity, extreme temperatures and other abiotic stresses. In this section, we examine how ncRNAs interact in these networks, focusing on their interactions with TFs, crosstalk with hormonal pathways and integration with other regulatory molecules.

5.11.1 Interaction with Transcription Factors

TFs are central regulators of gene expression and act as molecular switches that control numerous plant physiological processes, including stress responses. By binding to specific regions of DNA, they control the transcription of target genes, often in response to environmental signals. ncRNA, especially small (miRNA) and lncRNA, directly and indirectly affect the activity of the TF, which affects the transcription in stress conditions. Module miRNA-TF: miRNA often targets TFs involved in stress reactions. lncRNA–TF interactions: lncRNAs act as decoys or scaffolds to sequester TFs or recruit them to specific genomic regions. For example, the natural antisense transcript asHSFB2a in *Arabidopsis* regulates HSFB2a, a TF involved in heat stress response. By controlling the localisation and activity of HSFB2a, lncRNAs help fine-tune plant responses to high temperature.

5.11.2 Crosstalk with Hormonal Pathways

Plant hormones, such as ABA, auxin, gibberellin (GA), cytokinin and ethylene, are key signalling molecules in plant responses to abiotic stress. Each hormone affects specific stress-responsive genes and pathways, forming a tightly interconnected network that enables plants to adapt to complex environmental changes. ncRNAs are involved in hormone signalling cascades, adding another layer of regulation to hormone-mediated stress responses.

ABA plays a central role in responses to drought and salinity stress. The lncRNA DRIR is highly responsive to ABA and modulates the expression of ABA-regulated genes involved in stomatal closure, osmotic adjustment and other stress responses (Qin et al., 2017). Through this interaction, DRIR enhances ABA-mediated drought and salt tolerance. Ethylene and axin: MIR393 TIR1/AFB receptor miRNA etc. miRNA binds oxins with drought and physiological salt water. The MIR393, which regulates the transmission of axin signal, has an indirectly affected genetic network.

In the rice, a specific lncRNA adjusts the transmission of the Gibberlin signal, which affects the process, such as inhibiting drought growth. This regulation of GA levels by ncRNAs helps balance plant growth and stress responses, thereby increasing drought tolerance without compromising overall development.

5.11.3 Integration with Other Regulatory Molecules

Apart from TFs and hormones, other molecules like kinases, phosphatases and chromatin remodellers promote stress responses. ncRNAs are known to interact with these regulatory molecules, allowing plants to mount a coordinated response to stress. lncRNAs often serve as scaffolds, bringing together multiple proteins to form regulatory complexes. For instance, during heat stress, certain lncRNAs recruit chromatin-modifying complexes to activate heat-responsive genes.

miRNAs also regulate stress-dependent kinases and phosphatases, for example, miRNAs targeting SnRK1 (non-fermentable sucrose-related protein kinase) indirectly affect energy metabolism under flooding conditions by stabilising stress-related mRNAs. Epigenetic regulation: ncRNAs such as the lncRNAs COLDWRAP and COLDAIR mediate chromatin modification and target histone modifiers to control the expression of cold-responsive genes such as FLC, a key player in vernalisation. These interactions provide plants with mechanisms to 'remember' cold exposure and adjust flowering time accordingly (Azad et al., 2023).

ncRNAs play a key role in regulating signalling pathways that enable plants to sense and respond to abiotic stress. These pathways, which include calcium signalling, ROS signalling and hormone-induced responses such as ABA signalling, are essential for orchestrating adaptive responses at the cellular and molecular levels. Through complex interactions, ncRNAs influence gene expression patterns, regulate protein activities and modulate cross-communication among pathways, thereby enhancing plant resilience to environmental stressors (Das et al., 2023).

Calcium ions (Ca^{2+}) act as second messengers in various plant processes, including responses to drought, cold and salt stress. Fluctuations in cytosolic calcium levels are detected by calcium-binding proteins, such as calmodulin (CaM) and calcium-dependent protein kinases (CDPKs), which send signals to downstream targets. Recent studies suggest that miRNAs and lncRNAs modulate these calcium signalling responses under stress (Naz et al., 2024).

For example, under salt stress, miR398 targets the Cu/Zn superoxide dismutase (CSD) genes that are closely tied to calcium signalling. miR398 indirectly affects calcium signalling under high salinity conditions by regulating ROS detoxification pathways. Furthermore, certain ncRNAs, such as those upregulated in response to salinity and drought, interact with CDPK to fine-tune calcium-dependent responses and ensure that plants trigger appropriate cellular responses to prevent stress damage (Liu et al., 2023).

5.11.4 Reactive Oxygen Species (ROS) Signalling

ROS are generated as by-products of metabolic activity, dramatically increased under stress and function as signal transmission molecules that activate stress-responsive genes. However, excess ROS can damage cell components and requires a strictly adjusted balance. ncRNAs, especially miRNAs, play a key role in maintaining this balance by influencing antioxidant enzymes and ROS-generating proteins. For example, miR398 controls the expression of ROS-scavenging enzymes such as SOD and ascorbate peroxidase (APX) under drought and salt stress conditions. By controlling SOD levels, miR398 helps modulate ROS levels, enabling plants to mitigate oxidative damage without compromising ROS signalling. Moreover, under heat stress, the lncRNA Heat-responsive lncRNA (HRL) has been shown to interact with ROS-related TFs, providing a mechanism for modulating ROS levels through ncRNA-mediated gene regulation (Juan et al., 2021).

ABA is an important hormone in non-biological stress, especially in plant responses to drought and salt. ABA regulates processes such as the closure of the stomata, the expression of the genes and the permeating pressure adjustment so that plants can maintain water, so that they can survive in a hostile environment. ncRNAs affect ABA signalling at multiple levels, including direct regulation of ABA receptors and indirect regulation of ABA-responsive genes. In *Arabidopsis*, the miRNA miR159 targets MYB TFs that control ABA signalling, thereby reducing ABA sensitivity and preventing excessive stomatal closure, improving stress tolerance. Furthermore, dried lncRNA (lncRNA derived by drought) indicates that the ABA-sensitive gene is involved in the transmission of the ABA signal, regulates the expression of a sensitive ABA, improves drought and salt resistance and corrects the ABA cell level. It is done (Muhammad Aslam et al., 2022).

5.11.5 Other Hormonal Signalling Pathways

A network of hormonal crosstalk that synchronises plant adaptation to environmental stressors is created by ncRNAs, which also affect other hormones involved in stress responses, such as auxins, ethylene and GAs (Jiang et al., 2022). miRNAs such as miR393 affect auxin sensitivity and downstream gene networks linked to root architecture and growth suppression under stress by targeting auxin receptors like TIR1. Under water constraints, this miRNA-mediated control aids plants in maximising their growth (Chen et al., 2011). Ethylene is essential for controlling reactions to flooding and mechanical stress. Targeting AP2 TFs in the ethylene signalling pathway, miR172 modifies downstream reactions to maintain a balance between growth and stress response, especially in hypoxic or submerged environments (Khan et al., 2020).

GA pathways can also be impacted by ncRNAs, such as the genes in rice that are involved in GA production. Plants can alter their development and stress responses as a result of this control. This is particularly true during droughts, when reduced GA levels aid in energy and water conservation (Cao et al., 2024). Through these interactions, ncRNAs enable precise and intricate regulation of many pathways, contributing to hormone-mediated stress responses. The fact that ncRNAs modulate a variety of signal transduction pathways emphasises how crucial they are for plants to detect, react to and endure abiotic stress (Jha et al., 2020).

5.12 APPLICATIONS AND FUTURE PROSPECTS

ncRNAs, including miRNAs, small interfering RNAs (siRNAs) and lncRNAs, have emerged as important regulators of plant responses to abiotic stresses such as drought, salinity, extreme temperature and heavy metal toxicity. Their roles in gene regulation, signalling and metabolic pathways provide opportunities to improve stress tolerance in crop plants (Nemeth et al., 2024). Applications of ncRNAs in abiotic stress tolerance include the following:

i. Several specific ncRNAs are known to regulate genes involved in responses to drought, salinity and temperature. By upregulating or downregulating stress-responsive miRNAs or ncRNAs, scientists can develop crops that are better adapted to harsh environmental conditions. For example, miR393, which regulates ARFs, has been modified in rice to improve drought tolerance by altering root architecture and improving water uptake (Kim et al., 2024).
ii. Synthetic miRNAs designed to repress or activate stress-responsive genes may help plants tolerate environmental stresses. These synthetic ncRNAs could potentially be tailored to trigger adaptive responses such as increasing antioxidant production or altering stomatal aperture to improve water management under stress conditions (Pagano et al., 2021).
iii. Exogenous application of ncRNAs in biostimulant formulations is a novel non-transgenic strategy to enhance stress tolerance: spraying synthetic ncRNAs on crops can temporarily enhance specific stress responses. This application may be particularly useful in organic or non-GMO agricultural systems, flexibly and reversibly increasing stress tolerance without modifying the plant genome (Tiwari & Rajam, 2022).
iv. Stress-responsive ncRNAs can serve as biomarkers in breeding programmes, allowing breeders to identify plant varieties with natural stress tolerance, while miRNAs that respond strongly to stress (e.g., miR398, associated with oxidative stress resistance) can be used as indicators to select plants that are naturally resilient to certain environmental conditions (Gill et al., 2024).
v. By studying the expression profiles of ncRNAs under stress conditions, researchers can link specific ncRNA expression patterns to desirable phenotypes, such as drought-tolerant root structures or salt-tolerant ion transport systems. This linkage helps optimise breeding of resistant plant varieties (Bhogireddy et al., 2021).
vi. Stressed plants often show changes in ncRNA levels, and sensors that detect these changes in real time can be used to monitor plant stress conditions in the field. By following stress-related RNAs,

farmers can introduce timely interventions (for example, irrigation adjustments), reducing loss of agricultural crops and optimising resource use (Abdulraheem et al., 2024).

vii. With ncRNA biomarkers, plants under abiotic stress can be identified earlier than by visual symptoms, allowing preventive measures to be taken to avoid crop loss. For example, miRNAs upregulated during the early stages of drought stress can serve as early warning signals (Aina et al., 2024).

viii. CRISPR-Cas systems, especially the CRISPR-Cas13 variant that specifically targets RNA, can be used to edit or knock down stress-related ncRNAs with high precision. For example, modifying miRNAs involved in salt stress can enhance salt tolerance without affecting other traits, providing targeted resilience (Kordyś et al., 2022).

ix. ncRNA often participates in a crosstalk between different stress reactions. In other words, one ncRNA change can improve the resistance to some stress. CRISPR-CAS method enables scientists to selectively resist multipurpose stress. This is a very desirable feature for plants in climate change (Ahmad et al., 2024).

x. Breeders can focus on ARNC, including the ARNC regulatory network involved in stress reactions. ARNC interacts with specific routes such as ABA and ROS signals. By integrating these ARNCs into reproductive programmes, the accuracy of reproductive efforts increases, and especially resistant varieties with target stress can be developed (Cao et al., 2024).

xi. RNA adjusts the hormonal response (ABA, ethylene etc.) under restrictions, so that these ncRNA corrections can optimise the hormonal transmission route to improve the restoration of constraints. For example, miRNA, which regulates ABA signal transmission, can improve drought resistance by controlling the closure of the stomata and the loss of water (Zhang et al., 2022b).

xii. ncRNAs help plants adapt to long-term stress through epigenetic modifications and gene regulation that persist even after the stress subsides. This adaptability is critical for crops facing the unpredictable and persistent impacts of climate change (Xu et al., 2024).

xiii. ncRNAs with similar stress roles across species allow researchers to transfer stress tolerance traits between related plant species. For example, ncRNAs that confer drought tolerance in model plants like *Arabidopsis* could help develop similar strategies in food crops like rice and wheat (Ghorbanzadeh et al., 2022).

xiv. Advanced delivery methods, such as nanoparticles and biocompatible carriers, have been developed to deliver ncRNAs to plants to provide temporary or sustained protection to plants from stress. These systems aim to improve the uptake and stability of ncRNAs, thereby improving their efficacy in field applications. In sum, ncRNAs offer versatile, precise and sustainable tools for advancing plant stress tolerance and adapting crops to future agricultural challenges. As research progresses, these molecules hold the promise of transforming agricultural resilience through both biotechnological and natural breeding (Gil-Cabrerizo et al., 2024).

xv. Considering their roles in precise and efficient control of gene expression, signalling pathways and metabolic processes, the future prospects of ncRNAs in plant responses to abiotic stresses are broad and promising. As research progresses, ncRNAs are expected to play a key role in developing climate-tolerant crops to address challenges posed by increasing environmental stressors such as drought, salinity, extreme temperatures and heavy metal pollution (Yadav et al., 2024). Here are some potential directions for ncRNAs in abiotic stress response:

   a. ncRNAs that control overlapping responses to multiple stresses (e.g., drought and heat tolerance) may be valuable for breeding programmes aimed at producing plants that can thrive in diverse and unpredictable environmental conditions. These ncRNAs may be key to developing plants that can tolerate the complex effects of climate change (Talarico et al., 2024; Zhang et al., 2019).

   b. Some ncRNAs induce epigenetic changes that allow plants to 'remember' that they have been exposed to stress in the past. Future research could exploit this memory effect to develop plants that adapt more easily to repeated stress, resulting in more resilient crops in the next generation (Jin et al., 2024).

   c. Many ncRNAs exhibit conserved functions across plant species. As more ncRNAs are identified and their roles understood, cross-species applications will become possible, allowing stress tolerance traits to be transferred from model plants such as *Arabidopsis* to major crops such as rice, wheat and maize (Gill et al., 2024).

   d. Related wild plants often exhibit strong stress tolerance traits, and the study and isolation of ncRNAs from these species may provide a rich source of ncRNAs to enhance tolerance in cultivated crops, thereby expanding the genetic toolbox for sustainable agriculture (Zenda et al., 2021).

xvi. Combining AI (Artificial Intelligence) DRIR (drought induced lncRNA) and big data has enabled scientists to predict ncRNA functions and their interactions with other stress response pathways, accelerating the identification of potential targets for crop engineering. Machine learning algorithms could analyse ncRNA expression patterns under various stresses to identify the most effective molecules for promoting tolerance (Zhang et al., 2024). AI can also help model the behaviour of different ncRNAs under different environmental conditions, allowing researchers to develop crops that can adapt to specific climate zones and changing weather conditions (Balaska et al., 2023).

xvii. As climate change brings more frequent extreme weather events, developing crops that can withstand both sudden and persistent environmental changes will be critical. Modifications based on ncRNAs could help produce crops that not only survive, but thrive in high temperatures, saline soils and water-stressed regions (Ahmed et al., 2020).

xviii. Different regions face unique environmental challenges – for example, coastal regions may prioritise salt tolerance while semi-arid regions may require drought-tolerant crops – ncRNAs specific to these issues could contribute to the targeted development of climate-tolerant crops for different geographic regions (Raza et al., 2019).

5.13 CHALLENGES AND LIMITATIONS

Although the potential of ncRNAs to improve plant tolerance to abiotic stresses is widely recognised, there are significant challenges and limitations to using these molecules for practical applications. These obstacles include technical difficulties in research and development, as well as biological and regulatory considerations (Imaduwage & Hewadikaram, 2024).

Many ncRNAs have overlapping functions or interact with multiple signalling pathways. This redundancy can mask specific functional roles and make it difficult to determine which ncRNAs are essential for specific stress responses. Additionally, ncRNAs can interact with a network of signalling molecules, causing unexpected effects when manipulated. ncRNAs exhibit tissue-, stress- and even developmental-stage-specific expression. This vibration is complicated by the same ARNC, which can take different actions depending on the context, to normalise roles through various types and plants (Majumder et al., 2022).

Limited ncRNA databases prevent us from predicting and characterising ncRNA functions, making it difficult to identify promising ncRNAs for stress tolerance research. Unlike coding genes, functional characterisation of ncRNAs is more challenging because their mechanisms are often indirect and context-dependent. This limitation makes it difficult to identify ncRNAs with universal applications in stress tolerance (Micheel et al., 2021).

High-performance tools and tests are not yet available. This limits the speed that can be identified and verified by identifying the function ARNC. Some ARNCs cause epigenetic changes that may affect the signs and response of plants for a long time. However, the long-term impacts of these changes on plant health, growth and adaptability are not yet well understood, raising questions about the stability and reliability of ncRNA-induced tolerance. Companies and research institutions may be hesitant to invest in ncRNA technologies without clear pathways for bringing products to market (Fallet et al., 2023).

5.14 CONCLUSION

In recent years, ncRNAs have emerged as key regulators of plant adaptation to abiotic stresses such as drought, salinity, extreme temperatures and nutrient deficiencies. Unlike protein-coding genes, ncRNAs play a unique role in fine-tuning gene expression, allowing plants to respond rapidly and specifically to environmental challenges. This chapter reviewed various types of ncRNAs, such as miRNAs, siRNAs and lncRNAs and their involvement in gene regulation during abiotic stress responses. ncRNAs promote stress resistance by modulating critical molecular processes, including transcription, RNA processing and protein translation. They function within complex regulatory networks, interacting with TFs, hormones and other regulatory molecules to orchestrate plant responses. Specific pathways affected by ncRNAs, such as calcium, ROS and ABA signalling, highlight their universal role in orchestrating responses to different types of stress. However, the complex nature of these pathways also highlights the challenges of understanding and manipulating ncRNA function, as ncRNA molecules often exhibit redundancy, crosstalk and variable expression patterns depending on tissues, stresses and developmental stages. In the future, advances in high-throughput screening methods, computational modelling and functional validation are expected to broaden our understanding of ncRNAs in plant stress biology. Addressing current limitations will require not only technological innovations but also collaborative efforts between researchers, policymakers and industry stakeholders to develop frameworks for the safe and effective application of ncRNAs in agriculture. By overcoming these barriers, ncRNA-based approaches hold the potential to support sustainable agriculture and improve crop resilience, making them a promising frontier in plant biotechnology for abiotic stress tolerance.

REFERENCES

Abdulraheem, M. I., Xiong, Y., Moshood, A. Y., Cadenas-Pliego, G., Zhang, H., & Hu, J. (2024). Mechanisms of plant epigenetic regulation in response to plant stress: Recent discoveries and implications. *Plants*, *13*, 163.

Abideen, Z., Koyro, H.W., Huchzermeyer, B., Ahmed, M., Zulfiqar, F., Egan, T. and Khan, M.A., 2021. *Phragmites karka* plants adopt different strategies to regulate photosynthesis and ion flux in saline and water deficit conditions. *Plant Biosystems-An International Journal Dealing with all Aspects of Plant Biology*, *155(3)*,524–534.

Agrawal, N., Dasaradhi, P. V., Mohmmed, A., Malhotra, P., Bhatnagar, R. K., & Mukherjee, S. K. (2003). RNA interference: Biology, mechanism, and applications. *Microbiology and Molecular Biology Reviews*, *67*, 657–685.

Ahluwalia, O., Singh, P. C., & Bhatia, R. (2021). A review on drought stress in plants: Implications, mitigation and the role of plant growth promoting rhizobacteria. *Resources, Environment and Sustainability*, *5*, 100032.

Ahmad, Y., Haakim, Z., Iqbal, J., Abbasi, B. A., Mahmood, T., Iqbal, R., & Murtaza, G. (2024). Crosstalk between ncRNAs and stress signaling pathways unmasking the regulatory networks. In C. S. Prakash, M. Waseem, & S. Fiaz (Eds.), *ncRNAs: Mediated regulation: The unarguable biological function to environmental constrains*. Springer Nature Switzerland.

Ahmed, W., Xia, Y., Li, R., Bai, G., Siddique, K. H. M., & Guo, P. (2020). Non-coding RNAs: Functional roles in the regulation of stress response in Brassica crops. *Genomics*, *112*, 1419–1424.

Aina, O., Bakare, O. O., Fadaka, A. O., Keyster, M., & Klein, A. (2024). Plant biomarkers as early detection tools in stress management in food crops: A review. *Planta*, *259*, 60.

Anwar, K., Joshi, R., Dhankher, O. P., Singla-Pareek, S. L., & Pareek, A. (2021). Elucidating the response of crop plants towards individual, combined and sequentially occurring abiotic stresses. *International Journal of Molecular Sciences*, *22*, 6119.

Atta, K., Mondal, S., Gorai, S., Singh, A. P., Kumari, A., Ghosh, T., Roy, A., Hembram, S., Gaikwad, D. J., & Mondal, S. (2023). Impacts of salinity stress on crop plants: Improving salt tolerance through genetic and molecular dissection. *Frontiers in Plant Science*, *14*, 1241736.

Azad, M. F., Dawar, P., Esim, N., & Rock, C. D. (2023). Role of miRNAs in sucrose stress response, reactive oxygen species, and anthocyanin biosynthesis in *Arabidopsis thaliana*. *Frontiers in Plant Science*, *14*.

Balaska, V., Adamidou, Z., Vryzas, Z., & Gasteratos, A. (2023). Sustainable crop protection via robotics and artificial intelligence solutions. *Machines*, *11*, 774.

Banerjee, A., Roychoudhury, A., & Krishnamoorthi, S. (2016). Emerging techniques to decipher microRNAs (miRNAs) and their regulatory role in conferring abiotic stress tolerance of plants. *Plant Biotechnology Reports*, *10*, 185–205.

Bartas, M. (2024). Abiotic stresses in plants: From molecules to environment. *International Journal of Molecular Sciences*, *25*.

Bhogireddy, S., Mangrauthia, S. K., Kumar, R., Pandey, A. K., Singh, S., Jain, A., Budak, H., Varshney, R. K., & Kudapa, H. (2021). Regulatory non-coding RNAs: A new frontier in regulation of plant biology. *Functional & Integrative Genomics*, *21*, 313–330.

Bizzarri, A. R. & Cannistraro, S. (2021). Direct interaction of miRNA and circRNA with the oncosuppressor p53: An intriguing perspective in cancer research. *Cancers*, *13*.

Božić, M., Ignjatović Micić, D., Delić, N., & Nikolić, A. (2024). Maize miRNAs and their putative target genes involved in chilling stress response in 5-day old seedlings. *BMC Genomics*, *25*, 479.

Cao, W., Yang, L., Zhuang, M., Lv, H., Wang, Y., Zhang, Y., & Ji, J. (2024). Plant non-coding RNAs: The new frontier for the regulation of plant development and adaptation to stress. *Plant Physiology and Biochemistry*, *208*, 108435.

Chen, Z. H., Bao, M. L., Sun, Y. Z., Yang, Y. J., Xu, X. H., Wang, J. H., Han, N., Bian, H. W., & Zhu, M. Y. (2011). Regulation of auxin response by miR393-targeted transport inhibitor response protein 1 is involved in normal development in Arabidopsis. *Plant Molecular Biology*, *77*, 619–629.

Chinnusamy, V., Zhu, J., Zhou, T., & Zhu, J.-K. (2007). Small RNAs: Big role in abiotic stress tolerance of plants. In *Advances in molecular breeding toward drought and salt tolerant crops* (pp. 223–260). Dordrecht: Springer Netherlands.

Contreras-Cubas, C., Palomar, M., Arteaga-Vázquez, M., Reyes, J. L., & Covarrubias, A. A. (2012). Non-coding RNAs in the plant response to abiotic stress. *Planta*, *236*, 943–958.

Das, P., Agarwala, N., Gill, S. S., & Varshney, R. K. (2023). Emerging role of plant long non coding RNAs (lncRNAs) in salinity stress response. *Plant Stress*, *10*, 100265.

Delarue, M., Benhamed, M., & Fragkostefanakis, S. (2025). The role of epigenetics in tomato stress adaptation. *New Crops*, *2*, 100044.

Ding, Y., Huang, L., Jiang, Q., & Zhu, C. (2020). MicroRNAs as important regulators of heat stress responses in plants. *Journal of Agricultural and Food Chemistry*, *68*, 11320–11326.

Dos Santos, T. B., Ribas, A. F., De Souza, S. G. H., Budzinski, I. G. F., & Domingues, D. S. (2022). Physiological responses to drought, salinity, and heat stress in plants: A review. *Stresses*, *2*, 113–135.

Dotto, M. C., Petsch, K. A., Aukerman, M. J., Beatty, M., Hammell, M., & Timmermans, M. C. (2014). Genome-wide analysis of leafbladeless1-regulated and phased small RNAs underscores the importance of the TAS3 ta-siRNA pathway to maize development. *PLoS Genetics*, *10*, e1004826.

Eerdekens, N., Vaughan-Hirsch, J., & Van De Poel, B. (2024). The role of cytokinins during flooding stress in plants. *Environmental and Experimental Botany*, *228*, 105990.

Eldem, V., Çelikkol Akçay, U., Ozhuner, E., Bakır, Y., Uranbey, S., & Unver, T. (2012). Genome-wide identification of miRNAs responsive to drought in peach (*Prunus persica*) by high-throughput deep sequencing. *PLoS One*, *7*, 5.

El Sabagh, A., Islam, M. S., Hossain, A., Iqbal, M. A., Mubeen, M., Waleed, M., Reginato, M., Battaglia, M., Ahmed, S., & Rehman, A. (2022). Phytohormones as growth regulators during abiotic stress tolerance in plants. *Frontiers in Agronomy*, *4*, 765068.

Erdmann, R. M. & Picard, C. L. (2020). RNA-directed DNA methylation. *PLoS Genet*, *16*.

Fallet, M., Blanc, M., Di Criscio, M., Antczak, P., Engwall, M., Guerrero Bosagna, C., Rüegg, J., & Keiter, S. H. (2023). Present and future challenges for the investigation of transgenerational epigenetic inheritance. *Environment International*, *172*, 107776.

Fang, Y., Zheng, Y., Lu, W., Li, J., Duan, Y., Zhang, S., & Wang, Y. (2021). Roles of miR319-regulated TCPs in plant development and response to abiotic stress. *The Crop Journal*, *9*, 17–28.

Feng, D., Wang, R., Sun, X., Liu, L. N., Liu, P., Tang, J., Zhang, C., & Liu, H. (2023). Heavy metal stress in plants: Ways to alleviate with exogenous substances. *Science of the Total Environment*, *897*, 165397.

Gámez-Valero, A., Guisado-Corcoll, A., Herrero-Lorenzo, M., Solaguren-Beascoa, M., & Martí, E. (2020). Non-coding RNAs as sensors of oxidative stress in neurodegenerative diseases. *Antioxidants*, *9*, 1095.

Gautam, T. Gupta, S. and Sachan, M., (2024) Epigenomics for abiotic stress tolerance in cereals. In *Omics and system biology approaches for delivering better cereals. Journal of Ovarian Research*, 17.83.

Gelaw, T. A. & Sanan-Mishra, N. (2021). Non-coding RNAs in response to drought stress. *International Journal of Molecular Sciences*, *22*.

George, T. P., Subramanian, S., & Supriya, M. H. (2024). A brief review of noncoding RNA. *Egyptian Journal of Medical Human Genetics*, *25*, 98.

Ghorbanzadeh, Z., Hamid, R., Jacob, F., Asadi, S., Salekdeh, G. H., & Ghaffari, M. R. (2022). Non-coding RNA: Chief architects of drought-resilient roots. *Rhizosphere*, *23*, 100572.

Gil-Cabrerizo, P., Simon-Yarza, T., Garbayo, E., & Blanco-Prieto, M. J. (2024). Navigating the landscape of RNA delivery systems in cardiovascular disease therapeutics. *Advanced Drug Delivery Reviews, 208*, 115302.

Gill, S. S., Khan, N. A., Agarwala, N., Singh, K., Sunkar, R., & Tuteja, N. (2024). ncRNAs in plant development and stress responses. *Plant Physiology and Biochemistry, 214*, 108950.

Gómez-Martín, C., Zhou, H., Medina, J. M., Aparicio-Puerta, E., Shi, B., & Hackenberg, M. (2023). Genome-wide analysis of microRNA expression profile in roots and leaves of three wheat cultivars under water and drought conditions. *Biomolecules, 13*.

Gonzales, L. R., Blom, S., Henriques, R., Bachem, C. W. B., & Immink, R. G. H. (2024). LncRNAs: The art of being influential without protein. *Trends in Plant Science, 29*, 770–785.

Guan, Q., Lu, X., Zeng, H., Zhang, Y., & Zhu, J. (2013). Heat stress induction of mi R 398 triggers a regulatory loop that is critical for thermotolerance in a rabidopsis. *The Plant Journal, 74*, 840–851.

Han, J. J., Wang, X. Q., & Zhang, X. A. (2022). Functional interactions between lncRNAs/circRNAs and miRNAs: Insights into rheumatoid arthritis. *Frontiers in Immunology, 13*.

Heo, JB, Sung S. (2011) Vernalization-mediated epigenetic silencing by a long intronic noncoding RNA. *Science. 331*:76–9.

Howell, M. D., Fahlgren, N., Chapman, E. J., Cumbie, J. S., Sullivan, C. M., Givan, S. A., Kasschau, K. D., & Carrington, J. C. (2007). Genome-wide analysis of the RNA-DEPENDENT RNA POLYMERASE6/DICER-LIKE4 pathway in arabidopsis reveals dependency on miRNA-and tasiRNA-directed targeting. *The Plant Cell, 19*, 926–942.

Imaduwage, I. & Hewadikaram, M. (2024). Predicted roles of long non-coding RNAs in abiotic stress tolerance responses of plants. *Molecular Horticulture, 4*, 024–00094.

Jha, U. C., Nayyar, H., Jha, R., Khurshid, M., Zhou, M., Mantri, N., & Siddique, K. H. M. (2020). Long non-coding RNAs: Emerging players regulating plant abiotic stress response and adaptation. *BMC Plant Biology, 20*, 466.

Jiang, C., Wang, Y., He, Y., Peng, Y., Xie, L., Li, Y., Sun, W., Zhou, J., Zheng, C., & Xie, X. (2024). Identification and characterization of miRNAs and lncRNAs associated with salinity stress in rice panicles. *International Journal of Molecular Sciences, 25*, 8247.

Jiang, J., Zhu, H., Li, N., Batley, J., & Wang, Y. (2022). The miR393-target module regulates plant development and responses to biotic and abiotic stresses. *International Journal of Molecular Sciences, 23*.

Jin, Q., Chachar, M., Ali, A., Chachar, Z., Zhang, P., Riaz, A., Ahmed, N., & Chachar, S. (2024). Epigenetic regulation for heat stress adaptation in plants: New horizons for crop improvement under climate change. *Agronomy, 14*, 2105.

Juan, C. A., Pérez De La Lastra, J. M., Plou, F. J., & Pérez-Lebeña, E. (2021). The chemistry of reactive oxygen species (ROS) revisited: Outlining their role in biological macromolecules (DNA, lipids and proteins) and induced pathologies. *International Journal of Molecular Sciences, 22*.

Kar, M. M. & Raichaudhuri, A. (2021). Role of microRNAs in mediating biotic and abiotic stress in plants. *Plant Gene, 26*, 100277.

Karimi, M., Jafari, M., Shahvali, R., Ravash, R., & Shiran, B. (2023). Almond miRNA expression and horticultural implications. In *The almond tree genome*. Springer.

Kaur, S., Seem, K., Kumar, D., Kumar, S., Kaundal, R., & Mohapatra, T. (2024). Biogenesis to functional significance of microRNAs under drought stress in rice: Recent advances and future perspectives. *Plant Stress, 12*, 100447.

Khan, M. I. R., Trivellini, A., Chhillar, H., Chopra, P., Ferrante, A., Khan, N. A., & Ismail, A. M. (2020). The significance and functions of ethylene in flooding stress tolerance in plants. *Environmental and Experimental Botany, 179*, 104188.

Khraiwesh, B., Zhu, J.-K., & Zhu, J. (2012). Role of miRNAs and siRNAs in biotic and abiotic stress responses of plants. *Biochimica et Biophysica Acta (BBA)-Gene Regulatory Mechanisms, 1819*, 137–148.

Kim, J.-S., Kidokoro, S., Yamaguchi-Shinozaki, K., & Shinozaki, K. (2024). Regulatory networks in plant responses to drought and cold stress. *Plant Physiology, 195*, 170–189.

Kordyś, M., Sen, R., & Warkocki, Z. (2022). Applications of the versatile CRISPR-Cas13 RNA targeting system. *Wiley Interdisciplinary Reviews: RNA, 13*, 22.

Kryovrysanaki, N., James, A., Tselika, M., Bardani, E., & Kalantidis, K. (2022). RNA silencing pathways in plant development and defense. *The International Journal of Developmental Biology, 66*, 163–175.

Kumar, S. (2020). Abiotic stresses and their effects on plant growth, yield and nutritional quality of agricultural produce. *International Journal of Food Science and Agriculture, 4*, 367–378.

Li, J., Song, Q., Zuo, Z. F., & Liu, L. (2022). MicroRNA398: A master regulator of plant development and stress responses. *International Journal of Molecular Sciences, 23*.

Li, S., Yu, X., Lei, N., Cheng, Z., Zhao, P., He, Y., Wang, W., & Peng, M. (2017). Genome-wide identification and functional prediction of cold and/or drought-responsive lncRNAs in cassava. *Scientific Reports, 7*, 45981.

Li, X., Li, S., Wang, J., & Lin, J. (2020). Exogenous abscisic acid alleviates harmful effect of salt and alkali stresses on wheat seedlings. *International Journal of Environmental Research and Public Health, 17*, 3770.

Liang, C., Liu, X., Sun, Y., Yiu, S.-M., & Lim, B. L. (2014). Global small RNA analysis in fast-growing Arabidopsis thaliana with elevated concentrations of ATP and sugars. *BMC Genomics, 15*, 1–13.

Liu, Q., Yan, S., Yang, T., Zhang, S., Chen, Y. Q., & Liu, B. (2017). Small RNAs in regulating temperature stress response in plants. *Journal of Integrative Plant Biology, 59*, 774–791.

Liu, S., Zenda, T., Tian, Z., & Huang, Z. (2023). Metabolic pathways engineering for drought or/and heat tolerance in cereals. *Frontiers in Plant Science, 14*, 1111875.

Liu, W., Cheng, C., Chen, F., Ni, S., Lin, Y., & Lai, Z. (2018). High-throughput sequencing of small RNAs revealed the diversified cold-responsive pathways during cold stress in the wild banana (*Musa itinerans*). *BMC Plant Biology, 18*, 1–26.

Liu, X. X., Yu, X., Wang, H., & He, Y. K. (2021). Comparative profiling of small RNAs originated from two varieties of Chinese cabbage (*Brassica rapa*) with distinct leaf traits.

Ma, B., Wang, S., Wu, W., Shan, P., Chen, Y., Meng, J., Xing, L., Yun, J., Hao, L., Wang, X., Li, S., & Guo, Y. (2023). Mechanisms of circRNA/lncRNA-miRNA interactions and applications in disease and drug research. *Biomedicine & Pharmacotherapy, 162*, 114672.

Magar, N. D., Shah, P., Barbadikar, K. M., Bosamia, T. C., Madhav, M. S., Mangrauthia, S. K., Pandey, M. K., Sharma, S., Shanker, A. K., & Neeraja, C. (2023). Long non-coding RNA-mediated epigenetic response for abiotic stress tolerance in plants. *Plant Physiology and Biochemistry, 206*,108165.

Majumder, R., Ghosh, S., Das, A., Singh, M. K., Samanta, S., Saha, A., & Saha, R. P. (2022). Prokaryotic ncRNAs: Master regulators of gene expression. *Current Research in Pharmacology and Drug Discovery, 3*, 100136.

Mansoor, S., Ali Wani, O., Lone, J. K., Manhas, S., Kour, N., Alam, P., Ahmad, A., & Ahmad, P. (2022). Reactive oxygen species in plants: From source to sink. *Antioxidants*, *11*.

Masuta, Y., Nozawa, K., Takagi, H., Yaegashi, H., Tanaka, K., Ito, T., Saito, H., Kobayashi, H., Matsunaga, W., & Masuda, S. (2017). Inducible transposition of a heat-activated retrotransposon in tissue culture. *Plant and Cell Physiology*, *58*, 375–384.

Mattick, J. S., Amaral, P. P., Carninci, P., Carpenter, S., Chang, H. Y., Chen, L.-L., Chen, R., Dean, C., Dinger, M. E., Fitzgerald, K. A., Gingeras, T. R., Guttman, M., Hirose, T., Huarte, M., Johnson, R., Kanduri, C., Kapranov, P., Lawrence, J. B., Lee, J. T., …, & Wu J. E.2023). Long noncoding RNAs: Definitions, functions, challenges and recommendations. *Nature Reviews Molecular Cell Biology*, *24*, 430–447.

Matzke, M. A. & Mosher, R. A. (2014). RNA-directed DNA methylation: An epigenetic pathway of increasing complexity. *Nature Reviews Genetics*, *15*, 394–408.

Micheel, J., Safrastyan, A., & Wollny, D. (2021). Advances in non-coding RNA sequencing. *Noncoding RNA*, *7*.

Muhammad Aslam, M., Waseem, M., Jakada, B. H., Okal, E. J., Lei, Z., Saqib, H. S. A., Yuan, W., Xu, W., & Zhang, Q. (2022). Mechanisms of abscisic acid-mediated drought stress responses in plants. *International Journal of Molecular Sciences*, *23*.

Naz, M., Afzal, M. R., Raza, M. A., Pandey, S., Qi, S., Dai, Z., & Du, D. (2024). Calcium (Ca^{2+}) signaling in plants: A plant stress perspective. *South African Journal of Botany*, *169*, 464–485.

Nemeth, K., Bayraktar, R., Ferracin, M., & Calin, G. A. (2024). Non-coding RNAs in disease: From mechanisms to therapeutics. *Nature Reviews Genetics*, *25*, 211–232.

Neumeier, J. & Meister, G. (2021). siRNA specificity: RNAi mechanisms and strategies to reduce off-target effects. *Frontiers in Plant Science*, *11*.

Nguyen, N. H., Vu, N. T., & Cheong, J. J. (2022). Transcriptional stress memory and transgenerational inheritance of drought tolerance in plants. *International Journal of Molecular Sciences*, *23*.

Pagano, L., Rossi, R., Paesano, L., Marmiroli, N., & Marmiroli, M. (2021). miRNA regulation and stress adaptation in plants. *Environmental and Experimental Botany*, *184*, 104369.

Panahirad, S., Morshedloo, M. R., Ali, S., Hano, C., & Kulak, M. (2023). Secondary metabolites and their potential roles in plant tolerance against abiotic and biotic stress. *Plant Stress*, *10*, 100292.

Qasim, M., Abideen, Z., Adnan, M.Y., Gulzar, S., Gul, B., Rasheed, M. and Khan, M.A. (2017). Antioxidant properties, phenolic composition, bioactive compounds and nutritive value of medicinal halophytes commonly used as herbal teas. *South African Journal of Botany*, 110, 240–250.

Qiao, M., Hong, C., Jiao, Y., Hou, S., & Gao, H. (2024). Impacts of drought on photosynthesis in major food crops and the related mechanisms of plant responses to drought. *Plants*, *13*, 1808.

Qin, T., Zhao, H., Cui, P., Albesher, N. and Xiong, L. (2017). A nucleus-localized long non-coding RNA enhances drought and salt stress tolerance. *Plant Physiology*, *175*, 1321–1336.

Ramakrishnan, M., Zhang, Z., Mullasseri, S., Kalendar, R., Ahmad, Z., Sharma, A., Liu, G., Zhou, M., & Wei, Q. (2022). Epigenetic stress memory: A new approach to study cold and heat stress responses in plants. *Frontiers in Plant Science*, *13*.

Rashid, A., Schutte, B. J., Ulery, A., Deyholos, M. K., Sanogo, S., Lehnhoff, E. A., & Beck, L. (2023). Heavy metal contamination in agricultural soil: Environmental pollutants affecting crop health. *Agronomy*, *13*, 1521.

Ratti, M., Lampis, A., Ghidini, M., Salati, M., Mirchev, M. B., Valeri, N., & Hahne, J. C. (2020). MicroRNAs (miRNAs) and long non-coding RNAs (lncRNAs) as new tools for cancer therapy: First steps from bench to bedside. *Targeted Oncology*, *15*, 261–278.

Raza, A., Razzaq, A., Mehmood, S. S., Zou, X., Zhang, X., Lv, Y., & Xu, J. (2019). Impact of climate change on crops adaptation and strategies to tackle its outcome: A review. *Plants*, *8*, 34.

Sachdev, S., Ansari, S. A., Ansari, M. I., Fujita, M., & Hasanuzzaman, M. (2021). Abiotic stress and reactive oxygen species: Generation, signaling, and defense mechanisms. *Antioxidants*, *10*.

Sadura, I. & Janeczko, A. (2024). Are heat shock proteins important in low-temperature-stressed plants? A minireview. *Agronomy*, *14*, 1296.

Saeed, F., Shah, T., Hussain, Z., Rasul, S., & Manzoor, H. (2024). Interaction between ncRNAs and their response network in plants under multi-stress environment. In C. S. Prakash, M. Waseem, & S. Fiaz (Eds.), *ncRNAs: Mediated regulation: The unarguable biological function to environmental constrains*. Springer Nature Switzerland.

Sahu, P. K., Jayalakshmi, K., Tilgam, J., Gupta, A., Nagaraju, Y., Kumar, A., Hamid, S., Singh, H. V., Minkina, T., & Rajput, V. D. (2022). ROS generated from biotic stress: Effects on plants and alleviation by endophytic microbes. *Frontiers in Plant Science*, *13*, 1042936.

Samarfard, S., Ghorbani, A., Karbanowicz, T. P., Lim, Z. X., Saedi, M., Fariborzi, N., Mctaggart, A. R., & Izadpanah, K. (2022). Regulatory non-coding RNA: The core defense mechanism against plant pathogens. *Journal of Biotechnology*, *359*, 82–94.

Sarropoulou, E. & Fernández, I. (2023). Epigenetic regulation of gene expression by noncoding RNAs. In *Epigenetics in aquaculture* (pp. 65–93) Hoboken, NJ: John Wiley and Sons.

Shah, A. & Smith, D. L. (2020). Flavonoids in agriculture: Chemistry and roles in, biotic and abiotic stress responses, and microbial associations. *Agronomy*, *10*, 1209.

Sharma, Y., Sharma, A., MadhuShumaylaSingh, K., & Upadhyay, S. K. (2022). Long non-coding RNAs as emerging regulators of pathogen response in plants. *Non-Coding RNA*, *8*, 4.

Shelden, M. C. & Munns, R. (2023). Crop root system plasticity for improved yields in saline soils. *Frontiers in Plant Science*, *14*, 1120583.

Shi, B.-J. & Hussain, S. S. (2020). Toward elucidating the functions of miRNAs in drought stress tolerance. In *Plant Small RNA*. Elsevier.

Shriram, V., Kumar, V., Devarumath, R. M., Khare, T. S., & Wani, S. H. (2016). MicroRNAs as potential targets for abiotic stress tolerance in plants. *Frontiers in Plant Science*, *7*, 817.

Shoukat, E., Abideen, Z., Ahmed, M.Z., Gulzar, S. and Nielsen, B.L. (2019). Changes in growth and photosynthesis linked with intensity and duration of salinity in *Phragmites karka*. *Environmental and Experimental Botany*, *162*, 504–514.

Shoukat, E., Ahmed, M.Z., Abideen, Z., Azeem, M., Ibrahim, M., Gul, B. and Khan, M.A. (2020). Short and long term salinity induced differences in growth and tissue specific ion regulation of *Phragmites karka*. *Flora*, *263*,151550.

Shubham, D. & Ayan, R. (2022). Abiotic stress in plants. In K. Josphert Ngui (Ed.), *Advances in plant defense mechanisms*. IntechOpen.

Singh, A., Jain, D., Pandey, J., Yadav, M., Bansal, K. C., & Singh, I. K. (2023). Deciphering the role of miRNA in reprogramming plant responses to drought stress. *Critical Reviews in Biotechnology*, *43*, 613–627.

Sun, Z., Shu, L., Zhang, W., & Wang, Z. (2020). Cca-miR398 increases copper sulfate stress sensitivity via the regulation of CSD mRNA transcription levels in transgenic *Arabidopsis thaliana*. *PeerJ*, *8*, e9105.

Talarico, E., Zambelli, A., Araniti, F., Greco, E., Chiappetta, A., & Bruno, L. (2024). Unravelling the epigenetic code: DNA methylation in plants and its role in stress response. *Epigenomes*, *8*, 30.

Tiwari, R. & Rajam, M. V. (2022). RNA- and miRNA-interference to enhance abiotic stress tolerance in plants. *Journal of Plant Biochemistry and Biotechnology*, *31*, 689–704.

Waititu, J. K., Zhang, C., Liu, J., & Wang, H. (2020). Plant non-coding RNAs: Origin, biogenesis, mode of action and their roles in abiotic stress. *International Journal of Molecular Sciences*, *21*, 8401.

Wan, Y., Liu, J., Zhuang, Z., Wang, Q., & Li, H. (2024). Heavy metals in agricultural soils: Sources, influencing factors, and remediation strategies. *Toxics*, *12*, 63.

Wu, S., Wang, Y., Wang, J., Li, X., Li, J., & Ye, K. (2021). Profiling of RNA ribose methylation in *Arabidopsis thaliana*. *Nucleic Acids Research*, *49*, 4104–4119.

Wu, Y., Wang, X., Zhang, L., Zheng, Y., Liu, X., & Zhang, Y. (2023). The critical role of biochar to mitigate the adverse impacts of drought and salinity stress in plants. *Frontiers in Plant Science*, *14*, 1163451.

Xu, W.-B., Cao, F., Liu, P., Yan, K., & Guo, Q.-H. (2024). The multifaceted role of RNA-based regulation in plant stress memory. *Frontiers in Plant Science*, *15*, 1387575.

Yadav, A., Mathan, J., Dubey, A. K., & Singh, A. (2024). The emerging role of non-coding RNAs (ncRNAs) in plant growth, development, and stress response signaling. *Non-Coding RNA*, *10*, 13.

Yang, L., Lang, C., Wu, Y., Meng, D., Yang, T., Li, D., Jin, T., & Zhou, X. (2022). ROS1-mediated decrease in DNA methylation and increase in expression of defense genes and stress response genes in Arabidopsis thaliana due to abiotic stresses. *BMC Plant Biology*, *22*, 104.

Yang, Y., Liang, Y., Wang, C., & Wang, Y. (2024). MicroRNAs as potent regulators in nitrogen and phosphorus signaling transduction and their applications. *Stress Biology*, *4*, 38.

Zangouei, A. S., Zangoue, M., Taghehchian, N., Zangooie, A., Rahimi, H. R., Saburi, E., Alavi, M. S., & Moghbeli, M. (2023). Cell cycle related long non-coding RNAs as the critical regulators of breast cancer progression and metastasis. *Biological Research*, *56*, 1.

Zenda, T., Liu, S., Dong, A., & Duan, H. (2021). Advances in cereal crop genomics for resilience under climate change. *Life*, *11*.

Zhang, C., Tang, G., Peng, X., Sun, F., Liu, S., & Xi, Y. (2018). Long non-coding RNAs of switchgrass (*Panicum virgatum* L.) in multiple dehydration stresses. *BMC Plant Biology*, *18*, 1–15.

Zhang, F., Yang, J., Zhang, N., Wu, J., & Si, H. (2022a). Roles of microRNAs in abiotic stress response and characteristics regulation of plant. *Frontiers in Plant Science*, *13*, 919243.

Zhang, J. W., Long, Y., Xue, M. D., Xiao, X. G., & Pei, X. W. (2017). Identification of microRNAs in response to drought in common wild rice (*Oryza rufipogon* Griff.) shoots and roots. *PLoS One*, *12*.

Zhang, P., Wu, W., Chen, Q., & Chen, M. (2019). Non-coding RNAs and their integrated networks. *Journal of Integrative Bioinformatics*, *16*, 20190027.

Zhang, X., Ibrahim, Z., Khaskheli, M. B., Raza, H., Zhou, F., & Shamsi, I. H. (2024). Integrative approaches to abiotic stress management in crops: Combining bioinformatics educational tools and artificial intelligence applications. *Sustainability*, *16*, 7651.

Zhang, Y., Xu, J., Li, R., Ge, Y., & Li, Y. (2023). Plants' response to abiotic stress: Mechanisms and strategies. *International Journal of Molecular Sciences*, *24*.

Zhang, Y., Zhou, Y., Zhu, W., Liu, J., & Cheng, F. (2022b). Non-coding RNAs fine-tune the balance between plant growth and abiotic stress tolerance. *Frontiers in Plant Science*, *13*.

Zhao, C., Zhang, H., Song, C., Zhu, J.-K., & Shabala, S. (2020). Mechanisms of plant responses and adaptation to soil salinity. *The Innovation*, *1*, 100017.

Zhao, S., Zhang, Q., Liu, M., Zhou, H., Ma, C., & Wang, P. (2021). Regulation of plant responses to salt stress. *International Journal of Molecular Sciences*, *22*.

Zhao, X., Li, F., Ali, M., Li, X., Fu, X., & Zhang, X. (2024). Emerging roles and mechanisms of lncRNAs in fruit and vegetables. *Horticulture Research*, *11*.

Zhao, Z., Zang, S., Zou, W., Pan, Y. B., Yao, W., You, C., & Que, Y. (2022). Long non-coding RNAs: New players in plants. *International Journal of Molecular Sciences*, *23*.

Zheng, J., Zeng, E., Du, Y., He, C., Hu, Y., Jiao, Z., Wang, K., Li, W., Ludens, M., & Fu, J. (2019). Temporal small RNS expression profiling under drought reveals a potential regulatory role of small nucleolar rnas in the drought responses of maize. *The Plant Genome*, *12*, 180058.

6 Regulatory Role of ncRNAs in Plant Abiotic Stress Tolerance

Anil Namdev Kale, Yogesh Dashrath Naik, Sanjay Kumar Sheshma, Kumari Anjani, Karma Landup Bhutia and Vinay Kumar Sharma

6.1 INTRODUCTION

Drought, heat, cold, salinity and other abiotic stresses adversely affect the development and productivity of crop plants (Yadav et al., 2020). The plants exhibit sensitivity to such stresses and evolve a variety of strategies to cope with these challenges. The strategies basically involve morphological, physiological, biochemical and molecular alterations within plants (Kapoor et al., 2020). Among these alterations, molecular changes based key adaptations generally include the establishment of drought stress memory, synthesis of stress-responsive proteins and hormones, epigenetic modifications and variations in the levels of non-coding RNAs (ncRNAs) like consequences (Akbari et al., 2022; Mladenov et al., 2021). Notably, ncRNAs have been demonstrated to play a critical role in enhancing tolerance to abiotic stresses. Within the ncRNA category, structural ncRNAs are essential for protein synthesis, post-transcriptional modifications and regulation of gene expression, while regulatory ncRNAs significantly contribute in improving abiotic stress tolerance in agricultural plants (Dykes & Emanueli, 2017).

Short ncRNAs (sncRNAs) and long ncRNAs (lncRNAs) are two categories of regulatory ncRNAs. The primary classes of sncRNAs include microRNAs (miRNAs) and small or short interfering RNAs (siRNAs), which are essential for gene expression regulation through the RNA interference mechanism (Wang & Chekanova, 2016). The roles of various miRNAs in response to abiotic stresses have been extensively studied across multiple crop species (Akbari et al., 2022; Kumar et al., 2023). LncRNAs, typically greater than 200 nucleotides in length and synthesised by various RNA polymerases, are crucial for modulating the regulatory networks of genes. The secondary structures of lncRNAs can interact with DNA, RNA and proteins, thereby regulating gene expression in both cis and trans manners (Statello et al., 2021).

lncRNAs enhance or silence the expression of target genes during abiotic stress through various mechanisms, including functioning as scaffolds, guides, signals and competing endogenous RNAs (Zhang et al., 2022). Specific lncRNAs, such as DRIR and IPS1, modulate gene expression levels via these mechanisms (Verma et al., 2021). The regulation of gene expression by lncRNAs in response to abiotic stresses is also influenced by several chromatin modifications, including DNA methylation, histone modification and nucleosome remodelling (Yang et al., 2023). LncRNAs can act as recruiters for enzymes that facilitate epigenetic changes in the chromatin mediated by processes such as methylation, demethylation, acetylation and deacetylation. Notably, COLDAIR and APOLO are lncRNAs that play critical roles in modulating gene expression by recruiting enzymes involved in epigenetic modifications in response to abiotic stresses (Lucero et al., 2020).

The important role of ncRNA in enhancing abiotic stress tolerance attracts the attention of researchers. The sncRNA, such as tae-miR399, osa-miR12477, miR398, TCONS_00021861, LncR9A, lncRNA77580, helps in enhancing tolerance to the abiotic stresses in various crops (Chen et al., 2021; Lu et al., 2020; Parmar et al., 2020; Peng et al., 2021). Some of the sncRNA such as miR408 play an important role in enhancing the tolerance for more than one abiotic stresses (Ma et al., 2015). The regulatory functions of ncRNAs in increasing tolerance to abiotic stresses have also been documented in various species, including *Arabidopsis thaliana*, rice, maize, soybean, cotton and wheat.

Since the emergence of molecular breeding, protein-coding genes linked to specific agronomic traits have been prioritised as primary targets for crop improvement (Bi et al., 2009; Zhou & Luo, 2013). However, a great deal of emphasis on individual genes introduces significant limitations, as many agronomic traits are regulated by complex networks involving multiple genes. As an effective solution to address these challenges, research emphasis in recent times has increasingly focused on identifying regulatory genes including transcription factors that can modulate diverse regulatory networks influencing multiple traits (Zhang et al., 2023). While lncRNAs are essential for regulating various genomic activities, regulatory ncRNAs specifically target and modulate gene expression through a range of molecular mechanisms.

6.2 TYPES OF RNAs IN PLANTS

Coding and ncRNAs are the two broad categories of RNAs present in the plant cell. While coding RNA includes only mRNA that plays an important role in transferring the genetic information for synthesis of proteins, ncRNA is divided into two types: (a) structural ncRNAs and (b) regulatory ncRNAs (Zhang et al., 2019a). Structural ncRNAs

DOI: 10.1201/9781003583257-6

TABLE 6.1
Biogenesis Level Differences in miRNA and siRNA

Characteristics	miRNA	siRNA
Precursor	Single-stranded RNAs	Double-stranded RNAs
Dicer	DCL1	DCL2, DCL3 and DCL4
Length	21-nt	22-nt, 24-nt and 21-nt
Target	Multiple target	Specific with only one mRNA

include tRNA, rRNA, snRNA and snoRNA possessing indirect role in the protein synthesis (Zhang et al., 2019b). Regulatory ncRNAs is divided into two categories: (a) small ncRNA (sncRNAs) and (b) lncRNA. sncRNAs is shorter than 200 nucleotides and includes micro RNA (miRNA) and small or siRNA holding important role in RNA interference (Svoboda, 2020). Contrarily, lncRNAs are longer than 200 nucleotides and do not code for any polypeptide (Jha et al., 2020).

6.2.1 Short Non-Coding RNAs in Plants

Short ncRNAs (sncRNAs), typically less than 200 nucleotides in length, play a critical role in gene regulation by modulating gene expression at the transcriptional level and are collectively referred to as regulatory RNAs. In plants, miRNAs and siRNAs are the two primary classes of sncRNAs, both of which are essential for the regulation of gene expression (Shan et al., 2020). These sncRNAs are derived from long precursor RNAs, either single-stranded or double-stranded, through the enzymatic action of dicer. Despite their shared characteristics, miRNAs and siRNAs exhibit distinct biogenetic pathways (Bologna & Voinnet, 2014).

The biogenesis of sncRNAs involves distinct processing pathways mediated by different sets of enzymes (Shin & Shin, 2016). Endogenous miRNAs are derived from hairpin structures formed from long single-stranded RNA in the cytoplasm, originating as primary miRNAs (pri-miRNAs) within the nucleus. In contrast, siRNAs, which can be either endogenous or exogenous, are generated from double-stranded RNA in the cytoplasm by the action of the dicer enzyme. A comparative overview of the biogenesis pathways of miRNAs and siRNAs is presented in Table 6.1. These sncRNAs bind to the mRNA transcripts of target genes, leading to target mRNA degradation via the RNA-induced silencing complex (RISC), which ultimately results in the silencing of gene expression (Li et al., 2013).

The RISC along with Argonaute (AGO) proteins is responsible for altering gene expression. RISC complex along with guide RNA binds to the complementary strand of messenger RNA (mRNA), resulting in gene expression modification through two primary mechanisms: cleavage and transcriptional gene silencing (Shin & Shin, 2016). Distinct small RNAs, along with various AGO proteins, play crucial roles in these mechanisms (Table 6.2). Additionally, sncRNAs exhibit several characteristics that differentiate them from mRNA (Table 6.3).

6.2.2 Long Non-Coding RNAs in Plants

lncRNAs) are defined as transcripts longer than 200 nucleotides that do not encode full-length proteins. In plants, the biogenesis of lncRNAs involves various RNA polymerases, including RNA polymerases II, III, IV and V. The majority of lncRNAs are transcribed by RNA polymerase II and undergo post-transcriptional modifications, such as 5′ end capping, alternative splicing and the addition of 3′ poly-A tails, similar to mRNAs (Marquardt et al., 2014). Despite being transcribed by RNA polymerase II, lncRNAs exhibit variations in expression levels, size and expression specificity (Table 6.3). Transcripts produced by RNA polymerases IV and V play a crucial role in driving RNA-mediated DNA methylation (Wierzbicki et al., 2021).

The lncRNAs typically display secondary structures that are essential for their biological functions. They contain two primary types of functional sites: (a) interacting sites that facilitate specific interactions with RNA-binding proteins and (b) structural sites that define their secondary and tertiary structures, directing interactions with other

TABLE 6.2
Molecular Mechanism of RISC Mediated Gene Silencing

RNA and AGO Involved	Molecular Mechanism
miRNA- or 21-nt siRNA-loading AGO1	Slicing and degradation of target endogenous mRNA
24-nt siRNA-loading AGO4 and AGO6	Transcriptional gene silencing through heterochromatin formation
miRNA-loaded AGO1 and AGO10	Translational repression

TABLE 6.3
Comparison of Typical Characteristics of mRNAs, sncRNAs and lncRNAs

mRNAs	sncRNAs	lncRNAs
Synthesised by RNA pol II	Synthesised by RNA pol II and pol IV	Synthesised by RNA pol II, pol III, pol IV and pol V
Longer in size	<200 bp in size	> 200 bp in size
Consisting of more number of exons	No exons	Consisting of less number of exons
More constitutive expression	Most specific expression	Most specific expression
Expression level is high	Expression level is low	Expression level is low
All having ORF	Does not contain ORF	Some having ORF

molecular partners (Fabbri et al., 2019). In plants, lncRNAs are predominantly localised in the nucleus (Deforges et al., 2019; Wu et al., 2020). Various types of lncRNAs have been identified that play important roles in response to different abiotic stresses in species such as *Arabidopsis*, rice and maize (Kim et al., 2017; Zhang et al., 2014). The lncRNAs alter the gene expression level by interacting with the DNA, RNA and protein while following cis and trans type of gene regulation.

6.3 CLASSIFICATION OF LONG NON-CODING RNAs

Based on genomic location and the direction of transcription, lncRNAs are classified into four major categories. First one is long intergenic ncRNAs (lincRNAs) found present between the region of two genes and can be transcribed in both directions. Second one is intronic ncRNAs (incRNAs) found present in the intronic region of the single gene and can be transcribed by both directions. Natural antisense transcripts (NATs) are observed in the protein coding region of single genes and transcribed from the downstream site of the gene. Sense lncRNAs are transcribed from upstream region of the single protein coding genes.

Taking functional basis into consideration, lncRNAs can be classified as signal, scaffold, guide, decoy and ceRNA.

i. *Signal*: lncRNA can bind to transcription factors as signal molecules to participate in various regulatory reactions or take part in signalling pathways to further regulate the spatio-temporal expression of protein-coding genes (Swiezewski et al., 2009).
ii. *Decoy*: This type of regulatory mechanism reduces the expression level of miRNA by adsorbing miRNA to reduce the inhibition of mRNA and dynamically regulating the translation speed and stability of downstream target genes (Salmena et al., 2011).
iii. *Guide*: Many lncRNAs are in chromatin and can interact with proteins to promote or inhibit their binding activity in the target genomic region (Statello et al., 2021). lncRNA can guide RNA–protein complexes to bind to specific locations or recruit chromatin modifying enzymes to target genes either in cis or trans form (Wang et al., 2018).
iv. *ceRNA*: It strands for competitive endogenous RNA. The lncRNAs can indirectly regulate the interaction between miRNAs and their target mRNAs via binding to miRNAs. When ceRNA is transcribed, it makes competition with the silencing complex mediated by miRNA (RISC) to repress miRNA function and increase the expression of target mRNAs. Thus, potentially positive regulation exists between ceRNA and miRNA pairs (Thomson & Dinger, 2016).

6.4 LONG NON-CODING RNA AS MAJOR PLAYER IN GENE REGULATION

Gene expression is regulated by ncRNAs across multiple levels of controls. The molecular mechanisms involved in the regulation of genes with the help of ncRNAs are DNA methylation, histone modification and chromatin remodelling (Heo & Sung, 2011). Through interactions with DNA, RNA and proteins, lncRNAs can modulate chromatin structure and function, thereby influencing the transcriptional activity of both proximal and distal genes (Statello et al., 2021). Additionally, lncRNAs play a critical role in RNA splicing, stability and translation.

The lncRNAs play a vital role in regulating the gene expression via interaction with the chromatin and with DNA. De-condensation of the closed chromatin network is important for the expression of particular gene. The lncRNAs interact with the histone-modifying enzyme involved in histone acetylation, histone methylation and histone de-acetylation or act as scaffold to recruit such enzymes that help in making epigenetic changes in the histone to open the closed condensed structure of the histone (Jain et al., 2016; Sun et al., 2016). Histone modification by histone methylation at H3K4, H3K36 and H3K79 promote gene transcription, whereas H3K9, H3K27 and H4K20 inhibit gene transcription (Kouzarides, 2002). During the vernalisation stress, two lncRNAs, namely, COOLAIR and COLDAIR, play an important role. These lncRNAs modulate the expression of Flowering Locus C (FLC) at the transcriptional level with the help of histone modification (Heo & Sung, 2011; Sun et al., 2013). The transcription of COOLAIR occurs in antisense direction with FLC, while COLDAIR is transcribed from the intronic region of FLC in sense direction. PHD-PRC2 (plant homeo domain Polycomb Repressive Complex 2) complex is recruited by these lncRNAs and this complex allows histone modification of FLC by epigenetic regulation in *Arabidopsis.*

Auxin-regulated promoter loop (APOLO) 80 is responsible for the activation of auxin-responsive genes in *A. thaliana*. APOLO and auxin target genes are normally silenced by H3K27me3 and the presence of chromatin loops maintained by the Polycomb factor like heterochromatin protein 1 (LHP1). Following transcriptional activation of APOLO in response to auxin, the lncRNA recognises specific motifs at the promoters of its target genes, where it binds and generates R-loops that act as decoys of LHP1, thereby allowing target-gene expression (Ariel et al., 2020). Histone acetylation is regulated by histone acetyltransferases (HATs) and histone deacetylases (HDACs), specifically at lysine residues such as H3K4, H3K9, H3K27, H3K56 and H4K16. The enzymatic activity of HATs promotes gene transcription activation by acetylating histone lysine residues, thereby neutralising their positive charges and decreasing the affinity of histones for DNA. In contrast, HDACs facilitate histone deacetylation by removing acetyl groups, which is typically correlated with transcriptional repression. The lncRNAs play an important role in such kinds of histone modification. Furthermore, lncRNAs also act as scaffold in histone modification that binds two or more protein molecules and convert inactive protein into activate form in order to perform subsequent specific biological functions (Wang et al., 2018). Some lncRNAs display major role in the RNA-directed DNA methylation (RdDM). In plants, the involvement of lncRNAs in the histone modification and DNA methylation has been observed.

6.5 ROLE OF ncRNA IN ABIOTIC STRESS TOLERANCE

The sncRNAs and lncRNAs play critical roles in abiotic stress tolerance. Specifically, sncRNAs such as miRNAs are essential for the regulation of gene expression in response to various abiotic stresses in crop plants. The involvement of various miRNAs in abiotic stress tolerance has been documented across multiple crop species (Table 6.4). Among the different types of abiotic stresses, salinity is a prominent factor that negatively affects overall plant growth by inducing ionic imbalance, osmotic stress, oxidative stress and inhibiting nutrient uptake (Atta et al., 2023). To elucidate the role of miRNAs in salt tolerance, a transcriptomic study of shoots and roots was conducted on contrasting rice genotypes, revealing that the novel up-regulated osa-miR12477 regulates the expression of its target gene, L-ascorbate oxidase, which is crucial for salt tolerance in rice (Parmar et al., 2020). Additionally, the novel miRNA species miR164a has been identified in maize, where it enhances the expression levels of target genes GRMZM2G114850 and GRMZM2G008819, contributing to increased salt tolerance (Shan et al., 2020).

Drought represents a major abiotic constraint on agricultural production, significantly reducing crop yields. As a principal limiting factor for plant growth, drought interrupts essential physiological processes, including respiration, photosynthesis and stomatal dynamics, ultimately

TABLE 6.4
Target of Different sncRNAs under Abiotic Stresses

Crop	sncRNA	Target Genes	Abiotic Stress	References
Arabidopsis	*miR408*	*Plantacyanin, Cupredoxin, Laccase, Uclacyanin*	Salinity, cold and oxidative stress tolerance	Ma et al. (2015)
	miR398	*CSD1, CSD2* and *CCS*	Heat tolerance	Guan et al. (2013)
Rice	*osa-miR12477*	*L-ascorbate oxidase*	Salt tolerance	Parmar et al. (2020)
	miR531, miR827, miR8175, miR977, miR6300, miR1861, miR440, miR9773, miR3982 and *miR1876*	*NAC, MYB, MYC, ARF*	Drought stress tolerance	Singh et al. (2020)
	Osa-MIR2919	*LOC_Os01g72834, LOC_Os01g73090, LOC_Os01g69990, LOC_Os01g61880, LOC_Os01g69920, LOC_Os01g65920*	Negative regulator of drought tolerance	Kumar et al. (2023)
	miR169r-5p	*LOC_Os12g42400*	Heat tolerance	Liu et al. (2017)
Winter wheat	tae-miR399	*TaUBC24*	Freezing tolerance	Peng et al. (2021)
Chickpea	*miR319, miR393, miR394* and *miR159*	-	Cold tolerance	Akbari et al. (2022)
Maize	miR164a	*GRMZM2G114850* and *GRMZM2G008819*	Salt tolerance	Shan et al. (2020)

compromising plant growth and metabolic functions (Yang et al., 2021). Various studies have been conducted to analyse the effect of snRNA in crop plants (Table 6.4). A small RNA sequencing in rice under normal and drought stress condition at panicle initiation stage identified 10 new miRNAs for drought tolerance which targeted transcription factor such as NAC, MYB, MYC and ARF, helping rice plants in enhancing drought tolerance (Singh et al., 2020). Small RNA sequencing for drought stress in rice has identified Osa-MIR2919 as the negative regulator of drought tolerance (Kumar et al., 2023).

Heat stress in also a major abiotic stress with negative impact that adversely affects the yield of crop plants. A small RNA sequencing in two contracting genotype of rice identified the negative correlation between a new miR169r-5p targeting LOC_Os12g42400 gene.

The over-expression of miR169r-5p at flowering stage in rice helps the plants in improving heat stress tolerance (Liu et al., 2017). Another study indicates that the miR398 up-regulation during heat stress reduces expression level of target genes (CSD1, CSD2 and CCS) that helps in enhancing the thermo-tolerance in *Arabidopsis* plants (Guan et al., 2013).

Cold stress responses in plants involve sophisticated biochemical processes that alter cellular composition to protect against damage induced by low temperatures. Furthermore, cold stress markedly impacts plant morphology, resulting in growth inhibition and reduced yields (Eremina et al., 2016). The regulatory roles of various sncRNAs have been studied in several crops (Table 6.4). Chilling stress is a major abiotic stress in the winter wheat. The enhanced expression level of the tae-miR399 and its targeting gene TaUBC24 has been found playing an important role in enhancing freezing tolerance in the winter wheat (Peng et al., 2021). Chickpea is another winter crop in which yield reduction occurs due to cold stress. Four promising miRNAs, namely, miR319, miR393, miR394 and miR159 playing important role in cold tolerance have been identified during cold stress in the two contracting genotypes of chickpea (Akbari et al., 2022). Some of the miRNAs also play an important role in enhancing tolerance to various abiotic stress at a time. The role of miR408 has been studied under various abiotic stresses in the *Arabidopsis* and it has found that the enhanced level of miR408 plays an important role in enhancing the tolerance to salinity, cold and oxidative stress (Ma et al., 2015).

Several lncRNAs are known to play significant roles in enhancing abiotic stress tolerance in crop plants (Table 6.5). One such drought-responsive lncRNA, designated as drought-induced lncRNA (DRIR), regulates the FUT4 gene, which encodes fucosyl transferase in *A. thaliana*, thereby contributing to improved tolerance to both drought and salt stress (Qin et al., 2017). Additionally, the lncRNA TCONS_00021861 has been identified as crucial for drought tolerance in rice (Chen et al., 2021). The lncRNA MSTRG.28732.3 interacts with the genes Os02g0662700, Os02g0663100 and Os06g0105350 in response to drought stress in contrasting rice genotypes, playing a vital role in drought tolerance through its interaction with miR171 (Yang et al., 2022). To further elucidate the function of drought-responsive lncRNAs, a transgenic soybean line over-expressing lncRNA77580 has been developed. This genetic modification is associated with increased yield in soybean, primarily due to a higher number of seeds per plant (Chen et al., 2023).

The ceRNA-based mechanism of gene regulation involving lncRNAs plays a crucial role in the tolerance to various abiotic stresses. Numerous triplet interactions between lncRNAs, miRNAs and target genes have been known across different crops. The ceRNA mechanism is particularly significant in enhancing cold tolerance in crop plants. Transgenic approaches have been employed to investigate the function of lncR9A in *Brachypodium distachyon*, demonstrating that lncR9A operates through a ceRNA mechanism. This interaction ultimately results in improved cold tolerance in this crop species (Lu et al., 2020). Additionally, a novel lncRNA, CIL1, has been identified as a regulator of multiple cold stress-related genes, playing a key role in enhancing cold tolerance in *Arabidopsis*

TABLE 6.5
Role of lncRNA in Abiotic Stress Tolerance in Crop Plants

Crop	lncRNA	Target Gene	Abiotic Stress	References
Arabidopsis thaliana	*DRIR*	*FUT4*	Salt and drought tolerance	Qin et al. (2017)
	CIL1	-	Cold tolerance	Liu et al. (2022)
Rice	*NONOSAT000455*	*osa_miR5809b-LOC_Os11g01210*	Alkaline salt tolerance	Rehman et al. (2023)
	TCONS_00021861	*Indole-3-pyruvate monooxygenase*	Drought tolerance	Chen et al. (2021)
	MSTRG.28732.3	*Os02g0662700*, *Os02g0663100* and *Os06g0105350*	Drought tolerance	Yang et al. (2022)
Soybean	*lncRNA77580*	-	Drought tolerance	Chen et al. (2023)
Wheat	*lnc_521*, *lnc_593* and *lnc_743*	-	Salt tolerance	YANϴ et al. (2020)
	L0760, *L6247*, *L0208*, *L2098* and *L3065*	-	Alkaline salt tolerance	Wei et al. (2022)
	LncR9A	*CSD1*	Cold tolerance	Lu et al. (2020)
Cotton	*XH123*	-	Cold tolerance	Cao et al. (2021)

plants (Liu et al., 2022). Another lincRNA, XH123, associated with cold tolerance, has been discovered in cotton during the seeding stage (Cao et al., 2021).

The roles of five different lncRNAs, namely, L0760, L6247, L0208, L2098 and L3065, in salinity tolerance have been investigated in wheat by using a knockdown approach. This study revealed that the knockdown of L0760 and L2098 increased the sensitivity of wheat to alkaline condition, while the knockdown of L6247, L0208 and L3065 enhanced the plants' ability to tolerate alkaline stress (Wei et al., 2022). Additionally, a triplet consisting of NONOSAT000455, osa_miR5809b and LOC_Os11g01210 has been identified as critical for improving rice yield under alkaline salt stress. Furthermore, three additional lncRNAs, namely, lnc_521, lnc_593 and lnc_743, have been identified in a salt-tolerant wheat genotype, underscoring their significant contributions to salt tolerance. Additionally, lncRNA973 has been recognised as an important role player in enhancing salt tolerance in cotton (Zhang et al., 2019).

6.6 FUNCTIONS OF lncRNAs IN PLANT STRESS PRIMING

Crop plants frequently encounter multiple abiotic stresses throughout their life cycles. The abiotic stresses negatively impact crop yields and to cope with recurrent abiotic stress, plants establish a form of memory during the initial stress event, commonly referred to as priming (Kushawaha et al., 2021). This priming enhances the plant's ability to respond promptly to subsequent abiotic stresses, ultimately improving crop yield (Abid et al., 2016). Additionally, the priming established during one type of abiotic stress can facilitate a more rapid response to other stresses, a phenomenon known as cross-tolerance, which has been documented in various crops (Ru et al., 2023). The ncRNAs play a significant role in these mechanisms (Table 6.6).

Prior exposure to heat stress in rice has been shown to enhance tolerance to subsequent heat stress events. Small RNA sequencing conducted on heat-primed versus non-heat-primed plants during recurrent heat stress identified seven distinct miRNAs that were either up-regulated or down-regulated under heat stress conditions. These miRNAs target genes coding for heat shock activators and repressors, contributing to improved thermo-tolerance in rice (Kushawaha et al., 2021). Research on the effects of drought priming on heat stress in chickpeas has revealed that miR156, miR166, miR319, miR171 and miR5213 target small heat shock proteins (sHsps), heat shock factors (Hsfs) and Hsps. Stress priming induced by drought condition prior to heat stress results in reduced membrane damage and improved leaf water content (Juneja et al., 2023a). Another study examining the impact of drought priming on miRNA expression during recurrent heat stress in chickpeas identified miR172 and miR396, which were down-regulated due to drought priming. This down-regulation led to the up-regulation of ABF3 and GRF14, assisting chickpeas in coping with recurrent heat stress (Juneja et al., 2023). Additionally, heat stress priming in rice has identified osa-miR169 as an miRNA targeting nuclear factor Y, playing a crucial role in recurrent heat stress tolerance (Kushawaha et al., 2019).

6.7 ROLE OF lncRNA IN CROP IMPROVEMENT

Among various ncRNAs, which have been identified across multiple crops in response to different abiotic stresses, lncRNAs are critical for regulating gene expression associated with complex traits (Qin et al., 2017). Mutations in these lncRNAs can significantly affect crop improvement. A notable example is the long-day-specific male-fertility-associated RNA (LDMAR), which is 1236 bases in length and regulates photoperiod-sensitive male sterility (PSMS) in rice. Sufficient levels of the LDMAR transcript are essential for normal pollen development in plants exposed to long-day conditions. A spontaneous mutation resulting in a single nucleotide polymorphism (SNP) between the wild-type and mutant allele altered the secondary structure of LDMAR. This modification led to increased methylation in the putative promoter region of LDMAR, thereby reducing its transcription specifically under long-day conditions. Consequently, this reduction triggered premature programmed cell death in developing anthers, ultimately resulting in PSMS (Ding et al., 2012). Another example is the early flowering-completely dominant lncRNA, known as Ef-cd lncRNA, which is linked to early maturity in plants. While early maturity typically results in a yield

TABLE 6.6
Role of Stress Priming and ncRNA in Abiotic Stress Tolerance in Crop Plants

Crop	Priming Stress	Recurrent Stress	NcRNA	Target Genes	References
Rice	Heat	Heat	*osa-miR531b*, *osa-miR5149*, *osa-miR168a-5p*, *osa-miR1846d-5p*, *osa-miR5077*, *osa-miR156b-3p*, *osa-miR167e-3p*	Heat shock activator and repressor	Kushawaha et al. (2021)
	Heat	Heat	*osa-miR169*	*nuclear factor Y*	Kushawaha et al. (2019)
Chickpea	Drought	Heat	*miR156*, *miR166*, *miR319*, *miR171* and *miR5213*	*sHsps*, *Hsfs* and *Hsps*	Juneja et al. (2023)
	Drought	Heat	*miR172* and *miR396*	*ABF3* and *GRF14*	Juneja et al. (2023)

penalty, the Ef-cd lncRNA alleviates this issue. It is transcribed from the antisense strand of the flowering activator OsSOC1, resulting in a reduction of maturity duration. This facilitates early maturity in rice without compromising overall yield (Fang et al., 2019).

6.8 CONCLUSION

The different types of ncRNAs interacting with DNA, RNA and protein in cis and trans manner plays an important role in the gene regulation during various abiotic stresses. Differential mechanisms, such as ceRNA and triplet of the miRNA-lncRNA-target gene act as recruiter of the chromatin modifier enzymes using ncRNAs to deal with abiotic stress. Several ncRNAs, such as COLDAIR, function epigenetically to modulate the expression of their target genes by modulating histone modification status and chromatin organisation. Examples of beneficial lncRNAs include LDMAR and Ef-cd. These cases demonstrate that harnessing knowledge about the functional roles of plant lncRNAs can advance plant breeding research, potentially resulting in enhancements in crop quality and performance.

REFERENCES

Abid, M., Tian, Z., Ata-Ul-Karim, S. T., Liu, Y., Cui, Y., Zahoor, R., & Dai, T. (2016). Improved tolerance to post-anthesis drought stress by pre-drought priming at vegetative stages in drought-tolerant and-sensitive wheat cultivars. *Plant Physiology and Biochemistry*, *106*, 218–227. https://doi.org/10.1016/j.plaphy.2016.05.003

Akbari, A., Ismaili, A., Shobbar, Z. S., Amirbakhtiar, N., & Pouresmael, M. (2022). Identification of candidate microRNAs involved in cold tolerance of Chickpea. *Genetic Engineering and Biosafety Journal*, *11*(1), 42–53. https://doi.org/10.1038/s41598-023-33398-3

Ariel, F., Lucero, L., Christ, A., Mammarella, M. F., Jegu, T., Veluchamy, A., & Crespi, M. (2020). R-loop mediated trans action of the *APOLO* long noncoding RNA. *Molecular Cell*, *77*(5), 1055–1065. https://doi.org/10.1016/j.molcel.2019.12.015

Atta, K., Mondal, S., Gorai, S., Singh, A. P., Kumari, A., Ghosh, T., & Jespersen, D. (2023). Impacts of salinity stress on crop plants: Improving salt tolerance through genetic and molecular dissection. *Frontiers in Plant Science*, *14*, 1241736. https://doi.org/10.3389/fpls.2023.1241736

Bi, Y. M., Kant, S., Clark, J., Gidda, S., Ming, F., Xu, J., & Rothstein, S. J. (2009). Increased nitrogen-use efficiency in transgenic rice plants over-expressing a nitrogen-responsive early nodulin gene identified from rice expression profiling. *Plant, Cell & Environment*, *32*(12), 1749–1760. https://doi.org/10.1111/j.1365-3040.2009.02032.x

Bologna, N. G., & Voinnet, O. (2014). The diversity, biogenesis, and activities of endogenous silencing small RNAs in *Arabidopsis*. *Annual Review of Plant Biology*, *65*(1), 473–503. http://dx.doi.org/10.1146/annurev-arplant-050213-035728

Cao, Z., Zhao, T., Wang, L., Han, J., Chen, J., Hao, Y., & Guan, X. (2021). The lincRNA *XH123* is involved in cotton cold-stress regulation. *Plant Molecular Biology*, *106*, 521–531. https://doi.org/10.1007/s11103-021-01169-1

Chen, J., Zhong, Y., & Qi, X. (2021). LncRNA TCONS_00021861 is functionally associated with drought tolerance in rice (*Oryzasativa* L.) via competing endogenous RNA regulation. *BMC Plant Biology*, *21*(1), 12. https://doi.org/10.1186/s12870-021-03195-z

Chen, X., Jiang, X., Niu, F., Sun, X., Hu, Z., Gao, F., & Jiang, Q. (2023). Overexpression of lncRNA77580 regulates drought and salinity stress responses in soybean. *Plants*, *12*(1), 181. https://doi.org/10.3390/plants12010181

Deforges, J., Reis, R. S., Jacquet, P., Sheppard, S., Gadekar, V. P., Hart-Smith, G., & Poirier, Y. (2019). Control of cognate sense mRNA translation by cis-natural antisense RNAs. *Plant Physiology*, *180*(1), 305–322. https://doi.org/10.1104/pp.19.00043

Ding, J., Lu, Q., Ouyang, Y., Mao, H., Zhang, P., Yao, J., & Zhang, Q. (2012). A long noncoding RNA regulates photoperiod-sensitive male sterility, an essential component of hybrid rice. *Proceedings of the National Academy of Sciences*, *109*(7), 2654–2659. https://doi.org/10.1073/pnas.1121374109

Dykes, I. M., & Emanueli, C. (2017). Transcriptional and post-transcriptional gene regulation by long non-coding RNA. *Genomics, Proteomics and Bioinformatics*, *15*(3), 177–186. https://doi.org/10.1016/j.gpb.2016.12.005

Eremina, M., Rozhon, W., & Poppenberger, B. (2016). Hormonal control of cold stress responses in plants. *Cellular and Molecular Life Sciences*, *73*, 797–810. https://doi.org/10.1007/s00018-015-2089-6

Fabbri, M., Girnita, L., Varani, G., & Calin, G. A. (2019). Decrypting noncoding RNA interactions, structures, and functional networks. *Genome Research*, *29*(9), 1377–1388. https://doi.org/10.1101/gr.247239.118

Fang, J., Zhang, F., Wang, H., Wang, W., Zhao, F., Li, Z., & Chu, C. (2019). Ef-cd locus shortens rice maturity duration without yield penalty. *Proceedings of the National Academy of Sciences*, *116*(37), 18717–18722. https://doi.org/10.1073/pnas.1815030116

Guan, Q., Lu, X., Zeng, H., Zhang, Y., & Zhu, J. (2013). Heat stress induction of *miR398* triggers a regulatory loop that is critical for thermos-tolerance in *Arabidopsis*. *The Plant Journal*, *74*(5), 840–851. https://doi.org/10.1111/tpj.12169

Heo, J. B., & Sung, S. (2011). Vernalization-mediated epigenetic silencing by a long intronic noncoding RNA. *Science*, *331*(6013), 76–79. https://doi.org/10.1126/science.1197349

Jain, A. K., Xi, Y., McCarthy, R., Allton, K., Akdemir, K. C., Patel, L. R., & Barton, M. C. (2016). *LncPRESS1* is a p53-regulated LncRNA that safeguards pluripotency by disrupting SIRT6-mediated de-acetylation of histone H3K56. *Molecular Cell*, *64*(5), 967–981. https://doi.org/10.1016/j.molcel.2016.10.039

Jha, U. C., Nayyar, H., Jha, R., Khurshid, M., Zhou, M., Mantri, N., & Siddique, K. H. (2020). Long non-coding RNAs: emerging players regulating plant abiotic stress response and adaptation. *BMC Plant Biology*, *20*, 1–20. https://doi.org/10.1186/s12870-020-02595-x

Juneja, S., Saini, R., Adhikary, A., Yadav, R., Khan, S. A., Nayyar, H., & Kumar, S. (2023a). Drought priming evokes essential regulation of *Hsp* gene families, *Hsfs* and their related miRNAs and induces heat stress tolerance in chickpea. *Plant Stress*, *10*, 100189. https://doi.org/10.1016/j.stress.2023.100189

Juneja, S., Saini, R., Mukit, A., & Kumar, S. (2023b). Drought priming modulates ABF, GRFs, related microRNAs and induce metabolic adjustment during heat stress in chickpea. *Plant Physiology and Biochemistry*, *203*, 108007. https://doi.org/10.1016/j.plaphy.2023.108007

Kapoor, D., Bhardwaj, S., Landi, M., Sharma, A., Ramakrishnan, M., & Sharma, A. (2020). The impact of drought in plant metabolism: How to exploit tolerance mechanisms to increase crop production. *Applied Sciences*, *10*(16), 5692. https://doi.org/10.3390/app10165692

Kim, E. D., Xiong, Y., Pyo, Y., Kim, D. H., Kang, B. H., & Sung, S. (2017). Spatio-temporal analysis of coding and long noncoding transcripts during maize endosperm development. *Scientific Reports*, *7*(1), 3838. https://doi.org/10.1038/s41598-017-03878-4

Kouzarides, T. (2002). Histone methylation in transcriptional control. *Current Opinion in Genetics & Development*, *12*(2), 198–209. https://doi.org/10.1016/S0959-437X(02)00287-3

Kumar, D., Ramkumar, M. K., Dutta, B., Kumar, A., Pandey, R., Jain, P. K., & Sevanthi, A. M. (2023). Integration of miRNA dynamics and drought tolerant QTLs in rice reveals the role of miR2919 in drought stress response. *BMC Genomics*, *24*(1), 526. https://doi.org/10.1186/s12864-023-09609-6

Kushawaha, A. K., Khan, A., Sopory, S. K., & Sanan-Mishra, N. (2019). Light regulated Osa-miR169e is implicated during priming under high temperature stress in rice. *American Journal of Plant Sciences*, *10*(09), 1662. https://doi.org/10.4236/ajps.2019.109118

Kushawaha, A. K., Khan, A., Sopory, S. K., & Sanan-Mishra, N. (2021). Priming by high temperature stress induces microRNA regulated heat shock modules indicating their involvement in thermos priming response in rice. *Life*, *11*(4), 291. https://doi.org/10.3390/life11040291

Li, S., Liu, L., Zhuang, X., Yu, Y., Liu, X., Cui, X., & Chen, X. (2013). MicroRNAs inhibit the translation of target mRNAs on the endoplasmic reticulum in *Arabidopsis*. *Cell*, *153*(3), 562–574. https://doi.org/10.1016/j.cell.2013.04.005

Liu, G., Liu, F., Wang, Y., & Liu, X. (2022). A novel long noncoding RNA CIL1 enhances cold stress tolerance in *Arabidopsis*. *Plant Science*, *323*, 111370. https://doi.org/10.1016/j.plantsci.2022.111370

Liu, Q., Yang, T., Yu, T., Zhang, S., Mao, X., Zhao, J., & Liu, B. (2017). Integrating small RNA sequencing with QTL mapping for identification of miRNAs and their target genes associated with heat tolerance at the flowering stage in rice. *Frontiers in Plant Science*, *8*, 43. https://doi.org/10.3389/fpls.2017.00043

Lu, Q., Guo, F., Xu, Q., & Cang, J. (2020). LncRNA improves cold resistance of winter wheat by interacting with *miR398*. *Functional Plant Biology*, *47*(6), 544–557. https://doi.org/10.1071/FP19267

Lucero, L., Fonouni-Farde, C., Crespi, M., & Ariel, F. (2020). Long noncoding RNAs shape transcription in plants. *Transcription*, *11*(3–4), 160–171. https://doi.org/10.1080/21541264.2020.1764312

Ma, C., Burd, S., & Lers, A. (2015). *miR408* is involved in abiotic stress responses in *Arabidopsis*. *The Plant Journal*, *84*(1), 169–187. https://doi.org/10.1111/tpj.12999

Marquardt, S., Raitskin, O., Wu, Z., Liu, F., Sun, Q., & Dean, C. (2014). Functional consequences of splicing of the antisense transcript COOLAIR on FLC transcription. *Molecular Cell*, *54*(1), 156–165. https://doi.org/10.1016/j.molcel.2014.03.026

Mladenov, V., Fotopoulos, V., Kaiserli, E., Karalija, E., Maury, S., Baranek, M., Segal, N. A., Testillano, P. S., Vassileva, V., Pinto, G., & Nagel, M. (2021). Deciphering the epigenetic alphabet involved in transgenerational stress memory in crops. *International Journal of Molecular Sciences*, *22*(13), 7118. https://doi.org/10.3390/ijms22137118

Parmar, S., Gharat, S. A., Tagirasa, R., Chandra, T., Behera, L., Dash, S. K., & Shaw, B. P. (2020). Identification and expression analysis of miRNAs and elucidation of their role in salt tolerance in rice varieties susceptible and tolerant to salinity. *Public Library of Science One*, *15*(4), e0230958. https://doi.org/10.1371/journal.pone.0230958

Peng, K., Tian, Y., Sun, X., Song, C., Ren, Z., Bao, Y., & Cang, J. (2021). *Tae-miR399-UBC24* module enhances freezing tolerance in winter wheat via a CBF signaling pathway. *Journal of Agricultural and Food Chemistry*, *69*(45), 13398–13415. https://doi.org/10.1021/acs.jafc.1c04316

Qin, T., Zhao, H., Cui, P., Albesher, N., & Xiong, L. (2017). A nucleus-localized long non-coding RNA enhances drought and salt stress tolerance. *Plant Physiology*, *175*(3), 1321–1336. https://doi.org/10.1104/pp.17.00574

Rehman, O. U., Uzair, M., Farooq, M. S., Saleem, B., Attacha, S., Attia, K. A., & Khan, M. R. (2023). Comprehensive insights into the regulatory mechanisms of lncRNA zin alkaline-salt stress tolerance in rice. *Molecular Biology Reports*, *50*(9), 7381–7392. https://doi.org/10.1007/s11033-023-08648-2

Ru, C., Hu, X., Chen, D., & Wang, W. (2023). Droughts and thermo-priming enhance acclimation to later drought and heat stress in maize seedlings by improving leaf physiological activity. *Agronomy*, *13*(4), 1124. https://doi.org/10.3390/agronomy13041124

Salmena, L., Poliseno, L., Tay, Y., Kats, L., & Pandolfi, P. P. (2011). A *ceRNA* hypothesis: the Rosetta Stone of a hidden RNA language? *Cell*, *146*(3), 353–358. https://doi.org/10.1016/j.cell.2011.07.014

Shan, T., Fu, R., Xie, Y., Chen, Q., Wang, Y., Li, Z., & Wang, B. (2020). Regulatory mechanism of maize (*Zea mays* L.) miR164 in salt stress response. *Russian Journal of Genetics*, *56*, 835–842. https://doi.org/10.1134/S1022795420070133

Shin, S. Y., & Shin, C. (2016). Regulatory non-coding RNAs in plants: potential gene resources for the improvement of agricultural traits. *Plant Biotechnology Reports*, *10*, 35–47. https://doi.org/10.1007/s11816-016-0389-4

Singh, S., Kumar, A., Panda, D., Modi, M. K., & Sen, P. (2020). Identification and characterization of drought responsive miRNAs from a drought tolerant rice genotype of Assam. *Plant Gene*, *21*, 100213.https://doi.org/10.1016/j.plgene.2019.100213

Statello, L., Guo, C. J., Chen, L. L., & Huarte, M. (2021). Gene regulation by long non-coding RNAs and its biological functions. *Nature Reviews Molecular Cell Biology*, *22*(2), 96–118. https://doi.org/10.1038/s41580-020-00315-9

Sun, Q., Csorba, T., Skourti-Stathaki, K., Proudfoot, N. J., & Dean, C. (2013). R-loop stabilization represses antisense transcription at the Arabidopsis FLC locus. *Science*, *340*(6132), 619–621. https://doi.org/10.1126/science.1234848

Sun, T. T., He, J., Liang, Q., Ren, L. L., Yan, T. T., Yu, T. C., & Fang, J. Y. (2016). LncRNA GClnc1 promotes gastric carcinogenesis and may act as a modular scaffold of WDR5 and KAT2A complexes to specify the histone modification pattern. *Cancer Discovery*, *6*(7), 784–801. https://doi.org/10.1158/2159-8290.CD-15-0921

Svoboda, P. (2020). Key mechanistic principles and considerations concerning RNA interference. *Frontiers in Plant Science*, *11*, 1237. https://doi.org/10.3389/fpls.2020.01237

Swiezewski, S., Liu, F., Magusin, A., & Dean, C. (2009). Cold-induced silencing by long antisense transcripts of an *Arabidopsis* Polycomb target. *Nature*, *462*(7274), 799–802. https://doi.org/10.1038/nature08618

Thomson, D. W., & Dinger, M. E. (2016). Endogenous microRNA sponges: evidence and controversy. *Nature Reviews Genetics*, *17*(5), 272–283. https://doi.org/10.1038/nrg.2016.20

Verma, S., Verma, P. K., & Chakrabarty, D. (2021). Roles of long noncoding RNA during stress response in maize. In *Long noncoding RNAs in plants* (pp. 93–105). Academic Press. https://doi.org/10.1016/B978-0-12-821452-7.00007-6

Wang, H. L. V., & Chekanova, J. A. (2016). Small RNAs: essential regulators of gene expression and defenses against environmental stresses in plants. *Wiley Interdisciplinary Reviews: RNA*, *7*(3), 356–381. https://doi.org/10.1002/wrna.1340

Wang, Y., Luo, X., Sun, F., Hu, J., Zha, X., Su, W., & Yang, J. (2018). Overexpressing lncRNA LAIR increases grain yield and regulates neighbouring gene cluster expression in rice. *Nature Communications*, *9*(1), 3516. https://doi.org/10.1038/s41467-018-05829-7

Wei, L., Zhang, R., Zhang, M., Xia, G., & Liu, S. (2022). Functional analysis of long non-coding RNAs involved in alkaline stress responses in wheat. *Journal of Experimental Botany*, *73*(16), 5698–5714. https://doi.org/10.1093/jxb/erac211

Wierzbicki, A. T., Blevins, T., & Swiezewski, S. (2021). Long noncoding RNAs in plants. *Annual Review of Plant Biology*, *72*(1), 245–271. https://doi.org/10.1146/annurev-arplant-093020-035446

Wu, Z., Fang, X., Zhu, D., & Dean, C. (2020). Autonomous pathway: *FLOWERING LOCUS C* repression through an antisense-mediated chromatin-silencing mechanism. *Plant Physiology*, *182*(1), 27–37. https://doi.org/10.1104/pp.19.01009

Yadav, S., Modi, P., Dave, A., Vijapura, A., Patel, D., & Patel, M. (2020). Effect of abiotic stress on crops. *Sustainable Crop Production*, *3*(17), 5–16. https://doi.org/10.5772/intechopen.88434

Yang, H., Cui, Y., Feng, Y., Hu, Y., Liu, L., & Duan, L. (2023). Long non-coding RNAs of plants in response to abiotic stresses and their regulating roles in promoting environmental adaption. *Cells*, *12*(5), 729. https://doi.org/10.3390/cells12050729

Yang, X., Liu, C., Niu, X., Wang, L., Li, L., Yuan, Q., & Pei, X. (2022). Research on lncRNA related to drought resistance of Shanlan upland rice. *BMC Genomics*, *23*(1), 336. https://doi.org/10.1186/s12864-022-08546-0

Yang, X., Lu, M., Wang, Y., Wang, Y., Liu, Z., & Chen, S. (2021). Response mechanism of plants to drought stress. *Horticulturae*, *7*(3), 50. https://doi.org/10.3390/horticulturae7030050

Zhang, L., Lin, T., Zhu, G., Wu, B., Zhang, C., & Zhu, H. (2023). LncRNAs exert indispensable roles in orchestrating the interaction among diverse noncoding RNAs and enrich the regulatory network of plant growth and its adaptive environmental stress response. *Horticulture Research*, *10*(12), 234. https://doi.org/10.1093/hr/uhad234

Zhang, P., Wu, W., Chen, Q., & Chen, M. (2019). Non-coding RNAs and their integrated networks. *Journal of Integrative Bioinformatics*, *16*(3), 20190027. https://doi.org/10.1515/jib-2019-0027

Zhang, X., Dong, J., Deng, F., Wang, W., Cheng, Y., Song, L., & Shen, F. (2019). The long non-coding RNA *lncRNA973* is involved in cotton response to salt stress. *BMC Plant Biology*, *19*, 1–16. https://doi.org/10.1186/s12870-019-2088-0

Zhang, Y., Zhou, Y., Zhu, W., Liu, J., & Cheng, F. (2022). Non-coding RNAs fine-tune the balance between plant growth and abiotic stress tolerance. *Frontiers in Plant Science*, *13*, 965745. https://doi.org/10.3389/fpls.2022.965745

Zhang, Y. C., Liao, J. Y., Li, Z. Y., Yu, Y., Zhang, J. P., Li, Q. F., & Chen, Y. Q. (2014). Genome-wide screening and functional analysis identify a large number of long non-coding RNAs involved in the sexual reproduction of rice. *Genome Biology*, *15*, 1–16.https://doi.org/10.1186/s13059-014-0512-1

Zhou, M., & Luo, H. (2013). MicroRNA-mediated gene regulation: potential applications for plant genetic engineering. *Plant Molecular Biology*, *83*, 59–75.

7 Exploring the Role of Non-Coding RNAs in Plant Adaptation to Abiotic Stresses

Sadia Zafar, Javed Iqbal, Muhammad Arslan Ashraf, Rizwan Rasheed, Areeba Ijaz, Anis Ali Shah and Banzeer Ahsan Abbasi

7.1 INTRODUCTION

The regulation of gene expression is by far the most crucial aspect of the intriguing science of plant molecular biology, despite the fact that there are still many unanswered questions (Dhar et al., 2021). Globally, researchers studying plants are working tirelessly to synthesise the available data in an effort to unravel the riddles surrounding the gene regulation and reprogramming mechanisms that underlie plants' ability to respond to various biotic and abiotic stressors that they may encounter in their lifetime (Haider et al., 2022; Zhang, Ma et al., 2024). Plants must carefully balance their development, growth and reproduction in a spatiotemporal fashion as a result of these numerous stressors (Ahmad, Haakim, et al. (2024) Ahmad, Haider, et al. (2022); Osnato et al., 2022). But it does not seem like plants can overcome biotic and abiotic stress limits in and of themselves. Numerous resistance genes that reduce stress damage and thus give the plant protection against stress have been found and defined through extensive research (Jiang et al., 2022). Numerous biotechnology techniques have been used to alter the genetic composition of commercially significant crops to produce transgenic plants with better control against biotic and abiotic stress (Ricroch et al., 2022). But since genes are multifunctional, because they can have multiple functions, and because defence signalling networks are intricate and can interact with one another, using genetically modified plants has been severely limited. Thus, when using genetically engineered organisms, there is always a chance of unforeseen environmental impacts (Ali et al., 2022, Asghar et al., 2024, Zhang & Ning, 2021).

The dark intergenic spaces between genes that code for proteins have been better understood; thanks to developments in transcriptome analysis and sequencing technologies (Aerts et al., 2021).These regions have been previously labelled 'Junk DNA', transcriptional 'noise', or a product of experimentation. Despite not encoding proteins, a large proportion of eukaryotic genomes is extensively transcribed into RNA, and many of these transcripts may have important functions. These RNAs are known as non-coding RNAs for proteins (Havstad & Palazzo, 2022).

These days, non-coding RNAs (ncRNAs) are the most oftenusedtermfornon-protein-codingRNAs(Polisenoetal., 2024). They are categorised according to their pathway of action and genetic source. Together with regulatory and housekeeping ncRNAs, they form a variety of transcripts (Najafi et al., 2022). Small nuclear RNAs (snRNAs; 50–200 nt), small nucleolar RNAs (snoRNAs: 50–200 nt), transfer RNAs (tRNAs: 50–200 nt) and ribosomal RNAs (rRNAs) are examples of housekeeping ncRNAs that support ribosomal and cellular processes (Bai et al., 2021). The regulatory ncRNA family includes piwi-interacting RNAs (piRNAs: 24–32 nt), short RNAs (miRNAs: 20–24 nt), long non-coding RNAs (lncRNAs: >200 nt) and small interfering RNAs (siRNAs: 20–24 nt) (Statello et al., 2021). Non-coding in regulations transcribed from DNA, RNA molecules – also referred to as ribo-regulators – do not change into proteins. The transcriptional, post-transcriptional and epigenetic functions of non-coding RNAs influence the expression of genes (Morselli & Dieci, 2022).

Many short RNAs regulate gene expression, stress response and adaptability (Verma et al., 2022). They are also responsive to many environmental cues. Numerous studies and reviews have been undertaken on the role of short RNAs in growth and development, as well as their responsiveness to various stimuli (Statello et al., 2021). This study focuses on the role of lncRNAs in coordinating the plant immune response (Song et al., 2021). We begin by describing lncRNAs and providing a brief overview of their many roles. The next section discusses how long non-coding RNAs (lncRNAs) interact with short RNAs and transcription factors (TFs) to adapt to biotic and abiotic stresses in plants (Ijaz et al., 2024; Patra et al., 2023).

7.2 LONG RNAs WITHOUT CODING

Eukaryotic RNAs with an ORF of less than 100 amino acids, a length more lncRNAs are more than 200 nucleotides that lack the capacity to code for proteins (Wierzbicki et al., 2021). Like mRNAs, the majority of lncRNAs are 50-capped, spliced, polyadenylated (30-poly(A) tails) and transcribed using RNA polymerase II (Wiener et al., 2021). Post-transcriptional processing, secondary and tertiary structure creation, and similar epigenetic markers at their promoter regions, and expression that is unique to developmental stages, organs and cells albeit more so than mRNAs

DOI: 10.1201/9781003583257-7

TABLE 7.1
Different Types of Plant Interaction with Pathogen

Name of Plant	Name of Pathogen	Interaction	References
		lncRNA	
Mulberry (*Morus multicaulis*)	*Botrytis cinerea* and Pseudomonas syringae pv tomato DC3000	Inactivation of the calmodulin-like protein gene CML27 due to MuLnc1	Gai et al. (2018)
Cotton (*Gossypium hirsutum* L.)	Aphid	The CC-NBS-LRR family gene GhA01G0129 is a possible target of GhlncRNA149.1.	Zhang et al. (2023)
Arabidopsis thaliana	*Phytophthora capsici*	nalncFL7 negatively regulates *FORKED-LIKE7* (*FL7*)	Ai et al. (2023)
Tomato (*Solanum lycopersicum* L.)	*Phytophthora infestans*	lncRNA33732 interaction with *WRKY1*	Cui et al. (2019)
Wheat (*Triticum aestivum*)	*Blumeria graminis* f. sp. *tritici*	Non-coding RNA profiling	Xin et al. (2011)
Cotton (*G. hirsutum* L.)	*Verticillium dahliae* and *B. cinerea*	GhlncNAT-*ANX2* and GhlncNAT-RLP7 control the expression of *LOX1* and *LOX2*	Zhang et al. (2018)
Wheat (*T. aestivum*)	*Rhizoctonia cerealis*	In fungal growth retardation, MSTRG.4380.1	Yi et al. (2023)
		circRNA	
Legume crops	*P. infestans*	DNA, RNA, and protein interactions, as well as target protein regulation	Chand et al. (2021)
A. thaliana	*P. syringae* and *B. cinerea*	Resistance to *P. syringae* is mediated by circR194 and circR4022, whereas protection against B. cinerea is provided by circR11208	Wang et al. (2021)
Tea plant (*Camellia sinensis*)	*Helopeltis theivora*	Secondary metabolite activation and endogenous target mimics (eTMs) of target genes, such as phospholipase, lectin receptor, and aspartyl protease	Raja et al. (2023)
Tomato (*S. lycopersicum* L.)	Planticine®-induced defence responses	Upregulation of circRNA whitefly-induced gp91	Borodoloi et al. (2023)
Tomato (*S. lycopersicum* L.)	*P. infestans*	circRNA45 and positive regulators of tomato resistance response, circRNA47	Rakoczy-Lelek et al. (2023)

are some of their similarities. In contrast, transcripts of lncRNA are shorter and include fewer exons than those of mRNA (Staněk, 2021). Furthermore, the transcription of RNA polymerase III and other polymerases other than RNA polymerase II are responsible for the production of non-polyadenylated lncRNAs (poly(A)-) (Ahmad et al., 2024; Nojima & Proudfoot, 2022). Compared to mRNAs, lncRNAs exhibit no interspecies sequence conservation, lower expression levels, higher natural expression variation and poor evolutionary conservation across species. This is corroborated by a recent study of the non-coding transcriptome in maize and rice, which discovered that lncRNAs' positional conservation is significantly greater than their sequence conservation. Because there is no sequence conservation, it was suggested that the rapid evolution of the lncRNA sequence has occurred between maize and rice (Zhou et al., 2021).

7.2.1 Localisation of Cells

lncRNAs are eukaryotic RNAs that are longer than 200 nucleotides, have ORFs of fewer than 100 amino acids and lack the capacity to code for proteins (Wierzbicki et al., 2021). Most lncRNAs are 50-capped, spliced, polyadenylated (30-poly(A) tails) and transcribed using RNA polymerase II, exactly like mRNAs (Wiener et al., 2021). Their parallels to mRNAs include post-transcriptional processing, secondary and tertiary structure synthesis, comparable epigenetic markers at their promoter regions, and expression that is specific to developmental phases, tissues and cells albeit to a greater extent than mRNAs.

7.2.2 Classification

Relationships with known protein-coding genes, transcript length and other DNA areas with known functions, and repetitions are some of the characteristics that have been used to classify lncRNAs thus far (Ross & Ulitsky, 2022). In the context of biological states, subcellular localisation, function, targeting strategy, metabolic route or stability, sequence or structural conservation and similarities between DNA sequences and mRNA, the human genome demonstrates various impacts. However, they are mainly divided into five groups according to their positions in the genome relative to nearby protein-coding loci: (i) bidirectional lnRNAs (BI-lncRNAs), (ii) overlapping lnRNAs (OT-lncRNAs), (iii) intronic (incRNAs), (iv) long intergenic ncRNAs (lincRNAs) and (v) natural antisense transcripts (lncNATs) (Pecorino et al., 2021).

7.3 FUNCTIONS OF LONG NON-CODING RNAs

In many fundamental biological processes, such as the regulation of mRNA processing and transcription mechanisms for translation and localisation, lncRNAs are essential modulators, and employing a cis-mediated approach to activate their neighbouring genes turnover, splicing and editing, according to an increasing amount of recent research (Herman et al., 2022). Furthermore, among the three main epigenetic functions of lncRNAs are chromatin-mediated repression, transcriptional interference and histone modification, chromatin structural remodelling, imprinting regulation, and epigenetic gene regulation. Additionally, they contribute to evolution, RNA modification, inheritance and cell differentiation and development (Bhadouriya et al., 2021).

Further research on lncRNAs has confirmed that they play a role in a number of crucial biological functions in plants that are similar to those in mammals. For instance, it has been suggested that lncRNAs influence the expression of genes that code for proteins, control transcriptional gene silencing, and govern chromatin structure and epigenetic changes. Additionally, they function as scaffolds, enhancers, decoys, activators, signals, byproducts, or guides of the promoters of genes that code for proteins. Additionally, they use a range of different methods to control their targets in a cis or trans manner at several levels (Figure 7.1) (Morselli & Dieci, 2022).

In particular, the flattening of rice leaves depends on the activity of cis-natural antisense lncRNA overexpression of the OsMYB60 gene, which codes for the endogenous lncRNA TWISTED LEAF (TL), and downregulation of TL may result in twisted leaves in transgenic rice plants (Yadav et al., 2024). According to reports, overexpressing LAIR, rice grain yield is increased by several LRK genes' expression levels are also raised by the lncRNA that is transcribed from the antisense strand of the next gene LRK cluster (Nadarajah & Abdul Rahman, 2021). XLOC_0063639 and XLOC_007072, respectively, regulate two rice miRNAs that govern the growth of seeds and blooms: OsmiR160 and OsmiR164 (Hazra et al., 2023). Furthermore, the high expression of these two lncRNAs in the pistil, anther, and early spikelet seeds suggests that miRNA-lncRNA functional couples play a part in the formation of seeds and/or flowers (Tu et al., 2022). Moreover, XLOC_057324 mediates panicle development and reproduction. Another antisense lncRNA (Os01g0356951) derived from a pseudogene is the ESP gene as a result of epigenetic changes, is thought to have an impact on the shape of rice panicles (Beretta et al., 2023).

7.3.1 Multiple Roles for ncRNAs under Different Conditions

The identification of many stress-responsive non-coding RNAs (ncRNAs) in rice has been facilitated by genome-wide transcriptome sequencing (Xu et al., 2021). On the other hand, nothing is known about how those ncRNAs affect how plants react to environmental stressors. In order to reduce the harm caused by biological stresses that seriously jeopardise rice production, rice has developed a number of defence mechanisms. lncRNAs have a major impact on rice plants' defence systems and biological stress reactions. Rice's lncRNAs are believed to be impacted by heat, drought, cold, salt, heavy metals, nutrient shortages, frost and radiation, much like by other abiotic stressors (Avramova, 2015; Haier et al., 2021).

One major risk to human health and a class II metalloid that causes cancer is arsenic (As). More significantly it is spreading throughout Southeast Asia, contaminating soil and groundwater in nations, including Bangladesh, China, India and Thailand (Ozturk et al., 2022). The primary crop in these areas, rice, needs a lot of water, which is contaminated with a high concentration of arsenic, which is very concerning (Rokonuzzaman et al., 2022). This leads to excessive grain accumulation and As uptake by rice seedlings, which creates significant problems for food safety. Furthermore, As causes oxidative stress, depresses and interferes with glucose metabolism, all of which have a negative impact on agricultural production. Plant growth and reproduction are at risk from each of these. To lessen the buildup of As in rice grains, researchers were consequently compelled to employ finding rice cultivars with poor absorption systems, via molecular breeding, or genetic engineering (Geng et al., 2024). An array of whole-transcriptome studies at the mRNA or miRNA level have been carried out to profile the molecular responses in As-stressed rice. When arsenic seeps into flooded soil for rice production, it primarily takes the form of As(III). As a result, As(III) stress therapy has been the subject of several research (Hu et al., 2024).

Many lncRNAs, including those for arsenic (As) and cadmium (Cd), are functional in heavy metal stress. When comparing rice treated with Cd to rice grown under normal conditions, deep sequencing revealed differential expression of lncRNAs (75 downregulated and 69 upregulated). Important changes in genes linked under Cd stress to photosynthesis and the manufacture of phenylalanine, cysteine, methionine, and phenylpropanoids were also found by pathway analysis of pertinent target genes (Ghosh et al., 2022). The findings indicate that lncRNAs modulate the levels of Cd toxicity via trans-regulating secondary metabolites and photosynthesis, and by cis-regulating genes involved in the metabolism of cysteine-rich peptides. This activation of different physiological and biochemical responses is the outcome of these mechanisms (Nandi et al., 2021).

In response to drought stress, many lncRNAs change the amount of ABA. DPA of lncRNAs causes the activation of many sets of stress response genes related to ABA

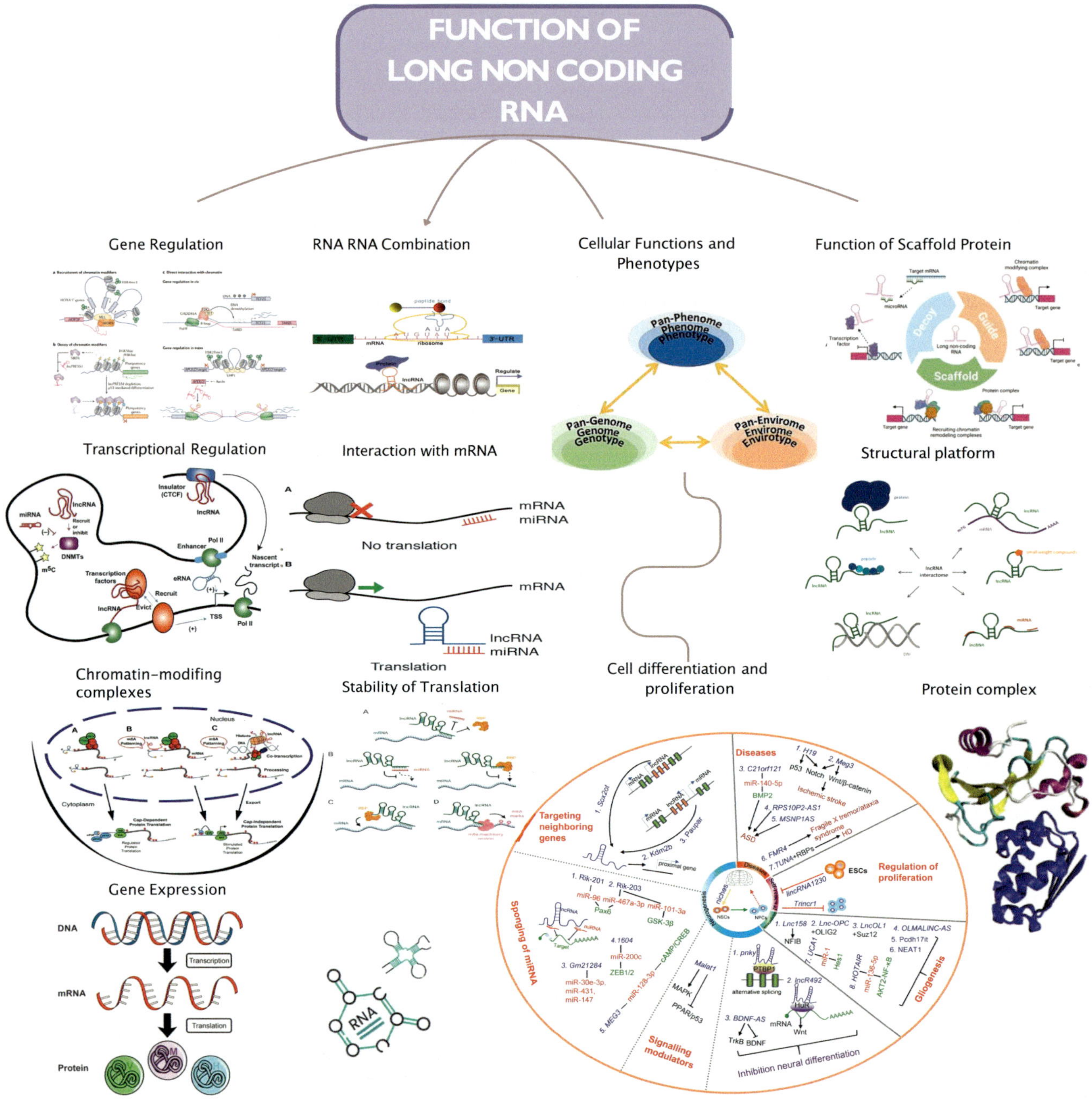

FIGURE 7.1 Long non-coding RNA's role.

production, transport and metabolism (Magar et al., 2023). Furthermore, several drought-associated DPA lncRNAs are connected to ABA trafficking and cell wall thickening under drought stress. However, non-DPA lncRNAs often participate in post-transcriptional regulation processes, in contrast to DPA lncRNAs (Li et al., 2024) that revealed rice's drought memory responses involved 238 lncRNAs and 6885 transcripts. A novel long non-coding RNA (lncRNA) called TCONS_00028567 has been discovered to regulate the SAPK10 (stress/ABA-activated protein kinase 10) gene and a precursor of osa-MIR1428e. This gene is involved in ABA signalling, which controls the creation of short circuits (Imaduwage & Hewadikaram, 2024).

When exposed to high temperatures, such as cold or heat, certain DPA lncRNAs negatively affect enzyme function. Researchers observed Chr03G0008, a putative lncRNA, to be expressed exclusively in seedlings that had received cold treatment, indicating the coordinated activity of several lncRNAs (Magar et al., 2023).

7.4 THE FUNCTION OF miRNAs IN VARIOUS STRESS REACTIONS

Wheat ncRNAs, like those in rice, are important for adaptive reactions to different metal stressors. Through the expression of transporters, the process controls the intake and movement of metals, such as heavy metals. Previously, several gene transporter families, such as the ZiFL gene family and ZIP transporters, members, NRAMP family, VITs and HMA family, were identified as regulating the accumulation of heavy metals. Reactive sites for ncRNAs like miR159 and miR167 are also found in these transporters (Gill et al., 2024). The absorption, transportation and vacuolar sequestration of hazardous metal ions are regulated by these miRNAs. TFs such as WRKY, MYB and others, as well as regulatory proteins, hormones and other biological processes whose gene expression fluctuates significantly, demonstrate how plants respond to toxic metals (Yang et al., 2022).

Stress from heavy metals (Cd, Fe or Zn) causes substantial changes in the transcriptome of plants, including wheat. Most of the information on these modifications has already been provided by the RNA-seq analysis. Recently, spl7 knockout mutants in Arabidopsis were used to finish the research of miRNA responses to heavy metals (Cd and Cu). MiRNAs such as miR397, miR398 and miR408 primarily target proteins involved in the positive regulation of zinc and copper. The Cu-limiting condition caused these miRNAs to be expressed (Tiwari, 2024).

Changes in climate are predicted to cause higher evaporation rates and lower precipitation rates in many significant wheat-growing regions, exacerbating the severity of drought stress. The transcriptional or translational regulation of gene expression governs all of the metabolic and physiological alterations brought on by drought; therefore, the regulation of miRNA is essential for the water stress. It has been shown that certain miRNA families are expressed differently in diverse species of the wheat family, including barley. In the year 2005, the wheat computational prediction was used to predict 16 miRNAs from nine families of conserved miRNAs. Wheat has so far been revealed to have 232 miRNAs, of which 83 were determined to belong to 18 families. Among them, all of the terrestrial plants under investigation appeared to have significant conservation of the miR156, miR319 and miR390 families, pointing to a potential early evolutionary history (Crescente et al., 2022).

Conserved wheat miRNAs, including other monocot plants, such as rice, maize, sorghum, barley and *Festuca arundinacea*, share TamiR2019, TamiR506, TamiR510, TamiR514 and TamiR516. Examining the expression patterns of drought-responsive miRNAs and how they vary depending on the kind of tissue under drought stress has also yielded further significant information. Since osmotic pressure and high water potential are directly maintained by the roots and leaves, these are the plant portions most susceptible to drought (Alshammari, 2022).

Microcomputers are sensitive to temperature. The detrimental consequences of heat stress primarily affect respiration, protein turnover, photosynthesis, nitrogen metabolism. The metabolism of lipids and nucleic acids has been thoroughly investigated in connection with RNA. MiR169-SBP, miR159-MYB, miR156-SBP/SPL, miR172-AP2/ERF, miR164-NAC, miR166-HD zip, miR396-GRF and miR5381-SAC2 are regulatory miRNA-target modules associated with heat stress and they have been demonstrated to be present in a variety of sensitive and resistant wheat cultivars (Moharana et al., 2024). Plants can withstand heat thanks to miR156, the first miRNA found in plants that is highly stimulated by high temperatures. It is also known that high temperatures cause transcriptional downregulation of the SPL TF, the miR156 target gene. Wheat was revealed to contain over 132 mature miRNAs and 37 unique miRNAs, six of which were proven to be heat-responsive (Saroha et al., 2024).

Prior to approving an experimental strategy, a metadata analysis, or a computational evaluation of non-coding transcripts, it is usually preferable to check for some basic information about the transcripts. The perfect model crop is maize (*Zea mays*), which effectively supports the favoured method of ncRNA identification by computational evaluations in conjunction with the available genomics datasets (Wang et al., 2023). According to computational research, the recently discovered lncRNAs in maize may regulate the expression of other genes through several RNA-mediated pathways. Comprehensive research is being done to determine how ncRNAs function in maize under various abiotic stressors (Figure 7.2).

7.5 FUNCTION OF NON-CODING RNAs AGAINST DIVERSE ABIOTIC STRESSORS

The growth and productivity of maize can be considerably reduced by drought stress, and the variety of maize needs to be modified in order to increase resistance to this abiotic stress. Reports also show that ncRNA actively participates in plant regulation as a major product. It was discovered that 664 lncRNAs directly correlated with the drought response. Precursors of microRNAs involved in the maize drought stress response, including miR172c, miR169d, miR167j, miR399b, miR169h, miR399e and miR827, comprise a subset of these lncRNAs. It has been discovered that a variety of lncRNAs attached to these CDS-derived proteins' expression is controlled by CDS sequences, which helps to suppress transcription or post-transcriptional processes under stress. These lncRNAs can induce the expression of genes that react to drought stress by endogenous target mimicking, antisense transcription-based regulation, chromatin remodelling and TF control (Cheng et al., 2024).

One abiotic stressor that can have a major impact on maize growth is heat, which can change the plant's physiology, productivity and metabolism. To assist plants withstand heat stress, many crops generate a variety of heat-shock proteins, also known as chaperones, phytohormones and antioxidant enzymes. There are DE-lncRNAs that respond to heat stress; 423 lncRNAs show downregulation, whereas 570 lncRNAs show upregulation. The amount of heat-responsive lncRNA and its expression in maize can be influenced by genetic variation and inbreeding (Sharma et al., 2024). Analysis of structural properties and expression levels has revealed a major difference between lncRNAs and microRNAs: reduced expressions of lncRNA content

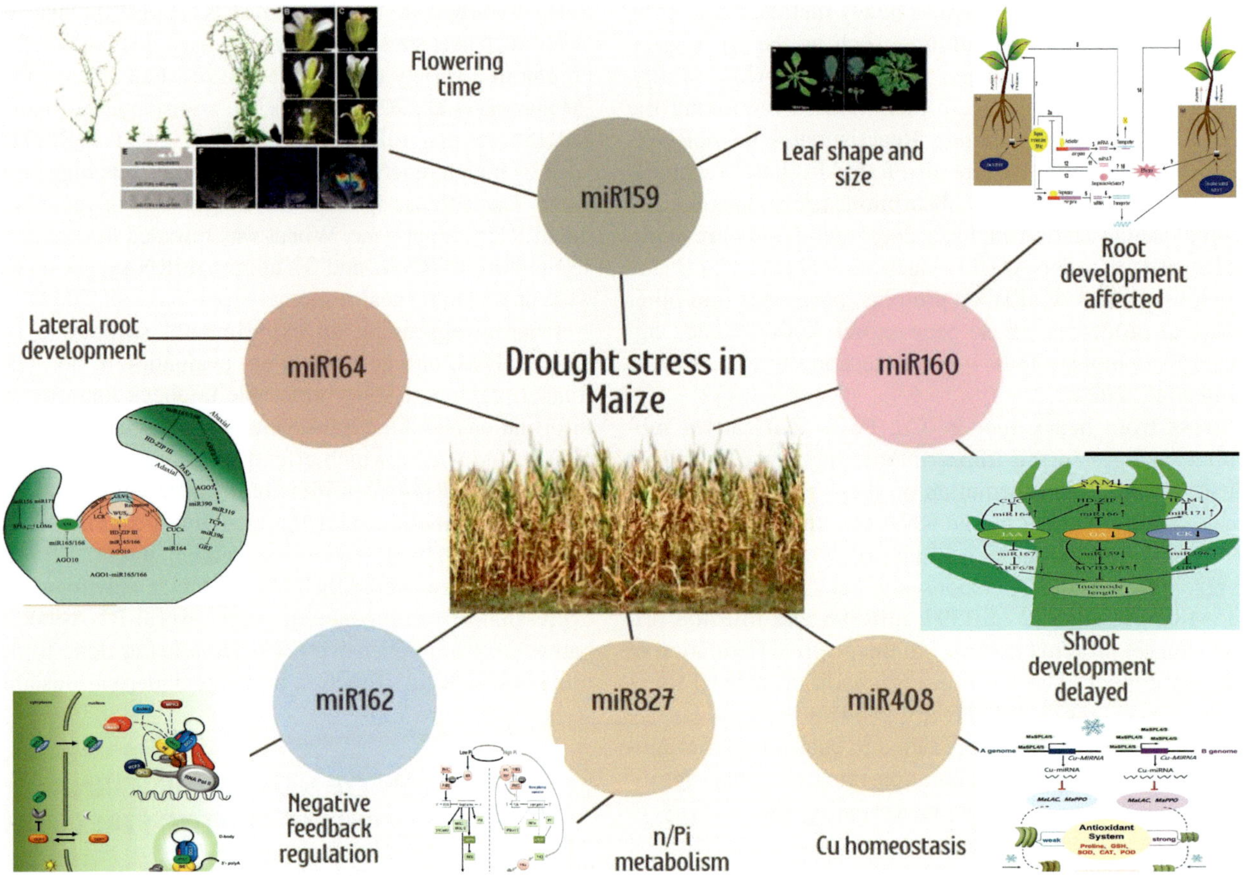

FIGURE 7.2 Drought stress in Maize.

during heat stress. Since most target genes cannot produce any proteins, these lncRNAs help them regulate heat stress through cis- or trans-mediated pathways. According to the GO and KEGG studies, heat stress may be regulated by DE-lncRNAs and other related gene homologs and have varying degrees of expressivity. To lessen the stress impact, target gene expression is directed by intricately linked regulatory networks composed of mRNAs, lncRNAs and miRNAs. These networks produce hormones and enzymes that respond to heat (Shi et al., 2023).

Nutrient availability is the most important factor in determining crop health, stress tolerance and yield when growing in soil. Stress is caused by a deficiency of phosphorus, a macronutrient that plants utilise. A lncRNA called Pi-deficiency-induced lncRNA1 (PILINCR1) can successfully affect the miR399-controlled breaking of PHO2 in maize with reduced Pi tolerance. A portion of microRNAs are connected to long non-coding RNAs (lncRNAs) that are essential for poor nutrition control. An lncRNA associated with the miR159 regulates GAMYB-like gene expression in plant seeds and anthers (Pei et al., 2024). When combined, these cause cellular processes that are frequently linked to low nutritional stress, such as GA-induced aleurone production. Low nitrogen stress in maize is regulated by a few lncRNAs, such as Zm00001d048998, Zm00001d051804, Zm00001d049380 and Zm00001d051666, which have been demonstrated to control neighbouring and remote genes through trans- and cis-processes (Liu et al., 2022).

It was found that a few selected legume crops' lncRNA sequences were similar, such as *Medicago truncatula*, chickpea and soybean, as well as in a model plant. Furthermore, in order to ascertain the functional roles of ncRNAs, ncRNA databases for numerous bean plants, such as SoyKB, have been constructed (Mishra et al., 2023). Twenty-four families of novel miRNAs were very recently discovered. Long and small ncRNAs frequently modulate stress tolerance in any plant (Zhang et al., 2024). In soybean, 256 unique miRNAs engaged under both biotic and abiotic stressors (rust-susceptible and resistant soybeans and drought-sensitive and tolerant soybean) were found. It has also been found that six families of conserved miRNAs exist in other plant species. In soybeans, 22 families have already been documented in earlier research (Aleem et al., 2022).

The regulation of germination, growth, maturity and stages of seed development of chickpeas has been discovered to be influenced by a number of miRNAs. In situations of drought stress, miR408 was shown to enhance or increase chickpeas' resilience to drought stress. Both shoots and roots showed significantly higher levels of miR398a,b, and miR408 under drought stress; however, it was found that roots had downregulated production of another M. truncatula miRNA with varied expression in different tissues was miR169. Chickpeas exhibited DE-lncRNA under drought and salt stress. Additionally, chickpeas exposed to salt and dryness showed an increase in heme-3-lysine-9 (H3K9) acetylation (Khemka et al., 2022). Soybeans are a well-known crop of high-protein legumes that can tolerate more water stress than maize. Growing research suggests that drought-responsive miRNAs may play a part in soybeans and other plant species. A total of 5372 circular RNAs have been identified by research from the root, shoot and leaf of soybean plants. The majority of these circular RNAs (around 80%) come from paralogous genes (Rasheed et al., 2022). Histone deacetylase genes were found to express more when common beans were subjected to cold stress (Yung et al., 2023). In common beans, the expression of the miRNAs has been observed to change in response to a variety of nutrient deficits, such as phosphorus (miR399, upregulated) and nitrogen (miR399, downregulated) (Islam et al., 2022).

7.6 CONCLUSION

With the use of cutting-edge biotechnological instruments and methods, research on plant stress tolerance based on ncRNA has accelerated recently. Significant epigenetic regulatory roles of lncRNAs include histone changes, chromatin-mediated repression, transcriptional interference-regulated imprinting, remodelling of chromatin structure and epigenetic gene control. Additionally, lncRNAs are essential for RNA modification, cell development and differentiation, inheritance, and evolution. Furthermore, by functioning target genes can be cis- or trans-regulated by lncRNAs as byproducts, signals, decoys, guides, scaffolds, enhancers or promoter activators. However, more details regarding the ways in which lncRNAs regulate are anticipated to become available soon given the steady advancement of research and development on the subject.

REFERENCES

Abbas, A., Shah, A. N., Tanveer, M., Ahmed, W., Shah, A. A., Fiaz, S., & Ullah, S. (2022). MiRNA fine tuning for crop improvement: Using advance computational models and biotechnological tools. *Molecular Biology Reports*, *49*(6), 5437–5450.

Aerts, N., Pereira Mendes, M., & Van Wees, S. C. (2021). Multiple levels of crosstalk in hormone networks regulating plant defense. *The Plant Journal*, *105*(2), 489–504.

Ahmad, Y., Haakim, Z., Iqbal, J., Abbasi, B. A., Mahmood, T., Iqbal, R., & Murtaza, G. (2024). Crosstalk between ncRNAs and stress signaling pathways unmasking the regulatory networks. In *ncRNAs: Mediated regulation: The unarguable biological function to environmental constrains* (pp. 265–282). Springer Nature Switzerland.

Ahmad, Y., Haider, S., Iqbal, J., Abbasi, B. A., Yaseen, T., & Mahmood, T. (2022). The mechanisms of genome editing technologies in crop plants. In *Principles and practices of OMICS and genome editing for crop improvement* (pp. 295–313). Springer International Publishing.

Ahmad, Y., Haider, S., Iqbal, J., Naseer, S., Attia, K. A., Mohammed, A. A., & Mahmood, T. (2024). In-silico analysis and transformation of OsMYB48 transcription factor driven by CaMV35S promoter in model plant–*Nicotiana tabacum* L. conferring abiotic stress tolerance. *GM Crops & Food*, *15*(1), 130–149.

Aleem, M., Aleem, S., Sharif, I., Wu, Z., Aleem, M., Tahir, A., & Ahmad, P. (2022). Characterization of SOD and GPX gene families in the soybeans in response to drought and salinity stresses. *Antioxidants*, *11*(3), 460.

Ali, I., Salah, K. B. H., Sher, H., Ali, H., Ullah, Z., Ali, A., & Mahmood, T. (2022). Drought stress enhances the efficiency of floral dip method of Agrobacterium-mediated transformation in *Arabidopsis thaliana*. *Brazilian Journal of Biology*, *84*, e259326.

Alshammari, M. (2022). *Characterisation of miRNA during drought stress in wheat plants* (Doctoral dissertation, University of East Anglia).

Asghar, A., Ullah, Z., Ali, T., Iqbal, J., Abbasi, B. A., Ijaz, S., & Mahmood, T. (2024). Introduction to nc-RNA and their crucial role in biological regulation under environmental stress. In *ncRNAs: Mediated regulation: The unarguable biological function to environmental constrains* (pp. 1–46). Springer Nature Switzerland.

Avramova, Z. (2015). Transcriptional 'memory' of a stress: Transient chromatin and memory (epigenetic) marks at stress-response genes. *The Plant Journal*, *83*(1), 149–159.

Bai, R., Wan, R., Wang, L., Xu, K., Zhang, Q., Lei, J., & Shi, Y. (2021). Structure of the activated human minor spliceosome. *Science*, *371*(6535), eabg0879.

Beretta, V. M., Franchini, E., Ud Din, I., Lacchini, E., Van den Broeck, L., Sozzani, R., … & Kater, M. M. (2023). The ALOG family members OsG1L1 and OsG1L2 regulate inflorescence branching in rice. *The Plant Journal*, *115*(2), 351–368.

Bhadouriya, S. L., Mehrotra, S., Basantani, M. K., Loake, G. J., & Mehrotra, R. (2021). Role of chromatin architecture in plant stress responses: An update. *Frontiers in Plant Science*, *11*, 603380.

Cheng, B., Pei, W., Wan, K., Pan, R., & Zhang, W. (2024). LncRNA cis-and trans-regulation provides new insight into drought stress responses in wild barley. *Physiologia Plantarum*, *176*(4), e14424.

Crescente, J. M., Zavallo, D., Del Vas, M., Asurmendi, S., Helguera, M., Fernandez, E., & Vanzetti, L. S. (2022). Genome-wide identification of MITE-derived microRNAs and their targets in bread wheat. *BMC Genomics*, *23*(1), 154.

Dhar, G. A., Saha, S., Mitra, P., & Nag Chaudhuri, R. (2021). DNA methylation and regulation of gene expression: Guardian of our health. *The Nucleus*, *64*(3), 259–270.

Fontemaggi, G. (2023). Non-coding RNA regulatory networks in post-transcriptional regulation of VEGFA in cancer. *IUBMB Life*, *75*(1), 30–39.

Geng, A., Lian, W., Wang, Y., Liu, M., Zhang, Y., Wang, X., & Chen, G. (2024). The molecular mechanism of the response of rice to arsenic stress and effective strategies to reduce the accumulation of arsenic in grain. *International Journal of Molecular Sciences*, *25*(5), 2861.

Ghosh, S., Adhikari, S., Adhikari, A., & Hossain, Z. (2022). Contribution of plant miRNAome studies towards understanding heavy metal stress responses: Current status and future perspectives. *Environmental and Experimental Botany*, *194*, 104705.

Gill, S. S., Khan, N. A., Agarwala, N., Singh, K., Sunkar, R., & Tuteja, N. (2024). ncRNAs in plant development and stress responses. *Plant Physiology and Biochemistry*, 108950. doi: 10.1016/j.plaphy.2024.108950.

Haider, S., Raza, A., Iqbal, J., Shaukat, M., & Mahmood, T. (2022). Analyzing the regulatory role of heat shock transcription factors in plant heat stress tolerance: A brief appraisal. *Molecular Biology Reports*, *49*(6), 5771–5785.

Havstad, J. C., & Palazzo, A. F. (2022). Not functional yet a difference maker: Junk DNA as a case study. *Biology & Philosophy*, *37*(4), 29.

Hazra, S., Moulick, D., Mukherjee, A., Sahib, S., Chowardhara, B., Majumdar, A., … & Dey, A. (2023). Evaluation of efficacy of non-coding RNA in abiotic stress management of field crops: Current status and future prospective. *Plant Physiology and Biochemistry*, *203*, 107940.

Herman, A. B., Tsitsipatis, D., & Gorospe, M. (2022). Integrated lncRNA function upon genomic and epigenomic regulation. *Molecular Cell*, *82*(12), 2252–2266.

Hossain, A., Skalicky, M., Brestic, M., Maitra, S., Ashraful Alam, M., Syed, M. A., … & Islam, T. (2021). Consequences and mitigation strategies of abiotic stresses in wheat (*Triticum aestivum* L.) under the changing climate. *Agronomy*, *11*(2), 241.

Hu, L., Wang, X., Guo, S., Cao, M., Kang, Y., Ding, Z., & Guo, X. (2024). Whole-transcriptome sequencing analysis to identify key circRNAs, miRNAs, and mRNAs in the development of yak testes. *BMC Genomics*, *25*(1), 824.

Ijaz, S., Iqbal, J., Abbasi, B. A., Yaseen, T., Ullah, Z., Alqahtani, T., & Mahmood, T. (2024). nc-RNA: A key player in stress response pathways, a molecular perspective. In *ncRNAs: Mediated regulation: The unarguable biological function to environmental constrains* (pp. 139–163). Springer Nature Switzerland.

Imaduwage, I. U. H., & Hewadikaram, M. (2024). Predicted roles of long non-coding RNAs in abiotic stress tolerance responses of plants. *Molecular Horticulture*, *4*(1), 20.

Islam, W., Tauqeer, A., Waheed, A., & Zeng, F. (2022). MicroRNA mediated plant responses to nutrient stress. *International Journal of Molecular Sciences*, *23*(5), 2562.

Jiang, Q., Feng, M., Ye, C., & Yu, X. (2022). Effects and relevant mechanisms of non-antibiotic factors on the horizontal transfer of antibiotic resistance genes in water environments: A review. *Science of The Total Environment*, *806*, 150568.

Khemka, N., Rajkumar, M. S., Garg, R., & Jain, M. (2022). Genome-wide analysis suggests the potential role of lncRNAs during seed development and seed size/weight determination in chickpea. *Planta*, *256*(4), 79.

Li, S., Zhao, Y., Tan, S., & Li, Z. (2024). Non-coding RNAs and leaf senescence: Small molecules with important roles. *Plant Physiology and Biochemistry*, 108399.

Liu, P., Zhang, Y., Zou, C., Yang, C., Pan, G., Ma, L., & Shen, Y. (2022). Integrated analysis of long non-coding RNAs and mRNAs reveals the regulatory network of maize seedling root responding to salt stress. *BMC Genomics*, *23*(1), 16.

Magar, N. D., Shah, P., Barbadikar, K. M., Bosamia, T. C., Madhav, M. S., Mangrauthia, S. K., … & Sundaram, R. M. (2023). Long non-coding RNA-mediated epigenetic response for abiotic stress tolerance in plants. *Plant Physiology and Biochemistry*, Jan;206: 108165.

Mishra, G. P., Dikshit, H. K., Devi, J., Aski, M. S., & Durgesh, K. (2023). Epigenetics of abiotic stress tolerance in legumes. In *Legumes: Physiology and molecular biology of abiotic stress tolerance* (pp. 89–114). Springer Nature Singapore.

Moharana, M., Khan, F., & Pattanayak, S. K. (2024) Strategic synthesis of diagnostic novel materials against infectious diseases: Past, present, and future. In *Nanomaterials and point of care technologies* (pp. 158–173). CRC Press.

Morselli, M., & Dieci, G. (2022). Epigenetic regulation of human non-coding RNA gene transcription. *Biochemical Society Transactions*, *50*(2), 723–736.

Nadarajah, K. K., & Abdul Rahman, N. S. N. (2021). The role of non-coding RNA in rice immunity. *Agronomy*, *12*(1), 39.

Najafi, S., Tan, S. C., Raee, P., Rahmati, Y., Asemani, Y., Lee, E. H. C., … & Aghamiri, S. (2022). Gene regulation by antisense transcription: A focus on neurological and cancer diseases. *Biomedicine & Pharmacotherapy*, *145*, 112265.

Nandi, S., Das, A., Singh, S., Kumari, K., Kumar, B., & Chaudhary, K., … & Dey, S. S. (2021) Deciphering differential mRNA and lncRNA expression profiles in response to drought stress in cucumber (*Cucumis sativus* L.). *Available at SSRN 4949270.*

Nojima, T., & Proudfoot, N. J. (2022). Mechanisms of lncRNA biogenesis as revealed by nascent transcriptomics. *Nature Reviews Molecular Cell Biology*, *23*(6), 389–406.

Osnato, M., Cota, I., Nebhnani, P., Cereijo, U., & Pelaz, S. (2022). Photoperiod control of plant growth: Flowering time genes beyond flowering. *Frontiers in Plant Science*, *12*, 805635.

Ozturk, M., Metin, M., Altay, V., Bhat, R. A., Ejaz, M., Gul, A., … & Kawano, T. (2022). Arsenic and human health: Genotoxicity, epigenomic effects, and cancer signaling. *Biological Trace Element Research*, Mar;200(3):988–1001..

Patra, G. K., Gupta, D., Rout, G. R., & Panda, S. K. (2023). Role of long non coding RNA in plants under abiotic and biotic stresses. *Plant Physiology and Biochemistry*, *194*, 96–110.

Pecorino, L. (2021). *Molecular biology of cancer: Mechanisms, targets, and therapeutics*. Oxford University Press.

Pei, L., Gao, X., Tian, X., Liu, N., Chen, M., Fernie, A. R., & Li, H. (2024). A microRNA528-ZmLac3 module regulates low phosphate tolerance in maize. *The Plant Journal*, Jun;118(6):2233–2248.

Poliseno, L., Lanza, M., & Pandolfi, P. P. (2024). Coding, or non-coding, that is the question. *Cell Research*, *34*(9), 609–629.

Rasheed, A., Mahmood, A., Maqbool, R., Albaqami, M., Sher, A., Sattar, A., … & Wu, Z. (2022). Key insights to develop drought-resilient soybean: A review. *Journal of King Saud University-Science*, *34*(5), 102089.

Ricroch, A. E., Martin-Laffon, J., Rault, B., Pallares, V. C., & Kuntz, M. (2022). Next biotechnological plants for addressing global challenges: The contribution of transgenesis and new breeding techniques. *New Biotechnology*, *66*, 25–35.

Rokonuzzaman, M., Li, W. C., Man, Y. B., Tsang, Y. F., & Ye, Z. (2022). Arsenic accumulation in rice: Sources, human health impact and probable mitigation approaches. *Rice Science*, *29*(4), 309–327.

Ross, C. J., & Ulitsky, I. (2022). Discovering functional motifs in long noncoding RNAs. *Wiley Interdisciplinary Reviews: RNA*, *13*(4), e1708.

Saroha, M., Arya, A., Singh, G., & Sharma, P. (2024). Genome-wide expression analysis of novel heat-responsive microRNAs and their targets in contrasting wheat genotypes at reproductive stage under terminal heat stress. *Frontiers in Plant Science*, *15*, 1328114.

Sharma, N., Sharma, L., Onkarappa, D., Yogendra, K., Bose, J., & Sharma, R. A. (2024). Molecular basis and engineering strategies for transcription factor-mediated reproductive-stage heat tolerance in crop plants. *Agronomy*, *14*(1), 159.

Shi, R., Ma, R., Jiang, X., Tang, X., Gong, Y., Yu, Z., & Shi, Y. (2023). Implications of lncRNAs and circRNAs in psoriasis: A review. *RNA Biology*, *20*(1), 334–347.

Song, L., Fang, Y., Chen, L., Wang, J., & Chen, X. (2021). Role of non-coding RNAs in plant immunity. *Plant Communications*, *2*(3).

Staněk, D. (2021). Long non-coding RNAs and splicing. *Essays in Biochemistry*, *65*(4), 723–729.

Statello, L., Guo, C. J., Chen, L. L., & Huarte, M. (2021). Gene regulation by long non-coding RNAs and its biological functions. *Nature Reviews Molecular Cell Biology*, *22*(2), 96–118.

Tiwari, M. (2024). The role of microRNA in stress signaling and adaptive response in plants. In *Stress biology in photosynthetic organisms: Molecular insights and cellular responses* (pp. 87–106). Springer Nature Singapore.

Tu, Z., Xia, H., Yang, L., Zhai, X., Shen, Y., & Li, H. (2022). The roles of microRNA-long non-coding RNA-mRNA networks in the regulation of leaf and flower development in Liriodendron chinense. *Frontiers in Plant Science*, *13*, 816875.

Verma, S., Negi, N. P., Pareek, S., Mudgal, G., & Kumar, D. (2022). Auxin response factors in plant adaptation to drought and salinity stress. *Physiologia Plantarum*, *174*(3), e13714.

Wang, W., Min, L., Qiu, X., Wu, X., Liu, C., Ma, J., … & Zhu, L. (2021). Biological function of long non-coding RNA (LncRNA) Xist. *Frontiers in Cell and Developmental Biology*, *9*, 645647.

Wang, Z., Cui, Q., Su, C., Zhao, S., Wang, R., Wang, Z., … & Luan, Y. (2023). Unveiling the secrets of non-coding RNA-encoded peptides in plants: A comprehensive review of mining methods and research progress. *International Journal of Biological Macromolecules*, *242*, 124952.

Wiener, D., Antebi, Y., & Schwartz, S. (2021). Decoupling of degradation from deadenylation reshapes poly (A) tail length in yeast meiosis. *Nature Structural & Molecular Biology*, *28*(12), 1038–1049.

Wierzbicki, A. T., Blevins, T., & Swiezewski, S. (2021). Long noncoding RNAs in plants. *Annual Review of Plant Biology*, *72*(1), 245–271.

Xu, Y. C., Zhang, J., Zhang, D. Y., Nan, Y. H., Ge, S., & Guo, Y. L. (2021). Identification of long noncoding natural antisense transcripts (lncNATs) correlated with drought stress response in wild rice (*Oryza nivara*). *BMC Genomics*, *22*(1), 424.

Yadav, A., Mathan, J., Dubey, A. K., & Singh, A. (2024). The emerging role of non-coding RNAs (ncRNAs) in plant growth, development, and stress response signaling. *Non-Coding RNA*, *10*(1), 13.

Yang, Y., Huang, J., Sun, Q., Wang, J., Huang, L., Fu, S., … & He, Y. (2022). microRNAs: Key players in plant response to metal toxicity. *International Journal of Molecular Sciences*, *23*(15), 8642.

Yung, W. S., Huang, C., Li, M. W., & Lam, H. M. (2023). Changes in epigenetic features in legumes under abiotic stresses. *The Plant Genome*, *16*(4), e20237.

Zhang, B., Fei, Y., Feng, J., Zhu, X., Wang, R., Xiao, H., … & Huang, J. (2022). RiceNCexp: A rice non-coding RNA co-expression atlas based on massive RNA-seq and small-RNA seq data. *Journal of Experimental Botany*, *73*(18), 6068–6077.

Zhang, D., Ma, Y., Naz, M., Ahmed, N., Zhang, L., Zhou, J. J., … & Chen, Z. (2024). Advances in circRNAs in the past decade: Review of circRNAs biogenesis, regulatory mechanisms, and functions in plants. *Genes*, *15*(7), 958.

Zhang, J., Lv, J., Qin, J., Zhang, M., He, X., Ma, B., … & Hong, Z. (2024). Unraveling the mysteries of early embryonic arrest: Genetic factors and molecular mechanisms. *Journal of Assisted Reproduction and Genetics*, Dec;41(12):3301–3316.

Zhang, Y. Y., & Ning, B. T. (2021). Signaling pathways and intervention therapies in sepsis. *Signal Transduction and Targeted Therapy*, *6*(1), 407.

Zhou, R., Sanz-Jimenez, P., Zhu, X. T., Feng, J. W., Shao, L., Song, J. M., & Chen, L. L. (2021). Analysis of rice transcriptome reveals the lncRNA/circRNA regulation in tissue development. *Rice*, *14*, 1–16.

8 ncRNAs and Their Impact on Stress Tolerance in Plants

Saleem Ur Rahman, Ishtiaq Ali, Nasir Ali and Ehtisham Shafique

8.1 INTRODUCTION

An important feature of plants as sessile organisms is that they always exist in an environment that presents varied potential stresses that can influence their growth, development and reproduction. Of these stresses, drought, salinity, temperature stresses and toxic heavy metals are the most damaging (Zhu, 2016). These stresses not only reduce plant output but also endanger world food security, particularly due to climate change. It is therefore important to know how plants perceive, respond and adapt to these abiotic stresses in order to use this knowledge to improve stress tolerance in crops. In this regard, the current development in molecular biology has unfolded the importance of non-coding RNAs (ncRNAs) in plant responses to abiotic stress and has given new perspectives into the complex gene regulatory networks that are involved in stress tolerance in plants (Zhu, 2016).

The long non-coding RNAs (lncRNAs) are a large group of short RNA molecules that do not code for proteins but are involved in the regulation of gene expression and appear to be involved in transcriptional, post-transcriptional and epigenetic regulation. Of these, microRNAs (miRNAs), small interfering RNAs (siRNAs) and lncRNAs have been well documented in plants. These ncRNAs play a role in the regulation of stress-responsive genes to enable plants to respond to dynamic condition in the environment. As transcriptional regulators, they play an indispensable role in regulating plant responses to abiotic stress, including water and nutrients transportation, distribution of ions, scavenging of reactive oxygen species (ROS) and signalling of hormones etc. (Jones-Rhoades et al., 2006).

The identification of miRNAs in plants revolutionised the understanding of this level of gene regulation. The miRNAs are usually 21–24 nucleotides molecules that, through the RNA-induced silencing complex (RISC), provide target mRNAs for degradation or translational repression. This mechanism of action makes it possible for the miRNAs to regulate the expression of a number of genes into stress responses. For example, miR159, miR398 and miR393 are involved in important stress-associated networks; abscisic acid (ABA) signalling, detoxification of ROS and root morphogenesis (Law & Jacobsen, 2010).

Besides the miRNAs, siRNAs are also known to play roles in plant stress responses. siRNAs are derived from long dsRNA precursors and are mainly involved in RNAi pathway. This pathway is involved in immune response to viruses and transposable elements, and also in regulation of endogenous genes under abiotic stress conditions: siRNAs, nat-siRNAs and phasiRNAs are involved in regulation of genes related to salt stress, temperature and heavy metal toxicity.

Another class of ncRNA is the lncRNAs, which are generally greater than 200 nucleotides in length and have recently received much consideration. While miRNAs and siRNAs operate mainly at the post-transcriptional level, lncRNA are involved in diverse directions of regulations such as chromatin remodelling, transcriptional regulation and miRNA regulation among others, lncRNA acts as molecular sponges, endogenous competitors or channellers, binding to proteins, DNA and other RNA molecules. For instance, DRIR (drought-induced lncRNA) and COLDAIR (cold-assisted intronic RNA) have been reported to regulate the stress response to drought and cold through the regulation of hormone signalling and chromatin remodelling, respectively.

Ability of ncRNAs to modulate responses to abiotic stress involves their interoperation with other species of molecules that belong to the stress response network: transcription factors, signalling molecules and other ncRNAs. For example, effects of miRNAs on the stress-responsive genes may be mediated by the modulation of the transcription factors. In the same way lncRNA can interact directly with miRNA and facilitate their absorption thus controlling their actions on target mRNAs.

In addition, the current research has also shown that ncRNAs are involved in epigenetic regulation under abiotic stress conditions. Imprinting and small ncRNAs are involved in DNA methylation and histone modifications that respond to environmental stress by adjusting gene expression. siRNAs are major components of RNA-directed DNA methylation (RdDM), which is specifically responsible for TE silencing and, thereby, genome stability. Further, lncRNAs have been shown to target chromatin-modifying complexes to the desired genomic location and affect chromatin structure and gene expression under stress (Xie et al., 2017).

The functional versatility of ncRNAs and their participation in various regulatory networks support their role in plant stress biology. With increased research in this field, new ncRNAs and their target genes are discovered and consequently new regulatory mechanisms that play a role in stress tolerance. These discoveries have important implications for agriculture because they offer new objectives for

DOI: 10.1201/9781003583257-8

the genetic manipulation and breeding of crops to enhance their resistance to abiotic stress (Liu et al., 2014).

The prospect for ncRNA research in crop enhancement is enormous. The study shows that by regulating the levels of certain ncRNAs, it is possible to increase stress tolerance in crops and, therefore, increase yields and productivity of crops in conditions of unfavourable environmental influences. For instance, the study has demonstrated that rice with overexpression of miR156 enhances drought tolerance due to alteration in SPL transcription factors. Likewise, reduced expression of miR398 in Arabidopsis improves the ability of the plant to withstand oxidative damage due to up-regulation of CSDs that are involved in the removal of ROS.

This is also useful in breeding programmes as we try to establish stress-responsive ncRNAs for selection of stress-tolerant varieties. These traits can be introduced into the elite crop varieties using the marker-assisted selection (MAS) and genomic selection (GS) application with the aim of improving stress-tolerant crop breeding. In addition, uncovering the functions of ncRNA genes enables precise modification via genome editing tools, including CRISPR/Cas9, opening up a new avenue for improving stress tolerance in species of plants (Zhang et al., 2018).

However, there are still many unknowns regarding the participation of ncRNAs in abiotic stress response mechanisms. The functional annotation of ncRNAs in various plant species and under diverse stress conditions remains limited, and much more effort is required to unravel the entire range of ncRNA-dependent regulatory pathways. However, the relationships between ncRNAs and other molecules including proteins, metabolites and hormones have to be studied further to understand how plants use multiple signals to develop an efficient stress response (Liu et al., 2014).

ncRNAs are an interesting and dynamically developing area of study in plant biology. They are involved in abiotic stress tolerance in a complex manner, including post-transcriptional regulation, epigenetic regulation and transcriptional regulation. The more we learn about the molecular roles of ncRNA, the closer we get to finding new approaches to enhancing crop performance in the context of growing environmental pressures. The information derived from ncRNA studies offers a great potential for increasing the sustainability of agriculture and guaranteeing food safety in the twenty-first century.

8.2 OVERVIEW OF ncRNAs

ncRNAs are broadly classified based on their length and function.

8.2.1 MicroRNAs (miRNAs)

miRNAs are endogenous, non-coding small RNA (sRNA) species derived from plants that are critically involved in gene expression regulation to control plant growth, development and stress tolerance. These miRNAs are normally between 20 and 24 nucleotides in size and act through complementary base pairing with mRNAs with issues of either mRNA cleavage or translation repression. Plant miRNAs were initially identified in *Arabidopsis thaliana*, a model plant species for plant molecular biology research, and thousands of different miRNAs have been subsequently described in numerous plant species. These molecules are required for modulating gene expression in response to both internal developmental signals as well as external environmental stimuli (Jones-Rhoades et al., 2006).

Plant miRNAs are largely involved in controlling the temporal and spatial expression of genes that are involved in development processes, including leaf shape, root branching and flowering. For instance, the miR156 family acts on SPL, which are important for transition from juvenile to adult phase in plants. The miR164 family, in contrast, targets NAC transcription factors that play a role in leaf and flower formation, such as senescence or organ abscission. Thus, miRNAs regulate those developmental genes to timely suppress or promote their expression to make sure that plants' morphology and structure well fit for the current growth environment and developmental stage (Wu, Fang, et al., 2009).

Plant miRNAs also have important functions in stress responses and enable plants to cope with biotic and abiotic stress. Certain miRNAs are overexpressed or underexpressed to alter the cluster of genes helpful for combating stress that include drought, salinity and nutrient deficiency. For instance, miR393 was reported to improve resistance to pathogens through genes controlling auxin signalling – a growth-responsive hormone signalling that is vital during defence processes, although it antagonises growth. On the other hand, miR398 is reported to target superoxide dismutase (SOD) genes that facilitate the removal of O_2 during stress conditions. These stress-responsive miRNAs allow the plants to physiologically modify in order to survive and perform optimally under stress (Sunkar et al., 2012).

Besides their individual functions, plant miRNAs are components of intricate signalling networks through which plants can transduce multiple signals and respond accordingly. A significant number of plant miRNAs are controlled by transcription factors that are sensitive to environmental and developmental cues, thus, feedback and crosstalk between the pathways occur. For instance, miR172 targets the three genes belonging to the APETALA2-like family to regulate flowering time and photoperiod and temperature. This integration ability of plant miRNAs enables them to regulate complex developmental processes such as flowering, which needs the plant to perceive environmental changes. In general, miRNAs are considered as essential components of the regulatory system in plants and help them to accurately and modally adapt the expression of target genes in response to their environment.

8.2.2 Small Interfering RNAs (siRNAs)

Plant siRNAs are 21–24 nucleotides long small, double-stranded RNA molecules that mediate gene regulation and defence. These species are derived from longer dsRNAs by DCL proteins and incorporated into Argonaute (AGO) proteins to form an RISC. Subsequently this complex transports the siRNA to other complementary mRNA molecules and causes their degradation or inhibition of their translation. This RNA interference (RNAi) pathway is central for transcriptional regulation in plants and can also degrade viral RNA and protect the plants from viral infections (Baulcombe, 2004).

Transposon silencing and regulation of genome stability are considered to be the major function of siRNAs in plants. Transposons are mobile genetic elements, MGEs which can disrupt the regulation and function of genes within the genome. Both observations suggest that plants use a class of siRNAs coined repeat-associated siRNAs (rasiRNAs) to repress these ITEs, especially through a process called RNA-directed DNA methylation (RdDM). During this process, siRNAs help in recruiting methylation machinery towards the transposon areas that put methyl groups on DNA and thus downregulates transposon. This silencing system is crucial for genome defence from fragmentation due to transposon mobility (Zilberman et al., 2007).

The siRNAs are also used by plants as an effective way to fight viral diseases that may affect them. Once a virus has entered the plant and taken up its residence within the plant cell, the viral genome is denatured and its double-stranded RNA is detected by DCL proteins that digest the RNA into a virus-sourced siRNA or vsiRNAs. These vsiRNAs are incorporated into the RISC complex in order to target the viral RNA so that it will not be allowed to replicate and spread within the plant. This antiviral response referred as post transcriptional gene silencing (PTGS) is an integral component of plant defence and PTGS is known to occur in many systems. In addition, there are certain viral suppressor proteins that shield them from this defence mechanism that again shows that there is an on-going battle between plants and their viruses (Ding & Voinnet, 2007).

siRNAs in plants are also used in the conversation between cells and even between organisms. Studies have revealed that siRNAs are mobile and functional intercellular and between tissues to manage gene expression in distinct parts of the plant, so that the plant responds uniformly to stress signals. Furthermore, specific plant-derived siRNAs can be transferred to neighbouring organisms, for example fungi or insects and can regulate gene expression within these organisms to deter pests or halt pathogen development. For instance, cotton plants are capable of manufacturing siRNAs that can be ingested by the attacking fungi and render them incompetent for growth. Perhaps, this peculiar interspecies siRNA highlights the fact that, while siRNAs are involved as regulatory RNA within plant species, they may also operate as some sort of external tool for recognition and survival within its environment (Zhang et al., 2016).

8.2.3 Long Non-Coding RNAs (lncRNAs)

In plants, lncRNAs are RNA more than 200 nucleotide in length that do not code for proteins, but are involved in the regulation of gene expression at the transcriptional, post-transcriptional and epigenomic levels. In contrast to other small formatted RNAs, lncRNAs are more diverse in terms of their architecture and function, are often scaffold or decoy which directs the protein complexes towards the target DNA strands or traps other RNAs modulates various cellular activities. In plants, lncRNAs are reported to play various fundamental cellular processes, including growth, development and stress interactions. Research has shown that lncRNAs can originate from any of the genome locations that include intergenic, antisense and intronic regions notwithstanding the fact that the lncRNAs plant transcriptome is relatively complex.

It is evidenced that lncRNAs have an essential regulatory role in the epigenetic regulation of plants, including DNA methyltransferase and histone modification. Some of the lncRNAs interact with chromatin-modifying enzymes, which in turn produce changes in chromatin that result in gene repression or activation. For example, the lncRNA COLDAIR takes part in *A. thaliana* processes that leads to vernalisation, the process by which plant flowers after prolonged exposure to cold. COLDAIR binds to a polycomb repressive complex and guides it to the FLOWERING LOCUS C (FLC) gene repressing its expression through histone methylation and enables flowering in the spring. This example shows that lncRNAs can be regulatory molecules that function in response to environmental stimuli in order to regulate the change of seasons in plants (Heo & Sung, 2011).

lncRNAs are also participated in plant stress responses, in which they appeared to be involved in the regulation of the genes related to biotic and abiotic stress tolerance. LDMAR (Long-Day-Specific Male Sterility-Associated RNA) is a lncRNA that is particularly active in the photoperiod-sensitive male sterility of rice, which is fundamental for producing hybrid seeds. It showed that LDMAR plays a role in reproductive adaptation to environmental change under long-day conditions given that its expression levels influence the development of pollen. Moreover, under drought stress some lncRNA became active to control stress-responsive genes for the growth of plant under stress conditions. These findings further support the role of lncRNAs in allowing plants to functionally plasticise the physiological and developmental aspects of their biology in response to stress.

In addition to direct gene regulation, plant lncRNAs also include molecular sponges that suppress miRNAs; it affects mRNA targets of such miRNAs. These are termed competitive endogenous RNAs (ceRNAs), whereby lncRNAs interact with miRNA and block their action to repress their target mRNAs thus controlling plant genes expression networks. For instance, the lncRNA PtCENL1 can control the muffled of *Populus trichocarpa* by sequestering miR156 thus regulating the genes concerned in growth regulation. This ceRNA functionality brings an extra layer of regulation, making it possible for the plants to modulate post-transcriptionally the levels of gene expression in the response to internal as well as outside stimuli. Because of its roles as miRNA sponges, lncRNAs are essential molecular entities that enhance the resistance and flexibility of plants in gene regulation networks.

8.3 BIOGENESIS AND FUNCTION OF ncRNAs

Biogenesis of the ncRNAs in plants starts with transcription, which can be done by RNA polymerase II or RNA polymerase III depending on the natures of the ncRNA. A broad classification of ncRNAs includes the small ncRNAs such as miRNAs and siRNAs, and the lncRNAs. These mature miRNAs are incorporated into the RISC with AGO proteins for post-transcriptional gene silencing (Bartel, 2004). siRNAs, in contrast, is derived from double-stranded RNAs and are essential in the process of defence against viruses and transposable element through the RNAi pathway (Baulcombe, 2004).

A larger class is lncRNAs and is usually greater than 200 nucleotides and do not have coding potential. They originate from all parts of the genome located intergenic and intronic regions and are less uniform in their processing than small ncRNAs. SOME lncRNAs are processed by the dicer-like proteins, but other lncRNAs undergo other post-transcriptional modifications. Many of these lncRNAs function as molecular tethers, recruiting certain proteins to particular genomic loci, or forming large RNA-protein particles that regulate gene expression. In contrast to the regulatory roles of miRNAs and siRNAs, lncRNA can be either positive or negative regulators with multiple tasks in plant growth, development or stress response (Liu et al., 2012).

In terms of role, ncRNAs in plants are involved in various genes expression regulation processes such as gene silencing, mRNA stability, translation and others miRNAs for example bind to target mRNAs introducing complementary sequences and cause mRNA degradation or preventing translation afterwards reducing the expression of certain genes. This regulatory mechanism is fundamental to the regulation of diverse developmental processes, including leaf and root initiation, flowering and stress responses. In Arabidopsis, the post-transcriptional regulation of flowering time revealed that the miRNA miR156 is capable of modulating developmental transitions in response to internal signals through targeting SPL gene family (Wu, Park, et al., 2009). On the other hand, siRNA's role includes defence mechanisms of plant cells involving RNA silencing against foreign invades elements like virus through the process of mRNA degrading.

Next to the influence of post-transcriptional regulation, ncRNAs are also involved in epigenetic changes in plants. By recruiting the methylation gears, via the RdDM machinery the siRNAs home in on loci of interests thereby orchestrating transcriptional gene silencing of transposons together with repetitive sequences. This process requires at some levels formaldehyde cross-linking is essential for genome integrity and stability not permitting transposon movement. For example, siRNAs from transposons are used in *A. thaliana* to regulate such transposons through directing DNA methylation so that their activity is inhibited and the genome is stabilised (Zilberman et al., 2007). This function of ncRNAs shows the general ability of ncRNAs to modulate genome structure and function at epigenetic level.

They also work as 'miRNA sponges' hence they can control the mRNA target of these miRNAs through an indirect mechanism. This ceRNA mechanism has been identified in a variety of plant species: lncRNAs which interact with cognate miRNAs and thus inhibit the repression of their target genes. As a consequence, lncRNAs direct the availability of active miRNA to modulate the gene expression networks. For instance, some lncRNAs in *P. trichocarpa* regulate the target genes mediating the developmental process by sponging miR156. Such interaction represents other layers of regulation of ncRNAs within plant cells – not only do they regulate specific gene, but they also manage larger globules of regulation.

In addition, a number of plant ncRNAs play a role in interactions between different kingdoms, which may modulate the gene expression of the interacting organisms themselves. Research also reveals that some plant-derived siRNAs can be transported into the fungal pathogens or insect pests and alter the expression of certain genes that reduce virulence or inhibit growth. For instance, siRNA is transcribed from cotton plants and ingested by fungal pathogens in order to inhibit disease development. This cross-kingdom RNAi is the proof of adaptive roles of plant ncRNAs enabling them not only to control own genes but also to expand their biological functions to intercellular communication, providing plants with a complex mechanism for increasing their resistance to various biotic stresses.

8.4 ROLE OF ncRNAs IN ABIOTIC STRESS TOLERANCE

8.4.1 Drought Stress

Plant ncRNAs including miRNAs, siRNAs and lncRNAs are significant for gene expression regulation related to drought stress, allowing plants to acclimate within water limitation environments. For instance, it has been discovered that miR156 and miR159 tracts are involved in controlling transcription factors that control stomatal characteristic, root structure and a stress reaction pathway that needed for increasing drought tolerance (Zhou et al., 2010). These miRNAs help the plant to fine tune changes in growth pattern and in water use, thus improving the plants' ability to cope with water shortage.

Aside from the miRNAs, the second type of small ncRNAs is the siRNAs that are involved in the drought stress signalling epigenetically main mediators of RdDM that leads to DNA methylation of particular gene loci. During drought conditions, the targeted siRNAs can direct the methylation of stress-inducible genes, therefore switch 'on' or 'off' to reduce their utilisation of resources or always remain active for survival. The drought-related siRNAs have been found to alter the expression of genes related to osmotic adjustment and cellular protection in the rice (*Oryza sativa*) which explains how long-term response to dry conditions works in the plant. Thus, the stable epigenetic modifications involving siRNAs play an important role in more precise regulation of plant's reaction on stress, such as drought, on the genomic level.

lncRNAs include a large category of ncRNAs that are involved in stress response of plants under drought stress condition. Some lncRNAs have roles in miRNAs, mainly through the mechanism of the so-called miRNA sponges, which compete with target mRNAs for binding to miRNAs and thereby modulating gene expression. For example, in maize (*Zea mays*) there is a devoted lncRNA that, through acting as a 'sponge' for miR156, regulates a family of transcription elements significant for drought tolerance (Xin et al., 2011). This lncRNA sequesters miR156 and ensures higher accumulation of these transcription factors under drought conditions which in turn allows the plant to cope with water deficit stress. Interactions mentioned earlier demonstrate how lncRNAs regulate drought tolerance through controlling miRNA levels and activities.

Besides, lncRNAs also have profound role in the direct regulation of transcriptional events during drought stress through their function as protein matrices or through direct interacting with DNA. Specific lncRNAs can influence the activity of drought-responsive genes by recruiting chromatin modifiers that change the chromatin structure. In *A. thaliana*, some lncRNAs have been described to regulate the chromatin state of the stress-related genes and facilitate or repress transcription under drought stress. Thus, lncRNAs are able to regulate the structure of chromatin and, therefore, perform the singular regulatory function for drought stress complementary to the actions of sRNAs.

Besides, ncRNAs in plants are presented as being connected with interacting along the hormone signalling pathways, which is crucial for drought response. Drought, in turn, evokes the elevation of the plant hormone called ABA that controls many protective processes. A lot of drought-responsive miRNAs and lncRNAs are implicated in ABA signalling indicating that these ncRNAs have roles as second messengers harmonising hormonal signals to gene expression remodelling. For instance, miR168 and miR169 are regulated by ABA and participate in gating the stomatal movement and water use efficiency which are two important components of drought stress tolerance (Sunkar & Zhu, 2004). Here we display hormonal regulation of ncRNAs that elucidate the role of these molecules in the organisation of multi-signalling pathways during drought stress.

Last but not least, plant ncRNAs are also engaged in inter-organismal interactions that allow improving drought tolerance. It has been established that some of the drought-responsive ncRNAs are cellular mobile in that they can move from one plant cell to another or they are secreted out to the rhizosphere where they can influence symbiotic endophytes. It has been reported that drought-imposed miRNAs can upregulate symbiotic nitrogen fixation in some legumes by regulating the expression of genes responsible for root nodulation in them to help the plant to deal with nutrient stresses resulting from water deficit (Mabo et al., 2012). These results indicate that plant ncRNAs serve not only functions inside the plant but also can regulate external environment, which offers further precise regulation for improving drought tolerance.

8.4.2 Salinity Stress

Plant ncRNAs involved in 'Salinity Stress Responses' control the genes directly involved in ionic balance, osmotic stress tolerance and stress signalling pathways. It is, therefore, significant since salinity stress causes damage through impossible ionic toxicity and osmotic stress in plants, affecting normal cell functions as well as growth. Of ncRNAs, miRNAs have been studied thoroughly relating to the regulation of stress response genes under salinity stress condition. For instance, miR398 is expected to reduce genes that have the capacity to code for SODs which help to overcome the effects of high salinity levels on the stress signals. In preventing or suppressing SOD activity, miR398 allows the plants in the regulation of ROS in that they prevent the buildup of ROS in cellular structures, thereby affecting the survivability of the plants under salt-stressed conditions.

Besides miRNAs, siRNAs play crucial role in gene regulation in response to salinity stress by altering epigenetic changes. The siRNAs are involved in RNA-directed DNA methylation (RdDM) process that leads to DNA methylation in gene locus and thus repression or activation of the genes. Depending on the type of stress that is salinity stress in the present case, siRNAs can regulate genes that are not required during non-stress situations or can induce stress genes which help in ionic and osmotic regulation. For example, salinity-generated siRNAs have been found that direct DNA methylation modification in rice (*O. sativa*) involved with Na^+ and K^+ transporter genes. This regulation is helpful in keeping up ionic balance within cells which is important for plant survival especially in conditions of high salinity.

lncRNAs are involved in salinity tolerance through the process of miRNA sponge, by which the expression of the target genes of the miRNAs is regulated. In *A. thaliana*, several lncRNAs have been identified to regulate the target miR399 of phosphate homeostasis and stress tolerance. Warning that salinity stress provokes modulation of miR399 levels through these lncRNAs enables increased expression of genes that support ion homeostasis. It also allows plants to give preference to stress-responsive gene expression thus affording resource mean to increase salinity stress resilience while regulating STMN, LBM, LNRs/miRNAs interactions. Thus, the example of interactions between different ncRNAs shows that stress adaptation is regulated by intricate networks.

Additionally, lncRNAs can modulate salinity stress-induced transcriptional changes either through its interaction with proteins that control chromatin accessibility or by being attached to DNA regions that are linked to stress-responsive genes. For instance, several lncRNAs may target chromatin-modifying enzymes to salinity-responsive genes that regulate ions transport and osmotic balance. For example, a certain salt-responsive lncRNA in maize (*Z. mays*) has been identified to control the genes involved in the exclusion of Na^+ from cells, thereby protecting the plant from ionic toxicity potential under salt stress condition. The roles described above in modulating chromatin structure and gene transcription also underline a novel and immediate function of lncRNAs in promoting salinity tolerance.

These ncRNAs are also involved with hormone signalling pathways concerning the regulation of salinity stress response, especially ABA dependent. ABA is the primary hormone involved in stress signalling, and many of the identified salinity responsive miRNAs and lncRNAs are regulated by ABA. For instance, miR169 that targets the gene family encoding the NF-YA transcription factors involved in regulating genes needed for stomatal closure under saline conditions and under ABA control. Thus, it is revealed that miR169 plays a role in stomatal regulation and assists plant to cope with water stress due to both salinity and drought implications; thus, the examples show that ncRNAs are involved in variety roles of stress tolerance (Zhou et al., 2010). This hormonal regulation of ncRNAs also signifies the implication of these ncRNAs into more complex signal transduction pathways connecting extrinsic stress stimuli with intrinsic cellular adjustments.

It has been reported that ncRNAs are involved in symbiotic relations under salinity stress in terms of nutrient uptake and stress endurance. For instance, the legume endogenous miRNAs controlling symbiotic nitrogen fixation are affected by high salinity by targeting genes for root nodulation. In *Medicago truncatula* the miR396 is regulated in both mycorrhizal symbiosis and root development, which are important for stressed environments' nutrient uptake. This also means that plants can coordinate responses to salinity stress by combinations of both ncRNAs and benefits acquired from mutualistic soil microorganisms for a more extensive stress adaptation strategy.

8.4.3 Temperature Stress

ncRNAs are specifically useful in regulating the changes that help plants adapt well to the effects of temperature stress. Heat and cold shocks cause changes in cell membrane permeability, distort metabolic processes and decrease growth and yield in crop plants. With regard to the ncRNAs, miRNAs are particularly important for the regulation of genes associated with temperature stress. For instance, miR398 is well studied in the context of heat stress and involved in the regulation of the genes encoding copper/zinc superoxide dismutases (CSD1 and CSD2), which have antioxidant function to eliminate ROS produced under heat stress conditions. This regulation enhances oxidative damage to be reduced, thus providing cell safety and maintaining stability during variations in the level of heat (Guan et al., 2013).

Besides miRNAs, siRNAs are significant for regulation of plant response to temperature stress targeting gene expression due to transcriptional and post-transcriptional RNA silencing. There is RNA-directed DNA methylation (RdDM) which is involved in gene silencing by adding methyl groups to the DNA sequence. Moreover, importantly siRNAs may modulate or induce the key genes for stress tolerance at temperature stress, for examples, genes related to heat shock proteins and signal transduction pathway. For example, *A. thaliana* is involved in temperature-sensitive siRNAs regulation of heat shock proteins (HSPs) that deter protein denaturation at high temperatures (Ito et al., 2011).

Other type of ncRNA involved in temperature stress tolerance regulation are lncRNAs that can function as molecular sponge, platform or template for other RNA molecules and proteins. lncRNAs can function as ceRNAs to explain that RNA molecules bind and thereby retain miRNAs which should otherwise target and degrade mRNAs encoding stress-response proteins. For instance, under cold stress conditions, some lncRNAs in a model plant Arabidopsis recruit specific miRNAs targeting mRNAs of cold-responsive transcription factors, which thus lets these factors for stress-adaptive pathways accumulation. This miRNA–lncRNA connection is useful for plants so as to respond effectively to a system of protection once they are exposed to low temperatures (Zhang & Swanson, 2014).

Moreover, it has been illustrated that lncRNAs can directly associate with chromatin-modifying enzymes to regulate chromatin states at the temperature stress-responsive genes. Interestingly in rice (*O. sativa*), there is one lncRNA that interacts with chromatin-remodelling complexes that remodify histone markers for genes that are related to heat stress responses. It enables more stable and versatile gene regulation which endows plants a capacity to sense gradual changes in temperature over time and possibly the memory of previous temperature stress, which can improve subsequent stress resistance (Wunderlich & Vollmer, 2017).

ncRNAs also regulate hormonal signalling pathways associated with temperature stress conditions. Stress responsive plant growth hormones are ABA, ethylene and gibberellin hormones all of which have effects on the plant growth. Specific class of miRNAs is elicited to ABA signalling and in managing water use efficiency and stomata closure under heat stress to minimise water loss. As example, miR159 in Arabidopsis is described to target MYB proteins, that are involved in ABA-mediated high temperature response. Through targeting hormone-responsive gene networks, miR159 and other comparable ncRNAs aid the plants adapt to the temperature stress by regulating growth and survival under quite unfavourable environmental conditions (Reyes & Chua, 2007).

ncRNAs are also involved in the interaction of two or maybe more stress response pathways in order to allow the plant to overcome two or maybe more stress types. Stress by temperature is frequently coupled with drought or salinity and some ncRNAs are capable to coordinate different signals of stress. For example, Byrne and colleagues elucidated a role for miR169 as having a function in both temperature and drought stress through targeting NF-YA genes that code for proteins implicated in stomatal closure and water retention. This kind of cross talking no doubt enables the plants to effectively mount an integrated response to dynamic environmental stressors where ncRNAs exhibit quite evident and diverse role in plant stress biology.

8.4.4 Heavy Metal Stress

Plant ncRNAs have been known to play important contributions to heavy metal response mechanisms to regulate genes on heavy metal uptake, transport and detoxification. Cadmium (Cd), lead (Pb), arsenic (As) are systemic heavy metals that can bioaccumulate in tissues and organs of plants and disrupt normal intra-cellular metabolism, leading to oxidative stress as well as cell morbid transformations. These stress responses are controlled by another type of ncRNA called miRNAs that evidence feedback regulation by degrading target mRNAs or repressing their translation. For instance, miR393 is proven to rise in cadmium toxicity in *A. thaliana* and is involved in signalling, particularly auxin signalling pathway that involves TIR1/AFB auxin receptor mRNAs. This regulation assists in controlling root architecture hence minimising cadmium accumulation and its prejudice impact.

Another studied ncRNA is miR398, which participates in the regulation of oxidative stress derived from metal stress. When plants are exposed to metals including copper and cadmium, miR398 is generally downregulated to permit the genes it targets to be expressed and these genes are CSD1 and CSD2 which encode for copper/zinc superoxide dismutases (Cu/Zn SODs). These enzymes assist in detoxification of ROS produced under metal stress and guard the plant cells from oxidative stress. This regulation is crucial because high ROS levels cause lipid peroxidation, protein degradation, DNA damage, and may result in the plant death (Fang et al., 2016).

siRNA has also played pivotal role in heavy metal stress tolerance in plants through a developmentally regulated process called RNA-directed DNA methylation (RdDM). This epigenetic regulation mechanism helps regulate and contain the genes of metal transport and sequestration to enable plants to take control and deposit hazardous metals. For instance, siRNAs have been identified to control the expression of genes related to metal transport in rice (*O. sativa*). The siRNAs are involved in teasing the genes responsible for the limitation of heavy metals in the sensitive tissues in order to reduce tolerance of the plant to heavy metal stresses.

Of the best-characterised ncRNAs involved in plant responses to heavy metal stress, lncRNAs are another type. lncRNAs may have an apt for acting as molecular sponges which inhibit the miRNAs that should otherwise degrade the mRNAs of metal-stress-responsive genes. This mechanism permits the accumulation of key mRNAs underlying metal detoxification and stress response. For instance, in *M. truncatula*, a number of well-identified lncRNAs are reported to be involved in regulating interactions with miRNAs in the plant when exposed to aluminium toxicity thereby contributing to cellular homeostasis under metal stress conditions (Chekanova, 2015).

lncRNAs can also guide chromatin-modifying complexes for the change of stress-associated genes promoter chromatin state and control the gene expression under the condition of metal exposure. In Arabidopsis, it was shown that lncRNA take part in histone modification and DNA methylation processes, which are responsible for the regulation of the transcription of the genes involved in metal detoxification, including metallothionein and phytochelatin synthase genes. It is identified that for complexes formation and accumulation of heavy metals in vacuoles, metallothioneins and phytochelatins play a vital role and minimise the cytotoxicity. Thus, it is not surprising to have identified that lncRNAs act as regulators of detoxifying agents to enable plants to survive in environments containing high levels of heavy metals.

ncRNA are also involved in crosstalk of different stress signalling pathways in order to enable plants to respond to complex combined stresses, including heavy metal and drought or salinity stress for example. Some of these miRNAs and lncRNA emerged to be equally involved in handling both heavy metal and abiotic stress so as to mobilise a completeness that can minimise resource wastage and that will enhance growth in response to both facets of stress. For illustration, miR156 that is implicated in both heavy metal stress and drought has been identified to regulate SPL transcription factors which have developmental significance. By acting as regulators of both metabolic and developmental processes across these regions, these ncRNAs provide plants with the means to experience multiple stresses simultaneously and stay in balance (Paul et al., 2016).

8.5 MECHANISMS OF ncRNA-MEDIATED ABIOTIC STRESS TOLERANCE

8.5.1 Post-Transcriptional Regulation

Plant ncRNAs are involved in the post-transcription regulation in which they selectively stabilised or destabilised, translated or delayed, processed or unprocessed mRNAs. Two of the ncRNA classes, miRNA and siRNA are particularly well recognised to be involved in this process. These small ncRNAs interact with target messages through sequence complementary and often cause message degradation or translational silence. For instance, miRNAs in Arabidopsis in stress response genes by subscribing to their mRNAs for cleavage and therefore limiting the synthesis of stress-related proteins when stress condition is not present as this would otherwise drain energy and materials. This form of control enables plants to quickly adapt to their surrounding when the products of some genes are of no use.

The role of miRNAs in post-transcriptional regulation is quite robust and was described as 'miRNA-guided cleavage'. This is especially the case where miRNAs hydrogen bond with their target mRNAs with a level of affinity that approaches totality meaning the target mRNA gets cleaved and then degraded, For instance, miR156 controlling plant development and flowering by binding SPL (SQUAMOSA promoter binding protein-like) genes. The SPLs encode for transcription factors that play a role in flowering and identity of leaves in Arabidopsis; the, cleavage of SPL mRNAs by miR156 sets the timing for flowering and development of, additional leaves in response to environmental signals (Wu, Suhasini, et al., 2009). Thus, miRNAs directly regulate a wide range of target-genes and indirectly coordinate developmental changes to occur in proper response to the environmental cues that are crucial for plant survival.

In addition to the cleavage of target mRNAs, there are other ways that miRNAs can repress translation, through partially complementary sites that are more typical of animals but seen in plants as well. This repression without cleavage avoids the translation of the mRNA into protein, and they provide a reversible way to modulate the gene expression. For example, the BMI-21 has been reported to control the translation of CSD1 and CSD2, genes that in some way participate in oxidative stress response, they encodes copper/zinc superoxide dismutases. While the plant is under normal conditions, miR398 inhibits the synthesis of the crop's antioxidant enzymes but, when the plant is stressed the action of the miRNA is deactivated so that the antioxidants are produced to neutralise the destructive ROS. This mechanism is an important component of plant adaptation to changing across environmental stresses because it affords a rapid and efficient means of regulating protein synthesis according to the cell's requirements.

Another kind of ncRNA that plays role in regulation in post-transcriptional level is the siRNA. In plants, siRNAs take a part in RNAi paths, that is necessary for degradation of definite mRNAs. One of the best known cases is the participation of siRNAs in the transcriptional gene silencing of transposable elements, which, if not controlled, can significantly interfere with gene expression. But giving the guidance to degradation of transposable element-derived mRNAs, siRNAs play crucial role in homeostasis to genome stability and integrity. This function is especially needed for plant species with a huge genome number, where there are many transposable elements, which can lead to mutation or affect the expression of the genes located nearby (Slotkin & Martienssen, 2007).

lncRNAs are also involved in the post-transcriptional regulation via molecular siphoning or, in other words, acting as miRNA sponges. These lncRNAs display molecular mimicry and act as competing endogenous RNAs by sequestering miRNAs hence allowing the miRNA to bind other molecules instead of their target mRNAs. For example, some lncRNAs in *M. truncatula* sequester a set of miRNA that regulate nutrient signalling and therefore the mRNAs of nutrient-responsive genes can be sustained and translated into proteins required for uptake and metabolisms of nutrient (Heo & Sung, 2011). Such type of interaction demonstrates that lncRNAs for miRNA sponging can control miRNA function and thus impact the post-transcriptional regulation of other genes.

Besides being passive, lncRNAs can selectively bind on RNA-binding proteins to stabilise a particular mRNA or to regulate its splicing. Sometimes lncRNAs associate with RNA-binding proteins and bind to target mRNAs and stabilise them or change their splicing pattern necessary for producing functional protein from a specific mRNA. This regulation can be crucial in stress responses, because usually plants generate different splice variants of stress-related genes in order to survive the unfavourable conditions. Through the regulation of these complexes during their assembly, lncRNAs offer an added layer of post-transcriptional regulation that endows plants flexibility in adjusting to the stressors within their environment.

8.5.2 Epigenetic Regulation

Plant ncRNAs are involved in gene regulation at epigenetic level meaning that they affect gene expression through changes that do not involve alterations to the DNA sequence. The small ncRNA including siRNA and miRNA are involved in various cellular processes including DNA methylation, histone modification and chromatin remodelling. RNA-directed DNA methylation is one of the major epigenetic regulating processes in which siRNAs target DNA methylation to certain DNA sequences particularly transposons and repetitive sequences which would otherwise be active and can sometimes compromise genome stability. For instance, in *A. thaliana*, siRNAs are involved in RNA-directed DNA methylation (RdDM) which is very important in the plant organism since it silences transposons and other repetitive elements in plant genome.

In RdDM, plant siRNAs are derived from double-stranded RNAs by DCL enzymes and incorporate with AGO proteins to form RISCs. These complexes then direct the siRNAs to their cognates DNA sequences and this attracts DNA methyltransferases to the targeted cytosine residues (Matzke & Mosher, 2014). This methylation inhibits gene activity through direct occupation of binding sites for transcription factors, or through alteration of chromatin structure and making the DNA difficult for transcription. Thus, siRNAs have a direct function for genome stability through inhibition of activation of potentially dangerous sequences and control of their expression.

lncRNAs also participated in epigenetic mechanisms through interaction with chromatin-related enzymes to modify epigenome status. It also showed that in plants, lncRNAs can recruit histone modification complexes to particular genomic loci leading to alterations in chromatin organisation and gene regulation. For example, COLDAIR lncRNA in Arabidopsis positions PRC2 to modulate histone methylation at FLC gene, which is responsible for flowering process. The COLDAIR-PRC2 complex can also establish H3K27me3 mark, which is generally associated with gene repression, to regulate FLC and enable the plant to flower when exposed to cold stress for an extended period. Such regulatory mechanism reflects how lncRNAs regulate plant response to environmental fluctuations through epigenetic works.

For instance, apart from being responsible for DNA methylation, histone modification, plant ncRNAs participate in chromatin modification which alters the chromatin structure to regulate gene expression. The ncRNAs influence the gene expression patterns dependent upon the environmental stress by associating with chromatin remodellers. For instance, in salt stress, there are known lncRNAs that interact with chromatin-modifying complexes, changes in chromatin architecture ensue that promotes expression of stress-responsive genes (Kim et al., 2017). Through the action of ncRNAs, the chromatin structure is dynamically regulated to help plants to respond effectively to environmental pressures: changing the gene expression to favour growth and development.

8.5.3 Interaction with Transcription Factors

It revealed that various ncRNAs are involved in regulations of gene at different levels like chromatin remodelling, transcriptional regulation and alterations in post-transcriptional modifications. Of these ncRNAs, the ability of these molecules to bind transcription factors (TFs) has recently received considerable attention. Transcription factors are proteins that can interact with DNA sequence, that is, control the process of transcription of the genes situated near them. Emerging data also indicate that ncRNAs can regulate TFs in a variety of ways, either directly or indirectly, including the modification of TF activity, changes to the affinity of the TF to DNA, and through ceRNA networks.

To date, the direct binding is one of the most common ways through which ncRNAs can influence the activity of TFs. Some lncRNAs have been demonstrated to bind TFs and function as co-activators or inhibitors, because they gobble up or deliver bivalent proteins to their targets. For example, lncRNA HOTAIR physically associates with the polycomb repressive complex 2 (PRC2) which is well known for involvement in gene repression. This interaction also impinges on the ability of TFs to bind to their target genes and control chromatin statuses and transcriptional repression (Rinn et al., 2007). Opposing influence is the lncRNA MALAT1 which associates with the transcriptional regulator proteins such as those of TFs including SR proteins to modulate the regulation of TPM and pre-mRNA splicing in context of different cellular compartments (Gao et al., 2018).

However, it is also possible ncRNA can affect transcription factors activity 'indirectly', through control of the TF and/or its co-factor synthesis. Post-transcriptional regulation of TFs through gap mRNAs including miRNAs has however been reported as important sRNAs. For instance, miR155 has a function in the modulation of the immune TFs, for instance the NF-κB route through the recognition and binding to the 3′ UTR region of a subset of TFs and their regulators (O'Connell et al., 2010). This post-transcriptional regulation can further modulate the response of cells to various stimuli that require miRNAs and other sRNAs as components of the regulation of transcription.

In addition, competition with cognate TFs for the same sequence motifs is another important process in which ncRNA affects transcriptional regulation. The ceRNA hypothesis: the ceRNA affects the amount and availability of certain TFs because the ceRNA and the target mRNA contain overlapping MREs. This competition may therefore alter the state of cell transcribe that relates to diseases such as cancer. For instance, PTENP1 binds to miRNAs that target PTEN thereby controlling PTEN mRNA's access to ribosomes and elemental proteins such as TFs that are absolutely fundamental to carcinogenesis (Salmena et al., 2011).

8.6 APPLICATIONS OF ncRNA RESEARCH IN CROP IMPROVEMENT

8.6.1 Genetic Engineering

Because of the ability of ncRNA to modulate gene expression and control cellular processes, ncRNAs have received significant interest in genetic engineering. The most popular area of using ncRNAs in genetic engineering is probably the usage of siRNAs and miRNAs as molecular switches to control gene expression levels. siRNAs are used to knock down the genes of interest by causing the degradation of their complementary mRNA. This has particular implications for functional genomics that demonstrate gene function through gene knock down techniques; miRNAs are used to control the post-transcriptional expression of genes that are implicated in diseases such as cancer. For instance, synthetic miRNAs have been invented to inhibit oncogenes and thereby inhibit tumour progression in the contexts of xenografts (Agarwal et al., 2018).

Besides the gene silencing function, artificial ncRNAs, for instance, lncRNA have been employed due to their functions in gene activation The chromatin structure of genes can be modified by lncRNAs and provide a binding site for transcription machinery, and subsequently ranscribing target genes, which makes lncRNA vital instruments in typical genetic manipulation for gene activation. Over the years, synthetic lncRNAs have been designed to regulate the genes of interest by binding to transcription factors or chromatin remodellers to the promoters. For instance, site-directing engineered lncRNAs have been employed to initiate the function of target endogenous genes, making great potential in gene therapy for diseases with gene inactivation or gene mutation.

ncRNAs also serve for synthetic biology and help to create new genetic circuits. Since ncRNAs can control the gene expression in a programmable manner, it is ideal for programmable synthetic biology where one needs to programme cell's gene networks. For instance, there are normal RNAs that switch depending on a certain metabolite; these are called riboswitches that can be used to govern gene expression based on a signal. This application of ncRNAs has been realised in metabolic engineering; engineered microbes are developed to synthesise useful chemicals or bio fuels upon certain stimuli. These applications show how using ncRNAs can help in building multilayered gene regulatory networks.

Thus, in regenerative medicine, ncRNAs are under investigation to evaluate their ability to transform somatic cells to pluripotential stem cells. For most cell types the process of inducing pluripotency involves the use of a set of transcription factors known as Yamanaka factors. Consequently, researchers have discovered that lncRNAs and miRNAs are able to play an important role in this reprogramming process. For example, miRNAs are found to play a significant role in maintenance of pluripotency and adding miRNAs to the list of transcription factors that help in generation of iPSCs. This approach could have revolutionary consequences in converting tissues and formation of individualised medications for different diseases (Lujan et al., 2016).

The significance of ncRNAs in genetic engineering lies in their potential application in gene editing. New developments in this technology have given rise to more functions which include CRISPRi and CRISPRa using RNA molecules to regulatively degrade or synthesise gene expression, respectively; further, ncRNAs act as the target recognition molecules to direct the CRISPR-Cas9 system to the specific locus in the genome that needs to be knocked down or activated. These techniques have been successfully employed in both prokaryotic and eukaryotic systems making it a powerful tool in the genome-wide screening, disease modelling and in the discovery of therapeutic agents. Combined with CRISPR systems, ncRNAs allow the generation of enhanced genetic modifications with relatively little off-target activities (Konermann et al., 2015).

ncRNAs, having important roles in gene regulation, stress response and crop traits enhancement, are believed to be potential molecular tools in plant breeding. The most fulfilling area of ncRNAs in plant breeding is the use of microRNA (miRNA) for increasing resistance against biotic and abiotic stresses. For instance, they wrote crops to generate miRNAs towards genes that are associated with disease susceptibility, such as Fusarium wilt or leaf rust, leading to disease resistance improvement. In Arabidopsis, Json has shown that miR393 helps a plant defend against root-knot nematodes and similar strategies are being tried in rice and wheat as well.

8.6.2 Breeding Programmes

Another major area that has been observed to benefit from use of ncRNAs in plant breeding is on the control of growth and development. Several lncRNAs also exert their functions on different developmental aspects such as seed development, flowering time and root development. For example, it was reported that in rice lncRNAs can control the development of roots and interact with complexes that modify chromatin. Through regulating these lncRNAs, breeders can increase positive traits like improved root systems which give higher nutrient and water uptake ability useful for crops to be grown in areas of low water availability. The modulation of the lncRNAs could also help in enhancing seed size and leading to enhanced yield of important food crops like maize and soybean.

Apart from enhancing stress tolerance and developmental characters, ncRNAs are employed in controlling flowering and fruiting times in plants, which are important indicators in agricultural production. For instance, in plants miRNAs have been evidenced to form the foundation of regulation of flowering time in plants through targeting genes in the flowering pathway, including FLC gene in Arabidopsis. Likewise, in tomato, miR172 regulates the time of flowering and fruit ripening targeting the AP2 gene. Because of these ncRNAs, breeders can alter the growth patterns of plants and ensure that most of them flower and fruit at the same time hence increasing production and maximising production cycles. This regulation of flowering and ripening can be especially important to fruits and vegetables where timing is crucial for market and shelf worth.

In addition, ncRNAs are involved in controlling the reaction of the plant to the environmental stress, which is significant given climate change tendencies recently, for instance, miRNAs involved in drought and salt tolerance were targeted in breeding. For instance, miR156 on the maize and rice plant was observed to control genes that enhance root development under drought, enhancing water absorption. Likewise, warmth stable miR396 has been identified linked to the regulation of genes encoding for salt tolerance in Arabidopsis as well as in other plants. Breeders, therefore, need to unlock this potential of these ncRNAs by engineering crops that express such molecules, which will better cope with the detrimental effects of climate change in the world food bowls (Jia et al., 2017).

ncRNAs present one of the biggest strengths in technology for transgenic and gene editing in plant breeding. The availability and innovation of manipulating techniques specifically CRISPR-Cas9 has provided a platform to directly target some of the free flowing ncRNAs with a point of interest in the down/control/up regulation of gene expression. As a result, the possibility to produce genetically engineered plants with desirable characteristics such as increased yields, improved nutritive value and resistance to pests and diseases arises. As an example, Arabidopsis plants with deficient or knocked-down miRNAs through the employment of CRISPR-Cas9 exhibit improved plant growth besides better stress resistance. This approach can be used on crops to enhance the breeding processes by enhancing the control of genes and the departure from standard breeding techniques.

8.7 CONCLUSION AND FUTURE PERSPECTIVES

sRNAs are ncRNA molecules that have been identified to play important functions in plants concerning their growth, reproduction, regulation of stress and overall yield, which they dictate, which makes them important in plant breeding and biotechnology. The current molecular concept of miRNAs, lncRNAs and other ncRNAs regulating gene expression has revolutionised plant breeding. ncRNAs help in regulating some vital cellular and molecular processes such as disease resistance, stress tolerance, and growth and development making ncRNAs useful tools in view of climate change in increasing crop productivity. With further refinements in understanding, functional roles of several ncRNAs would give fresh picture of plants system and offer new opportunities to enhance crop characteristics.

There is enormous possibility in the application of ncRNA in plant breeding in the future especially in precision gene editing and synthetic biology. As gene-editing methods improve, all CRISPR/Cas9 platforms and other platforms involved can increase the visibility of ncRNA to achieve a specific gene, thereby helping develop plants with favourable characteristics. This greater control over gene expression also implies that new plant varieties can be bred with less halting crossbreeding and more speed. In addition, with the possibilities to engineer ncRNA opener to target and regulate multiple genes concurrently, the opportunity to genetically define crops that not only increase production yield but are also adaptable to environmental changes has emerged (Zhu et al., 2016).

Nevertheless, there are several issues to address in the process of bringing ncRNA findings into widespread clinical use. Among the challenges facing researchers, there is one major concern – the intricate intertwining of ncRNA with the plant's genetic regulation network. lncRNAs are themselves implicated in complex allosteric systems, and their action tends to be tissue-specific and subtlety influenced by context. Thus, to comprehend all these interactions, fully expanded research needs to be made on the function and regulation mechanisms of different ncRNA species in different plant species. Furthermore, the impact of off-targeted engineered ncRNAs and other bead fall consequences of manipulating the expression and function of ncRNA in plants requires extensive assessment to enable the determination of safety and sustainability of the formulated streams of plants. Consequently, a focus in future studies should be to enhance the comprehension on ncRNA functions, and to establish methods for purposeful and tightly controlled utilisation in plant breeding (Axtell, 2013).

As practical applications, the combinatorial use of ncRNA technologies with the current high-throughput genomic tools may be fundamental to the future of plant breeding. Incorporation of RNA sequencing, bioinformatics and functional genomics will allow for improved studies towards the identification of ncRNAs in various plants and under different conditions. This will give a more profound insight into how ncRNAs modulate traits like yield, stress and disease tolerance, and help take more targeted approaches to crop improvement. The discovery of ncRNA-based biomarkers can also help with the identification of plants with the desirable traits at an early stage and therefore help promote breeding, increasing crop production.

Further effective research into these ncRNAs will afford more developments in plant biotechnology that can unlock challenges such as food security and climate change. Furthering our understanding on how to manipulate ncRNAs in plants, it is possible to grow crops that not only are resistant to biotic stresses as well as abiotic stresses but also are suitable for growth under adverse, or extreme conditions, important for producing food in arid, flooded and otherwise climatically precarious regions of the world. Also, the application of ncRNAs in the agriculture sector may limit the usage of chemical fertilisers and pesticides, therefore encouraging eco-friendly agriculture. Thus, it is possible to say that the further development of ncRNA research in plants may open great perspectives to enhance crop resistance, productivity and sustainability.

REFERENCES

Agarwal, G., Angeles, R., Pirrie, M., Mcleod, B., Marzanek, F., Parascandalo, J., & Thabane, L. (2018). Evaluation of a community paramedicine health promotion and lifestyle risk assessment program for older adults who live in social housing: A cluster randomized trial. *CMAJ*, *190*, E638–E647.

Axtell, M. J. (2013). Classification and comparison of small RNAs from plants. *Annual Review of Plant Biology*, *64*, 137–159.

Bartel. D. P. (2004). MicroRNAs: Genomics, biogenesis, mechanism, and function. *Cell*, *116*(2), 281–297. https://doi.org/10.1016/s0092-8674(04)00045-5.

Baulcombe, D. (2004). RNA silencing in plants. *Nature*, *431*, 356–363.

Chekanova, J. A. (2015). Long non-coding RNAs and their functions in plants. *Current Opinion in Plant Biology*, *27*, 207–216.

Ding, S. W., & Voinnet, O. (2007). Antiviral immunity directed by small RNAs. *Cell*, *130*, 413–426.

Fang, E. F., Kassahun, H., Croteau, D. L., Scheibye-Knudsen, M., Marosi, K., Lu, H., Shamanna, R. A., Kalyanasundaram, S., Bollineni, R. C., Wilson, M. A., Iser, W. B., Wollman, B. N., Morevati, M., Li, J., Kerr, J. S., Lu, Q., Waltz, T. B., Tian, J., Sinclair, D. A., Mattson, M. P., Nilsen, H., & Bohr, V. A. (2016). NAD(+) replenishment improves lifespan and healthspan in ataxia telangiectasia models via mitophagy and DNA repair. *Cell Metabolism*, *24*, 566–581.

Gao, J., Langemeyer, L., Kümmel, D., Reggiori, F., & Ungermann, C. (2018) *Molecular mechanism to target the endosomal Mon1-Ccz1 GEF complex to the pre-autophagosomal structure*. 7.

Guan, Q., Lu, X., Zeng, H., Zhang, Y., & Zhu, J. (2013). Heat stress induction of miR398 triggers a regulatory loop that is critical for thermotolerance in Arabidopsis. *The Plant Journal*, *74*, 840–851.

Heo, J. B., & Sung, S. (2011). Vernalization-mediated epigenetic silencing by a long intronic noncoding RNA. *Science*, *331*, 76–79.

Ito, S., Shen, L., Dai, Q., Wu, S. C., Collins, L. B., Swenberg, J. A., He, C., & Zhang, Y. (2011). Tet proteins can convert 5-methylcytosine to 5-formylcytosine and 5-carboxylcytosine. *Science*, *333*, 1300–1303.

Jia, B., Raphenya, A. R., Alcock, B., Waglechner, N., Guo, P., Tsang, K. K., Lago, B. A., Dave, B. M., Pereira, S., Sharma, A. N., Doshi, S., Courtot, M., Lo, R., Williams, L. E., Frye, J. G., Elsayegh, T., Sardar, D., Westman, E. L., Pawlowski, A. C., Johnson, T. A., Brinkman, F. S., Wright, G. D., & Mcarthur, A. G. (2017). CARD 2017: Expansion and model-centric curation of the comprehensive antibiotic resistance database. *Nucleic Acids Research*, *45*, D566–D573.

Jones-Rhoades, M. W., Bartel, D. P., & Bartel, B. (2006). MicroRNAS and their regulatory roles in plants. *Annual Review of Plant Biology*, *57*, 19–53.

Kim, J., Sung, D. J., & Lee, J. (2017). Therapeutic effectiveness of instrument-assisted soft tissue mobilization for soft tissue injury: Mechanisms and practical application. *Journal of Exercise Rehabilitation*, *13*, 12–22.

Konermann, S., Brigham, M. D., Trevino, A. E., Joung, J., Abudayyeh, O. O., Barcena, C., Hsu, P. D., Habib, N., Gootenberg, J. S., Nishimasu, H., Nureki, O., & Zhang, F. (2015). Genome-scale transcriptional activation by an engineered CRISPR-Cas9 complex. *Nature*, *517*, 583–588.

Law, J. A., & Jacobsen, S. E. (2010). Establishing, maintaining and modifying DNA methylation patterns in plants and animals. *Nature Reviews Genetics*, *11*, 204–220.

Liu, J., Liu, X., Wang, W., Mccauley, L., Pinto-Martin, J., Wang, Y., Li, L., Yan, C., & Rogan, W. J. (2014). Blood lead concentrations and children's behavioral and emotional problems: a cohort study. *JAMA Pediatrics*, *168*, 737–45.

Liu, X., Ramirez, S., Pang, P. T., Puryear, C. B., Govindarajan, A., Deisseroth, K., & Tonegawa, S. (2012). Optogenetic stimulation of a hippocampal engram activates fear memory recall. *Nature*, *484*, 381–5.

Lujan, S. A., Williams, J. S., & Kunkel, T. A. (2016). DNA Polymerases Divide the Labor of Genome Replication. *Trends in Cell Biology*, *26*, 640–654.

Mabo, P., Victor, F., Bazin, P., Ahres, S., Babuty, D., Da Costa, A., Binet, D., & Daubert, J. C. (2012). A randomized trial of long-term remote monitoring of pacemaker recipients (the COMPAS trial). *European Heart Journal*, *33*, 1105–1111.

Matzke, M. A., & Mosher, R. A. (2014). RNA-directed DNA methylation: An epigenetic pathway of increasing complexity. *Nature Reviews Genetics*, *15*, 394–408.

O'connell, R. M., Rao, D. S., Chaudhuri, A. A., & Baltimore, D. (2010). Physiological and pathological roles for microRNAs in the immune system. *Nature Reviews Drug Discovery*, *10*, 111–122.

Paul, J., Modi, A., & Patel, J. (2016). Predicting green product consumption using theory of planned behavior and reasoned action. *Journal of Retailing and Consumer Services*, *29*, 123–134.

Reyes, J. L., & Chua, N. H. (2007). ABA induction of miR159 controls transcript levels of two MYB factors during Arabidopsis seed germination. *The Plant Journal*, *49*, 592–606.

Rinn, J. L., Kertesz, M., Wang, J. K., Squazzo, S. L., Xu, X., Brugmann, S. A., Goodnough, L. H., Helms, J. A., Farnham, P. J., Segal, E., & Chang, H. Y. (2007). Functional demarcation of active and silent chromatin domains in human HOX loci by noncoding RNAs. *Cell*, *129*, 1311–1323.

Salmena, L., Poliseno, L., Tay, Y., Kats, L., & Pandolfi, P. P. (2011). A ceRNA hypothesis: the Rosetta Stone of a hidden RNA language? *Cell*, *146*, 353–358.

Slotkin, R. K., & Martienssen, R. (2007). Transposable elements and the epigenetic regulation of the genome. *Nature Reviews Genetics*, *8*, 272–285.

Sunkar, R., Li, Y. F., & Jagadeeswaran, G. (2012). Functions of microRNAs in plant stress responses. *Trends in Plant Science*, *17*, 196–203.

Sunkar, R., & Zhu, J. K. (2004). Novel and stress-regulated microRNAs and other small RNAs from Arabidopsis. *Plant Cell*, *16*, 2001–2019.

Wu, G., Park, M. Y., Conway, S. R., Wang, J. W., Weigel, D., & Poethig, R. S. (2009). The sequential action of miR156 and miR172 regulates developmental timing in Arabidopsis. *Cell*, *138*, 750–759.

Wu, P., Fang, Y., Guan, Z., Fan, B., Kong, J., Yao, Z., Liu, X., Fuller, C. J., Susser, E., Lu, J., & Hoven, C. W. (2009). The psychological impact of the SARS epidemic on hospital employees in China: exposure, risk perception, and altruistic acceptance of risk. *The Canadian Journal of Psychiatry*, *54*, 302–311.

Wu, Y., Suhasini, A. N., & Brosh, R. M. Jr. (2009). Welcome the family of FANCJ-like helicases to the block of genome stability maintenance proteins. *Cellular and Molecular Life Sciences*, *66*, 1209–1222.

Wunderlich, K. L., & Vollmer, T. R. (2017). Effects of serial and concurrent training on receptive identification tasks: A systematic replication. *Journal of Applied Behavior Analysis*, *50*, 641–652.

Xie, D., Miller, E., Sharpe, P., Jackson, E., & Zhu, Q. (2017). Omega-3 production by fermentation of *Yarrowia lipolytica*: From fed-batch to continuous. *Biotechnology and Bioengineering*, *114*, 798–812.

Xin, M., Wang, Y., Yao, Y., Song, N., Hu, Z., Qin, D., Xie, C., Peng, H., Ni, Z., & Sun, Q. (2011). Identification and characterization of wheat long non-protein coding RNAs responsive to powdery mildew infection and heat stress by using microarray analysis and SBS sequencing. *BMC Plant Biology*, *11*, 61.

Zhang, D. D., & Swanson, L. A. (2014). Linking social entrepreneurship and sustainability. *Journal of Social Entrepreneurship*, *5*, 175–191.

Zhang, X., Xiao, W., & Hu, X. (2018). Predicting essential proteins by integrating orthology, gene expressions, and PPI networks. *PLoS One*, *13*, e0195410.

Zhang, Y., Chen, J., Qiu, J., Li, Y., Wang, J., & Jiao, J. (2016) Intakes of fish and polyunsaturated fatty acids and mild-to-severe cognitive impairment risks: A dose-response meta-analysis of 21 cohort studies. *The American Journal of Clinical Nutrition*, *103*, 330–340.

Zhou, T., Lu, Y., & Wang, B. (2010). Integrating TTF and UTAUT to explain mobile banking user adoption. *Computers in Human Behavior*, *26*, 760–767.

Zhu, J. K. (2016). Abiotic stress signaling and responses in plants. *Cell*, *167*, 313–324.

Zhu, Z., Piao, S., Myneni, R. B., Huang, M., Zeng, Z., Canadell, J. G., Ciais, P., Sitch, S., Friedlingstein, P., Arneth, A., Cao, C., Cheng, L., Kato, E., Koven, C., Li, Y., Lian, X., Liu, Y., Liu, R., Mao, J., Pan, Y., Peng, S., Peñuelas, J., Poulter, B., Pugh, T. a. M., Stocker, B. D., Viovy, N., Wang, X., Wang, Y., Xiao, Z., Yang, H., Zaehle, S. & Zeng, N. (2016) Greening of the Earth and its drivers. *Nature Climate Change*, *6*, 791–795.

Zilberman, D., Gehring, M., Tran, R. K., Ballinger, T., & Henikoff, S. (2007). Genome-wide analysis of *Arabidopsis thaliana* DNA methylation uncovers an interdependence between methylation and transcription. *Nature Genetics*, *39*, 61–69.

9 ncRNAs as Regulators of Abiotic Stress Responses in Plants

Anamish Tyagi, Priyanka Jaiswal, Banshidhar Jha, Nitin Kumar, Santosh Kumar and Charupriya Chauhan

9.1 INTRODUCTION

Plants, due to their immobility, are unable to evade stress and must, therefore, develop mechanisms to tolerate or mitigate its effects to sustain growth and survival in dynamic environments. Abiotic stressors, including drought, salinity, cold, heat and nutrient deficiencies, can lead to considerable yield losses, profoundly impacting plant growth and productivity (Hirayama & Shinozaki, 2010). Food security concerns are anticipated to escalate as a result of alterations in soil quality, the loss of arable land driven by urbanisation, the ever-increasing global population and the impacts of climate change (Mittler, 2006). In recent times, non-coding RNAs (ncRNAs) have emerged as significant entities that determine the genetic complexity and the resultant phenotype of an organism. Despite RNA constituting up to 90% of the eukaryotic genome, merely approximately 2% of them are transcribed and further translated to proteins (Rai et al., 2019). Ariel et al. (2015) previously regarded the remaining ncRNAs as having limited utility, given their lack of protein-coding potential and poorly conserved sequences. Advanced sequencing methods and experimental studies have demonstrated that ncRNAs are integral to various genetic and epigenetic regulations (Cech & Steitz, 2014). The identification of ncRNAs in plants has been significantly facilitated by advancements in next-generation sequencing (NGS) technology alongside homology-based and experimental methodologies (Eddy, 2001). Since their discovery, ncRNAs have been classified into various groups according to their origin, biogenesis, mechanism of action and other characteristics. Regulatory ncRNAs, *viz.* short interfering RNAs (siRNA), long ncRNAs (lncRNAs), piwi-interacting RNAs (piRNAs), microRNAs (miRNAs) and represent significant categories among the various ncRNA transcripts. Though these are transcribed from DNA, they do not produce protein. However, it plays an important role in the regulation of transcriptional and post-transcriptional modification, influencing the response of a plant to environmental stimuli, its growth and development. The other class of ncRNA, popularly known as housekeeping ncRNAs, which includes rRNAs, snRNAs, transfer RNA (tRNAs) and snoRNAs (small nucleolar RNAs), play essential roles in ribosomal and cellular functions (Table 9.1).

9.2 BIOGENESIS OF REGULATORY ncRNAs

9.2.1 Transfer RNA Derived Fragments (tRFs)

tRNA, being the second most abundant RNA in a cell after rRNA, is associated with some special kind of fragments that are generated during their transcription by RNA polymerase III. At first, the precursor tRNAs are first transcribed by RNA polymerase III in the nucleus. After the removal of the 3′ trailer and 5′ leader sequences, a CCA (cytidine–cytidine–adenosine) tail is added to the mature tRNA 3′ terminus. Additionally, this mechanism produces tRFs (tRNA-derived fragments) from (13–20 bp) that can potentially come from the 5′ or 3′ end. Additional cleavage may occur at the 5′ or 3′ ends of mature tRNAs, yielding fragments of 31–40 nt each. In the selection of biological processes associated with the trans-synaptic neuronal cytoarchitecture, transcription factors (tRFs) are recognised as potential components or effectors of modulation interfering with gene expression at the transcriptional level.

9.2.2 Small Interfering RNAs (SiRNAs)

siRNAs are double-stranded ncRNA molecules of 21–24 bp involved in gene silencing by degrading messenger RNA (mRNA). They are commonly found in both plant and mammalian systems. dsRNA duplexes are synthesised from ssRNA, which essentially requires RNA-dependent RNA polymerases (RDR1-6) (Voinnet, 2009). The former (dsRNA duplexes) are cleaved by Dicer-like proteins (DCL1-4) into short RNA (sRNA) duplexes. DCL1 generates sRNAs of 18–21 nt in length, whereas the other three (DCL2, DCL4 and DCL3) produce sRNAs of lengths of 22, 21 and 24 nt, respectively. Further, these sRNA duplexes are moved to cytoplasm facilitated by the integration of RNA-induced silencing complex (RISC) with ARGONAUTE (AGO) proteins. Inside the cytoplasm, they associated themselves with post-transcriptional gene silencing (PTGS), otherwise confined within the nucleus itself and perform chromatin-related functions (Brosnan & Voinnet, 2011). Instances of both external and internal origins that can produce siRNAs encompass transposons, retro-elements and genomic regions abundant in repeats (Patil et al., 2014). Instances of external sources encompass viral RNA and transgenes. The endogenous siRNAs may be present in several forms and can be distinguished by their

DOI: 10.1201/9781003583257-9

TABLE 9.1
Classification of Non-Coding RNAs (ncRNAs)

Category	Type	Nucleotide Length
Housekeeping ncRNAs	Transfer RNAs (tRNAs)	50–200 nt
	Small nucleolar RNAs (snoRNAs)	50–200 nt
	Small nuclear RNAs (snRNAs)	50–200 nt
Regulatory ncRNAs	MicroRNAs (miRNAs)	20–24 nt
	Piwi-interacting RNAs (piRNAs)	24–32 nt
	Short interfering RNAs (siRNAs)	20–24 nt
	Long non-coding RNAs (lncRNAs)	>200 nt

biogenesis and the processing enzymes that are involved in their origin and maturation. These are discussed as follows.

9.2.3 Heterochromatic siRNAs (hc siRNAs) and cis-acting siRNAs (casiRNAs)

These are specific types of endogenous siRNAs having a stretch of 24 nt. The production process of these siRNAs is primarily capitalised on two important factors, namely RDR2 and DCL3, which collectively account for 70% of the overall composition. They play an important role in epigenetic modifications, maintaining structural as well as functional integrity of the genome and facilitating plant reproduction (Sanan-Mishra et al., 2021).

9.2.4 Cis-NAT siRNAs

These siRNAs are 24 nt long and constitute only 6–16% of the overall composition of endogenous siRNAs. This siRNA undergoes amplification and processing by RDR6 and DCL1, respectively. They crucially affect the growth of plants and support their adaptation to environmental changes.

9.2.5 PhasiRNAs

Phased small interfering RNAs (phasiRNAs) represent a distinct class of small RNAs, typically measuring 21 or 24 nt in length, that are derived from the precise cleavage of RNA transcripts at specific loci known as PHAS loci. This cleavage process is orchestrated by miRNAs, particularly 22-nt species such as miR2118, which directs the production of 21-nt phasiRNAs, and miR2275, which facilitates the generation of 24-nt phasiRNAs. PHAS loci encompass a diverse array of genes, including those encoding transcription factors such as MYBs and NACs, hormone-responsive genes like auxin response factors (ARFs) and nucleotide-binding leucine-rich repeat (NLR) genes. The biogenesis of phasiRNAs involves the conversion of cleaved RNA fragments into double-stranded RNA intermediates, mediated by RNA-DEPENDENT RNA POLYMERASE 6 (RDR6). Subsequently, these double-stranded RNAs are processed into phasiRNAs by Dicer-like proteins. Functionally, phasiRNAs act as negative regulators, guiding the RNA-induced silencing complex (RISC) to target complementary mRNA sequences, thereby promoting their degradation. This silencing mechanism is integral to numerous biological processes, including plant development and responses to environmental stresses (Liu et al., 2020).

9.2.6 Repeat-Associated siRNAs (rasiRNAs)

The repeat-associated siRNAs (rasiRNAs) consist of 24 nt. They contribute to the silencing of transcriptional genes to preserve the stability of the genome and are derived from transposons and repetitive genomic sequences. The differences in rasiRNAs between the hybrids of maize inbred lines B73 and Mo17 were found to enhance heterosis significantly (Barber et al., 2012).

9.2.7 Long siRNAs (lsiRNAs)

The occurrence of pathogen infection or particular growth conditions leads to the synthesis of long siRNAs (lsiRNAs), which consist of 30–40 nt. The biogenesis of lsiRNAs predominantly relies on proteins such as DCL1, HEN1, AGO7, HYL1 and HST, with a lesser dependency on Pol IV and RDR6 (Katiyar-Agarwal et al., 2007; Katiyar-Agarwal & Jin, 2010). For example, the effector avrRpt2 in *Pseudomonas syringae* disrupts RAP-domain proteins associated with disease resistance in Arabidopsis, leading to the production of AtlsiRNA-1 to AtlsiRNA-5 (Katiyar-Agarwal et al., 2007).

9.3 PLANT microRNAs (miRNAs)

Typically measuring 20–24 nt in length, these ncRNAs are present across a diverse range of organisms, encompassing both plants and mammals. Deng et al. (2019) report that the initial discovery of miRNA occurred in *Caenorhabditis elegans*, where it interacts with specific mRNA targets to regulate developmental timing. miRNAs are grouped into classes of 'intergenic' or 'intronic' based on their stem-loop structure. miRNA genes are transcribed by RNA polymerase II, producing single-stranded primary transcripts known as pri-miRNAs. In Arabidopsis, the RNase III enzyme Dicer-like 1 (DCL1) processes primary microRNA (pri-miRNA) transcripts into mature miRNAs within the nucleus. This processing is enhanced by the RNA-binding protein (RBP) HYPONASTIC LEAVES 1 (HYL1) and the zinc finger protein SERRATE (SE), which interact with DCL1 to ensure precise and efficient cleavage of pri-miRNAs. Upon loading with this miRNA, the AGO1 protein translocate to the cytoplasm, where it gets assembled with target mRNAs, thereby forming the miRNA-induced silencing complex (miRISC), which initiates PTGS process (Ambros, 2011). The generation of miRNA (~24 nt) facilitates the AGO4-mediated RISC in targeting DNA methylation at the transcriptional level, establishing a connection to epigenetic modifications (Deng et al., 2019). Two prevalent mechanisms by which miRNAs attenuate the production of target transcripts are translation inhibition and transcript cleavage. The activity of exoribonuclease facilitates the cleavage of transcripts, whereas the targeting of the 5′ untranslated region inhibits ribosome recruitment, leading to a suppression of translation (Carthew & Sontheimer, 2009). Short open reading frames (sORFs) present in the 5′ leader sequence of primordial miRNAs (pri-miRNAs) translate into miRNA-encoded peptides (miPEPs) (Feng et al., 2023a; Yadav et al., 2021). By increasing the transcription of their respective pri-miRNAs, these peptides control a number of biological processes. For example, miPEP165a and miPEP171b support optimum to good growth of primary root (PR) in *Medicago* and Arabidopsis while suppressing target genes that help in the proliferation of lateral root (LR) (Lauressergues et al., 2015). Furthermore, vvi-miPEP171d1 is known to have a part in controlling the development of adventitious roots in grapevines. MiPEP858a has been shown to affect auxin signalling and phenylpropanoid pathway gene expression, which in turn affects plant development and secondary metabolite synthesis in Arabidopsis (Chen et al., 2020a).

9.4 LONG NON-CODING RNAs (lncRNAs)

These ncRNAs exceed 200 nt in length and are found in various biological forms, such as plants, animal species, fungal organisms and prokaryotic organisms. While most lncRNAs are found in the nucleus, they do not code for any proteins. These entities are associated with chromatin and may also demonstrate activity in the cytoplasm (Wang & Chekanova, 2017). The main categories of strive intergenic RNAs include natural anti-sense transcripts, intronic ncRNA, and long intergenic RNAs (Mattick et al., 2023). NATs originate from the anti-sense strand of coding genes, whereas IncRNAs can be produced in both orientations. In eukaryotic cells, three specialised RNA polymerases – RNA polymerase I, II and III – regulate the transcription of nuclear genomes, producing a variety of coding and ncRNA molecules. Among these, RNA polymerase II is primarily responsible for synthesising mRNAs and various lncRNAs. On the other hand, the synthesis of lncRNAs is typically regulated by Pol I and Pol III. Additionally, plants utilise two specialised enzymes, Pol IV and Pol V, to produce lncRNAs. These RNAs play a crucial role in identifying and silencing transposons (Wierzbicki et al., 2021). lncRNAs regulate gene expression through diverse mechanisms, functioning at various levels in either cis (local) or trans (distant) modes, reflecting their multifaceted roles. Distinct sequence motifs or structures characterise these entities, and their functional activity is enhanced through sequence mutual complementarity with DNA or RNA (Chen et al., 2020b). Additionally, the repression of targets mediated by miRNAs can be affected by lncRNAs through the mechanism of competitive endogenous RNA (ceRNA), enabling these lncRNAs to interact with mRNAs for the recognition of miRNA molecules (Bouba et al., 2019). Endogenous target mimics, including lncRNAs, can compete for different micro RNAs and act as precursors for both siRNAs and miRNAs. Recent advances in technology have enabled the identification of various lncRNAs. EARLY NODULIN 40 (ENOD40), identified as the first plant lncRNA, was initially isolated from nodule primordia in *Medicago* species and shown to play a role in symbiotic nodule organogenesis, with the potential to influence the subcellular localisation of the nuclear RBP MtRBP1 (Crespi et al., 1994; Waseem et al., 2021). A variety of lncRNAs have been identified for their involvement in development and reactions to stress. In the context of vernalisation, it is noted that COLD-ASSISTED INTRONIC NON-CODING RNA (COLDAIR) contributes to the silencing of the FLOWERING LOCUS C (FLC) gene. Furthermore, the DOG1 anti-sense lncRNA plays a role in the regulation of seed dormancy, whereas the lncRNA SVALKA is associated with the freezing response (Kindgren et al., 2018). In modern studies, single-cell RNA sequencing has emerged as a widely used technique for pinpointing lncRNAs that are unique to specific cell types, along with the gene regulatory networks that govern these molecules (Zhao et al., 2022). Liu et al. (2020) identified the single-cell-based plant lncRNA atlas database (scPLAD) in Arabidopsis seedlings, which is helpful in studying plant lncRNA functions due to its cell-type-specific fingerprints.

9.5 CIRCULAR RNAs (circRNAs)

In recent decades, several circular molecules of RNA (circRNAs) have been discovered and characterised by their unique single-stranded closed-loop structure, indicating their widespread presence and abundance across all eukaryotic organisms (Zhou et al., 2020). The generation of these entities is accomplished via a technique referred to as 'back-splicing', which involves the covalent linkage of the 5′ and 3′ ends. Unlike linear RNAs, circular RNAs demonstrate more excellent stability and lower vulnerability to deterioration by Ribonuclease R (RNase R) due to their closed-loop configuration. CircRNAs typically exhibit lower expression levels, and the efficiency of back-splicing is usually less than that of linear splicing. Nonetheless, their intrinsic stability facilitates the gradual accumulation within particular cell types or tissues (Liu et al., 2023). In the early investigations of circRNAs, genome-wide profiling was frequently accomplished using rRNA-depleted total RNA sequencing. Nevertheless, a method employing linear RNase R for library generation demonstrates greater efficacy in enriching circRNAs, considering their low expression levels (Feng et al., 2023b). Circular RNAs can be classified into various types according to their genomic location, including exonic circRNAs, intronic circRNAs, intergenic circRNAs and UTR circRNAs (found in untranslated regions). The phases of gene expression, both transcriptional and post-transcriptional, are regulated by their mechanisms. CircRNAs serve as reservoirs for RBPs and miRNAs, highlighting their importance as potential biomarkers (Lukiw, 2013).

9.6 IMPACT OF ncRNAs ON PLANT GROWTH AND DEVELOPMENT

ncRNAs in plants have emerged as key regulators of chromatin dynamics, crucial for transcriptional regulation, and play vital roles in various developmental processes such as root, leaf, flower, endosperm and seed formation. ncRNAs are an integral component of epigenomics of a cell/organism and ensure accurate developmental outcomes in plants by modulating the expression of various important genes.

9.6.1 Root Growth and Development

Roots serve as subterranean structures for plants, offering stability to the aerial parts while enabling the uptake of water and vital nutrients required for development. Tian et al. (2014) detail the root system of Arabidopsis, emphasising the existence of distinct LRs in conjunction with PRs. After multiple rounds of asymmetrical cell divisions and subsequent expansions, stem cells situated near the quiescent centre generate PRs, which mainly support the development of meristematic tissues in the root. The pericycle cells create a tissue that surrounds the cells involved in the formation of LR primordia, located directly opposite each xylem pole, in conjunction with quiescent cells found at the phloem poles. The cells serve as the essential specification point for the development of LRs. A range of plant species has indicated that ncRNAs, such as circRNAs, siRNAs, miRNAs and lncRNAs, play a vital role in root development in plants (Table 9.2). Nuclear speckle RBPs play a crucial role in regulating alternative splicing and facilitating auxin-regulated growth of LRs. ASCO, as an alternative spliced mRNA target, is involved in interactions with these proteins. Zhang et al. (2018) define root hairs (RHs) as specialised extensions of individual cells that develop from epidermal trichoblast cells. Recent studies indicate that long-chain ncRNA AUXIN-REGULATED PROMOTER LOOP (APOLO) can identify various genes through the formation of DNA–RNA duplexes, known as R-loops, by employing sequence complementarity (Ariel et al., 2020). The establishment of the plant, its relationship with rhizospheric microbes and the absorption of nutrients through roots depend on these factors.

9.6.2 Leaf Development

The intricate mechanisms underlying leaf formation are essential for plant growth. Leaf development encompasses several stages, including initiation from the shoot apical meristem, establishment of polarity, progression through distinct developmental phases, morphological changes and, finally, senescence. The development of leaves is influenced by the role of ncRNAs, including circRNAs, miRNA-transcription factor regulating modules, siRNAs, miRNAs and lncRNAs (Table 9.2). The miR165/166 gene family in Arabidopsis consists of nine distinct members. The target sites of HD-ZIP III transcription factors show limited conservation across terrestrial plant species, and even small sequence differences—such as single-nucleotide variations—can influence their recognition (Satterlee & Scanlon, 2019). During leaf development, miR166 accumulates on the abaxial side of immature leaf primordia, while HD-ZIPIII is expressed in the meristematic region and on the adaxial surface of developing leaves (Chitwood et al., 2009; Pekker et al., 2005; Satterlee & Scanlon, 2019; Skopelitis et al., 2017). The impact of miR319 on the regulation of TCP transcription factors, such as TCP2, TCP3, TCP4, TCP10 and TCP24, is significant. High levels of miR319 cause the downregulation of TCP genes, leading to crinkled leaves and significant changes in leaf morphology in Arabidopsis. The reduction in leaf size exemplifies functional redundancy, potentially arising from the overexpression of miR396 or mutations affecting multiple GRFs. The presence of larger leaves is associated with the overexpression of GRF or mutations occurring in the miR396-binding region of GRF2, GRF7 or GRF9 (Bresso et al., 2018). As plant leaves mature, senescence takes place, serving as a crucial mechanism for enhancing offspring production and increasing their chances of survival.

9.6.3 Development of Seed, Endosperm and Seed Nutrients

Grain development is a crucial biological process that greatly influences plant productivity and quality. The weight of a thousand kernels, the number of spikes or panicles generated by each plant and the tally of grains per spike or panicle are various aspects that affect grain yield. The structure of the spike or panicle significantly affects these parameters. A range of ncRNAs have been recognised for their impact on traits related to seeds. Zhang et al. (2014) discovered around 65 lncRNAs linked to rice reproduction by utilising strand-specific RNA sequencing techniques. The reproductive stage showed significant overexpression of three lncRNAs, namely XLOC_0063639, XLOC_057324 and XLOC_007072. He et al. (2006) identified a quantitative trait locus containing several leucine-rich repeat receptor kinase (LRK) genes, which were associated with improved rice yield. Cao et al. (2021) discovered 2753 differentially expressed lncRNAs in wheat, emphasising their significance in spike development. Oil plays a crucial role in

TABLE 9.2
Involvement of ncRNAs in Plant Development and Stress Responses

Trait	siRNAs	miRNAs and Corresponding Targets	lncRNAs	circRNAs
Root development	–	RLD1/2 and miR165-SPL10 are miRNAs	APOLO, lncWOX11a, lncWOX5	DE-circRNAs
Leaf development	–	HR-miR165, miR319-TCP and miR 390–TAS3	longevity-related lncRNAs, twisting leaves	Circuits AT1C294965, AT4C38080 and AT5C18990
Seed, endosperm, nutrient development	Maternal p4-siRNAs	miR156, miR156-SPL14/SPL16, miR172, miR319, miR396	LAIR, XLOC_0063639, XLOC_057324, XLOC_007072, TCONS_0000891, TCONS_0000874, MSTRG.22563, MSTRG.86004	Fatty acid metabolism is aided by DE-circRNAs
Drought stress	The effects of dehydration on siRNAs	miR169-NFYA5, miR408-laccase	DRIR	circcirc, CPKtests for circulating RNA, CKX and circThe cork
Salt stress	OsSIDP301, SRO5 and P5CDH NATs	osa-miR1427f-LAO	Salt-responsive DE-lncRNAs	circ00260, circ_0013624, circ_0017320, circRNA Chr10:245,344–120,346,873
Flooding/hypoxia	Hypoxia-induced tasiRNA	miR156-SPL	Response of *Medicago truncatula* to waterlogging stress through DE-lncRNAs and genome-wide lncRNAs	
Heat stress	Targeted siRNA, heat-induced TAS1,	miR156-NAC	TCONS_00084391, TCONS_00010866, HILinc1	1538 heat-specific contexts RNAs
Cold stress	Cold-induced tasiRNA	miR319-TCP, miR319-OsCP6, OsTCP21	COLDAIR, COLDWRAP, COOLAIR, CIL1	1453 chilling-responsive circRNAs, 1830 low-temperature-induced circRNAs

seeds, offering beneficial effects on human health. During seed development and maturation, which includes morphogenesis, cell growth, reserve accumulation and desiccation, many oilseed crops synthesise and store oil. The production of seed oil encompasses four essential processes: the de novo synthesis of fatty acids, the chemical alteration of acyl groups, the assembly of glycerol and the creation of oil droplets. The expression patterns of 13 lncRNAs were strongly correlated with lipid-related genes in the developing seeds of *Brassica napus*, indicating a potential role for these lncRNAs in oil synthesis. Li et al. (2023) identified 8094 expressed lncRNAs in two *B. napus* accessions, ZS11, known for its high oil content, and WH5557, characterised by its low oil content. The relationship between seed storage of oil and the lncRNAs MSTRG.22563 as well as MSTRG.86004 is notably important. A 2.0% increase in oil content was observed with MSTRG.86004; however, this was linked to negative impacts on seed growth. Overexpression of MSTRG.22563 led to a decrease in seed oil content by 3.1–3.9% and changed the composition of seed fatty acids. The development of endosperm in cereal grains plays a crucial role in human nutrition, requiring an in-depth exploration of the metabolic and molecular processes at play. Before cellularisation, a syncytial phase takes place where endosperm nuclei undergo several rounds of mitosis without cytokinesis, leading to the formation of a multinucleated cell. The precise measurements of the seed are established during this transitional phase. Both glutamine and gliadin, as storage proteins, are of significant relevance within the endosperm. Essential amino acids such as proline, glutamine, asparagine and arginine play a crucial role in regulating the quality of baked goods. In wheat, a coexpression between storage proteins and several lncRNAs is observed. The anti-sense lncRNA TraesLNC1D26001.1 is consistently integrated with TaGLU-1Dy and is translated from the promoter associated with the high molecular weight glutelin subunit (HMW-GS) (Zhang et al., 2023). Functional analysis revealed no significant changes in protein levels, suggesting that this lncRNA does not affect the transcription or translation of the neighbouring gene.

9.7 THE ROLE OF ncRNAs IN THE RESILIENCE TO VARIOUS ABIOTIC STRESSES

Plants, being stationary organisms, are exposed to various abiotic stresses such as heat, drought, salinity, flooding and low temperatures. These stresses can significantly impact plant growth and productivity, with global data showing reductions ranging from 50 to 70%. In many regions, abiotic stressors, particularly those linked to climate change, have a widespread and damaging effect on plant life. The increasing occurrence of these stresses, along with the challenges posed by global warming, presents considerable challenges for plant ecosystems and agriculture globally.

9.7.1 Drought Stress

Given the significant challenges that drought stress presents to agricultural systems, numerous global initiatives are actively working to create crop varieties capable of enduring these challenging conditions. Plants exposed to this stress encounter osmotic imbalance, degradation of the membrane system, reduced respiration and a decline in photosynthesis. Throughout their life cycle, these challenges impede plant growth and metabolism, consequently diminishing agricultural output and quality. Gelaw and Sanan-Mishra (2021) discovered a variety of drought-responsive ncRNAs that interact with genes associated with calcium signalling, starch and sucrose production, ethylene and ABA synthesis or signalling, as well as numerous other metabolic processes. A significant quantity of lncRNAs was identified in the ear, tassel and leaf tissues of maize under both well-watered and drought conditions, covering four developmental stages (V12, V14, V18 and R1).

The identification of drought-responsive long ncRNAs (DRIR) highlights their crucial role in boosting plant resilience to salt and drought stress. In 2019, Zhang et al. conducted large-scale circRNA profiling, identifying 2174 and 1354 high-confidence circRNAs in maize and Arabidopsis, respectively, with the majority showing differential expression in response to drought. Two circular RNAs were found to be linked with the GRMZM2G032852 and GRMZM5G817173 genes, which encode cytokinin oxidase/dehydrogenase (CKX) and calcium-dependent protein kinase (CPK), respectively. The back-splicing events occurring between exons resulted in allele insertions and deletions within the maize panel, leading to the formation of circRNAs, specifically circCKX and circCPK. Plants exhibiting insertion alleles showed superior performance compared to those with deletion alleles, highlighting increased circCPK expression levels and better survival rates under both drought and normal conditions.

9.7.2 Salt Stress

Salt stress affects around 33% of agricultural land, significantly hindering plant growth and development. The rise in ionic and osmotic pressures in plants leads to toxicity, oxidative damage and osmotic stress (Hasanuzzaman & Fujita, 2022).

The role of ncRNAs in regulating salt stress responses in plants remains inadequately explored. For example, rice varieties that exhibit salt tolerance and are exposed to NaCl treatment possess over 75 conserved miRNAs. The plant subjected to salt treatment exhibited an accumulation of oxidative stress associated with osa-miR12477, which explicitly targets L-ascorbate oxidase. In a related context, it was observed that the roots of the salt-tolerant (Horkuch) and salt-sensitive (IR-29) rice genetic variants contained 1529 and 2103 putative lncRNAs, respectively. The study emphasised TCONS_00008914 along with TCONS_00008749, two lncRNAs with differing expression levels, as potential target mimics of established rice miRNAs (Jain et al., 2021).

9.7.3 Flooding/Hypoxia Stress

Flooding describes the temporary immersion of land by water, while waterlogging refers to a condition where soil remains fully or nearly saturated for prolonged periods. Both conditions negatively impact various physiological processes in plants, such as ionic balance, photosynthesis and energy production at the cellular level. In hypoxic conditions, marked by reduced oxygen availability, the difficulties encountered by plants become more evident, frequently arising in natural settings. The metabolic processes occurring within the plant are essential for its growth and development, and these processes can be adversely affected by inadequate oxygen levels in the root zone. ncRNAs have been identified in various plant species, including rice, maize, Arabidopsis, Populus, *Medicago*, tomato and cucumber, under conditions of waterlogging or hypoxia. Studies have identified approximately 291 and 58 miRNAs in the internode and node of rice, accordingly, that respond to deep-sea stress (Panda et al., 2022). In their study, Cho et al. (2012) investigate sucrose-derived non-fermenting-related enzyme (SnRK1), a protein kinase that assesses the impact of flooded areas on the energy metabolism of plants. Stief et al. (2014) showed that increasing the expression of miR156 can enhance abiotic stress resilience in various plant species by affecting the activity of SQUAMOSA-promoter linking protein-like (SPL) transcription factors.

Feyissa et al. (2021) conducted a study investigating various genetic constructs associated with the flood tolerance of lucerne, focusing specifically on SPL6RNAi, SPL13_RNAi, miR156OE, wild-type and empty vector plants. The research delved deeper into the traits of the flood-tolerant lucerne variety AAC-Trueman, contrasting it with the flood-sensitive variety AC-Caribou.

9.7.4 Heat and Cold Stress

Low temperatures, particularly those falling below 0°C, as well as moderate temperatures between 0°C and 15°C, significantly influence plant development, growth and physiological processes. Temperature-responsive regulatory genes, such as heat shock factors and C-repeat/dehydration-responsive element-binding proteins (CBF/DREB1s), modulate the regulatory components of genes involved in heat stress response and cold-regulated genes (COR). These factors are crucial in enabling plants to endure both cold and heat stress conditions. Strand-specific sequencing and chip-based microarray techniques were employed to identify lncRNAs associated with thermal stress conditions, encompassing both cold and heat. Throughout various stages of development, these lncRNAs exhibit tissue specificity in Table 9.2. An example case is that lnc-173 connects explicitly with the heat-responsive SUCROSE SYNTHASE ENZYME4 gene, in contrast to the lncRNA (Yuan et al., 2017). The capacity of eukaryotic organisms to adjust and recuperate from thermal stress is fundamentally linked to the function of heat shock transcription factors (HSFs). Heat stress triggers the production of the natural anti-sense transcript of HSFB2a (asHSFB2a), which helps regulate the expression of HSFB2a. Recent findings highlight the presence of heat-induced long intergenic ncRNA 1 (HILincl) in pears. This comprehensive ncRNA interacts with heat stress factor (HSF), leading to an augmentation of HSF's transcriptional activity and consequently facilitating an increase in multiprotein bridging factor 1c. Plants exhibit a response to cold stress through a mechanism called vernalisation, which effectively suppresses flowering during the winter months and, under favourable conditions, promotes flowering in the reproductive season. The FLC gene is essential for the regulation of vernalisation. A number of lncRNAs are involved in chromatin modification to reduce FLC expression, such as COLDAIR, COOLAIR and COLDWRAP (Jampala et al., 2021). COLD-INDUCED lncRNA1 (CIL) plays a role in low-temperature stress response by modulating the ROS pathway and molecules involved in osmotic adjustment (Table 9.2) (Liu et al., 2022).

9.8 SUMMARY AND FUTURE PERSPECTIVES

Advancements in high-throughput sequencing technologies, such as single-cell RNA sequencing, have significantly improved our ability to detect and study ncRNAs. The implementation of various validation techniques, such as mass spectrometry, quantitative real-time PCR, CRISPR/Cas9 and northern blotting, has contributed significantly to this progress. The flexibility of these ncRNAs, coupled with their common evolutionary backgrounds and preserved mechanisms across different plant species, highlights their importance in a wide range of biological processes. Unique expression patterns across different tissues play a crucial role in shaping plant development and growth. Plant responses to abiotic stresses are regulated at the molecular level by intricate and multifaceted systems, which include transcriptional, post-transcriptional and epigenetic mechanisms. Recent studies often utilise deep learning algorithms to predict new ncRNAs that are customised for the unique tissues and environments of various plant species. The importance of laboratory validation in clarifying the biological roles of ncRNAs has been highlighted. This validation provides crucial resources for the advancement of plants that can better withstand, adapt to and thrive in changing climatic conditions.

REFERENCES

Ambros, V. (2011). MicroRNAs and developmental timing. *Current Opinion in Genetics and Development*, *21*, 511–517. https://doi.org/10.1016/j.gde.2011.04.003

Ariel, F., Lucero, L., Christ, A., Mammarella, M. F., Jegu, T., Veluchamy, A., Mariappan, K., Latrasse, D., Blein, T., Liu, C., Benhamed, M., & Crespi, M. (2020). R-loop mediated trans action of the APOLO long noncoding RNA. *Molecular Cell*, *77*, 1055–1065.e4. https://doi.org/10.1016/j.molcel.2019.12.015

Ariel, F., Romero-Barrios, N. Jégu, T., Benhamed, M., & Crespi, M. (2015). Battles and hijacks: Non-coding transcription in plants. *Trends in Plant Science*, *20*, 362–371. https://doi.org/10.1016/j.tplants.2015.03.003

Barber, W. T., Zhang, W., Win, H., Varala, K. K., Dorweiler, J. E., Hudson, M. E., & Moose, S. P. (2012). Repeat associated small RNAs vary among parents and following hybridization in Maize. *Proceedings of the National Academy of Sciences of the United States of America*, *109*, 10444–10449. https://doi.org/10.1073/pnas.1202073109

Bouba, I., Kang, Q., Luan, Y. S.,& Meng, J. (2019). Predicting miRNA-lncRNA interactions and recognising their regulatory roles in stress response of plants. *Mathematical Biosciences*, *312*, 67–76. https://doi.org/10.1016/j.mbs.2019.04.006

Bresso, E. G., Chorostecki, U., Rodriguez, R. E., Palatnikm, J. F., & Schommer, C. (2018). Spatial control of gene expression by miR319-regulated TCP transcription factors in leaf development. *Plant Physiology*, *176*(2), 1694–1708. https://doi.org/10.1104/pp.17.00823

Brosnan, C. A. & Voinnet, O. (2011). Cell-to-cell and long-distance siRNA movement in plants: Mechanisms and biological implications. *Current Opinion in Plant Biology*, *14*, 580–587. https://doi.org/10.1016/j.pbi.2011.07.011

Cao, P., Fan, W., Li, P., & Hu, Y. (2021). Genome-wide profiling of long non-coding RNAs involved in wheat spike development. *BMC Genomics*, *22*, 493. https://doi.org/10.1186/s12864-021-07851-4

Carthew, R.W. & Sontheimer, E. J. (2009). Origins and mechanisms of miRNAs and siRNAs. *Cell*, *136*, 642–655. https://doi.org/10.1016/j.cell.2009.01.035

Cech, T. R. & Steitz, J. A. (2014). The non-coding RNA revolution-trashing old rules to forge new ones. *Cell*, *157*, 77–94. https://doi.org/10.1016/j.cell.2014.03.008

Chen, L., Zhu, Q. H., & Kaufmann, K. (2020b). Long non-coding RNAs in plants: Emerging modulators of gene activity in development and stress responses. *Planta*, *252*, 92. https://doi.org/10.1007/s00425-020-03480-5

Chen, Q. J., Deng, B. H., Gao, J., Zhao, Z. Y., Chen, Z. L., Song, S. R., Wang, L., Zhao, L. P., Xu, W. P., Zhang, C. X., et al. (2020a). A miRNA encoded small peptide, Vvi-MiPEP171d1, regulates adventitious root formation. *Plant Physiology*, *183*, 656–670. https://doi.org/10.1104/pp.20.00197

Chitwood, D. H., Nogueira, F. T. S., Howell, M. D., Montgomery, T. A., Carrington, J. C., & Timmermans, M. C. P. (2009). Pattern formation via small RNA mobility. *Genes & Development*, *23*, 549–554. https://doi.org/10.1101/gad.1770009

Cho, Y. H., Hong, J. W., Kim, E. C., & Yoo, S. D. (2012). Regulatory functions of SnRK1 in stress-responsive gene expression and in plant growth and development. *Plant Physiology*, *158*, 1955–1964. https://doi.org/10.1104/pp.111.189829

Crespi, M. D., Jurkevitch, E., Poiret, M., d'Aubenton-Carafa, Y., Petrovics, G., Kondorosi, E., & Kondorosi, A. (1994). enod40, a gene expressed during nodule organogenesis, codes for a non-translatable RNA involved in plant growth. *The EMBO Journal*. https://doi.org/10.1002/j.1460-2075.1994.tb06839.x

Deng, Y., Luo, K. L., Shaye, D. D., & Greenwald, I. (2019). A screen of the conserved kinome for negative regulators of LIN-12 negative regulatory region ("NRR")-Missense activity in *Caenorhabditis elegans*. *G3 Genes|Genomes|Genetics*, *9*(11), 3567–3574. https://doi.org/10.1534/g3.119.400471

Eddy, S. R. (2001). Non-coding RNA genes and the modern RNA world. *Nature Reviews Genetics*, *2*, 919–929. https://doi.org/10.1038/35103511

Feng, X. Y., Zhu, S. X., Pu, K. J., Huang, H. J., Chen, Y. Q., & Wang, W. T. (2023a). New insight into CircRNAs: Characterisation, strategies, and biomedical applications. *Experimental Hematology & Oncology*, *12*, 91.

Feng, Y. Z., Zhu, Q. F., Xue, J., Chen, P., & Yu, Y. (2023b). Shining in the dark: The big world of small peptides in plants. *aBIOTECH*, *4*, 238–256.

Feyissa, B. A., Amyot, L.,Nasrollahi, V., Papadopoulos, Y., Kohalmi, S. E., & Hannoufa, A. (2021). Involvement of the MiR156/SPL module in flooding response in *Medicago sativa*. *Scientific Reports*, *11*, 3243.

Gelaw, T. A. & Sanan-Mishra, N. (2021). Non-coding RNAs in response to drought stress. *International Journal of Molecular Sciences*, *22*(22), 12519. https://doi.org/10.3390/ijms222212519

Hasanuzzaman, M. & Fujita, M. (2022). Plant responses and tolerance to salt stress: Physiological and molecular interventions. *International Journal of Molecular Sciences*, *23*, 15740. https://doi.org/10.3390/ijms23094810

He, G., Luo, X., Tian, F., Li, K., Zhu, Z., Su, W., Qian, X., Fu, Y., Wang, X., Sun, C., & Yang, J. (2006). Haplotype variation in structure and expression of a gene cluster associated with a quantitative trait locus for improved yield in rice. *Genome Research*, *16*(5), 618–626. https://doi.org/10.1101/gr.4814006

Hirayama, T. & Shinozaki, K. (2010). Research on plant abiotic stress responses in the post genome era: Past, present, and future. *The Plant Journal*, *61*, 1041–1052. https://doi.org/10.1111/j.1365-313X.2010.04124.x

Jain, P., Hussian, S., Nishad, J., Dubey, H., Bisht, D. S., Sharma, T. R., & Mondal, T. K. (2021). Identification and functional prediction of long non-coding RNAs of rice (*Oryza sativa* L.) at reproductive stage under salinity stress. *Molecular Biology Reports*, *48*(3), 2261–2271. https://doi.org/10.1007/s11033-021-06246-8

Jampala, P., Garhewal, A., & Lodha, M. (2021). Functions of long non-coding RNA in *Arabidopsis thaliana*. *Plant Signaling & Behavior*, *16*, 1925440. https://doi.org/10.1080/15592324.2021.1925440

Katiyar-Agarwal, S., Gao, S., Vivian-Smith, A., & Jin, H. (2007). A novel class of bacteria induced small RNAs in Arabidopsis. *Genes & Development*, *21*(23), 3123e3134. https://doi.org/10.1101/gad.1595107

Katiyar-Agarwal, S. & Jin, H. (2010). Role of small RNAs in host-microbe interactions. *Annual Review of Phytopathology*, *48*, 225e246. DOI: 10.1146/annurev-phyto-073009-114457

Kindgren, P., Ard, R., Ivanov, M., & Marquardt, S. (2018). Transcriptional read-through of the long non-coding RNA SVALKA governs plant cold acclimation. *Nature Communications*, *9*, 4561.

Lauressergues, D., Couzigou, J. M., San Clemente, H., Martinez, Y., Dunand, C., Bécard, G., & Combier, J. P. (2015). Primary transcripts of microRNAs encode regulatory peptides. *Nature*, *520*, 90–93. https://doi.org/10.1038/nature14346

Li, Y., Tan, Z., Zeng, C., Xiao, M., Lin, S., Yao, W., Li, Q., Guo, L., & Lu, S. (2023). Regulation of seed oil accumulation by lncRNAs in *Brassica Napus*. *Biotechnology for Biofuels and Bioproducts*, *16*, 22.

Liu, G., Liu, F., Wang, Y., & Liu, X. (2022). A novel long non-coding RNA CIL1 enhances cold stress tolerance in Arabidopsis. *Plant Science*, *323*, 111370. https://doi.org/10.1016/j.plantsci.2022.111370

Liu, S., Guo, X. Y., Shang, Q. J., & Gao, P. (2023). The biogenesis, biological functions and modification of circular RNAs. *Experimental and Molecular Pathology*, *131*, 104861.

Liu, Y., Teng, C., Xia, R., & Meyers, B. C. (2020). PhasiRNAs in plants: Their biogenesis, genic sources, and roles in stress responses, development, and reproduction. *Plant Cell*, *32*, 3059–3080. https://doi.org/10.1105/tpc.20.00335

Liu, Z., Zhou, Y., Guo, J., Li, J., Tian, Z., Zhu, Z., Wang, J., Wu, R., Zhang, B., Hu, Y., Sun, Y., Shangguan, Y., Li, W., Li, T., Hu, Y., Guo, C., Rochaix, J. D., Miao, Y., & Sun, X. (2020). Global dynamic molecular profiling of stomatal lineage cell development by single-cell RNA sequencing. *Molecular Plant*, *13*(8), 1178–1193. https://doi.org/10.1016/j.molp.2020.06.010

Lukiw, W. J. (2013). Circular RNA (CircRNA) in Alzheimer's disease (AD). *Frontiers in Genetics*, *4*, 307. https://doi.org/10.3389/fgene.2013.00307

Mattick, J. S., Amaral, P. P., Carninci, P., Carpenter, S., Chang, H. Y., Chen, L. L., Chen, R., Dean, C., Dinger, M. E., Fitzgerald, K. A., Gingeras, T. R., Guttman, M., Hirose, T., Huarte, M., Johnson, R., Kanduri, C., Kapranov, P., Lawrence, J. B., Lee, J. T., …, & Wu, J. E. (2023). Long non-coding RNAs: Definitions, functions, challenges and recommendations. *Nature Reviews Molecular Cell Biology*, *24*(6), 430–447. https://doi.org/10.1038/s41580-022-00566-8

Mittler, R. (2006). Abiotic stress, the field environment and stress combination. *Trends in Plant Science*, *11*, 15–19. https://doi.org/10.1016/j.tplants.2005.11.002

Panda, A. K., Rawal, H. C., Jain, P., Mishra, V., Nishad, J., Chowrasia, S., Sarkar, A. K., Sen, P., Naik, S. K., & Mondal, T. K. (2022). Identification and analysis of miRNAs-LncRNAs-mRNAs modules involved in stem-elongation of deepwater rice (*Oryza sativa* L.). *Physiologia Plantarum*, *174*, 13736. https://doi.org/10.1111/ppl.13736

Patil, V. S., Zhou, R., & Rana, T. M. (2014). Gene regulation by non-coding RNAs. *Critical Reviews in Biochemistry and Molecular Biology*, *49*, 16–32. https://doi.org/10.3109/10409238.2013.844092

Pekker, I., Alvarez, J. P., & Eshed, Y. (2005). Auxin response factors mediate Arabidopsis organ asymmetry via modulation of KANADI activity. *Plant Cell*, *17*, 2899–2910. https://doi.org/10.1105/tpc.105.034876

Rai, M. I., Alam, M., Lightfoot, D. A., Gurha, P., & Afzal, A. J. (2019). Classification and experimental identification of plant long non-coding RNAs. *Genomics*, *111*, 997–1005. https://doi.org/10.1016/j.ygeno.2018.04.014

Sanan-Mishra, N., Abdul Kader Jailani, A., Mandal, B., & Mukherjee, S. K. (2021). Secondary siRNAs in plants: Biosynthesis, various functions, and applications in virology. *Frontiers in Plant Science*, *12*, 610283. https://doi.org/10.3389/fpls.2021.610283

Satterlee, J. W. & Scanlon, M. J. 2019. Coordination of leaf development across developmental axes. *Plants*, *8*(10), 433. https://doi.org/10.3390/plants8100433

Skopelitis, D. S., Benkovics, A. H., Husbands, A. Y., & Timmermans, M. C. P. (2017). Boundary formation through a direct threshold-based readout of mobile small RNA gradients. *Developmental Cell*, *43*, 265–273.e6. https://doi.org/10.1016/j.devcel.2017.10.003

Stief, A., Altmann, S., Hoffmann, K., Pant, B. D., Scheible, W. R., & Bäurle, I. (2014). Arabidopsis MiR156 regulates tolerance to recurring environmental stress through SPL transcription factors. *Plant Cell*, *26*, 1792–1807. https://doi.org/10.1105/tpc.114.123851

Tian, H., Jia, Y., Niu, T., Yu, Q., & Ding, Z. (2014). The key players of the primary root growth and development also function in lateral roots in Arabidopsis. *Plant Cell Reports*, *33*, 745–753. https://doi.org/10.1007/s00299-014-1575-x

Voinnet, O. (2009). Origin, biogenesis, and activity of plant microRNAs. *Cell*, *136*, 669–687. https://doi.org/10.1016/j.cell.2009.01.046

Wang, H. L. V., & Chekanova, J. A. (2017). Long noncoding RNAs in plants. *Long non coding RNA biology*, Advances in Experimental Medicine and Biology (AEMB, volume 1008) (pp. 133–154). Springer. https://doi.org/10.1007/978-981-10-5203-3_5

Waseem, M., Liu, Y., & Xia, R. (2021). Long non-coding RNAs, the dark matter: An emerging regulatory component in plants. *International Journal of Molecular Sciences*, *22*(1), 86. https://doi.org/10.3390/ijms22010086

Wierzbicki, A. T., Blevins, T., & Swiezewski, S. (2021). Annual review of plant biology long non-coding RNAs in plants. *Annual Review of Plant Biology*, *72*. https://doi.org/10.1146/annurev-arplant-093020-035446

Yadav, A., Sanyal, I., Rai, S. P., & Lata, C. (2021). An overview on miRNA-encoded peptides in plant biology research. *Genomics*, *113*, 2385–2391. https://doi.org/10.1016/j.ygeno.2021.05.013

Yuan, J., Chen, S., Jiao, W., Wang, L., Wang, L., Ye, W., Lu, J., Hong, D., You, S., & Cheng, Z. (2017). Both maternally and paternally imprinted genes regulate seed development in rice. *New Phytologist*, *216*, 373–387. https://doi.org/10.1111/nph.14510

Zhang, D. J.,Yang, Y. J., Liu, C. Y., Zhang, F., & Wu, Q. S. (2018). Root hair growth and development in response to nutrients and phytohormones. *Root Biology*, *52*, 65–84. https://doi.org/10.1007/978-3-319-75910-4_3

Zhang, Y. C., Liao, J. Y., Li, Z. Y., Yu, Y., Zhang, J. P., Li, Q. F., Qu, L. H., Shu, W. S., & Chen, Y. Q. (2014). Genome-wide screening and functional analysis identify a large number of long noncoding RNAs involved in the sexual reproduction of rice. *Genome Biology*, *15*, 512. https://doi.org/10.1186/s13059-014-0512-1

Zhang, Z., Zhang, R., Meng, F., Chen, Y., Wang, W., Yang, K., Gao, Y., Xin, M., Du, J., Hu, Z., & Ni, Z. (2023). A comprehensive atlas of long non-coding RNAs provides insight into grain development in wheat. *Seed Biology*, *2*(12). https://doi.org/10.48130/SeedBio-2023-0012

Zhao, X., Lan, Y., & Chen, D. (2022). Exploring long non-coding RNA networks from single cell omics data. *Computational and Structural Biotechnology Journal*, *20*, 4381–4389. https://doi.org/10.1016/j.csbj.2022.08.003

Zhou, W. Y., Cai, Z. R., Liu, J., Wang, D. S., Ju, H. Q., & Xu, R. H. (2020). Circular RNA: Metabolism, functions and interactions with proteins. *Molecular Cancer*, *19*, 172. https://doi.org/10.1186/s12943-020-01286-3

10 Harnessing the Power of ncRNAs
Enhancing Abiotic Stress Tolerance in Plants

Taufiq Nawaz, Shah Fahad, Sirmast Faiz, Nitish Joshi, Rehan Aslam, Muhammad Ikram, Shah Saud, Imran Khan and Riffat Latif

10.1 INTRODUCTION

Abiotic and biotic stresses affect plant growth, productivity and quality. They pose huge threat to agriculture productivity and global food security. In order to survive, plants must adapt to many abiotic stresses. These extreme temperatures, droughts and excessive salinity may have a negative effect on plant growth, development and productivity and have potential to harm global food security (Zhang et al., 2022). These stresses can induce disruption on cellular homeostasis, excessive production of reactive oxygen species (ROS) and can potentially lead to the molecular and physiological damage (Nadarajah, 2020). The plants have developed sophisticated tolerance mechanisms through gene expression, signalling and biochemical defence modulation to combat these adverse effects (Sachdev et al., 2021).

The plants response to abiotic stress via cellular homeostasis, from membrane stress perception to highly complex gene regulatory networks within the cell. Stress signals are transmitted and elicited via key signalling molecules, such as calcium ions (Ca^{2+}), abscisic acid (ABA), nitric oxide (NO) and ROS (Ravi et al., 2023). For instance, ABA signalling pathway is upregulated in plants during drought conditions to reduce water loss via stomatal closure activating genes related to osmosis and cell protection (Hussain et al., 2021). Saline environments reorient ion transport mechanisms to prevent buildup of toxic sodium and to clear the way for antioxidant scavenging of ROS to reduce oxidative damage (Jiang et al., 2024). Temperature extremes evoke unique responses on either side of the temperature spectrum. In heat stress, for example, plants increased protein production of heat shock proteins (HSPs) that are molecular chaperones that stabilise proteins to protect cellular structures from heat induced denaturation of proteins (Khan & Shahwar, 2020). On the other hand, cold stress increases synthesis of the antifreeze proteins that promote membrane fluidity, stabilise cell structure during freezing. These adaptations highlight the fact that the plant stress response is both dynamic and multifaceted and critically dependent on finely tuned gene expression networks (Jahed et al., 2023).

Due to the negative impact of climate change on agricultural productivity, there is an urgent need to develop crops that can withstand abiotic stresses. Drought, salinity and extreme temperatures are estimated to reduce crop yields by 15–35 per cent, reinforcing the need to improve plant resilience (Hasnain et al., 2023). Breeding techniques were used to select traits that would make a plant resilient in response to stress, but new developments in molecular biology and biotechnology offer more precise, more targeted and more efficient breeding methods. Recently, non-coding RNAs (ncRNAs) have been shown as promising tools in crop breeding because they are involved in stress response pathways. Understanding the role of ncRNAs in stress tolerance can alter the current and future approaches to crop improvement by providing new potential genetic engineering targets (Summanwar et al., 2020).

Generally, RNA molecules that are not translated into proteins but play some regulatory functions in the cell are known as non-coding RNAs (ncRNA) (Bhatti et al., 2021). Both microRNAs (miRNAs) and long non-coding RNAs (lncRNAs) are well-studied and involved in epigenetic modification and transcription regulation (Mudasir Ahmad et al., 2022). miRNAs are 20–24 nucleotides long; they are involved in mRNA degradation, or translation repression. They bind to their complementary sequence in their target mRNA. This mechanism allows for miRNA control of target expression in response to environmental conditions (Naeli et al., 2023).

On the other hand, lncRNAs are 200 nucleotides long, exhibit a highly diverse functions in cellular processes and its main functions include molecular scaffolds, decoys and chromatin remodellers (Nadhan et al., 2022). Interaction of lncRNAs with specific transcription factors and RNA-binding proteins allows them to regulate gene expression. Recently, lncRNAs are demonstrated to modify stress response pathways to regulate plant's response to abiotic stress. For example, specific lncRNAs increase tolerance to drought and high salinity by transducing antioxidants response and hormone signal transduction (Das et al., 2023; Ren et al., 2022). In addition, ncRNAs control epigenetic changes that enable plants to store the memory of previous stresses so as to react more quickly to the next stress. These ncRNAs act as mediators of DNA methylation and histone modifications to create epigenetic memory (Akhter et al., 2021).

In this chapter, we will discuss how ncRNAs perform different and changing functions in plant abiotic stress response. It starts with an overview of various ncRNAs and how they are involved in the regulatory mechanism for gene expression. The following sections elaborate on the role of miRNAs, lncRNAs and other ncRNAs in plant response to a range of abiotic stresses such as drought, salinity,

DOI: 10.1201/9781003583257-10

temperature and heavy metal toxicities. The next part of the chapter will present bioengineering approaches based on CRISPR-Cas, which are being utilised to reprogramme the ncRNA pathways in order to increase crop stress tolerance, reviewing the research of ncRNA (delivery systems, off-target effects and ethical aspects) and discussing possible challenges and limitations. In addition, the potential for ncRNAs to produce more resilient crops against climate change stress is discussed, and the future of ncRNAs in plant biotechnology is presented. In this chapter, we present an overview of previous research and recent technological trends and how ncRNAs are being used to improve stress-tolerant crops and to promote sustainable development in agriculture.

10.2 TYPES AND FUNCTIONS OF ncRNAs IN PLANTS

Non-coding RNAs (ncRNAs) of diverse complexity are found in plants but cannot translate into proteins, yet play pivotal roles in gene expression, especially in response to abiotic stress. This section focuses on the main classes of plant ncRNAs, such as microRNAs (miRNAs), small interfering RNAs (siRNAs), long non-coding RNAs (lncRNAs) and other newer classes such as tasiRNAs, phasiRNAs and circular RNAs (circRNAs). These ncRNA classes perform important roles in fine tuning such as gene regulation to increase resilience in the face of stress, and more general regulation of gene expression.

10.2.1 MicroRNAs (miRNAs): Structure, Biogenesis and General Functions

MicroRNAs (miRNAs) are small ncRNAs (~20–24 nt) that regulate gene expression post-transcriptionally. Primary miRNAs having length of ~70 nucleotides are transcribed by RNA polymerase II and form stem loop structures, which initiate miRNA processing. The mature miRNAs are processed from precursor miRNAs (pre-miRNAs) in the nucleus by DICER-LIKE 1 (DCL1) enzymes to finally generate the mature miRNAs (Leitão & Enguita, 2022). Specific mature miRNAs are then incorporated into the RNA-induced silencing complex (RISC) where they then guide the complex to target messenger RNA (mRNA) with similar sequences which results either in mRNA degradation or translational repression (Leitão & Enguita, 2022).

In plants, miRNAs control a diverse array of processes including development, hormone signalling and stress responses (e.g., to drought and salinity). For instance, miR398 is induced by oxidative stress and attacks Cu/Zn superoxide dismutase (Cu/Zn SOD) genes that are key for ROS detoxification (Ghosh et al., 2022). During stress conditions other miRNAs such as miR159 and miR319 are also important in the hormone signalling modulating gibberellin and jasmonate pathways to adjust growth in response to environmental changes (Singh et al., 2021). Therefore, miRNAs provide an important regulatory mechanism by which plants can rapidly adjust expression of genes that are critical in regulating the response to environmental fluctuation.

10.2.2 Small Interfering RNAs (siRNAs): Mechanism of Action and Roles in Gene Silencing

The second class of small ncRNAs are small interfering RNAs (siRNAs), which are mainly known for their role in RNA interference (RNAi), gene silencing, and are derived from double-stranded RNA (dsRNA) precursors through the activity of DICER–LIKE enzymes that process the dsRNA into 21–23 nt siRNA duplexes. Then these duplexes are then loaded onto the RISC complex where the guide strand directs the complex to the target complementary mRNA and cleaves and degrades the target mRNA (Ciechanowska et al., 2021). siRNAs play pivotal role by silencing expression in the defence against viral infections, transposable elements and other genomic parasites (Kong et al., 2022). Furthermore, siRNAs modulate transposons and maintain genome stability, particularly under stress conditions during which the activity of transposons may increase (Gill et al., 2021). In addition, stress-induced siRNAs can direct DNA methylation via RNA-directed DNA methylation (RdDM) to epigenetically regulate the expression of stress-responsive genes. In this way, siRNAs offer an additional means of controlling genes highly relevant for plant adaptation to abiotic stresses through their ability to manipulate epigenetic states (Ashapkin et al., 2020; Erdmann & Picard, 2020).

10.2.3 Long Non-Coding RNAs (lncRNAs): Structural and Functional Diversity

Long non-coding RNAs (lncRNAs) are a structurally diverse group of ncRNAs longer than 200 nucleotides with various regulatory roles in gene regulation (Statello et al., 2021). Unlike miRNAs and siRNAs, lncRNAs don't have specific length or have a unique biogenesis pathway, which makes them diverse in their function. As molecular scaffolds, decoys, guides or competitors of miRNA binding sites, they enable multiple layers of gene regulation (Li et al., 2014).

Abiotic stress responses depend heavily on lncRNAs, which modulate expression of stress-related genes and signalling pathways. For example, some lncRNAs are induced under drought stress and enhance stress tolerance by regulating ABA signalling pathways or ion transport to maintain cellular osmotic balance (Wei et al., 2024). Some lncRNAs physically interact with chromatin-modifying complexes to control histone modifications, chromatin structure and result in quick and reversible stress response (Li & Fu, 2019). Recent studies have also demonstrated that lncRNAs form complex regulatory networks with miRNAs and mRNAs to generate competitive endogenous RNA (ceRNA) networks for fine-tuned gene expression during stress (Ala, 2020).

10.2.4 Other Emerging ncRNAs: Including tasiRNAs, phasiRNAs and circRNAs

In addition to the well-characterised miRNAs, siRNAs and lncRNAs, several other ncRNAs have recently emerged as important regulators in plants such as trans-acting small interfering RNAs (tasiRNAs), phased small interfering RNAs (phasiRNAs) and circRNAs enlisted in Table 10.1.

- TAS siRNAs (TasiRNAs) are a subset of siRNAs found in TAS genomic loci. Their biogenesis is initiated by the action of miRNAs and is processed in phasic manner. TasiRNAs selectively target distinct mRNAs and are essential in the processes of development and stress responses such as lateral root growth, leaf morphogenesis.
- Typically, miRNAs trigger PhasiRNAs from PHAS loci. They are expressed in a phased pattern and principally influence development of the reproductive system, responding to abiotic stresses. PhasiRNAs can regulate genes of antioxidant response and ion homeostasis to response stress.
- Circular RNAs (circRNAs) refer to covalently closed RNAs that exhibit a continuous loop structure with stability and resistance to exonuclease degradation. Less studied in plants, circRNAs are thought to function as miRNA sponges, regulatory hubs and transcription modulators, thereby facilitating stress responses through sequestration of miRNAs to prevent their action of repressing stress-responsive genes.

10.3 FUNCTIONAL OVERVIEW OF ncRNAs IN REGULATING GENE EXPRESSION AND STRESS RESPONSES

Non-coding ribonucleic acid (ncRNA) is a type of RNA which is not being translated into proteins but into genome. In addition to renowned RNA (rRNA, tRNA, mRNA), various classes of non-codindg RNA including small nuclear RNAs (snRNAs) with 18–30 nucleotide, medium-sized ncRNAs (31–200 nt), long non-coding RNAs (lncRNAs) (>200 nt), small nucleolar RNAs (snoRNAs) microRNAs (miRNAs), small interfering RNAs (siRNAs), long non-coding RNAs (lncRNAs) and circular RNAs (circRNAs) have been found in plants. They have promising role in modulating gene expression networks, act as the pivotal regulators in chromatin remodelling, epigenetic memory, transcriptional and post-transcriptional regulation of gene expression, RNA splicing, the modulation of RNA stability and translation (Hirayama & Shinozaki, 2010; Khraiwesh et al., 2012; Nejat & Mantri, 2018). In addition, they have key roles in retaining genomic stability, growth and development, senescence and plant responses to biological and abiotic stresses. LncRNA is the newly appearing class which has versatile competence and transcribed by RNA polymerase II without having significant open reading frame (ORF). In the start it was thought that lncRNA are not stable because of their low expression level (Chekanova, 2015; Jampala et al., 2021; Khraiwesh et al., 2012).

On the other hand, some have half-lives greater than 12h. All these classes of non-coding RNAs are specific in their biosynthesis action on specific target genes, performing specific cellular processes like gene expression regulation (miRNAs, piRNAs, lncRNAs), RNA maturation (snRNAs,

TABLE 10.1
Classification of ncRNAs

Classification	Composition	Function	Reference
Circular ncRNAs	Exonic circRNAs, intronic circRNAs, UTR circRNAs, intergenic circRNAs, Other circRNAs	They have a promising role in plant responses to abiotic stresses including drought, salt heat and cold stress	Bhar and Roy (2023), Ma et al. (2022)
Housekeeping ncRNAs	Ribosomal RNA (rRNA), transfer RNA (tRNA), small nuclear RNA (snRNA), small nucleolar RNA (snoRNA)	They are involved in fundamental cellular processes and indirectly impact the plant responses to abiotic stress	Baruah et al. (2021), Matsui et al. (2013), Nejat and Mantri (2018)
Linear ncRNAs	SRNAs, microRNAs (miRNAs), natural antisense transcript-derived small interfering RNAs (nat-siRNAs), trans-acting siRNAs (tasiRNAs), repeat-associated siRNAs (rasi-RNAs), heterochromatic small interfering RNAs (hc-siRNAs)	They play a key role in regulation of gene expression at various stages to deal with abiotic stress challenges in plants	Bhar and Roy (2023), Jha et al. (2020), Ma et al. (2022)
Regulatory ncRNAs	Long ncRNAs, intergenic ncRNAs, intron ncRNAs, sense ncRNAs (slncRNAs), antisense ncRNAs (ancRNAs), piwi-interacting RNAs	They play a critical role in plant responses to abiotic stresses. They regulate gene expression, affecting different cellular processes that resulting contribute to stress tolerance	Hazra et al. (2023), Panchal et al. (2023), Waititu et al. (2020)

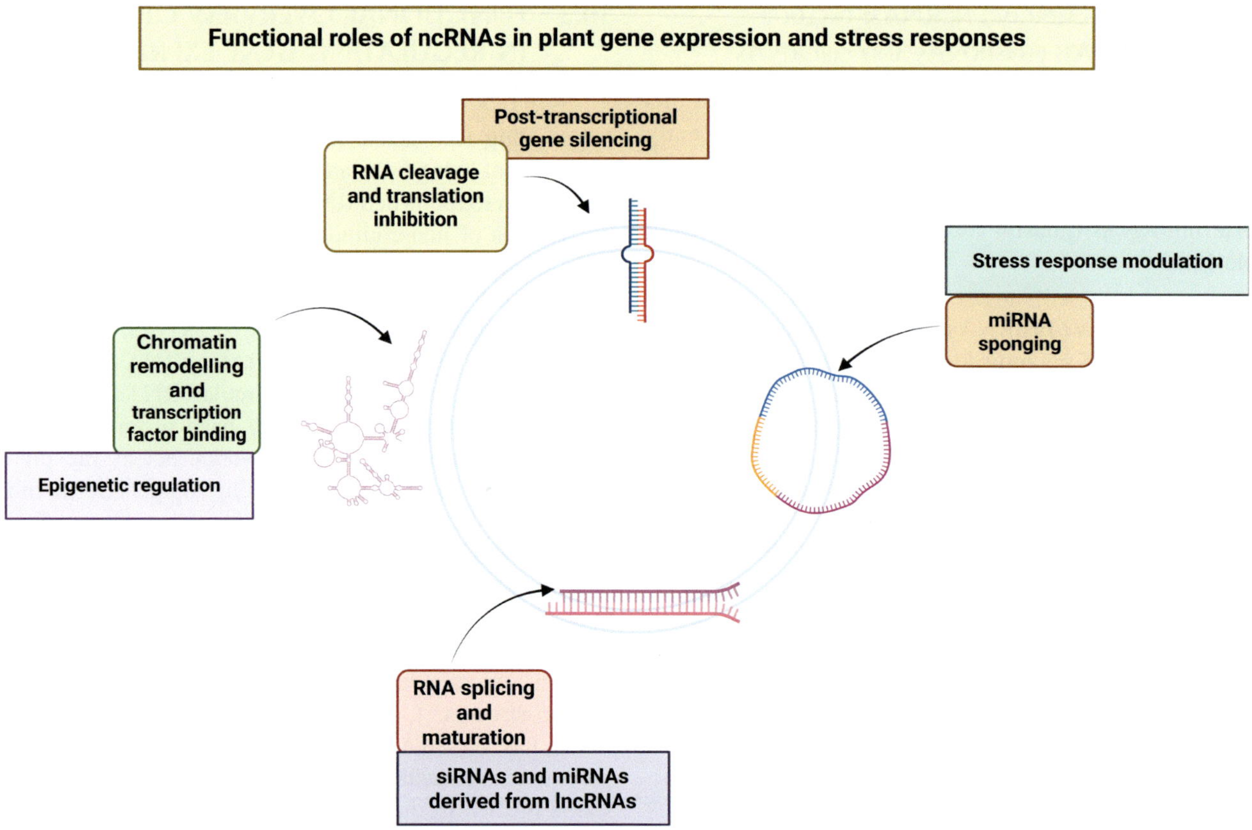

FIGURE 10.1 Mechanistic Roles of Non-Coding RNAs in Regulating Gene Expression and Stress Responses in Plants: An Integrated Overview of ncRNA Classes, Functions, and Biotechnological Applications.

snoRNAs) and protein synthesis (rRNAs, tRNAs). Such multifaceted regulation is made possible by virtue of their high versatility to interact with chromatin, functional proteins and different RNA species. Besides of all the differences either of their size, origin, synthesis or mode of action, the passages of regulation are collaboratively interrelated and connected either in the regulation of genes or during production of genome from harsh threats (Ghildiyal & Zamore, 2009).

In plants generally, ncRNAs operate by diverse means to fine tune gene expression under stress conditions, with miRNAs and siRNAs acting mostly at the post-transcriptional level by guiding translation or mRNA cleavage (Zhang et al., 2022, 2023). In contrast, lncRNAs are often involved in transcriptional or chromatin actions, which affect large gene networks through binding to transcription factors, RNA-binding proteins and chromatin-modifying complexes (Mishra & Kanduri, 2019). The emerging ncRNAs include tasiRNAs and phasiRNAs, both increase the level of complexity in regulation and influence distinct developmental processes and stress responses in a phased manner (Yu et al., 2019). With their unique structural stability, circRNAs regulate stress responses through sequestering regulatory miRNAs and contribute to the dynamic balance of gene regulation under fluctuating environmental conditions (van Wijk et al., 2022). Together, ncRNA-mediated mechanisms enable plant response to environmental stresses with high specificity and sensitivity by adjusting gene expression in a dynamic fashion. Therefore, this regulation allows plants adaptive flexibility to abiotic stresses, and the use of ncRNAs could be employed in biotechnological applications to improve crop stress tolerance (Figure 10.1).

10.4 MECHANISMS OF ncRNA-MEDIATED ABIOTIC STRESS TOLERANCE

Although plants have diverse regulatory mechanisms to deal with abiotic stresses such as severe drought, salinity, temperature change and heavy metal toxicity, ncRNAs play a key role as players in helping them survive these stresses. To achieve these goals, they control stress-responsive genes, modulate hormone signalling pathways, intervene with epigenetic modifications and influence antioxidant responses. In this chapter we explore how ncRNAs enable these mechanisms by which plants can adapt and resist the harmful effects of abiotic stress, collectively through tolerance to abiotic stress.

10.4.1 Post-Transcriptional Gene Silencing (PTGS)

One of the primary mechanisms of ncRNA regulation of stress-responsive genes is post-transcriptional gene silencing (PTGS), which involves mRNA degradation or translational repression. PTGS is mediated by major miRNAs and siRNAs to degrade or to prevent translation of complementary mRNA via RISCs (Tan et al., 2020). One example is miR398 that is critical in oxidative stress response, functioning to control levels of ROS via targeting copper/zinc superoxide dismutase (Cu/Zn SOD) mRNAs. When the plant is under stress, miR398 levels go down which allows the SOD to accumulate and scavenge the excess ROS protecting the plant from oxidative damage (Li et al., 2022). miR393 also modulates auxin signalling by modulating the auxin receptors to control drought-induced lateral root growth for resource conservation (Ghorbanzadeh et al., 2022).

10.4.2 Hormonal Crosstalk and Signal Transduction

ncRNAs play a key role in the complex network of hormonal signalling that is essential for plant stress responses(Gill et al., 2024). Levels of hormones such as ABA, ethylene, jasmonic acid (JA) and gibberellins are often altered under stress conditions and modify stress-responsive gene expression (Parwez et al., 2022). For example, like ABA levels increase with drought stress and initiate signal cascades that eventually cause stomatal closure, reduced transpiration and activate stress related genes (Hasnain et al., 2023; Mukherjee et al., 2023).

miR159, miR319 and other ncRNAs directly regulate ABA and JA pathways, affecting, for example, seed dormancy, root growth and stress-induced stomatal closure by modulating the response via MYB transcription factors targeted by miR159 (Gelaw & Sanan-Mishra, 2021). This ncRNA-regulated pathway allows plants to conserve water under drought stress. Similarly, miR319 regulates TCP transcription factors (**Teosinte Branched 1/Cycloidea/ Proliferating Cell Factor**) that interact with JA signalling for the control of leaf morphology and tolerance to abiotic stresses by altering growth in adverse conditions (Guo et al., 2021). Furthermore, long non-coding RNAs (lncRNAs) participate in hormonal regulation under stress through their activities such as a molecular decoy or guide to ensure precise control over hormonal signalling pathways (Jha et al., 2020). On the other hand, some lncRNAs function as miRNA sponges that regulate miRNAs related to ABA signalling, boosting the activity of the ABA pathway thereby allowing a more severe response to drought and salt stresses (Numan et al., 2024).

10.4.3 Epigenetic Modifications

Epigenetic modifications make it possible for plants to not only respond rapidly but also reversibly, to changes in the environment by modifying gene expression, and ncRNAs play fundamental roles in these processes and can guide chromatin modifying enzymes to specified genomic loci or control levels of DNA methylation (Akhter et al., 2021). One such mechanism that generates such interactions involves siRNAs binding to RNA-directed DNA methylation (RdDM) RNA that completes epigenetic silencing through the physical action of directing DNA methyltransferases to methylate DNA at targeted regions turning off the expression of genes in a heritable, yet flexible way (Kryovrysanaki et al., 2022).

In addition, lncRNA also functions in diverse roles in chromatin remodelling. For an example, lncRNAs can recruit chromatin modifying proteins, such as histone methyltransferases, to modulate histone and chromatin structure (Kim, 2019).This can lead to open or closed chromatin states enabling or blocking stress-responsive gene transcription. An example is the lncRNA COLDAIR which recruits the Polycomb Repressive Complex 2 (PRC2) to silence FLOWERING LOCUS C (FLC) through histone modification, to allow plants to flower after a lengthy cold period (Zhao et al., 2022). ncRNAs regulate epigenetic marks to set up 'stress memory' in plants, whereby plants that experienced a stress event before can better respond to further stress events (Yajnik et al., 2024). This epigenetic memory gives plants an adaptive advantage, allowing them to tune their responses to environmental challenges based on past experiences.

10.4.4 Antioxidant Mechanisms

Oxidative stress is an overproduction of ROS such as hydrogen peroxide (H_2O_2) and superoxide radicals experienced by plants under stress conditions (Hasanuzzaman et al., 2020; Sachdev et al., 2021). ROS participate in cell signalling, and in excess, their accumulation can damage proteins, lipids and DNA, thus damaging cellular functions (Juan et al., 2021). ncRNAs are essential to regulate these antioxidant pathways, by coordinating the antioxidant enzymes superoxide dismutase (SOD), catalase (CAT) and ascorbate peroxidase (APX) to detoxify ROS (Oyebamiji et al., 2024).

Several miRNAs known directly to modulate antioxidant defences are also included; for example, miR398 targets mRNAs encoding Cu/Zn SOD enzyme proteins that remove superoxide radicals. Normally miR398 represses SOD expression but during oxidative stress miR398 levels are downregulated freeing up expression of SOD to maintain ROS levels (Li et al., 2022). It modulates regulatory adjustment to maintain ROS homeostasis and the cells from oxidative damage during stress (Fujibayashi et al., 2021).

Furthermore, a variety of other ncRNAs including some of the lncRNAs and phasiRNAs have been shown to regulate redox balance and upregulate ROS scavenging capacity (Othman et al., 2024). These ncRNAs contribute to a coordinated antioxidant response by modulating the activity of transcription factors that control antioxidant genes, which depend on them to respond to stressful conditions (Kinoshita & Aoyama, 2021).

The study of the role of ncRNAs in mediating plant responses to abiotic stresses highlights their important role as regulatory molecules that influence the fine tuning of gene expression and optimal regulation of stress adaptation mechanisms. ncRNAs ensure the dynamic plant responses to environmental changes through direct regulation of stress-responsive genes, modulation of hormone signalling

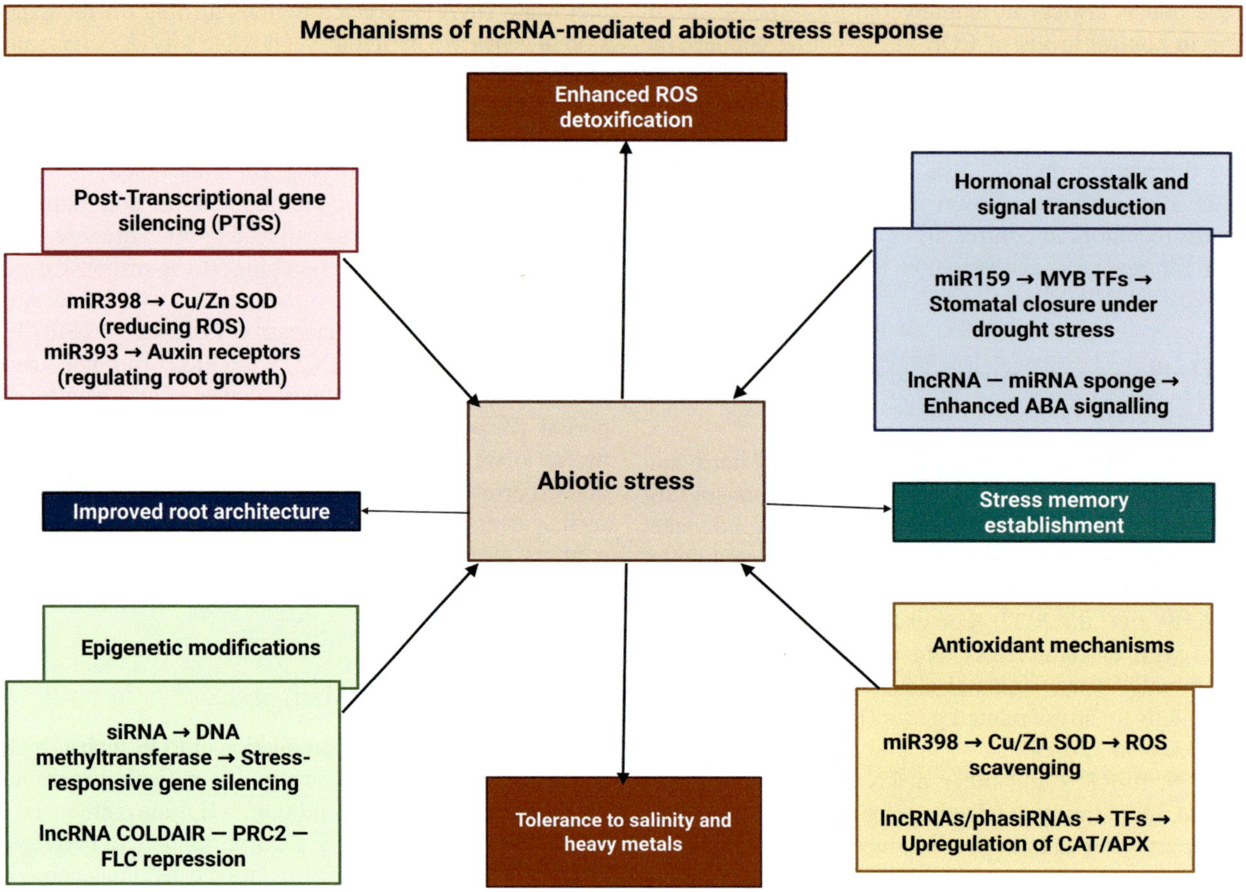

FIGURE 10.2 ncRNA-Mediated Abiotic Stress Tolerance ncRNAs help plants adapt to stress by regulating genes via PTGS, modulating hormone signaling, guiding epigenetic changes, and enhancing antioxidant defenses.

pathways, participating in epigenetic modifications and control over the antioxidant responses. Using ncRNA functions to advance crop stress tolerance is an avenue of future research to explore new ways to boost resilience in crops to global climate challenges in agricultural systems.

10.5 ncRNAs IN SPECIFIC ABIOTIC STRESS TOLERANCES

Increasing abiotic stress challenges in plants due to microclimate variations are major concerns for plant growth and yield. Environmental conditions including temperature, drought, freezing and salinity are ever-changing. Globally production of crops can be down due to abiotic stress (moisture stress, high salinity of soils, drought, extreme temperature conditions, oxidative damage and toxicity by heavy metals). All these abiotic stresses effect plants, resulting in decreased efficiency of most cellular functions. In recent years, the role of plant non-coding RNAs (ncRNAs) in plant response against abiotic stresses has been well described. It is this section that explores ncRNAs critical to stress tolerance of specific stresses such as drought, salinity, extreme temperatures (cold and heat) and so on, including those that facilitate tolerance to heavy metal toxicity and oxidative stress. Information about specific ncRNAs activated under these conditions, their mechanisms of action and their physiological effects on plants is provided by each case study (Figure 10.2).

10.5.1 Drought Tolerance

Drought stress is one of the most important abiotic stresses. It causes yield reduction and plants are exposed to it at least once in their life span. Specific ncRNAs, miRNAs and lncRNAs are activated in plants to regulate drought-tolerance pathways (Yang et al., 2022). For examples, miR166 and miR393 play important roles in drought response. miR166 regulates transcription factors involved in leaf morphology and root development, including HD-ZIP III genes that promote deeper root growth and thus water uptake during drought (Yadav et al., 2021).

miR393 is a conserved miRNA family present in many plants, which mainly targets genes encoding the transport inhibitor response1 (TIR1)/auxin signalling F-box (AFB) auxin receptors, and thus greatly affects the auxin signal perception (Xing et al., 2023).

Similarly, lncRNAs play an important in drought tolerance by regulating of ABA responsive genes. The lncRNA DRIR (drought-induced lncRNA in rice) increase drought tolerance. It is involved in the modification of ABA signalling pathways and expression of stress-responsive genes.

Induction of DRIR under drought conditions stimulates ABA-related genes expression and induce stomatal closure to reduce water loss (Xu et al., 2021).

10.5.2 Salinity Tolerance

Disruption of the ionic and osmotic balance caused by salinity stress affects plants. High levels of salt in soil decrease water absorption and can induce ion toxicity (Hayat et al., 2022). The ncRNAs regulate these stress responses by regulating genes for ion transport, osmoregulation and stress signalling (Sun et al., 2023).

A number of studies have demonstrated that microRNAs (miRNA) are pivotal to how plants respond to salt stress, particularly under salt stress affecting hormone signalling pathways. For instance, auxin signalling is linked to miR160, miR167 and miR393 in stress response mechanisms. In its normal state, these miRNAs are expressed at a low level, which only slightly inhibits AUXIN RESPONSE FACTORS (ARFs). But miR393 is upregulated under stress, suppressing auxin signalling by reducing TIR1 levels in favour of AUX/IAA-ARF complexes. Additionally, increased levels of miR160 and miR167 directly reduce ARF expression (Makkar et al., 2022; Zhang et al., 2021; Zhao et al., 2018).

Function of long non-coding RNAs (lncRNAs) has been implicated in salinity tolerance of rice (*Oryza sativa*). Recent work has identified the lncRNAs that bind specific transcription factors in response to salt to control genes encoding ion transport, which attenuates cytosolic Na^+ toxicity and maintains an optimal K^+/Na^+ ratio. For instance, a 2024 study identified variously expressed lncRNAs in the rice panicles under salt stress to play a role in stress response mechanisms

Research into the salt tolerant rice variety Pokkali also identified a set of lncRNAs potentially contributing to the tolerant phenotype in this variety (Jiang et al., 2024). These findings demonstrate the role of lncRNAs in helping rice tolerate salinity through regulation of ion homeostasis (Tiwari et al., 2023). These findings demonstrate the role of lncRNAs in helping rice tolerate salinity through regulation of ion homeostasis.

10.5.3 Cold and Heat Stress Tolerance

Temperature extremes (high temperatures, heat and low temperatures, cold) pose unique challenges for plants by disrupting cellular integrity and enzyme stability and cellular metabolic processes. ncRNAs are crucial for helping plants cope with temperature extremes by controlling genes eliciting heat and cold shock responses.

miR156 and miR398 are upregulated under heat stress to protect plants from thermal damage. miR156 targets SPL (SQUAMOSA PROMOTER BINDING PROTEIN-LIKE) transcription factors which promote expression of HSPs involved in heat stress tolerance that protect proteins from denaturation (Xu et al., 2016). miR398 regulates genes encoding the enzymes that produce ROS scavengers, thus preventing oxidative damage at elevated temperature (Zuo et al., 2021).

miR319 and miR393 are crucial in response to cold stress. The miR319 targets TCP transcription factors that are known to play a role in growth inhibition at low temperatures, so that the plant can save energy and resources if conditions are unfavourable (Lan & Qin, 2020). miR393 participates in auxin signalling and is critical to stabilise cell membranes and maintain cellular structure in the face of freezing temperatures (Rajwanshi et al., 2014). Similarly, lncRNAs like COOLAIR and COLDAIR also serve to be very important for cold tolerance. These lncRNAs regulate expression of FLOWERING LOCUS C (FLC) by the modulation of chromatin modifications, allowing plants to monitor flowering time and protect themselves from the elements of early frost (Jha et al., 2023).

10.5.4 Other Stresses (e.g., Heavy Metals, Oxidative Stress)

Cadmium, lead and mercury are some examples of heavy metals that have a negative impact on plants. These metals exert the damaging effects like oxidative stress and the interfere of key metabolic activities. It also revealed that plants employ ncRNAs to mitigate the effects of the heavy metals on cells and internal processes.

For example, miR167 and miR390 were found to be responsive to heavy metal stress. miR167 regulates auxin response factors that helps the plant to resist heavy metal uptake from the soil by influencing root structure (Ghosh et al., 2022; Yang et al., 2022). Similarly, miR390 regulates genes required to detoxify metal ions in order to compartmentalise and sequester heavy metals in the metal-tolerant compartment to limit their toxicity. Sometimes, oxidative stress in agreement with other abiotic stresses is correlated with high ROS productions (Ghosh et al., 2022). ncRNAs like miR398 and miR408 are key to ROS levels control. miR398 acts on the SOD enzymes that neutralise ROS, while miR408 regulates plants cyanin genes involved to photorespiration and redox imbalanced under stress conditions (Zhou et al., 2023).

Recent research has identified heavy metal-responsive lncRNAs, which impact genes involved in metal-binding proteins and antioxidants. These lncRNA play a role in metal tolerance by promoting the removal or sequestration of the toxic ions and antioxidant defence to reduce the oxidative damage caused by heavy metal (Zhou et al., 2023).

These case studies highlight the important roles of ncRNAs in plant tolerance to abiotic stress. Functionally, ncRNAs target specific stress-responsive genes; modulate hormonal pathways; and regulate ROS and ion balance to impart plants with a flexible and dynamic stress response that is essential for survival in extreme environmental conditions (Table 10.2). In fact, as our knowledge about these ncRNAs increases, their potential increases for biotechnological applications that can lead to the development of stress-tolerant crop varieties able to accommodate the growing challenges of a changing climate.

TABLE 10.2
Types of ncRNA and Their Role in Abiotic Stress

Types of ncRNA	Crop	Stress	Expression	Reference
miRNAs	*Chickpea, common beans, soybean, cowpea, pea*	Drought	Downregulate plantacyanin genes encoding and resulting regulate DREB expression under drought stress. Control drought stress by regulating transcription factors and protein kinases	Chand Jha et al. (2021), Hajyzadeh et al. (2015), Wu et al. (2017))
LTCONS	Medicago truncatula	Salinity	Upregulate the expression of genes involved in salinity stress adaptation	Pan et al. (2016), Wang et al. (2015)
lncRNA MtCIR1	*M. truncatula*	Salinity	Control cold stress	Zhao et al. (2020)
circRNAs	*Soybean*	Low phosphorus	Upregulate phosphorus use efficiency	Zhao et al. (2017)
Phosphate deficiency-induced lncRNAs (PDILs)	*M. truncatula*	Phosphate deficiency	Upregulate the phosphate uptake	Wang et al. (2017)
ENOD40	*M. truncatula, soybean*	Nitrogen-fixing nodules	Regulating gene expression, particularly in the development of nitrogen-fixing root nodules	Yang et al. (1993)
Long intergenic non-coding RNAs (lncRNAs)	*Soybean*	Salinity	Regulate the expression of genes involved in salinity stress adaptation	Chen et al. (2019), Ma et al. (2021)
siRNAs	*Yellow lupine*	Abscission of flowers	Upregulate flower development	Glazińska et al. (2019)
LRK antisense intergenic RNA	*Rice*	Yield	Upregulate the expression of *LRK* genes to enhance the yield	Wang et al. (2018)
LncNAT, TAR	*Fusarium oxysporum*	Drought, cold, high salt	Upregulate gene expressions to combat abiotic stress in plants	Wang et al. (2018)
LncRNAs	Camelina sativa	Drought	Upregulate the expression of genes to provide drought resistance	Zhu et al. (2013)
TalnRNA21	*Wheat*	Heat stress	Upregulate the siRNA expression to provide heat resistance	Xin et al. (2019)

10.6 APPROACHES TO ENHANCE ABIOTIC STRESS TOLERANCE USING ncRNAs

The key role of non-coding RNAs (ncRNAs) in plant stress tolerance has made crop improvement more exciting and innovative. To enhance abiotic stress tolerance in plants, scientists have been targeting ncRNAs with genetic engineering, CRISPR-Cas technologies, synthetic applications and novel gene-editing strategies. These approaches are precise and flexible, enabling plants to build in resilience to environmental threats. During abiotic stress, some of plant genes become active, as a result levels of several metabolites and proteins is enhanced to a greater extent. Some of which may be responsible for a certain degree of protection to these stresses. Hence, plants can withstand certain unfavourable conditions for normal growth. Plants have highly developed mechanisms to respond against several abiotic changes deteriorating the growth, yield and quality of the crops (Shriram et al., 2016; Wani et al., 2016). The survival of plants to these drastic environmental changes depends on the adjustment of many physiological and cellular processes achieved by perfecting gene expression and protein activities through transcriptional, post-transcriptional, translational and post-translational mechanisms. Modulation of gene expression is a key level of regulation during abiotic challenges.

10.6.1 Genetic Engineering and CRISPR-Cas for ncRNA Manipulation

The manipulation of miRNAs, siRNAs and lncRNAs that regulate stress genes, and some other ncRNAs associated with stress, is possible through genetic engineering. Notably, this approach has been promisingly useful for boosting plant resilience under abiotic stresses such as drought, salinity and temperature extremes. For example, the ability to manipulate miR398, a key regulator of oxidative stress, has been shown to increase drought and salinity tolerance by modulation of ROS scavenging enzymes, specifically copper/zinc superoxide dismutase (Cu/Zn SOD) (Zhang et al., 2024).

CRISPR-Cas technologies particularly CRISPR-Cas9 enable ncRNA gene editing by targeted editing (Mahato et al., 2024). Scientists now can use CRISPR to remove or replace much of the ncRNA to make more strong these stress-tolerance traits. For example, miRNAs that suppress drought-tolerant genes may be knocked down using CRISPR to allow survivability and water retention rates in plants to stay at a higher rate during prolonged drought conditions (Sami et al., 2021). In addition, the CRISPR-Cas technology was used to generate knock-out of miR159, an miRNA repressing ABA signalling, resulting in higher drought tolerance through increasing ABA sensitivity, and thus better water use efficiency (Premnath et al., 2022).

Reversible ncRNA manipulation without altering the genomic DNA is possible using advanced CRISPR tools such as CRISPR-Cas13, which target RNA. The ability to target ncRNAs at particular stress periods is particularly useful for achieving flexible and precise control over stress tolerance mechanisms (Wolter & Puchta, 2018).

10.6.2 Biotechnological Applications

Synthetic ncRNAs and ncRNA mimics have become invaluable tools to investigate stress-tolerance research. These synthetic molecules have natural ncRNA function and can temporarily be introduced into plants to activate stress response pathway without affecting genome (Ahmed et al., 2020; Othman et al., 2024). This approach is especially useful where the use of genetically modified organisms (GMOs) isn't permitted, especially in regulatory environments.

A second application involves synthetic miRNA mimics that can be used to regulate gene expression during stress. For example, if miR394 mimics are used to promote the antioxidant defences when they are necessary, plant responses to high salt or drought treatment are enhanced. This gives us stress responses that we can target without having to alter our genome permanently (Bernardi et al., 2022). In addition, synthetic lncRNAs are being constructed to regulate the association of specific transcription factors or chromatin modifying proteins responsible for abiotic stress responses. As an example, synthetic lncRNAs that mimic the function of COOLAIR can be used to regulate and improve cold tolerance in crop plants (Ijaz et al. 2024).

Synthetic ncRNAs could also be tailored to regulate hormonal pathways or ROS scavenging mechanisms during stress and are thus adaptable tools in regulating plant stress resilience in a variety of environments. Breeding and agricultural practices in which temporal control over stress responses is a need form an attractive basis for using ncRNA mimics to elicit transient effects on stress responses, which promises a useful tool for crop resilience enhancement without permanent genomic modification (Breunig et al., 2021).

10.6.3 Gene-Editing Strategies for Targeted ncRNA Modulation

Gene editing strategies have advanced rapidly and include methods for more precisely targeting ncRNAs that regulate stresses responses to abiotic factors (Ullah et al., 2022). In addition to conventional CRISPR-Cas9 applications, alternative gene-editing tools, including base editing and CRISPR prime editing, permit the editing of ncRNA gene sequences in a more localised manner through single nucleotide changes that refine ncRNA function without disrupting larger gene sequences (Zhu et al., 2020).

A recent advance is phasiRNA-based gene-editing systems. PhasiRNAs are ncRNAs that impose a sequential silencing of specific genes in response to stress. Researchers engineer phasiRNA-generating loci that activate stress-tolerance genes in a phasing fashion only in response to specific environmental conditions, minimising the unnecessary energy expenditure incurred by the plant and augmenting it with a flexibility of repertoire (Zhao et al., 2020). Another RNA-guided editing tool is CRISPR-Cas12a which can simultaneously edit multiple ncRNAs, involved in stress response. Most importantly, this tool has been particularly helpful when editing ncRNAs that interact with each other in intricate networks, enabling increased stress responses without undesired off-target effects (Zhang et al., 2021).

Additionally, without editing DNA sequences CRISPRa (CRISPR activation) and CRISPRi (CRISPR interference) approaches allow for modulation. In these approaches, CRISPR machinery is engineered to enhance or degrade expression of an ncRNA. This allows for fine tuning of genes that respond to particular stressors to enable only protective responses to be elicited when necessary while still maintaining plant resilience (Martella et al., 2019; Morelli et al., 2021).

As a result of genetic engineering, CRISPR-Cas innovations, synthetic ncRNA applications as well as new gene-editing strategies, scientists can now intervene in ncRNA roles with precision. They can make an important contribution to the development of more resilient crop varieties capable of withstanding often harsh environmental conditions and to sustainably managed agriculture and food security. As research continues, ncRNA will continue to contribute to crop improvement in biotechnology and equip crops to better address the stresses of climate change (Figure 10.3).

10.7 CHALLENGES AND LIMITATIONS IN ncRNA RESEARCH AND APPLICATION

Non-coding RNAs (ncRNAs) offer great promise to improve plant abiotic stress tolerance; however, there are a number of technical, practical and ethical challenges. For these ncRNAs to be fully exploited for agricultural biotechnology, these obstacles must be overcome. In this section, we review the challenges in ncRNA research, including detection and quantification problems, difficulties with delivery into crops and ethical concerns around delivery.

10.7.1 Technical Challenges in ncRNA Detection and Quantification

The accurate detection and quantification of ncRNA in complex plant tissues and under varying environmental conditions are one of the main challenges in ncRNA research (Budak et al., 2020; Hazra et al., 2023). Consequently, miRNAs, siRNAs and lncRNAs are typically characterised by low abundance, being extremely difficult to isolate and quantify (Wang & Chekanova, 2019). Additionally, ncRNA expression can be extremely dynamic: the change in ncRNA expression can be rapid over time in a time-dependent manner in response to stress, developmental stage or environmental stimuli (Dayal et al., 2024).

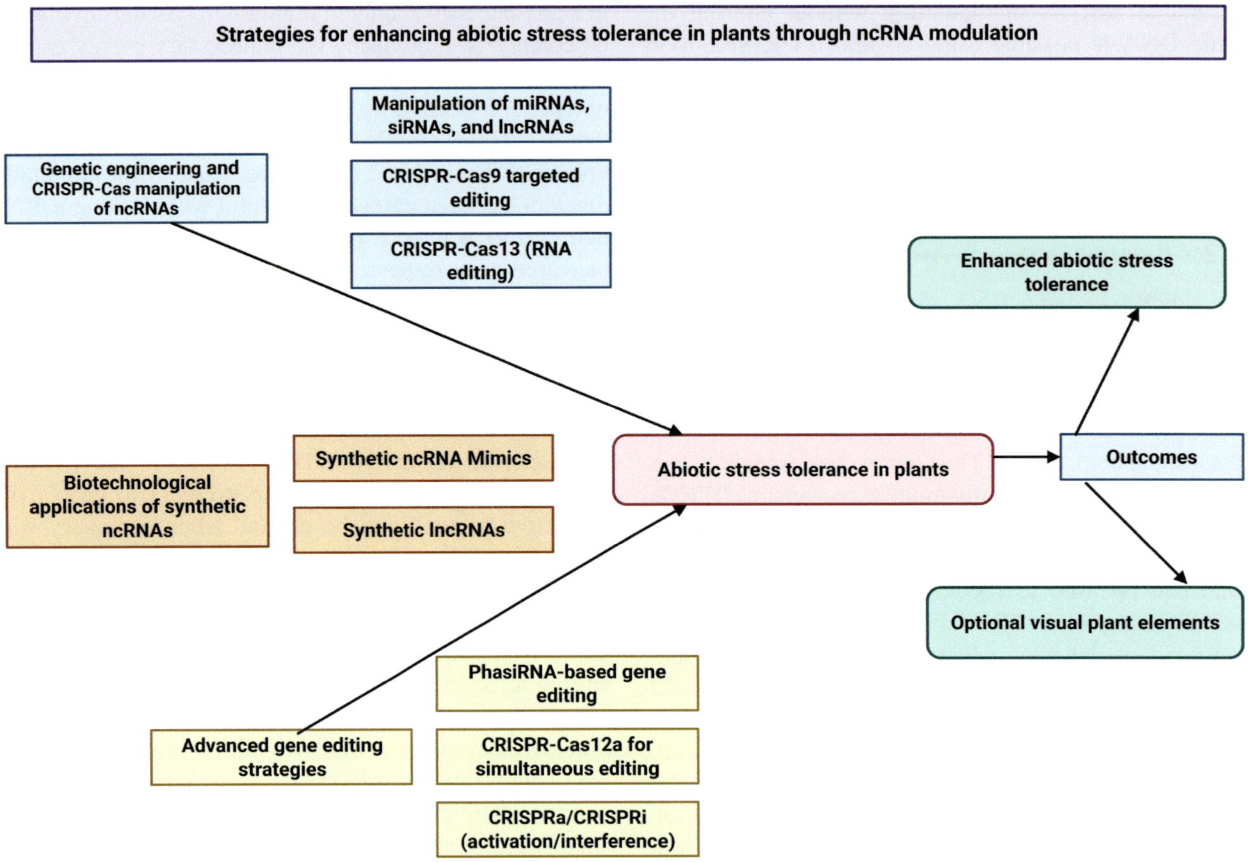

FIGURE 10.3 Strategies to enhance abiotic stress tolerance in plants using ncRNAs, including genetic engineering, CRISPR-Cas, synthetic ncRNAs, and advanced gene editing techniques for improved resilience.

Building the reliable expression profile and understanding the full repertoire of regulatory functions of plants is difficult because of the variability.

Quantitative PCR (qPCR), northern blotting or RNA sequencing (RNA-seq) are common ways to measure ncRNA expression but each of these can only be performed under certain conditions (Panda & Gorospe, 2018). For instance, qPCR involves precise primer design and may be insensitive to low abundance ncRNAs, while RNA-seq is broader but may miss ncRNA that is very low in abundance, expressed at extremely low abundance, or unstable under experimental conditions. However, new techniques, including single molecule sequencing and droplet digital PCR (ddPCR), provide enhanced sensitivity and specificity but are costly and dependent upon specialised equipment, making routine use of these techniques impractical.

The structural diversity of lncRNAs, and the lack of standardised annotation criteria complicates the identification of lncRNAs. Unlike protein-coding genes, ncRNAs do not possess well-defined coding sequences, thus making it impossible to delineate between functional ncRNAs and transcriptional noise. Improving ncRNA research and application in crop improvement requires more accurate detection technologies and robust databases for ncRNA annotation (Liao et al., 2018).

10.7.2 Limitations of ncRNA Delivery Systems in Crops

Delivery of ncRNAs into plant cells and tissues remains an important barrier to the application of ncRNA-based technologies to crops. Synthetic or engineered ncRNAs must reach target cells, exist stably within the plant and effectively express or modulate gene targets using delivery systems. Currently, ncRNA delivery in mature plant tissues, to generate transgenic plants, usually requires traditional DNA and RNA delivery methods such as Agrobacterium-mediated transformation and particle bombardment, which often yield low transformation efficiencies, and tend to deliver DNA or RNA specifically or unevenly across the plant (Mujtaba et al., 2021).

Moreover, RNA molecules are sensitive to degradation, including in the presence of nucleases present in plant cells. Therefore, delivery methods for ncRNAs need to be developed that will protect them from degradation while allowing them to be delivered to target cells. However, nanoparticle-based systems and viral vectors have emerged as promising options for delivering ncRNAs but need to be further optimised for effective delivery in an inexpensive and scalable manner for agricultural use. The second limitation resides in the regulatory and physical barriers that restrict the ncRNA movement within plants. For example, some ncRNAs are likely to travel long distances within the plant, e.g., from roots to leaves. These limitations must be addressed by developing delivery systems capable of delivering nanoparticles to specific tissues or throughout the plant via systemic delivery (Niazi & Magoola, 2024).

10.7.3 Potential Off-Target Effects and Ethical Considerations

When using CRISPR-Cas-based gene editing to target ncRNAs, great specificity is important. ncRNAs or edited sequences displaying off-target effects bind to unintended targets and induce unintended changes in gene expression both altering plant growth, development and stress responses, respectively. For example, miRNAs could potentially silence non-target genes by binding to non-complementary mRNAs. Yet other genes can be incorrectly activated or repressed, and this can disrupt cellular functions and deleteriously affect plant fitness in the absence of stress. Because ncRNA modifications induce off-target effects, both computational and experimental validation is necessary to minimise these effects. Potential off-target sites can be predicted by bioinformatics tools, but experimentation is needed to verify these predictions, which can be costly and time consuming. Developing ncRNA-based applications in crops to be both safe and effective requires specificity (Amirloo, 2020).

Ethical considerations, especially when it comes to the use of gene-editing technologies and synthetic ncRNAs in crops, are also important. The public acceptance of GMOs is mixed, but synthetic ncRNAs or the use of CRISPR-Cas to modify ncRNA genes will probably receive the same degree of scrutiny. Each country has different rules on how gene-edited crops are regulated, from mandatory testing for crops in some countries to virtually no rules for genes in others. In addition, there are ethical side issues concerning potential consequences for other potential environmental impacts of the release of the ncRNA-modified plants, particularly those of gene flow and ecological interaction with other organisms. Engaging stakeholders, communicating transparently and following closely to address ethical concerns through strictly following regulatory guidelines are all essential now to build public trust in ncRNA-based agricultural technologies. Despite these efforts to pursue the potential benefits from ncRNA-based interventions in agriculture, researchers continue to develop and refine their approaches, keeping mind the longer term implications before embarking on either (Cotter et al., 2020; Enríquez, 2021).

The potential for ncRNAs to improve crop abiotic stress tolerance is great, yet there are many challenges in detection, delivery, specificity and ethical considerations. However, what is lacking are these technologies developing further, more stringent validation processes and approaches to regulatory and ethical issues that are both responsible and humane. These improvements potentially enhance ncRNA-based applications in agriculture to help contribute to sustainable food production and environmental stress resilience (Seem et al., 2024).

10.8 FUTURE PROSPECTS AND CONCLUDING REMARKS

New ncRNA technologies and new insights are bringing new understanding of ncRNAs and new possibilities for getting plants more stress tolerant with advancing the field of non-coding RNA (ncRNA) research. These advances hold particularly promising prospects for addressing some of the most challenging problems in agriculture, improving crops and food security in response to global climate change. This concluding section presents emerging tools and techniques and the field of crop development where ncRNAs may be used in future research.

10.8.1 Emerging Technologies for ncRNA Research in Plants

However, ncRNA research has been enhanced by recent technological advancements that allow a greater understanding of the role of these molecules in plants and their use in stress tolerance. Recent advances in high-throughput sequencing have revolutionised ncRNA studies, enabling the identification and quantification of ncRNAs with great precision over a wide range of plant tissues and conditions. For example, small RNA sequencing, RNA immunoprecipitation (RIP)-seq and single-cell RNA sequencing (scRNA-seq) have been particularly useful techniques in identifying the diversity and dynamics of ncRNAs in response to abiotic stresses. The analysis of ncRNA interaction with specific proteins and identification of regulatory targets in plant cell using these tools (Sripathi et al., 2021).

Despite the advances of CRISPR-Cas technology, ncRNA manipulation abides new strategies of precise ncRNA manipulation. CRISPRa and CRISPRi CRISPR tools offer researchers control over the stress-response pathways, reducing the risk of off-target effects, without permanent DNA changes. As with CRISPR-Cas13, which can target RNA directly, which is a new approach capable of transiently modulating ncRNAs under specific stress conditions, this advance could be especially advantageous for the development of transient stress responses that do not require editing of the plant genome (Kordyś et al., 2022).

Furthermore, nanotechnology and viral vectors are also promising platforms for the delivery of synthetic ncRNA to plants. For example, nanoparticles can protect ncRNA from degradation and deliver stable ncRNA to specific plant tissues. In addition to delivering viral vectors that can be used to introduce synthetic ncRNAs into plants, this represents a scalable, cost-effective means of introducing ncRNAs into plants for agricultural applications. These delivery methods could eliminate traditional limitations for ncRNA applications (e.g., stability and systemic distribution) and therefore enable ncRNA-based interventions to become more feasible for crop improvement (Arjunan et al., 2024).

Application of ncRNAs to crop improvement represents one of the exciting opportunities to address the challenge of global food security. A critical need exists to enhance crop resilience to abiotic stresses, which will be required as a means to sustain yields and preserve food security in the face of increasing global population and effects of climate change. ncRNA-based breeding could be a game changer in this regard. As plant breeders can identify and select plants expressing advantageous ncRNA profiles for that stress, they can further develop crop varieties that grow naturally expressing beneficial ncRNAs as responses to a particular stress condition. By allowing for breeding of desired traits without direct genetic modification, this approach has lower regulatory hurdles and a higher public acceptability (Hazra et al., 2023).

Furthermore, synthetic ncRNAs and ncRNA mimics can be the sustainable alternative of classic chemical treatment. For example, instead of using external growth regulatory or stress-protective chemicals, farmers might use synthetic ncRNAs to enhance the plant's natural resilience mechanisms. The advantage in this method is an inexpensive and environmentally friendly way to handle stressed crops that are becoming increasingly chemical input free and towards more sustainable farming. As ncRNA advances continue, they may be essential players in precision agriculture, or a method of crop production where factors such as environmental conditions and ncRNA concentration determine the optimal states to implement gene expression. Real-time crop monitoring and management can be offered using ncRNA-based approaches combined with digital farming technologies (e.g., sensors, predictive models) and may offer an opportunity to increase agricultural productivity and resilience (Dávalos et al., 2019; Othman et al., 2024).

10.8.2 Summary of Key Points and Future Directions for Research

This work represents an exploration of ncRNAs in abiotic stress tolerance in which we observe how these molecules influence plant resilience through influencing gene expression, regulating hormone signalling, controlling epigenetic changes and supporting antioxidant defences. There is also different genetic engineering, CRISP-based editing, synthetic applications and gene-editing strategies, which together are valuable approaches to increase the functionality of ncRNA and make the plants better resist the stress.

Although ncRNAs show significant promise in this field, many issues remain unsolved regarding their detection, delivery, specificity and ethical considerations. Some of these issues have been addressed by the recent advances in high-throughput sequencing, CRISPR-Cas technology and nanotechnology, but more research is required to fully optimise ncRNA-based interventions. Improving ncRNA stability in plant cells, developing robust delivery systems and minimising off-target effects are key steps needed to make ncRNA applications viable and scalable in agriculture.

Looking ahead, key areas of research will likely include the following:

- Refining ncRNA delivery methods, with a focus on ensuring stable, tissue-specific distribution across different plant species.
- Exploring the interplay of ncRNAs with other regulatory molecules, such as transcription factors and small peptides, to gain a comprehensive understanding of how ncRNAs operate within broader stress response networks.
- Investigating ncRNA networks in crop plants beyond model organisms, such as Arabidopsis, to determine how these findings translate to staple crops like rice, maize and wheat.
- Advancing synthetic biology applications of ncRNAs, including the design of artificial ncRNAs tailored to specific stress scenarios, allowing for rapid adaptation to unpredictable climate conditions.

As these research avenues are explored, ncRNAs have the potential to become foundational tools in crop improvement. By enhancing the natural resilience of crops through targeted ncRNA manipulation, we can develop plant varieties that are better equipped to thrive in diverse environmental conditions, supporting food security and sustainable agriculture in a changing world.

REFERENCES

Ahmed, W., Xia, Y., Li, R., Bai, G., Siddique, K. H. M., & Guo, P. (2020). Non-coding RNAs: Functional roles in the regulation of stress response in Brassica crops. *Genomics*, *112*(2), 1419–1424. https://doi.org/10.1016/j.ygeno.2019.08.011

Akhter, Z., Bi, Z., Ali, K., Sun, C., Fiaz, S., Haider, F. U., & Bai, J. (2021). In response to abiotic stress, DNA methylation confers epigenetic changes in plants. *Plants*, *10*(6), 1096. https://www.mdpi.com/2223-7747/10/6/1096

Ala, f. (2020). Competing endogenous RNAs, non-coding RNAs and diseases: An intertwined story. *Cells*, *9*(7), 1574. https://www.mdpi.com/2073-4409/9/7/1574

Amirloo, B. (2020). *Design and development of novel therapeutic interventions:'Intelligent' catalytic systems for irreversible cleavage of disease-relevant RNA*. The University of Manchester. https://doi.org/https://www.proquest.com/openview/b23e31fa32df7b415fed26a8bb35d8cc/1?pq-origsite=gscholar&cbl=51922&diss=y

Arjunan, N., Thiruvengadam, V., & Sushil, S. N. (2024). Nanoparticle-mediated dsRNA delivery for precision insect pest control: A comprehensive review. *Molecular Biology Reports*, *51*(1), 355. https://doi.org/10.1007/s11033-023-09187-6

Ashapkin, V. V., Kutueva, L. I., Aleksandrushkina, N. I., & Vanyushin, B. F. (2020). Epigenetic mechanisms of plant adaptation to biotic and abiotic stresses. *International Journal of Molecular Sciences*, *21*(20), 7457. https://www.mdpi.com/1422-0067/21/20/7457

Baruah, P. M., Krishnatreya, D. B., Bordoloi, K. S., Gill, S. S., & Agarwala, N. (2021). Genome wide identification and characterization of abiotic stress responsive lncRNAs in *Capsicum annuum*. *Plant Physiology and Biochemistry*, *162*, 221–236. https://doi.org/10.1016/j.plaphy.2021.02.031

Bernardi, Y., Ponso, M. A., Belén, F., Vegetti, A. C., & Dotto, M. C. (2022). MicroRNA miR394 regulates flowering time in *Arabidopsis thaliana*. *Plant Cell Reports*, *41*(6), 1375–1388. https://doi.org/10.1007/s00299-022-02863-0

Bhar, A., & Roy, A. (2023). Emphasizing the role of long noncoding RNAs (lncRNA), circular RNA (circRNA), and micropeptides (miPs) in plant biotic stress tolerance. *Plants*, *12*(23), 3951. https://www.mdpi.com/2223-7747/12/23/3951

Bhatti, G. K., Khullar, N., Sidhu, I. S., Navik, U. S., Reddy, A. P., Reddy, P. H., & Bhatti, J. S. (2021). Emerging role of non-coding RNA in health and disease. *Metabolic Brain Disease*, *36*(6), 1119–1134. https://doi.org/10.1007/s11011-021-00739-y

Breunig, C. T., Köferle, A., Neuner, A. M., Wiesbeck, M. F., Baumann, V., & Stricker, S. H. (2021). CRISPR tools for physiology and cell state changes: Potential of transcriptional engineering and epigenome editing. *Physiological Reviews*, *101*(1), 177–211. https://doi.org/10.1152/physrev.00034.2019

Budak, H., Kaya, S. B., & Cagirici, H. B. (2020). Long noncoding RNA in plants in the era of reference sequences. *Frontiers in Plant Science*, *11*, 276. https://doi.org/10.3389/fpls.2020.00276

Chand Jha, U., Nayyar, H., Mantri, N., & Siddique, K. H. M. (2021). Non-coding RNAs in legumes: Their emerging roles in regulating biotic/abiotic stress responses and plant growth and development. *Cells*, *10*(7), 1674. https://www.mdpi.com/2073-4409/10/7/1674

Chekanova, J. A. (2015). Long non-coding RNAs and their functions in plants. *Current Opinion in Plant Biology*, *27*, 207–216. https://doi.org/10.1016/j.pbi.2015.08.003

Chen, R., Li, M., Zhang, H., Duan, L., Sun, X., Jiang, Q., Zhang, H., & Hu, Z. (2019). Continuous salt stress-induced long non-coding RNAs and DNA methylation patterns in soybean roots. *BMC Genomics*, *20*(1), 730. https://doi.org/10.1186/s12864-019-6101-7

Ciechanowska, K., Pokornowska, M., & Kurzyńska-Kokorniak, A. (2021). Genetic insight into the domain structure and functions of dicer-type ribonucleases. *International Journal of Molecular Sciences*, *22*(2), 616. https://www.mdpi.com/1422-0067/22/2/616

Cotter, J., Kawall, K., & Then, C. (2020). New genetic engineering technologies. *Report of the results from the RAGES project*, 2016–2019. https://doi.org/https://www.testbiotech.org/wp-content/uploads/2020/01/RAGES_report-new-genetic-engineering-techniques.pdf

Das, P., Agarwala, N., Gill, S. S., & Varshney, R. K. (2023). Emerging role of plant long non coding RNAs (lncRNAs) in salinity stress response. *Plant Stress*, *10*, 100265. https://doi.org/10.1016/j.stress.2023.100265

Dávalos, A., Henriques, R., Latasa, M. J., Laparra, M., & Coca, M. (2019). Literature review of baseline information on non-coding RNA (ncRNA) to support the risk assessment of ncRNA-based genetically modified plants for food and feed. *EFSA Supporting Publications*, *16*(8), 1688E. https://doi.org/10.2903/sp.efsa.2019.EN-1688

Dayal, S., Chaubey, D., Joshi, D. C., Ranmale, S., & Pillai, B. (2024). Noncoding RNAs: Emerging regulators of behavioral complexity. *Wiley Interdisciplinary Reviews: RNA*, *15*(3), e1847. https://doi.org/10.1002/wrna.1847

Enríquez, P. (2021). *Rewriting nature: The future of genome editing and how to bridge the gap between law and science*. Cambridge University Press. .

Erdmann, R. M., & Picard, C. L. (2020). RNA-directed DNA methylation. *PLoS Genetics*, *16*(10). https://doi.org/10.1371/journal.pgen.1009034

Fujibayashi, M., Suzuki, T., & Sano, T. (2021). Mechanism underlying potato spindle tuber viroid affecting tomato (*Solanum lycopersicum*): Loss of control over reactive oxygen species production. *Journal of General Plant Pathology*, *87*(4), 226–235. https://doi.org/10.1007/s10327-021-01000-1

Gelaw, T. A., & Sanan-Mishra, N. (2021). Non-coding RNAs in response to drought stress. *International Journal of Molecular Sciences*, *22*(22), 12519. https://www.mdpi.com/1422-0067/22/22/12519

Ghildiyal, M., & Zamore, P. D. (2009). Small silencing RNAs: An expanding universe. *Nature Reviews Genetics*, *10*(2), 94–108. https://doi.org/10.1038/nrg2504

Ghorbanzadeh, Z., Hamid, R., Jacob, F., Asadi, S., Salekdeh, G. H., & Ghaffari, M. R. (2022). Non-coding RNA: Chief architects of drought-resilient roots. *Rhizosphere*, *23*, 100572. https://doi.org/10.1016/j.rhisph.2022.100572

Ghosh, S., Adhikari, S., Adhikari, A., & Hossain, Z. (2022). Contribution of plant miRNAome studies towards understanding heavy metal stress responses: Current status and future perspectives. *Environmental and Experimental Botany*, *194*, 104705. https://doi.org/10.1016/j.envexpbot.2021.104705

Gill, R. A., Federico, S., & Liu, F. A. (2021). On the role of transposable elements in the regulation of gene expression and subgenomic interactions in crop genomes. *Critical Reviews in Plant Sciences*, *40*(2), 157–189. https://doi.org/10.1080/07352689.2021.1920731

Gill, S. S., Khan, N. A., Agarwala, N., Singh, K., Sunkar, R., & Tuteja, N. (2024). ncRNAs in plant development and stress responses. *Plant Physiology and Biochemistry*, *214*, 108950. https://doi.org/10.1016/j.plaphy.2024.108950

Glazińska, P., Kulasek, M., Glinkowski, W., Wojciechowski, W., & Kosiński, J. (2019). Integrated analysis of small RNA, transcriptome and degradome sequencing provides new insights into floral development and abscission in yellow lupine (*Lupinus luteus* L.). *International Journal of Molecular Sciences*, *20*(20), 5122. https://www.mdpi.com/1422-0067/20/20/5122

Guo, M., Li, H., Zhu, L., Wu, Z., Li, J., & Li, Z. (2021). Genome-wide identification of microRNAs associated with osmotic stress and elucidation of the role of miR319 in *Medicago ruthenica* seedlings. *Plant Physiology and Biochemistry*, *168*, 53–61. https://doi.org/10.1016/j.plaphy.2021.09.033

Hajyzadeh, M., Turktas, M., Khawar, K. M., & Unver, T. (2015). miR408 overexpression causes increased drought tolerance in chickpea. *Gene*, *555*(2), 186–193. https://doi.org/10.1016/j.gene.2014.11.002

Hasanuzzaman, M., Bhuyan, M. H. M. B., Zulfiqar, F., Raza, A., Mohsin, S. M., Mahmud, J. A., Fujita, M., & Fotopoulos, V. (2020). Reactive oxygen species and antioxidant defense in plants under abiotic stress: Revisiting the crucial role of a universal defense regulator. *Antioxidants*, *9*(8), 681. https://www.mdpi.com/2076-3921/9/8/681

Hasnain, M., Munir, N., Abideen, Z., Zulfiqar, F., Koyro, H. W., El-Naggar, A., Caçador, I., Duarte, B., Rinklebe, J., & Yong, J. W. H. (2023). Biochar-plant interaction and detoxification strategies under abiotic stresses for achieving agricultural resilience: A critical review. *Ecotoxicology and Environmental Safety*, *249*, 114408. https://doi.org/10.1016/j.ecoenv.2022.114408

Hayat, K., Hafeez, M., Bilal, K., & Shabbir, M. S. (2022). Interactive effects of organizational structure and team work quality on project success in project based non profit organizations. *iRASD Journal of Management*, *4*(1), 84–103. https://doi.org/10.52131/jom.2022.0401.0064

Hazra, S., Moulick, D., Mukherjee, A., Sahib, S., Chowardhara, B., Majumdar, A., Upadhyay, M. K., Yadav, P., Roy, P., Santra, S. C., Mandal, S., Nandy, S., & Dey, A. (2023). Evaluation of efficacy of non-coding RNA in abiotic stress management of field crops: Current status and future prospective. *Plant Physiology and Biochemistry*, *203*, 107940. https://doi.org/10.1016/j.plaphy.2023.107940

Hirayama, T., & Shinozaki, K. (2010). Research on plant abiotic stress responses in the post-genome era: Past, present and future. *The Plant Journal*, *61*(6), 1041–1052.

Hussain, Q., Asim, M., Zhang, R., Khan, R., Farooq, S., & Wu, J. (2021). Transcription factors interact with ABA through gene expression and signaling pathways to mitigate drought and salinity stress. *Biomolecules*, *11*(8), 1159. https://www.mdpi.com/2218-273X/11/8/1159

Ijaz, S., Iqbal, J., Abbasi, B. A., Ullah, Z., Yaseen, T., Munir, F., Hyder, S., Kanwal, S., Sahito, Z. A., & Mahmood, T. (2024). From single nc-RNAs to networks: Understanding the complexity of environmental adaption. In C. S. Prakash, M. Waseem, & S. Fiaz (Eds.), *ncRNAs: Mediated regulation: The unarguable biological function to environmental constrains* (pp. 283–310). Springer Nature Switzerland. https://doi.org/10.1007/978-3-031-69354-0_11

Jahed, K. R., Saini, A. K., & Sherif, S. M. (2023). Coping with the cold: Unveiling cryoprotectants, molecular signaling pathways, and strategies for cold stress resilience [Review]. *Frontiers in Plant Science*, *14*. https://doi.org/10.3389/fpls.2023.1246093

Jampala, P., Akanksha, G., & Lodha, M. (2021). Functions of long non-coding RNA in *Arabidopsis thaliana*. *Plant Signaling & Behavior*, *16*(9), 1925440. https://doi.org/10.1080/15592324.2021.1925440

Jha, U. C., Nayyar, H., Jha, R., Khurshid, M., Zhou, M., Mantri, N., & Siddique, K. H. M. (2020). Long non-coding RNAs: Emerging players regulating plant abiotic stress response and adaptation. *BMC Plant Biology*, *20*(1), 466. https://doi.org/10.1186/s12870-020-02595-x

Jha, U. C., Nayyar, H., Roychowdhury, R., Prasad, P. V. V., Parida, S. K., & Siddique, K. H. M. (2023). Non-coding RNAs (ncRNAs) in plant: Master regulators for adapting to extreme temperature conditions. *Plant Physiology and Biochemistry*, *205*, 108164. https://doi.org/10.1016/j.plaphy.2023.108164

Jiang, C., Wang, Y., He, Y., Peng, Y., Xie, L., Li, Y., Sun, W., Zhou, J., Zheng, C., & Xie, X. (2024). Identification and characterization of miRNAs and lncRNAs associated with salinity stress in rice panicles. *International Journal of Molecular Sciences*, *25*(15), 8247. https://www.mdpi.com/1422-0067/25/15/8247

Jiang, S., Lan, Z., Zhang, Y., Kang, X., Zhao, L., Wu, X., & Gao, H. (2024). Mechanisms by which exogenous substances enhance plant salt tolerance through the modulation of ion membrane transport and reactive oxygen species metabolism. *Antioxidants*, *13*(9), 1050. https://www.mdpi.com/2076-3921/13/9/1050

Juan, C. A., Pérez de la Lastra, J. M., Plou, F. J., & Pérez-Lebeña, E. (2021). The chemistry of reactive oxygen species (ROS) revisited: Outlining their role in biological macromolecules (DNA, lipids and proteins) and induced pathologies. *International Journal of Molecular Sciences*, *22*(9), 4642. https://www.mdpi.com/1422-0067/22/9/4642

Khan, Z., & Shahwar, D. (2020). Role of heat shock proteins (HSPs) and heat stress tolerance in crop plants. In R. Roychowdhury, S. Choudhury, M. Hasanuzzaman, & S. Srivastava (Eds.), *Sustainable agriculture in the era of climate change* (pp. 211–234). Springer International Publishing. https://doi.org/10.1007/978-3-030-45669-6_9

Khraiwesh, B., Zhu, J.-K., & Zhu, J. (2012). Role of miRNAs and siRNAs in biotic and abiotic stress responses of plants. *Biochimica et Biophysica Acta (BBA) – Gene Regulatory Mechanisms*, *1819*(2), 137–148. https://doi.org/10.1016/j.bbagrm.2011.05.001

Kim, J.-H. (2019). Chromatin remodeling and epigenetic regulation in plant DNA damage repair. *International Journal of Molecular Sciences*, *20*(17), 4093. https://www.mdpi.com/1422-0067/20/17/4093

Kinoshita, C., & Aoyama, K. (2021). The role of non-coding RNAs in the neuroprotective effects of glutathione. *International Journal of Molecular Sciences*, *22*(8), 4245. https://www.mdpi.com/1422-0067/22/8/4245

Kong, X., Yang, M., Le, B. H., He, W., & Hou, Y. (2022). The master role of siRNAs in plant immunity. *Molecular Plant Pathology*, *23*(10), 1565–1574. https://doi.org/10.1111/mpp.13250

Kordyś, M., Sen, R., & Warkocki, Z. (2022). Applications of the versatile CRISPR-Cas13 RNA targeting system. *Wiley Interdisciplinary Reviews: RNA*, *13*(3), e1694. https://doi.org/10.1002/wrna.1694

Kryovrysanaki, N., James, A., Tselika, M., Bardani, E., & Kalantidis, K. (2022). RNA silencing pathways in plant development and defense. *The International Journal of Developmental Biology*, *66*(1-2-3), 163–175. https://doi.org/https://doi.org/10.1387/ijdb.210189kk

Lan, J., & Qin, G. (2020). The regulation of CIN-like TCP transcription factors. *International Journal of Molecular Sciences*, *21*(12), 4498. https://www.mdpi.com/1422-0067/21/12/4498

Leitão, A. L., & Enguita, F. J. (2022). A structural view of miRNA biogenesis and function. *Non-Coding RNA*, *8*(1), 10. https://www.mdpi.com/2311-553X/8/1/10

Li, J., Song, Q., Zuo, Z.-F., & Liu, L. (2022). MicroRNA398: A master regulator of plant development and stress responses. *International Journal of Molecular Sciences*, *23*(18), 10803. https://www.mdpi.com/1422-0067/23/18/10803

Li, X., & Fu, X.-D. (2019). Chromatin-associated RNAs as facilitators of functional genomic interactions. *Nature Reviews Genetics*, *20*(9), 503–519. https://doi.org/10.1038/s41576-019-0135-1

Li, X., Wu, Z., Fu, X., & Han, W. (2014). lncRNAs: Insights into their function and mechanics in underlying disorders. *Mutation Research/Reviews in Mutation Research*, *762*, 1–21. https://doi.org/10.1016/j.mrrev.2014.04.002

Liao, P., Li, S., Cui, X., & Zheng, Y. (2018). A comprehensive review of web-based resources of non-coding RNAs for plant science research. *International Journal of Biological Sciences*, *14*(8), 819–832. https://doi.org/10.7150/ijbs.24593

Ma, B., Zhang, A., Zhao, Q., Li, Z., Lamboro, A., He, H., Li, Y., Jiao, S., Guan, S., Liu, S., Yao, D., & Zhang, J. (2021). Genome-wide identification and analysis of long non-coding RNAs involved in fatty acid biosynthesis in young soybean pods. *Scientific Reports*, *11*(1), 7603. https://doi.org/10.1038/s41598-021-87048-7

Ma, X., Zhao, F., & Zhou, B. (2022). The characters of non-coding RNAs and their biological roles in plant development and abiotic stress response. *International Journal of Molecular Sciences*, *23*(8), 4124. https://www.mdpi.com/1422-0067/23/8/4124

Mahato, R. K., Bhattacharya, S., Khullar, N., Sidhu, I. S., Reddy, P. H., Bhatti, G. K., & Bhatti, J. S. (2024). Targeting long non-coding RNAs in cancer therapy using CRISPR-Cas9 technology: A novel paradigm for precision oncology. *Journal of Biotechnology*, *379*, 98–119. https://doi.org/10.1016/j.jbiotec.2023.12.003

Makkar, H., Arora, S., Khuman, A. K., & Chaudhary, B. (2022). Target-mimicry-based miR167 diminution confers salt-stress tolerance during in vitro organogenesis of tobacco (*Nicotiana tabacum* L.). *Journal of Plant Growth Regulation*, *41*(4), 1462–1480. https://doi.org/10.1007/s00344-021-10376-5

Martella, A., Firth, M., Taylor, B. J. M., Göppert, A., Cuomo, E. M., Roth, R. G., Dickson, A. J., & Fisher, D. I. (2019). Systematic evaluation of CRISPRa and CRISPRi modalities enables development of a multiplexed, orthogonal gene activation and repression system. *ACS Synthetic Biology*, *8*(9), 1998–2006. https://doi.org/10.1021/acssynbio.8b00527

Matsui, A., Nguyen, A. H., Nakaminami, K., & Seki, M. (2013). Arabidopsis non-coding RNA regulation in abiotic stress responses. *International Journal of Molecular Sciences*, *14*(11), 22642–22654. https://www.mdpi.com/1422-0067/14/11/22642

Mishra, K., & Kanduri, C. (2019). Understanding long noncoding RNA and chromatin interactions: What we know so far. *Non-Coding RNA*, *5*(4), 54. https://www.mdpi.com/2311-553X/5/4/54

Morelli, E., Gulla', A., Amodio, N., Taiana, E., Neri, A., Fulciniti, M., & Munshi, N. C. (2021). CRISPR interference (CRISPRi) and CRISPR activation (CRISPRa) to explore the oncogenic lncRNA network. In A. Navarro (Ed.), *Long non-coding RNAs in cancer* (pp. 189–204). Springer US. https://doi.org/10.1007/978-1-0716-1581-2_13

Mudasir Ahmad, S., Ahmad Dar, M. A., Ahmad, F., Majeed Shah, I., Ahmad Bhat, K., & Ahmad Dar, T. (2022). Long non-coding RNAs: Biogenesis, mechanism of action and role in different biological and pathological processes. In L. Tutar (Ed.), *Recent advances in noncoding RNAs*. IntechOpen. https://doi.org/10.5772/intechopen.104861

Mujtaba, M., Wang, D., Carvalho, L. B., Oliveira, J. L., Espirito Santo Pereira, A. d., Sharif, R., Jogaiah, S., Paidi, M. K., Wang, L., Ali, Q., & Fraceto, L. F. (2021). Nanocarrier-mediated delivery of miRNA, RNAi, and CRISPR-Cas for plant protection: Current trends and future directions. *ACS Agricultural Science & Technology*, *1*(5), 417–435. https://doi.org/10.1021/acsagscitech.1c00146

Mukherjee, A., Dwivedi, S., Bhagavatula, L., & Datta, S. (2023). Integration of light and ABA signaling pathways to combat drought stress in plants. *Plant Cell Reports*, *42*(5), 829–841. https://doi.org/10.1007/s00299-023-02999-7

Nadarajah, K. K. (2020). ROS homeostasis in abiotic stress tolerance in plants. *International Journal of Molecular Sciences*, *21*(15), 5208. https://www.mdpi.com/1422-0067/21/15/5208

Nadhan, R., Isidoro, C., Song, Y. S., & Dhanasekaran, D. N. (2022). Signaling by LncRNAs: Structure, cellular homeostasis, and disease pathology. *Cells*, *11*(16), 2517. https://www.mdpi.com/2073-4409/11/16/2517

Naeli, P., Winter, T., Hackett, A. P., Alboushi, L., & Jafarnejad, S. M. (2023). The intricate balance between microRNA-induced mRNA decay and translational repression. *The FEBS Journal*, *290*(10), 2508–2524. https://doi.org/10.1111/febs.16422

Nejat, N., & Mantri, N. (2018). Emerging roles of long non-coding RNAs in plant response to biotic and abiotic stresses. *Critical Reviews in Biotechnology*, *38*(1), 93–105. https://doi.org/10.1080/07388551.2017.1312270

Niazi, S. K., & Magoola, M. (2024). MicroRNA Nobel Prize: Timely recognition and high anticipation of future products—A prospective analysis. *International Journal of Molecular Sciences*, *25*(23), 12883.

Numan, M., Sun, Y., & Li, G. (2024). Exploring the emerging role of long non-coding RNAs (lncRNAs) in plant biology: Functions, mechanisms of action, and future directions. *Plant Physiology and Biochemistry*, *212*, 108797. https://doi.org/10.1016/j.plaphy.2024.108797

Othman, S. M. I. S., Mustaffa, A. F., Mohd Zahid, N. I. I., Che-Othman, M. H., Samad, A. F. A., Goh, H.-H., & Ismail, I. (2024). Harnessing the potential of non-coding RNA: An insight into its mechanism and interaction in plant biotic stress. *Plant Physiology and Biochemistry*, *207*, 108387. https://doi.org/10.1016/j.plaphy.2024.108387

Oyebamiji, Y. O., Adigun, B. A., Shamsudin, N. A. A., Ikmal, A. M., Salisu, M. A., Malike, F. A., & Lateef, A. A. (2024). Recent advancements in mitigating abiotic stresses in crops. *Horticulturae*, *10*(2), 156. https://www.mdpi.com/2311-7524/10/2/156

Pan, W. J., Tao, J. J., Cheng, T., Bian, X. H., Wei, W., Zhang, W. K., Ma, B., Chen, S. Y., & Zhang, J. S. (2016). Soybean miR172a improves salt tolerance and can function as a long-distance signal. *Molecular Plant*, *9*(9), 1337–1340. https://doi.org/10.1016/j.molp.2016.05.010

Panchal, A., Maurya, J., Seni, S., Singh, R. K., & Prasad, M. (2023). An insight into the roles of regulatory ncRNAs in plants: An abiotic stress and developmental perspective. *Plant Physiology and Biochemistry*, *201*, 107823. https://doi.org/10.1016/j.plaphy.2023.107823

Panda, A. C., & Gorospe, M. (2018). Detection and analysis of circular RNAs by RT-PCR. *Bio-Protocol*, *8*(6), e2775–e2775. https://doi.org/https://en-cdn.bio-protocol.org/pdf/bio-protocol2775.pdf?rel_link=YmlvLXByb3RvY29sLm9yZy9lbi9icGRldGFpbD9pZD0yNzc1JnR5cGU9MA==

Parwez, R., Aftab, T., Gill, S. S., & Naeem, M. (2022). Abscisic acid signaling and crosstalk with phytohormones in regulation of environmental stress responses. *Environmental and Experimental Botany*, *199*, 104885. https://doi.org/10.1016/j.envexpbot.2022.104885

Premnath, A., Ramalingam, A. P., Ramasamy, S. P., Karnatam, K. S., & Ramadoss, B. R. (2022). Genome editing and miRNA-based approaches in cereals under abiotic stress. In A. A. H. Abdel Latef (Ed.), *Sustainable remedies for abiotic stress in cereals* (pp. 647–673). Springer Nature Singapore. https://doi.org/10.1007/978-981-19-5121-3_25

Rajwanshi, R., Chakraborty, S., Jayanandi, K., Deb, B., & Lightfoot, D. A. (2014). Orthologous plant microRNAs: Microregulators with great potential for improving stress tolerance in plants. *Theoretical and Applied Genetics*, *127*(12), 2525–2543. https://doi.org/10.1007/s00122-014-2391-y

Ravi, B., Foyer, C. H., & Pandey, G. K. (2023). The integration of reactive oxygen species (ROS) and calcium signalling in abiotic stress responses. *Plant, Cell & Environment*, *46*(7), 1985–2006. https://doi.org/https://doi.org/10.1111/pce.14596

Ren, J., Jiang, C., Zhang, H., Shi, X., Ai, X., Li, R., Dong, J., Wang, J., Zhao, X., & Yu, H. (2022). LncRNA-mediated ceRNA networks provide novel potential biomarkers for peanut drought tolerance. *Physiologia Plantarum*, *174*(1). https://doi.org/10.1111/ppl.13610

Sachdev, S., Ansari, S. A., Ansari, M. I., Fujita, M., & Hasanuzzaman, M. (2021). Abiotic stress and reactive oxygen species: Generation, signaling, and defense mechanisms. *Antioxidants*, *10*(2), 277. https://www.mdpi.com/2076-3921/10/2/277

Sami, A., Xue, Z., Tazein, S., Arshad, A., He Zhu, Z., Ping Chen, Y., Hong, Y., Tian Zhu, X., & Jin Zhou, K. (2021). CRISPR–Cas9-based genetic engineering for crop improvement under drought stress. *Bioengineered*, *12*(1), 5814–5829. https://doi.org/10.1080/21655979.2021.1969831

Seem, K., Kaur, S., Kumar, S., & Mohapatra, T. (2024). Epigenome editing for targeted DNA (de) methylation: A new perspective in modulating gene expression. *Critical Reviews in Biochemistry and Molecular Biology*, *59*(1-2), 69–98. https://doi.org/10.1080/10409238.2024.2320659

Shriram, V., Kumar, V., Devarumath, R. M., Khare, T. S., & Wani, S. H. (2016). MicroRNAs as potential targets for abiotic stress tolerance in plants. *Frontiers in Plant Science*, *7*. https://doi.org/10.3389/fpls.2016.00817

Singh, P., Dutta, P., & Chakrabarty, D. (2021). miRNAs play critical roles in response to abiotic stress by modulating cross-talk of phytohormone signaling. *Plant Cell Reports*, *40*(9), 1617–1630. https://doi.org/10.1007/s00299-021-02736-y

Sripathi, V. R., Anche, V. C., Gossett, Z. B., & Walker, L. T. (2021). Recent applications of RNA sequencing in food and agriculture. In *Applications of RNA-Seq in biology and medicine*. IntechOpen. https://doi.org/10.5772/intechopen.97500

Statello, L., Guo, C.-J., Chen, L.-L., & Huarte, M. (2021). Gene regulation by long non-coding RNAs and its biological functions. *Nature Reviews Molecular Cell Biology*, *22*(2), 96–118. https://doi.org/10.1038/s41580-020-00315-9

Summanwar, A., Basu, U., Rahman, H., & Kav, N. N. V. (2020). Non-coding RNAs as emerging targets for crop improvement. *Plant Science*, *297*, 110521. https://doi.org/10.1016/j.plantsci.2020.110521

Sun, X., Tang, M., Xu, L., Luo, X., Shang, Y., Duan, W., Huang, Z., Jin, C., & Chen, G. (2023). Genome-wide identification of long non-coding RNAs and their potential functions in radish response to salt stress. *Frontiers in Genetics*, *14*. https://doi.org/10.3389/fgene.2023.1232363

Tan, H., Li, B., & Guo, H. (2020). The diversity of post-transcriptional gene silencing mediated by small silencing RNAs in plants. *Essays in Biochemistry*, *64*(6), 919–930.

Tiwari, S., Jain, M., Singla-Pareek, S. L., Bhalla, P. L., Singh, M. B., & Pareek, A. (2023). Pokkali: A naturally evolved salt-tolerant rice shows a distinguished set of lncRNAs possibly contributing to the tolerant phenotype. *International Journal of Molecular Sciences*, *24*(14), 11677. https://www.mdpi.com/1422-0067/24/14/11677

Ullah, I., Kamel, E. A. R., Shah, S. T., Basit, A., Mohamed, H. I., & Sajid, M. (2022). Application of RNAi technology: A novel approach to navigate abiotic stresses. *Molecular Biology Reports*, *49*(11), 10975–10993. https://doi.org/10.1007/s11033-022-07871-7

van Wijk, N., Zohar, K., & Linial, M. (2022). Challenging cellular homeostasis: Spatial and temporal regulation of miRNAs. *International Journal of Molecular Sciences*, *23*(24), 16152. https://www.mdpi.com/1422-0067/23/24/16152

Waititu, J. K., Zhang, C., Liu, J., & Wang, H. (2020). Plant non-coding RNAs: Origin, biogenesis, mode of action and their roles in abiotic stress. *International Journal of Molecular Sciences*, *21*(21), 8401. https://www.mdpi.com/1422-0067/21/21/8401

Wang, H.-L. V., & Chekanova, J. A. (2019). An overview of methodologies in studying lncRNAs in the high-throughput era: When acronyms ATTACK! In J. A. Chekanova & H.-L. V. Wang (Eds.), *Plant long non-coding RNAs: Methods and protocols* (pp. 1–30). Springer New York. https://doi.org/10.1007/978-1-4939-9045-0_1

Wang, T., Zhao, M., Zhang, X., Liu, M., Yang, C., Chen, Y., Chen, R., Wen, J., Mysore, K. S., & Zhang, W.-H. (2017). Novel phosphate deficiency-responsive long non-coding RNAs in the legume model plant *Medicago truncatula*. *Journal of Experimental Botany*, *68*(21–22), 5937–5948. https://doi.org/10.1093/jxb/erx384

Wang, T.-Z., Liu, M., Zhao, M.-G., Chen, R., & Zhang, W.-H. (2015). Identification and characterization of long non-coding RNAs involved in osmotic and salt stress in *Medicago truncatula* using genome-wide high-throughput sequencing. *BMC Plant Biology*, *15*(1), 131. https://doi.org/10.1186/s12870-015-0530-5

Wang, Y., Luo, X., Sun, F., Hu, J., Zha, X., Su, W., & Yang, J. (2018). Overexpressing lncRNA LAIR increases grain yield and regulates neighbouring gene cluster expression in rice. *Nature Communications*, *9*(1), 3516. https://doi.org/10.1038/s41467-018-05829-7

Wani, S. H., Kumar, V., Shriram, V., & Sah, S. K. (2016). Phytohormones and their metabolic engineering for abiotic stress tolerance in crop plants. *The Crop Journal*, *4*(3), 162–176. https://doi.org/10.1016/j.cj.2016.01.010

Wei, J., Li, H., Huang, X., Zhao, Y., Ouyang, L., Wei, M., Wang, C., Wang, J., & Lu, G. (2024). Elucidating the regulatory role of long non-coding RNAs in drought stress response during seed germination in leaf mustard. *PeerJ*, *12*, e17661. https://doi.org/10.7717/peerj.17661

Wolter, F., & Puchta, H. (2018). The CRISPR/Cas revolution reaches the RNA world: Cas13, a new Swiss Army knife for plant biologists. *The Plant Journal*, *94*(5), 767–775. https://doi.org/10.1111/tpj.13899

Wu, J., Wang, L., & Wang, S. (2017). MicroRNAs associated with drought response in the pulse crop common bean (*Phaseolus vulgaris* L.). *Gene*, *628*, 78–86. https://doi.org/10.1016/j.gene.2017.07.038

Xin, M., Peng, H., Ni, Z., Yao, Y., Hu, Z., & Sun, Q. (2019). Wheat responses and tolerance to high temperature. In M. Hasanuzzaman, K. Nahar, & M. A. Hossain (Eds.), *Wheat production in changing environments: Responses, adaptation and tolerance* (pp. 139–147). Springer Singapore. https://doi.org/10.1007/978-981-13-6883-7_6

Xing, X., Cao, C., Xu, Z., Qi, Y., Fei, T., Jiang, H., & Wang, X. (2023). Reduced soybean water stress tolerance by miR393a-mediated repression of GmTIR1 and abscisic acid accumulation. *Journal of Plant Growth Regulation*, *42*(2), 1067–1083. https://doi.org/10.1007/s00344-022-10614-4

Xu, M., Hu, T., Zhao, J., Park, M.-Y., Earley, K. W., Wu, G., Yang, L., & Poethig, R. S. (2016). Developmental functions of miR156-regulated *SQUAMOSA PROMOTER BINDING PROTEIN-LIKE (SPL)* genes in *Arabidopsis thaliana*. *PLoS Genetics*, *12*(8), e1006263. https://doi.org/10.1371/journal.pgen.1006263

Xu, Y.-C., Zhang, J., Zhang, D.-Y., Nan, Y.-H., Ge, S., & Guo, Y.-L. (2021). Identification of long noncoding natural antisense transcripts (lncNATs) correlated with drought stress response in wild rice (*Oryza nivara*). *BMC Genomics*, *22*(1), 424. https://doi.org/10.1186/s12864-021-07754-4

Yadav, A., Kumar, S., Verma, R., Lata, C., Sanyal, I., & Rai, S. P. (2021). microRNA 166: An evolutionarily conserved stress biomarker in land plants targeting HD-ZIP family. *Physiology and Molecular Biology of Plants*, *27*(11), 2471–2485. https://doi.org/10.1007/s12298-021-01096-x

Yajnik, K. N., Singh, I. K., & Singh, A. (2024). lncRNAs and epigenetics regulate plant's resilience against biotic stresses. *Plant Physiology and Biochemistry*, *214*, 108892. https://doi.org/10.1016/j.plaphy.2024.108892

Yang, W., Katinakis, P., Hendriks, P., Smolders, A., De Vries, F., Spee, J., Van Kammen, A., Bisseling, T., & Franssen, H. (1993). Characterization of GmENOD40, a gene showing novel patterns of cell-specific expression during soybean nodule development. *The Plant Journal*, *3*(4), 573–585. https://doi.org/10.1046/j.1365-313x.1993.03040573.x

Yang, X., Liu, C., Niu, X., Wang, L., Li, L., Yuan, Q., & Pei, X. (2022). Research on lncRNA related to drought resistance of Shanlan upland rice. *BMC Genomics*, *23*(1), 336. https://doi.org/10.1186/s12864-022-08546-0

Yang, Y., Huang, J., Sun, Q., Wang, J., Huang, L., Fu, S., Qin, S., Xie, X., Ge, S., Li, X., Cheng, Z., Wang, X., Chen, H., Zheng, B., & He, Y. (2022). microRNAs: Key players in plant response to metal toxicity. *International Journal of Molecular Sciences*, *23*(15), 8642. https://www.mdpi.com/1422-0067/23/15/8642

Yu, Y., Zhang, Y., Chen, X., & Chen, Y. (2019). Plant noncoding RNAs: Hidden players in development and stress responses. *Annual Review of Cell and Developmental Biology*, *35*(1), 407–431.

Zhang, D., Zhang, Z., Unver, T., & Zhang, B. (2021). CRISPR/Cas: A powerful tool for gene function study and crop improvement. *Journal of Advanced Research*, *29*, 207–221. https://doi.org/10.1016/j.jare.2020.10.003

Zhang, H., Zhu, J., Gong, Z., & Zhu, J.-K. (2022). Abiotic stress responses in plants. *Nature Reviews Genetics*, *23*(2), 104–119. https://doi.org/10.1038/s41576-021-00413-0

Zhang, L., Lin, T., Zhu, G., Wu, B., Zhang, C., & Zhu, H. (2023). LncRNAs exert indispensable roles in orchestrating the interaction among diverse noncoding RNAs and enrich the regulatory network of plant growth and its adaptive environmental stress response. *Horticulture Research*, *10*(12). https://doi.org/10.1093/hr/uhad234

Zhang, X., Shen, J., Xu, Q., Dong, J., Song, L., Wang, W., & Shen, F. (2021). Long noncoding RNA lncRNA354 functions as a competing endogenous RNA of miR160b to regulate ARF genes in response to salt stress in upland cotton. *Plant, Cell & Environment*, *44*(10), 3302–3321. https://doi.org/10.1111/pce.14133

Zhang, X., Zhu, X., Chen, L., Fan, H., Liu, X., Yang, N., Duan, Y., & Wang, Y. (2024). MiR398b targets superoxide dismutase genes in soybean in defense against *Heterodera glycines* via modulating reactive oxygen species homeostasis. *Phytopathology®*, *114*(8), 1950–1962. https://doi.org/10.1094/PHYTO-09-23-0343-R

Zhang, Y., Zhou, Y., Zhu, W., Liu, J., & Cheng, F. (2022). Noncoding RNAs fine-tune the balance between plant growth and abiotic stress tolerance. *Frontiers in Plant Science*, *13*. https://doi.org/https://doi.org/10.3389/fpls.2022.965745

Zhao, J., Yuan, S., Zhou, M., Yuan, N., Li, Z., Hu, Q., Bethea, F. G., Liu, H., Li, S., & Luo, H. (2018). Transgenic creeping bentgrass overexpressing Osa-miR393a exhibits altered plant development and improved multiple stress tolerance. *Plant Biotechnology Journal*, *17*(1), 233–251. https://doi.org/https://doi.org/10.1111/pbi.12960

Zhao, M., Wang, T., Sun, T., Yu, X., Tian, R., & Zhang, W.-H. (2020). Identification of tissue-specific and cold-responsive lncRNAs in *Medicago truncatula* by high-throughput RNA sequencing. *BMC Plant Biology*, *20*(1), 99. https://doi.org/10.1186/s12870-020-2301-1

Zhao, T., Tao, X., Li, M., Gao, M., Chen, J., Zhou, N., Mei, G., Fang, L., Ding, L., Zhou, B., Zhang, T., & Guan, X. (2020). Role of phasiRNAs from two distinct phasing frames of GhMYB2 loci in cis- gene regulation in the cotton genome. *BMC Plant Biology*, *20*(1), 219. https://doi.org/10.1186/s12870-020-02430-3

Zhao, W., Cheng, Y., Zhang, C., You, Q., Shen, X., Guo, W., & Jiao, Y. (2017). Genome-wide identification and characterization of circular RNAs by high throughput sequencing in soybean. *Scientific Reports*, *7*(1), 5636. https://doi.org/10.1038/s41598-017-05922-9

Zhao, Z., Zang, S., Zou, W., Pan, Y.-B., Yao, W., You, C., & Que, Y. (2022). Long non-coding RNAs: New players in plants. *International Journal of Molecular Sciences*, *23*(16), 9301. https://www.mdpi.com/1422-0067/23/16/9301

Zhou, X.-Y., Wang, N.-H., Qiu, C.-W., Luo, L., Zhang, M., Zhang, S., Gao, Z.-F., Ahmed, I. M., & Wu, F. (2023). Transcriptome profiling uncovers the lncRNA-mediated regulatory networks associated with tolerance to cadmium stress in barley. *Environmental and Experimental Botany*, *206*, 105156. https://doi.org/10.1016/j.envexpbot.2022.105156

Zhu, H., Li, C., & Gao, C. (2020). Applications of CRISPR–Cas in agriculture and plant biotechnology. *Nature Reviews Molecular Cell Biology*, *21*(11), 661–677. https://doi.org/10.1038/s41580-020-00288-9

Zhu, H., Stephen, S., Taylor, J., Helliwell, C. A., & Wang, B. (2013). Long noncoding RNAs responsive to *Fusarium oxysporum* infection in *Arabidopsis thaliana*. *New Phytologist*, *201*(2), 574–584. https://doi.org/10.1111/nph.12537

Zuo, Z., He, W., Li, J., Mo, B., & Liu, L. (2021). Small RNAs: The essential regulators in plant thermotolerance. *Frontiers in Plant Science*, *12*. https://doi.org/10.3389/fpls.2021.726762

11 Stress-Responsive ncRNAs
Paving the Way for Resilient Plants

Rachna Shashikant Jadhav, Vinay Sharma, Himanshu Sharma, Chandresh Kumari, Pradeep Singh and Saurabh Kulshrestha

11.1 INTRODUCTION TO STRESS-RESPONSIVE NON-CODING RNAs (ncRNAs)

Deoxyribonucleic acid (DNA) and ribonucleic acid (RNA) play diverse roles in cellular processes in living organisms and are essential for transcription and translation, by which cells read out, or express, the genetic instructions in their genes. RNA plays an equally important role as DNA. RNA is composed of ribose sugar, phosphate groups and four nitrogenous bases (adenine, uracil, cytosine and guanine). It is structurally similar to DNA but is usually single-stranded in nature and consists of uracil as a replacement of thymine. Three main types of RNA that code and are involved in protein synthesis are mRNA, rRNA and tRNA. Apart from coding RNAs, snRNAs (small nuclear RNAs), circRNA (circular RNAs) and lnRNA (long non-coding RNAs) account for ncRNAs (non-coding RNAs) (Wang & Farhana, 2024). As essential coding RNAs are for life processes, ncRNAs also play crucial roles in several biological processes. Although ncRNAs do not encode proteins in plants, they are integral to gene regulation, development and responses to environmental stresses, which pave the way for building resilience in plants (Yadav et al., 2024). The current review discusses the different categories of ncRNAs that act as stress regulators, in gene silencing, hormone signalling pathways, epigenetic regulation, epigenetic and transcriptional regulation and stress tolerance, thereby contributing to stress response while paving the way for resilient plants. The chapter also focuses on highlighting the ncRNA-based biotechnological strategies for improving stress tolerance in plants and integrating ncRNA knowledge into sustainable agriculture.

11.1.1 Categorisation of Non-Coding RNAs and Their Functions

Long ncRNAs are produced as a result of transcription and RNA processing. The three main types of ncRNAs mentioned earlier are sRNA, lncRNA and circRNA. The further subtypes of ncRNAs are discussed below:

a. *Small RNAs (sRNAs)*: sRNAs are the first class of ncRNAs that range from 21 to 24 nucleotides. They include two major subtypes of RNAs: microRNA (miRNAs) and small interfering RNA (siRNAs). Altogether, these subtypes are known to act as stress-response regulators during plant stress.
b. *MicroRNAs (miRNAs)*: miRNAs are endogenous ncRNAs in plants that are about 21 nucleotides long, originating from a stem-loop structure from precursor RNAs. They are generated from introns of pre-mRNAs (Rodriguez et al., 2004). miRNAs employ various mechanisms to regulate gene expression in both the cytoplasm and the nucleus, including mediating gene silencing at the post-transcriptional level (Lagos-Quintana et al., 2001). They also control gene expression post-transcriptionally by targeting mRNAs for translational repression. In plants, miRNAs target a limited number of mRNA molecules, approximately 1% of protein-coding genes, whereas in animals, they regulate nearly 60% of protein-coding genes (Friedman et al., 2009; Li et al., 2010). Although only 1% of miRNAs target mRNAs in plants, their significant role in gene regulation is undeniable, as many of these target mRNAs encode transcription factors involved in key developmental processes, from seed germination to maturation. miRBase serves as a central public repository and online resource for miRNA sequences and annotations. The latest update on miRBase (v22) consists of 38,589 entries that represent hairpin precursor miRNAs from around 271 organisms, 326 hairpins of which belong to *Arabidopsis thaliana* (Kozomara et al., 2019).
c. *Small interfering RNAs (siRNAs)*: siRNAs are derivatives of dsRNA precursors that are around 25 nucleotides long, having both sense and antisense polarity. Similar to their molecules of origin, they are double-stranded in nature. They were first identified in plants that were undergoing cosuppression or virus-induced gene silencing. They are known to play a vital role in RNA silencing, genome stability by silencing transposons and defence mechanisms against viruses (Hammond et al., 2000). Plant siRNAs are categorised into trans-acting siRNAs (tas-siRNAs), heterochromatic siRNAs (hc-siRNAs), repeat-associated siRNAs (ra-siRNAs), endogenous siRNAs (endosiRNAs) and long siRNAs (lsiRNAs). In plants, tas-siRNAs regulate genic mRNAs through post-transcriptional target degradation or translational inhibition (Felippes & Weigel, 2009). Meanwhile, hc-siRNAs act on silenced genomic regions, including centromeric repeats and inactive

DOI: 10.1201/9781003583257-11

transposable elements, by inducing heterochromatic marks such as DNA or histone methylation (Law & Jacobsen, 2010). endosiRNAs target pseudogenes and genomic parasites, such as viruses and transposable elements in plants as well as animals via post-transcriptional target degradation (Ding, 2010). Whereas lsiRNAs are produced in response to pathogen infection or under specific growth conditions (Katiyar-Agarwal et al., 2007).

d. *Long Non-Coding RNAs (lncRNAs)*: lncRNAs are RNA molecules at least 200 nucleotides in length. They play roles in transcription modulation, chromatin remodelling and post-transcriptional regulation, functioning as regulators of RNA-binding proteins despite lacking protein-coding capability. lncRNAs are further divided into six subtypes: long intergenic ncRNAs (lincRNAs), long intronic ncRNAs, sense lncRNAs, antisense lncRNAs, bidirectional lncRNAs, cis-acting lncRNAs and trans-acting lncRNAs. lincRNAs are transcribed from both strands in intergenic regions. Long intronic ncRNAs are derived from introns of protein-coding genes. Sense lncRNAs originate from the sense strand, incorporating exons from protein-coding genes, whereas antisense lncRNAs are transcribed from the antisense strand, overlapping with exonic or intronic regions or spanning the protein-coding sequence via an intron. Bidirectional lncRNAs are situated near coding transcripts on the opposite strand. Cis-acting lncRNAs regulate gene expression in nearby genes, while trans-acting lncRNAs influence gene expression in distant genes (Kornienko et al., 2013; Ma et al., 2013; Kornienko et al., 2013
e. *Circular RNAs (circRNAs)*: circRNAs are endogenous single-stranded RNAs that form covalently closed RNA molecules. They were first identified in *A. thaliana* (Zhang & Dai, 2022). circRNAs are 100–1000 nucleotides long. They originate from exons, introns and untranslated regions (UTRs) and are also formed as a result of the linking of the 5′ terminus and the 3′ terminus after reverse splicing. They regulate alternative RNA splicing or transcription (Salzman, 2016; Ye et al., 2015; Calixto et al., 2019).

11.2 RESPONSE MECHANISMS OF STRESS PERCEPTION AND ROLES OF STRESS-RESPONSIVE ncRNAs

Plants face various environmental stresses depending on the environment in which they grow. Stresses or abiotic factors such as drought (water scarcity), salinity and extreme temperatures – high and low trigger sophisticated changes in plants that they respond to using several mechanisms. Plants detect different stress signals via receptors located on the plant cell's plasma membrane. The receptors can sense changes in temperature, excessive reactive oxygen species (ROS) etc. and respond using several mechanisms. Non-coding RNAs (ncRNAs), particularly small RNAs (sRNAs) and long non-coding RNAs (lncRNAs), are crucial regulators of plant responses to environmental stresses. These nucleic acid molecules influence stress perception, gene expression and overall stress adaptation by acting at transcriptional, post-transcriptional and epigenetic levels through various mechanisms, as outlined below.

11.2.1 Response to Drought Stress

In the absence of an optimum amount of water required for plant growth, or in drought-like conditions, the plant not only responds at the physiological, morphological and biochemical levels but also at a molecular level wherein the response is genetically programmed and the process is regulated in a sophisticated yet synchronised manner. The essential genetic regulation is brought about by ncRNAs that interact with other molecules to form potential networks to control multiple genes for responding to stress. Non-coding RNAs (ncRNAs), particularly small RNAs (sRNAs) and long non-coding RNAs (lncRNAs), are essential in regulating plant responses to environmental stresses. These molecules function at transcriptional, post-transcriptional and epigenetic levels, influencing stress perception, gene expression and overall adaptation to stress. As previously mentioned, miRNAs and siRNAs are two primary categories of small ncRNAs. Both types play significant roles at the molecular level, regulating multiple targets and acting as key regulators of processes such as photosynthesis, respiration, growth and stress responses in plants.

11.2.2 sRNAs' Role in Drought Stress Response

Small ncRNAs like miRNAs are known to play a vital role in promoting the plant's adaptation and tolerance to changing environmental conditions, including drought stress, wherein the miRNA controls the key genes and transcription factors (Hernández & Sanan-Mishra, 2017; Jatan & Lata, 2019). Zhang et al. (2016) found that miR169g, miR166h, miR172d and miR408 from the miRNA family were significantly upregulated in rice in response to drought stress, while miR164c, miR319b and miR1861d were downregulated under similar conditions. In a separate study, Zhao et al. (2007) observed that miR393 was transiently induced by drought stress in rice. Rice plants inoculated with a cultivable fungi – *Piriformospora indica* – exhibited increased expression of miR396, leading to the suppression of growth-regulating factor (GRF). This reduction in GRF activity decreased transpiration rates, allowing the plants to be resilient against drought conditions (Fard et al., 2017).

11.2.3 lncRNAs' Role in Drought Stress Response

In a study by Bardou et al. (2014) on lateral root development, two nuclear alternative splicing regulators, AtNSRa and AtNSRb, were known to be affected due to binding by an alternative splicing competitor (ASCO) long ncRNA. The same regulators are known to modulate gene expression in tomatoes when infected by *Phytophthora infestans* (Cui et al., 2017). Root system architecture (RSA) is the spatiotemporal arrangement of roots in a root system. In drought stress, transcription factors belonging to the NAC, WRKY, MYB and bHLH are regulated (Nykiel et al., 2022). Overexpression of NAC transcription factors OsNAC9, OsNAC10 and OsNAC045 specifically in the roots (Redillas et al., 2012; Seo Jeong et al., 2010; Yu et al., 2018) has demonstrated significant improvements in root development. Additionally, these transcription factors have been shown to enhance drought tolerance in C3 plants, highlighting their potential role in stress resilience. Qin et al. (2017) discovered a novel 755-nt-long drought-induced lincRNA (DRIR), localised in the nucleus, revealing the potential role of lncRNAs as positive regulators of drought stress tolerance in *Arabidopsis*. *Arabidopsis* lines with DRIR overexpression demonstrated increased drought tolerance compared to wild-type seedlings. This lincRNA was located in the nucleus and regulated the transcription of various drought-responsive genes, such as those involved in ABA signalling (ABI5, P5CS1, RD29A and RD29B), aquaporins (NIP1 and TIP4), annexin (ANNAT7), fucosyltransferase 4 (FUT4) and transcription factors (NAC3 and WRKY8).

Genome-wide transcription studies identified 19 potential lncRNAs in foxtail millet, 98 in rice, 664 in maize and 1597 in switchgrass (Chung et al., 2016; Qi et al., 2013; Zhang et al., 2014, 2018). Research in maize revealed that lncRNAs expressed during the silking stage play a vital role in conferring tolerance to drought stress (Pang et al., 2019). Ding et al. (2019) identified 124 differentially expressed lncRNAs (DElncRNAs) in cassava under drought stress, with 11 of these acting as target mimics for miR156, miR164, miR169 and miR172. Qin et al. (2017) also explored the role of lncRNAs in regulating drought stress tolerance in *Arabidopsis*, uncovering a novel nuclear-localised lncRNA, drought-induced lncRNA (DRIR). This 755-nt-long lncRNA regulates the transcription of several drought-responsive genes, including those related to ABA signalling (ABI5, P5CS1, RD29A and RD29B), aquaporin genes (NIP1 and TIP4), the annexin gene (ANNAT7), the fucosyltransferase 4 (FUT4) gene and transcription factor genes (NAC3 and WRKY8).

11.2.4 circRNAs' Role in Drought Stress Response

Dasmandlal et al. (2020) identified that circRNAs in chickpeas and soybeans are involved in crop transcriptomics and their differential expressions when faced with drought stress.

11.3 RESPONSE TO HIGH SALINITY

Salinity stress poses a significant challenge to plant growth and productivity, as it disrupts the ion homeostasis, inducing osmotic stress and generating oxidative damage. To deal with such salinity stress conditions, plants have evolved intricate mechanisms by which they not only perceive but also respond to high salinity, through complex signalling pathways and regulatory networks. Plants detect salinity-induced changes through various sensors that perceive osmotic stress, ion imbalance and mechanical alterations in cellular structures. Main initial signals include elevated levels of cytosolic calcium ions (Ca^{2+}), ROS accumulation and the production of cyclic nucleotides like cGMP (Ma et al., 2022).

11.3.1 sRNAs' Role in High Salinity Stress

The most prevalent RNA members of the sRNA family are microRNAs, which smartly regulate gene expression at the post-transcriptional and translational levels (Baulcombe, 2004). Salinity has been known to affect the sRNA population, indicating that the rise in the occurrence of unique sequences in roots under NaCl exposure suggests an accelerated process of sRNA accumulation in the presence of NaCl. However, not all of these sequences may correspond to miRNAs (De Paola et al., 2012). Small RNAs also play a role in regulating mRNA degradation, suppressing translation, modifying chromatin and responding to salt stress, including challenges such as oxidative stress, mineral nutrient deficiencies, dehydration and maintaining osmotic balance (Dutta et al., 2020).

Under salt stress conditions, miR156 expression was elevated, which improved seed germination rates by regulating early embryonic development in plants. This enhancement contributed to increased salt tolerance and higher germination rates in alfalfa seeds with miR156 overexpression (Muhammad et al., 2017). Salt stress also led to higher expression of miR169 in rice, which targets and degrades the NF-YA (nuclear factor-YA) gene, reducing ABA sensitivity. In *Arabidopsis*, overexpressing miR169 improved salt tolerance. Similarly, overexpression of miR393 in both rice and *A. thaliana* enhanced salt resistance by targeting TIR1 (transport inhibitor response protein 1). This interaction between miR393 and its target gene influenced seed germination by regulating the expression of salt tolerance-associated genes, P5CDH and SR05 (Gao et al., 2011). Additionally, miRNAs have been shown to boost plant salt resistance by regulating processes involved in the removal of ROS in *Phaseolus vulgaris* and cotton (Naya et al., 2014; Xie et al., 2015). Thus, understanding these sRNA mechanisms provides valuable insights into plant resilience to salinity stress and offers potential strategies for developing salt-tolerant crops through genetic and biotechnological interventions.

11.3.2 lncRNAs' Role in High Salinity Stress

lncRNAs control gene expression through multiple mechanisms, such as chromatin remodelling, transcriptional interference, post-transcriptional regulation and acting as molecular decoys or scaffolds (Rinn & Chang, 2012). Recent studies have highlighted the involvement of lncRNAs in fine-tuning plant responses to salinity stress by modulating stress-responsive genes, signalling pathways and metabolic processes.

Chen et al. (2023) found that drought-induced lncRNAs (DRIR) in *Arabidopsis* positively regulated several drought- and salinity-responsive transcripts. In cotton, lnc_388, lnc_883 and lnc_973 were linked to miRNAs that respond to salinity stress. Similarly, lncRNA77580 was recently identified as playing a role in soybeans' response to salinity stress. In *Chenopodium quinoa*, Luo et al. (2022) identified seven weighted gene co-expression modules highly associated with saline conditions. It was observed that quinoa lncRNAs may contribute to salinity stress responses by interacting with transcription factors such as HSF, WRKY, C2H2, ERF, MYB, NAC, bHLH, FAR1, GRASS, Dof and C3H.

Alternatively, these lncRNAs function independently by interacting directly with genes without the involvement of transcription factors. An RNA-seq analysis conducted by Kumar et al. (2021) on two salt-tolerant chickpea lines (ICCV 10 and JG 11) and two salt-sensitive lines (DCP 92–3 and Pusa 256) under salinity stress identified 3,373 lncRNAs functioning as cis-regulators of target genes, along with 80 lncRNAs serving as eTMs for 136 miRNAs.

lncRNAs therefore represent a new frontier in understanding plant stress biology. While significant progress has been made in elucidating their roles in salinity stress, several challenges remain. These include limited functional characterisation of lncRNAs, the complexity of their regulatory networks and their evolutionary conservation across species. Future research should focus on developing comprehensive databases, high-resolution functional assays and translational approaches to harness lncRNAs for sustainable agriculture.

11.3.3 circRNAs' Role in Salinity Stress Response

Circular RNAs (circRNAs) are another type of non-coding RNAs that have attracted attention for their regulatory functions in various biological processes, including responses to salinity stress in plants. The potential role of circRNAs in salinity stress response has been increasingly recognised, although research on this specific area is still emerging.

Liu et al. (2022) conducted a genome-wide association study (GWAS) to examine the survival rate of 300 maize accessions under salt stress, identifying 5 trait-associated SNPs and 86 candidate genes. RNA sequencing was performed on 28 transcriptome libraries from two maize lines with varying salt tolerance, both under normal and salt-treated conditions. This analysis uncovered a total of 1,217 highly expressed circRNAs, 371 of which were responsive to salt stress. Gao et al. (2024) reported that VvcircABH and VvABH (alpha-/beta-hydrolase) have distinct roles in salt tolerance and brassinosteroid signalling, with VvcircABH possibly regulating VvABH expression by inhibiting its splicing.

In conclusion, circular RNAs (circRNAs) play a crucial role in the regulation of plant responses to salinity stress by modulating gene expression, protein interactions and stress-responsive pathways. Through mechanisms such as miRNA interaction with transcription factors, circRNAs contribute to enhancing salt tolerance in plants. As research advances, understanding these roles could lead to innovative strategies for improving crop resilience to salinity, which is a growing challenge for global agriculture. Further studies are essential to explore the full potential of circRNAs in plant stress adaptation.

11.4 RESPONSE TO TEMPERATURE STRESS

Temperature stress encompasses both heat and cold stress. It poses a significant challenge to cellular homeostasis and survival in plants. Non-coding RNAs (ncRNAs), which include microRNAs (miRNAs), small interfering RNAs (siRNAs), long non-coding RNAs (lncRNAs) and circular RNAs (circRNAs), have emerged as key regulators in the cellular response to temperature fluctuations. These molecules modulate gene expression, protein synthesis and other physiological processes critical for stress adaptation. ncRNAs are pivotal in orchestrating cellular responses to temperature stress. Their ability to regulate gene expression and physiological pathways underscores their potential in adaptive strategies for stress resilience.

11.4.1 sRNAs' Role in Temperature Stress

miRNAs are part of sRNAs, and miR398 from the miRNA family has been shown to be rapidly induced in response to heat stress, accompanied by the downregulation of its target genes CSD1, CSD2 and CCS (copper/zinc superoxide dismutases and a copper chaperone for SOD) (Guan et al., 2013). miR398 and its target CSDs have also been implicated in the heat stress responses of *Brassica rapa* (Yu et al., 2012) and *Populus tomentosa* (Kotak et al., 2007), suggesting that the miR398-CSD/CCS pathway plays a broad role in plant responses to heat stress. miR172, another miRNA family member, is downregulated, and TOE2 (Target of EAT2) is upregulated by heat stress in *Helianthus annuus* (Khaksefidi et al., 2015). In wheat, miR164 can directly target the heat shock protein HSP17, which is upregulated during heat stress. Yu et al. (2013) conducted a differential expression analysis in *B. rapa* and identified that nat-siRNAs derived from 12 cis-NATs responded to heat stress, with most exhibiting strand bias. Furthermore, transcripts generating heat-responsive nat-siRNAs were mostly upregulated under heat stress, while transcripts from the opposite strands of the same loci were downregulated.

11.4.2 lncRNAs' Role in Temperature Stress

lncRNAs function through chromatin remodelling and transcriptional regulation and act as molecular sponges for miRNAs. In *B. rapa*, 34 uniquely expressed lncRNAs were identified under heat stress conditions. These lncRNAs regulated 192 target genes, the majority of which were associated with heat response functions (Song et al., 2016). lncRNAs are even known to stabilise mRNAs of heat-shock proteins (HSPs) during elevated temperatures (Song et al., 2019). Yu et al. (2013) performed a differential expression analysis in *B. rapa* and found that nat-siRNAs originating from 12 cis-NATs were responsive to heat stress, with the majority showing strand bias. Additionally, transcripts producing heat-responsive nat-siRNAs were generally upregulated under heat stress, whereas those from the opposite strands of the same loci were downregulated. Functional validation of lincRNA419, lincRNA207 and lincRNA234 demonstrated their upregulated expression under cold stress, highlighting their role in cold stress tolerance (Li et al., 2017).

11.4.3 CircRNAs' Role in Temperature Stress

Circular RNAs (circRNAs) are essential in a range of biological processes, including the response to environmental stresses like temperature stress. These single-stranded, covalently closed RNA molecules are stable because of their circular structure, which makes them well-suited for regulating stress responses. In tomatoes, circRNAs were suggested to be involved in the chilling response (Zuo et al., 2016). Gao et al. (2019) also found that circRNAs play a role in cold resistance in grapes. Findings by Wang et al. (2020) suggest that soybean circRNAs may encode proteins and participate in regulating low-temperature responses, offering insights into the molecular mechanisms involved in low-temperature responses in soybean. Yang et al. (2020) identified that circRNAs regulate genes involved in transcription and Ca^{2+} channel activity by targeting related proteins, including NAC, MYB, WRKY and calmodulin-binding proteins.

11.5 RESPONSE TO INFECTION BY MICROBES

Non-coding RNAs (ncRNAs) have gained significant attention in recent years for their role in regulating various biological processes, including host–pathogen interactions. In microbial infections, ncRNAs, including microRNAs (miRNAs), long non-coding RNAs (lncRNAs) and circular RNAs (circRNAs), have been found to influence both the microorganism's ability to infect a host and the host's defence mechanisms.

Plants detect pathogenic microbes through two main mechanisms:

1. *Pattern-Triggered Immunity (PTI)*: This involves the recognition of conserved microbial patterns, known as pathogen-associated molecular patterns (PAMPs), by pattern recognition receptors (PRRs) on the plant cell surface.
2. *Effector-Triggered Immunity (ETI)*: In this mechanism, plants detect specific pathogen effectors using intracellular nucleotide-binding leucine-rich repeat (NLR) proteins.

Once these patterns are recognised, defence responses are activated, which include the production of ROS, synthesis of antimicrobial compounds and activation of stress-responsive ncRNAs like miRNAs and siRNAs that regulate immune-related gene expression (Chisholm et al., 2006; Jones & Dangl, 2006; Peng et al., 2018). The signal-integration layer collects inputs from the recognition layer and conveys them to the defence-action layer, triggering various responses, such as the deposition of callose, ROS generation and activation of pathogenesis-related (PR) genes (Wang et al., 2019).

11.5.1 sRNAs' Role during Microbial Infection

Numerous miRNAs have been found to regulate the cleavage of R genes, ensuring precise control of their expression to modulate the immune response in *Arabidopsis*, *Leguminosae*, *Solanaceae* and other plant species (Fei et al., 2013). miR863-3P of the miRNA family is known to be involved in the cleavage and degradation of mRNA from two atypical receptor-like pseudokinase genes, ARLPK1 and ARLPK2, which act as negative regulators of plant immunity during the initial stages of *Pseudomonas syringae* pv. *tomato* (Pst) infection in *A. thaliana* (Niu et al., 2016). In another study on *Arabidopsis* by Soto-Suarez et al. (2017), miR396 expression is reduced during *Plectosphaerella cucumerina* infection, enabling its GRF transcription factor targets to initiate the reprogramming of the host's gene expression.

11.5.2 lncRNAs' Role During Microbial Infection

Numerous studies have demonstrated that certain lncRNAs act as precursors for miRNAs and siRNAs, facilitating the cleavage of target genes. In tomatoes, overexpression of lncRNA23468, which includes conserved endogenous target mimic sites for miR482b, reduces miR482b expression. This leads to increased expression of the target NLR gene *Solyc02g036270.2* and enhances resistance to *P. infestans* (Jiang et al., 2019). *Arabidopsis* plants overexpressing MuCML27 show improved resistance to *Botrytis cinerea* and *Pst*. Therefore, the miR3954–MuLncl–siRNAs–mRNAs module offers new insights into the mulberry defence response (Gai et al., 2018).

11.5.3 CircRNAs' Role during Microbial Infection

In tomatoes, circRNA45 and circRNA47 are believed to function as positive regulators of immunity against *P. infestans* by influencing the expression of miR477-3P and SpRLK (Hong et al., 2020). ncRNAs are central to both microbial infection mechanisms and host defence responses. The growing understanding of their roles in these processes could lead to the development of innovative, targeted pest management strategies. As more studies uncover the complex interactions between ncRNAs and pathogens, new molecular tools might emerge to

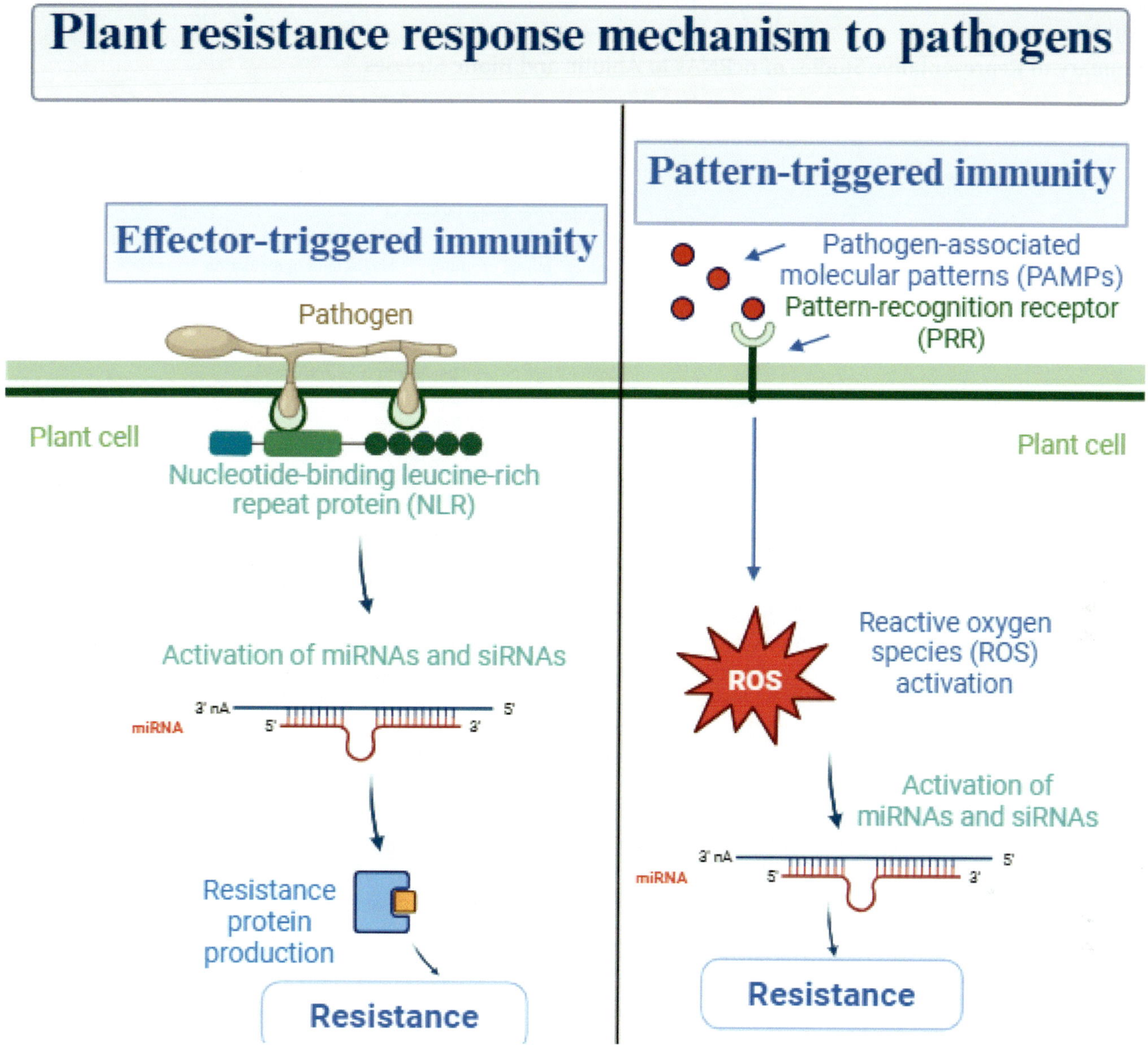

FIGURE 11.1 Plant resistance response mechanism to pathogens.

specifically control pest populations with minimal ecological impact (Figure 11.1).

11.6 CROSS-TALK BETWEEN ncRNAs AND HORMONAL SIGNALLING PATHWAYS

In recent years, the role of non-coding RNAs (ncRNAs) in plant stress responses has attracted significant attention. ncRNAs, especially microRNAs (miRNAs) and long non-coding RNAs (lncRNAs), are involved in regulating various physiological processes, including responses to environmental stresses such as drought, salinity and pathogen attacks, as mentioned earlier. A key aspect of this regulation is the interaction between ncRNAs and hormonal signalling pathways. Abscisic acid (ABA), jasmonic acid (JA) and ethylene are critical hormones that mediate stress responses, and their interaction with ncRNAs plays a crucial role in plant stress adaptation (Sun et al., 2006).

11.6.1 Interaction of Stress-Responsive ncRNAs with Hormones

Plant hormones play a crucial role in mediating stress responses when plants face challenges such as salinity, drought, heat, cold, heavy metals and bacterial or viral infections. Auxin is essential in improving plant tolerance to salt, drought, heat stress and microbial infections (Šečić et al., 2021; Waadt et al., 2022). miR160, miR167, miR390 and miR393, as primary regulators of auxin signalling, have been shown to be differentially expressed under salt and drought stresses (Kumar et al., 2014).

TABLE 11.1
Summary of Representative Studies of ncRNAs to Abiotic and Biotic Stresses

Stress Type	ncRNA Type	Plant Species	Key Findings	References
Drought	miR169	*Solanum lycopersicum*	The overexpression of Sly-miR169c, a member of the miR169 family, in tomato plants effectively suppresses the transcripts of its target genes, thereby negatively regulating stomatal movement.	Zhang et al. (2011)
Salinity	miR398	Oryza sativa	The downregulation of miR398 targets supports growth under optimal conditions, whereas their upregulation enhances rice's tolerance to salinity.	Lu et al. (2022)
Temperature	miR160a, 166a, 167h, 5175a	Hordeum vulgare	miR160a, miR166a, miR167h and miR5175a were identified as being upregulated by heat stress at both the mature miRNA and precursor pri-miRNA levels.	Kruszka et al. (2014)
Cucumber mosaic virus (CMV) infection	miR482	*Solanum lycopersicum*	The miR482-NBS pathway is crucial for regulating resistance to CMV in tomatoes.	Feng et al. (2014)
Sclerotinia stem rot (SSR) infection	miR319a	*Brassica rapa*	miR319a and its target gene BraTCP4 regulate stem rot resistance by influencing the pathways involving WRKY genes.	Dong et al. (2021)

Jasmonic acid (JA) is another critical hormone involved in plant defence responses, particularly in response to wounding and pathogen attacks. ncRNAs are known to play essential roles in modulating JA biosynthesis and signalling. Jasmonic acids (JAs) play species-specific roles in nodulation, either promoting or inhibiting the process. They influence the formation of infection threads and the expression of nodulation-related genes, such as ENOD11.

Ethylene (ET) plays an important role in the nodulation process as a negative regulator. In chickpeas, the expression of miR171b was inversely related to that of ERF (Ca_12975) and has been found to regulate nodule formation (De Luis et al., 2012; Garg et al., 2019).

Transcriptomic analyses by Aung et al. (2017) in *Medicago sativa* showed that overexpression of miR156 increased the expression of gibberellin (GA) receptors and the GA-regulated protein family, thereby enhancing nitrogen fixation activity. More recently, it was found that rhizobial infection in soybean roots induces the genes gibberellin 20-oxidase 1a (GA20ox1a), GA3ox1a and GA2ox1a and the activation of these genes is crucial for nodule organogenesis (Chu et al., 2022).

In *Lotus japonicus*, the CLE-RS1/2-HAR1 signalling pathway promotes the transport of shoot-derived cytokinins (CK) to the root, thereby inhibiting nodulation. Conversely, during rhizobial infection, the nodulation factor (NF) stimulates the accumulation of cytokinins (CK) (Sasaki et al., 2014; Van Zeijl et al., 2015) (Table 11.1).

11.7 CHALLENGES AND FUTURE PROSPECTS IN ncRNA RESEARCH FOR STRESS RESILIENCE

The role of non-coding RNAs (ncRNAs) in regulating various physiological and stress-response processes in plants is becoming increasingly recognised. ncRNAs, which include microRNAs (miRNAs), long non-coding RNAs (lncRNAs), small interfering RNAs (siRNAs) and other types, influence gene expression, chromatin remodelling and stress tolerance mechanisms. However, despite the growing understanding of their importance in stress resilience, several challenges remain in fully understanding the functions of ncRNAs in plants.

a. *Complexity of ncRNA Interactions*: ncRNA-mediated regulation involves complex networks of interactions, which include interactions between different classes of ncRNAs, as well as with mRNAs and proteins. Disentangling these networks in plants is difficult due to the high level of redundancy and pleiotropy within these regulatory systems (Jones-Rhoades et al., 2006). While some ncRNAs have been implicated in stress responses, the full scope of their roles remains unclear, as many ncRNAs are expressed under stress conditions, but their specific targets and mechanisms of action have yet to be fully characterised.
b. *Lack of Comprehensive ncRNA Databases for Plants*: While significant progress has been made in annotating plant genomes, comprehensive, well-curated databases specifically focused on ncRNAs are still limited. This hampers the identification of novel ncRNAs and their associated targets. Furthermore, cross-species comparisons are often limited due to the variability in ncRNA expression across plant species.
c. *Challenges in Functional Validation of ncRNAs*: One of the major hurdles in ncRNA research is the functional validation of their roles in plant stress tolerance. Traditional knockout or knockdown strategies for gene silencing are not always effective in plants due to the presence of gene redundancy or

the complex regulation of ncRNA expression (Lyu et al., 2023). Additionally, the functional characterisation of lncRNAs and other non-coding species is particularly challenging, as these transcripts often have weak or tissue-specific expression patterns.

d. *Technical Limitations in ncRNA Profiling*: Advances in high-throughput sequencing technologies have allowed for large-scale profiling of ncRNAs in plants, but detecting low-abundance ncRNAs and distinguishing them from background noise remains a challenge. Furthermore, the quantification of ncRNAs often requires sophisticated bioinformatic tools and statistical methods, which may not be widely accessible in all research settings.

11.8 FUTURE DIRECTIONS AND TECHNOLOGICAL ADVANCEMENTS IN ncRNA RESEARCH

a. *Advancements in CRISPR/Cas Systems for ncRNA Research*: The advent of CRISPR/Cas systems, particularly CRISPR/Cas9, has revolutionised genetic engineering by enabling precise editing of plant genomes. This technology has significant potential for investigating ncRNA functions. Recent innovations in CRISPR-based tools, such as CRISPR/Cas13 (which targets RNA directly), open up new avenues for studying the role of ncRNAs in stress responses (Abudayyeh et al., 2017). By knocking out specific ncRNAs or modifying their expression, CRISPR/Cas systems could facilitate a more detailed understanding of how these regulatory elements influence plant stress tolerance and adaptation.
b. *Single-Cell Transcriptomics and ncRNA Profiling*: With the development of single-cell RNA sequencing technologies, it is now possible to profile ncRNA expression at the single-cell level, providing insights into the spatial and temporal regulation of ncRNAs under stress conditions (Zhuo et al., 2019). This approach can also help identify specific cell types that are involved in stress responses, providing a more nuanced understanding of how ncRNAs contribute to plant resilience.
c. *Integration of Multi-Omics Approaches*: Future research in ncRNA functions will benefit from integrating transcriptomics, proteomics and metabolomics data. By combining ncRNA profiles with other omic data, researchers can gain a more comprehensive view of how ncRNAs influence gene networks and metabolic pathways involved in stress responses (Borges et al., 2015). For example, integrating ncRNA data with metabolomics could reveal how ncRNAs regulate the production of secondary metabolites that help plants cope with abiotic or biotic stresses.
d. *Artificial Intelligence and Machine Learning*: As the volume of ncRNA-related data grows, machine learning and artificial intelligence tools will be essential for identifying novel ncRNAs, predicting their functions and discovering new gene regulatory networks (Zhou et al., 2020). These technologies will also be useful in analysing large-scale datasets from various plant species to identify conserved ncRNA functions across species.
e. *Improved Functional Genomics Tools for ncRNA Studies*: New tools and techniques for functional genomics, such as the use of small molecule inhibitors or overexpression systems, could aid in the dissection of ncRNA functions in plant stress responses. Additionally, advances in transgenic plants and CRISPR/Cas systems for fine-tuning ncRNA expression will allow researchers to perform functional validation with higher precision.

Although significant progress has been made in understanding the role of ncRNAs in plant stress resilience, several challenges remain in fully exploring their potential. The integration of advanced technologies such as CRISPR/Cas systems, single-cell transcriptomics, multi-omics approaches and artificial intelligence will offer new insights into the functional roles of ncRNAs in stress responses. As these technologies continue to advance, they hold great promise for deepening our understanding of plant resilience mechanisms and aiding the development of crops with improved stress tolerance (Figure 11.2).

11.9 CONCLUSION

11.9.1 Integrating ncRNA Knowledge into Sustainable Agriculture

In recent years, non-coding RNAs (ncRNAs) have become recognised as essential regulators in plant biology, affecting various biological processes, including growth, stress response and development. Notable findings emphasise the diverse roles of ncRNAs, such as microRNAs (miRNAs) and long non-coding RNAs (lncRNAs), in regulating gene expression under various environmental stresses. These RNA molecules regulate the expression of genes involved in stress tolerance, pathogen resistance and growth adaptability. Their role in gene regulatory networks makes them crucial for enhancing crop resilience and productivity.

Research has also identified specific ncRNAs that regulate plant responses to abiotic stresses like drought, salinity and temperature fluctuations. These findings emphasise the potential of ncRNAs in modifying plant genomes for enhanced stress tolerance, ultimately contributing to sustainable agricultural practices.

11.9.2 Significance of ncRNA Research

The study of ncRNAs holds significant promise for advancing sustainable agriculture. Unlike traditional approaches that focus on protein-coding genes, ncRNAs provide a layer

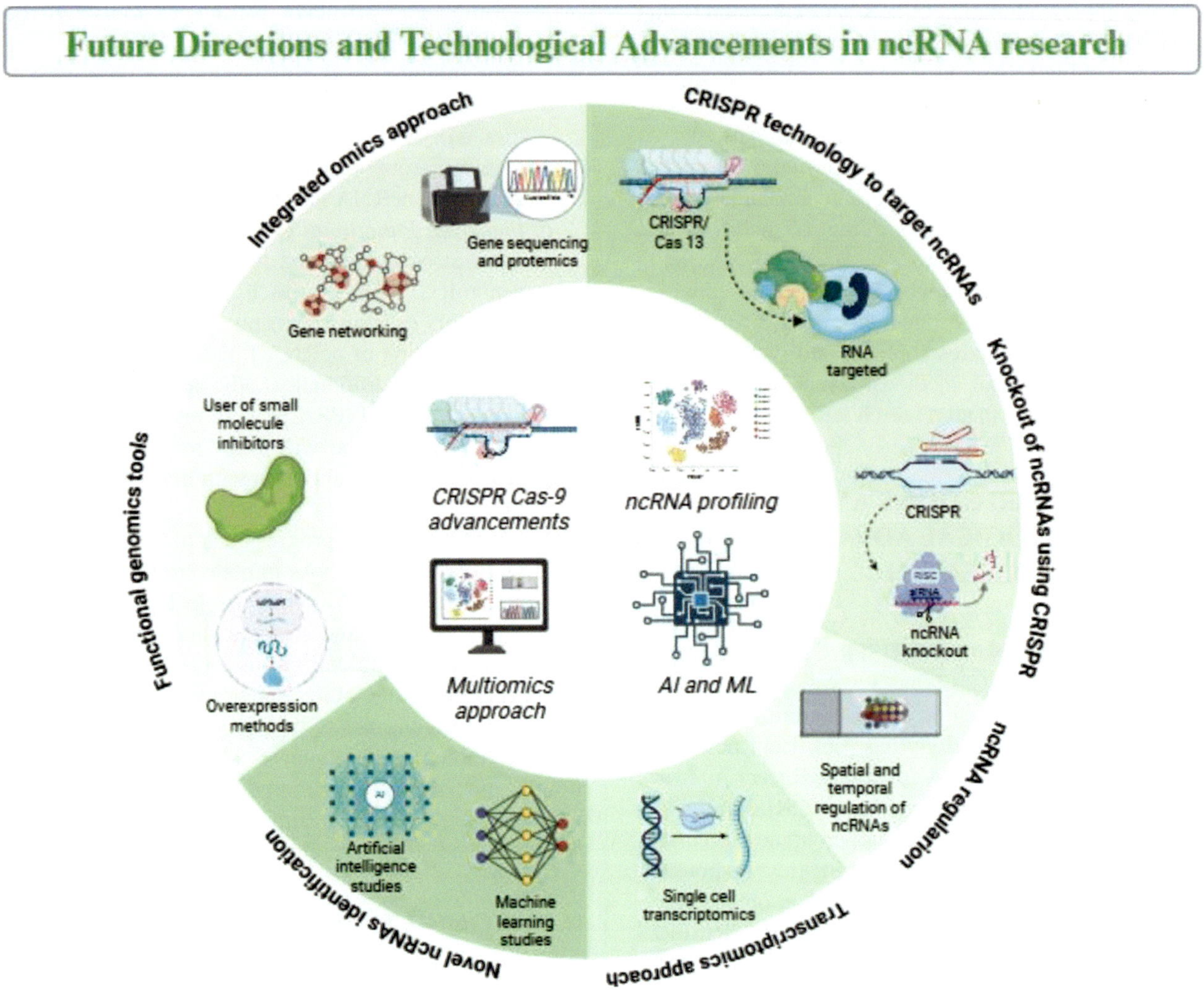

FIGURE 11.2 Future directions and technological advancements in ncRNA research.

of regulatory control that is often overlooked. Their ability to fine-tune gene expression in response to external stimuli positions ncRNAs as a valuable resource for crop improvement. By understanding how ncRNAs function in plants, researchers can design strategies that optimise crop traits like drought tolerance, nutrient use efficiency and resistance to pests and diseases.

Moreover, ncRNA-based technologies, such as gene silencing and overexpression techniques, have shown the potential to precisely control traits without introducing foreign DNA, offering a more sustainable and eco-friendly alternative to genetically modified organisms (GMOs). The ability to manipulate ncRNA expression provides a novel route to enhance crop resilience, thus contributing to food security in the face of climate change.

11.9.3 The Potential of ncRNA-Based Approaches in Developing Climate-Resilient Crops

With the accelerating impacts of climate change, the development of climate-resilient crops is of paramount importance for sustainable agriculture. ncRNA-based approaches offer a powerful tool for addressing climate-related challenges by enhancing plant tolerance to various environmental stresses. MicroRNAs and small interfering RNAs (siRNAs) have been shown to regulate critical pathways that control drought tolerance, salt stress and high-temperature resistance. The ability to modulate these ncRNAs could enable the creation of crops that can better withstand fluctuating environmental conditions, thus ensuring stable yields despite unpredictable climates.

Furthermore, long non-coding RNAs (lncRNAs) involved in epigenetic regulation and chromatin remodelling could be harnessed to modify plants' adaptive responses to stress, making them more resilient over time. The integration of ncRNA-based tools with traditional breeding techniques or CRISPR/Cas9 technology could fast-track the development of climate-resilient crops.

By focusing on ncRNA as a regulatory element, rather than simply modifying individual genes, it is possible to create crops with a broader range of stress-resilient traits. This approach would lead to more sustainable and resource-efficient agricultural practices, reducing the need for chemical interventions and improving food security in the face of a changing global climate.

REFERENCES

Abudayyeh, O. O., Gootenberg, J. S., Essletzbichler, P., Han, S., Joung, J., Belanto, J. J., Verdine, V., Cox, D. B. T., Kellner, M. J., Regev, A., Lander, E. S., Voytas, D. F., Ting, A. Y., & Zhang, F. (2017). RNA targeting with CRISPR-Cas13. *Nature, 550*(7675), 280–284. https://doi.org/10.1038/nature24049

Aung, B., Gao, R., Gruber, M. Y., Yuan, Z. C., Sumarah, M., & Hannoufa, A. (2017). MsmiR156 affects global gene expression and promotes root regenerative capacity and nitrogen fixation activity in alfalfa. *Transgenic Research, 26*, 541–557. https://doi.org/10.1007/s11248-017-0024-3

Bardou, F., Ariel, F., Simpson, C. G., Romero-Barrios, N., Laporte, P., Balzergue, S., Brown, J. W. S., & Crespi, M. (2014). Long noncoding RNA modulates alternative splicing regulators in *Arabidopsis. Developmental Cell, 30*, 166–176. https://doi.org/10.1016/j.devcel.2014.06.017

Baulcombe, D. (2004). RNA silencing in plants. *Nature, 431*(7006), 356–363.

Borges, F., & Martienssen, R. A. (2015). The expanding world of small RNAs in plants. *Nature reviews. Molecular Cell Biology, 16*(12), 727–741. https://doi.org/10.1038/nrm4085

Calixto, C. P. G., Tzioutziou, N. A., James, A. B., Hornyik, C., Guo, W., Zhang, R., Nimmo, H. G., & Brown, J. W. S. (2019). Cold-dependent expression and alternative splicing of arabidopsis long non-coding RNAs. *Frontiers in Plant Science, 10*, 235. https://doi.org/10.3389/fpls.2019.00235

Chen, X., Jiang, X., Niu, F., Sun, X., Hu, Z., Gao, F., Zhang, H., & Jiang, Q. (2023). Overexpression of *lncRNA77580* regulates drought and salinity stress responses in soybean. *Plants (Basel, Switzerland), 12*(1), 181. https://doi.org/10.3390/plants12010181

Chisholm, S. T., Coaker, G., Day, B., & Staskawicz, B. J. (2006). Host-microbe interactions: shaping the evolution of the plant immune response. *Cell, 124*, 803–814. https://doi.org/10.1016/j.cell.2006.02.008

Chu, X., Su, H., Hayashi, S., Gresshoff, P. M., & Ferguson, B. J. (2022). Spatiotemporal changes in gibberellin content are required for soybean nodulation. *New Phytologist, 234*, 479–493. https://doi.org/10.1111/NPH.17902

Chung, P. J., Jung, H., Jeong, D. H., Ha, S. H., Choi, Y. D., & Kim, J. K. (2016). Transcriptome profiling of drought responsive noncoding RNAs and their target genes in rice. *BMC Genomics, 17*(1). https://doi.org/10.1186/s12864-016-2997-3

Cui, J., Luan, Y., Jiang, N., Bao, H., & Meng, J. (2017). Comparative transcriptome analysis between resistant and susceptible tomato allows the identification of lncRNA16397 conferring resistance to *Phytophthora infestans* by co-expressing glutaredoxin. *The Plant Journal, 89*, 577–589. https://doi.org/10.1111/tpj.13408

Dasmandal, T., Rao, A. R., & Sahu, S. (2020). Identification and characterization of circular RNAs regulating genes responsible for drought stress tolerance in chickpea and soybean. *Indian J. Genet.*, 80(1), 1–8. https://doi.org/10.31742/IJGPB.80.1.1

De Luis, A., Markmann, K., Cognat, V., Holt, D. B., Charpentier, M., Parniske, M., Stougaard, J., & Voinnet, O. (2012). Two microRNAs linked to nodule infection and nitrogen-fixing ability in the legume *Lotus japonicus. Plant Physiology, 160*(4), 2137–2154. https://doi.org/10.1104/pp.112.204883

De Paola, D., Cattonaro, F., Pignone, D., & Sonnante, G. (2012). The miRNAome of globe artichoke: Conserved and novel micro RNAs and target analysis. *BMC Genomics, 13*, 41

Ding, S. W. (2010). RNA-based antiviral immunity. *Nature Reviews. Immunology, 10*(9), 632–644. https://doi.org/10.1038/nri2824

Ding, Z., Tie, W., Fu, L., Yan, Y., Liu, G., Yan, W., Li, Y., Wu, C., Zhang, J., & Hu, W. (2019). Strand-specific RNA-seq based identification and functional prediction of drought-responsive lncRNAs in cassava. *BMC Genomics, 20*(1), 214. https://doi.org/10.1186/s12864-019-5585-5

Dong, W., Ren, W., Wang, X., Mao, Y., & He, Y. (4 May 2021). MicroRNA319a regulates plant resistance to *Sclerotinia* stem rot. *Journal of Experimental Botany, 72*(10), 3540–3553. https://doi.org/10.1093/jxb/erab070

Dutta, T., Neelapu, N. R., Wani, S. H., & Surekha, C. (2020). *Salt stress tolerance and small RNA*.

Ebrahimi Khaksefidi, R., Mirlohi, S., Khalaji, F., Fakhari, Z., Shiran, B., Fallahi, H., Rafiei, F., Budak, H., & Ebrahimie, E. (2015). Differential expression of seven conserved microRNAs in response to abiotic stress and their regulatory network in *Helianthus annuus. Frontiers in Plant Science, 6*, 741. https://doi.org/10.3389/fpls.2015.00741

Fei, Q., Xia, R., & Meyers, B. C. (2013). Phased, secondary, small interfering RNAs in posttranscriptional regulatory networks. *Plant Cell, 25*, 2400–2415. https://doi.org/10.1105/tpc.113.114652

Felippes, F. F., & Weigel, D. (2009). Triggering the formation of tasiRNAs in *Arabidopsis thaliana*: the role of microRNA miR173. *EMBO Reports, 10*, 1–7.

Feng, J., Liu, S., Wang, M., Lang, Q., & Jin, C. (2014). Identification of microRNAs and their targets in tomato infected with *Cucumber mosaic virus* based on deep sequencing. *Planta, 240*, 1335–1352. https://doi.org/10.1007/s00425-014-2158-3

Friedman, R. C., Farh, K. K., Burge, C. B., & Bartel, D. P. (2009). Most mammalian mRNAs are conserved targets of microRNAs. *Genome Research, 19*(1), 92–105. https://doi.org/10.1101/gr.082701.108

Gai, Y. P., Yuan, S. S., Zhao, Y. N., Zhao, H. N., Zhang, H. L., & Ji, X. L. (2018). A novel LncRNA, MuLnc1, associated with environmental stress in mulberry (*Morus multicaulis*). *Frontiers in Plant Science, 9*, 669. https://doi.org/10.3389/fpls.2018.00669

Gao, P., Bai, X., Yang, L., Lv, D., Pan, X., Li, Y., Cai, H., Ji, W., Chen, Q., & Zhu, Y. (2011). osa-MIR393: A salinity- and alkaline stress-related microRNA gene. *Molecular Biology Reports, 38*, 237–242. https://doi.org/10.1007/s11033-010-0100-8

Gao, Z., Li, J., Luo, M., Li, H., Chen, Q., Wang, L., Song, S., Zhao, L., Xu, W., Zhang, C., Wang, S., & Ma, C. (2019). Characterization and cloning of grape circular RNAs identified the cold resistance-related *Vv-circATS1, Plant Physiology, 180*(2), 966–985. https://doi.org/10.1104/pp.18.01331

Gao, Z., Su, Y., Wang, Y., Li, Y., Wu, Y., Sun, X., Yao, Y., Ma, C., Li, J., & Du, Y. (2024). The antisense CircRNA VvcircABH controls salt tolerance and the brassinosteroid signaling response by suppressing cognate mRNA splicing in grape. *New Phytologist, 245*. https://doi.org/10.1111/nph.20306

Garg, V., Khan, A. W., Kudapa, H., Kale, S. M., Chitikineni, A., Qiwei, S., Sharma, M., Li, C., Zhang, B., Xin, L., Kishor, P. B. K., & Varshney, R. K. (2019). Integrated transcriptome, small RNA and degradome sequencing approaches provide insights into Ascochyta blight resistance in chickpea. *Plant Biotechnology Journal, 17*(5), 914–931. https://doi.org/10.1111/pbi.13026

Guan, Q. M., Lu, X. Y., Zeng, H. T., Zhang, Y. Y., & Zhu, J. H. (2013). Heat stress induction of miR398 triggers a regulatory loop that is critical for thermotolerance in *Arabidopsis*. *The Plant Journal*, *74*, 840–851. https://doi.org/10.1111/tpj.12169

Hammond, S. M., Berstein, E., Beach, D., & Hannon, G. J. (2000). An RNA-directed nuclease mediates post-transcriptional gene silencing in *Drosophila* cells. *Nature*, *404*, 293–296.

Hernández, Y., & Sanan-Mishra, N. (2017). miRNA mediated regulation of NAC transcription factors in plant development and environment stress response. *Plant Gene*, *11*, 190–198. https://doi.org/10.1016/j.plgene.2017.05.013

Hong, Y. H., Meng, J., Zhang, M., & Luan, Y. S. (2020). Identification of tomato circular RNAs responsive to *Phytophthora infestans*. *Gene*, *746*, 144652. https://doi.org/10.1016/j.gene.2020.144652

Jatan, R., & Lata, C. (2019). Role of microRNAs in abiotic and biotic stress resistance in plants. *Proceedings of the Indian National Science Academy*, *85*, 553–567. https://doi.org/10.16943/ptinsa/2019/49586

Jiang, N., Cui, J., Shi, Y., Yang, G., Zhou, X., Hou, X., Meng, J., & Luan, Y. (2019). Tomato lncRNA23468 functions as a competing endogenous RNA to modulate NBS-LRR genes by decoying miR482b in the tomato-*Phytophthora infestans* interaction. *Horticulture Research*, *6*, 28. https://doi.org/10.1038/s41438-018-0096-0

Jones, J. D., & Dangl, J. L. (2006). The plant immune system. *Nature*, *444*, 323–329. https://doi.org/10.1038/nature05286

Katiyar-Agarwal, S., Gao, S., Vivian-Smith, A., & Jin, H. (2007). A novel class of bacteria-induced small RNAs in *Arabidopsis*. *Genes & Development*, *21*(23), 3123–3134. https://doi.org/10.1101/gad.1595107

Kornienko, A. E., Guenzl, P. M., Barlow, D. P., & Pauler, F. M. (2013). Gene regulation by the act of long non-coding RNA transcription. *BMC Biology*, *11*, 59.

Kotak, S., Larkindale, J., Lee, U., Koskull-Doring, P. V., Vierling, E., & Scharf, K. D. (2007). Complexity of the heat stress response in plants. *Current Opinion in Plant Biology*, *10*, 310–316. https://doi.org/10.1016/j.pbi.2007.04.011

Kozomara, A., Birgaoanu, M., & Griffiths-Jones, S. (2019). miRBase: From microRNA sequences to function. *Nucleic Acids Research*, *47*(D1), D155–D162. https://doi.org/10.1093/nar/gky1141

Kruszka, K., Pacak, A., Swida-Barteczka, A., Nuc, P., Alaba, S., Wroblewska, Z., Karlowski, W., Jarmolowski, A., & Szweykowska-Kulinska, Z. (November 2014). Transcriptionally and post-transcriptionally regulated microRNAs in heat stress response in barley. *Journal of Experimental Botany*, *65*(20), 6123–6135. https://doi.org/10.1093/jxb/eru353

Kumar, N., Bharadwaj, C., Sahu, S., Shiv, A., Shrivastava, A. K., Reddy, S. P. P., Soren, K. R., Patil, B. S., Pal, M., Soni, A., Roorkiwal, M., & Varshney, R. K. (2021). Genome-wide identification and functional prediction of salt- stress related long non-coding RNAs (lncRNAs) in chickpea (*Cicer arietinum* L.). *Physiology and Molecular Biology of Plants: An International Journal of Functional Plant Biology*, *27*(11), 2605–2619. https://doi.org/10.1007/s12298-021-01093-0

Kumar, R. (2014). Role of microRNAs in biotic and abiotic stress responses in crop plants. *Applied Biochemistry and Biotechnology*, *174*(1), 93–115. https://doi.org/10.1007/s12010-014-0914-2

Kumar, R. R., Pathak, H., Sharma, S. K., Kale, Y. K., Nirjal, M. K., & Singh, G. P. (2014). Novel and conserved heat-responsive microRNAs in wheat (*Triticum aestivum* L.). *Functional & Integrative Genomics*, *15*, 1–26. https://doi.org/10.1007/s10142-014-0421-0

Lagos-Quintana, M., Rauhut, R., Lendeckel, W., & Tuschl, T. (2001). Identification of novel genes coding for small expressed RNAs. *Science*, *294*, 853–858. https://doi.org/10.1126/science.1064921

Law, J. A., & Jacobsen, S. E. (2010). Establishing, maintaining and modifying DNA methylation patterns in plants and animals. *Nature Reviews Genetics*, *11*, 1–17.

Li, S., Yu, X., Lei, N., Cheng, Z., Zhao, P., He, Y., Wang, W., & Peng, M. (2017). Genome-wide identification and functional prediction of cold and/or drought-responsive lncRNAs in cassava. *Scientific Reports*, *7*, 45981. https://doi.org/10.1038/srep45981

Li, Y. F., Zheng, Y., Addo-Quaye, C., Zhang, L., Saini, A., Jagadeeswaran, G., Axtell, M. J., Zhang, W., & Sunkar, R. (2010). Transcriptome-wide identification of microRNA targets in rice. *The Plant Journal: For Cell and Molecular Biology*, *62*(5), 742–759. https://doi.org/10.1111/j.1365-313X.2010.04187.x

Liu, P., Zhu, Y., Liu, H., Liang, Z., Zhang, M., Zou, C., Yuan, G., Gao, S., Pan, G., Shen, Y., & Ma, L. (2022). A combination of a genome-wide association study and a transcriptome analysis reveals circRNAs as new regulators involved in the response to salt stress in maize. *International Journal of Molecular Sciences*, *23*(17), 9755. https://doi.org/10.3390/ijms23179755

Lu, Y., Yao, K., Gong, Z., Zhang, Y., Meng, Y., & Liu, Q. (2022). Molecular manipulations of miR398 increase rice grain yield under different conditions. *Frontiers in Plant Science*, *13*, 1037604. https://doi.org/10.3389/fpls.2022.1037604

Luo, C., He, B., Shi, P., Xi, J., Gui, H., Pang, B., Cheng, J., Hu, F., Chen, X., & Lv, Y. (2022). Transcriptome dynamics uncovers long non-coding RNAs response to salinity stress in *Chenopodium quinoa*. *Frontiers in Plant Science*, *13*, 988845. https://doi.org/10.3389/fpls.2022.988845

Lyu, Q. R., Zhang, S., Zhang, Z., & Tang, Z. (2023). Functional knockout of long non-coding RNAs with genome editing. *Frontiers in Genetics*, *14*, 1242129. https://doi.org/10.3389/fgene.2023.1242129

Ma, L., Bajic, V. B., & Zhang, Z. (2013). On the classification of long non-coding RNAs. *RNA Biology*, *10*, 924–933.

Ma, L., Liu, X., Lv, W., & Yang, Y. (2022). Molecular mechanisms of plant responses to salt stress. *Frontiers in Plant Science*, *13*, 934877. https://doi.org/10.3389/fpls.2022.934877

Mohsenifard, E., Ghabooli, M., Mehri, N., & Bakhshi, B. (2017). Regulation of miR159 and miR396 mediated by Piriformospora indica confer drought tolerance in rice. *Journal of Plant Molecular Breeding*, 5. 10.22058/jpmb.2017.60864.1129.

Muhammad, A., Margaret, Y. G., & Ken, W. (2017). An insight into microRNA-156 role in salinity stress responses of alfalfa. *Frontiers in Plant Science*, *8*, 356–359. https://doi.org/10.3389/fpls.2017.00356

Naya, L., Paul, S., Valdés-López, O., Mendoza-Soto, A. B., Nova-Franco, B., Sosa-Valencia, G., Reyes, J. L., & Hernández, G. (2014). Regulation of copper homeostasis and biotic interactions by microRNA 398b in common bean. *PLoS ONE*, *9*, e84416. https://doi.org/10.1371/journal.pone.0084416

Niu, D., Lii, Y. E., Chellappan, P., Lei, L., Peralta, K., Jiang, C., Guo, J., Coaker, G., & Jin, H. (2016). miRNA863-3p sequentially targets negative immune regulator ARLPKs and positive regulator SERRATE upon bacterial infection. *Nature Communications*, *7*, 11324. https://doi.org/10.1038/ncomms11324

Nykiel, M., Gietler, M., Fidler, J., Prabucka, B., Rybarczyk-Płońska, A., Graska, J., Boguszewska-Mańkowska, D., Muszyńska, E., Morkunas, I., & Labudda, M. (2022). Signal transduction in cereal plants struggling with environmental stresses: From perception to response. *Plants*, *11*(8), 1009. https://doi.org/10.3390/plants11081009

Pang, J., Zhang, X., Ma, X., & Zhao, J. (2019). Spatio-temporal transcriptional dynamics of maize long non-coding RNAs responsive to drought stress. *Genes*, *10*, 138. https://doi.org/10.3390/genes10020138

Peng, Y., van Wersch, R., & Zhang, Y. (2018). Convergent and divergent signaling in PAMP-triggered immunity and effector-triggered immunity. *Molecular Plant-Microbe Interactions*, *31*, 403–409. https://doi.org/10.1094/MPMI-06-17-0145-CR

Qi, X., Xie, S., Liu, Y., Yi, F., & Yu, J. (2013). Genome-wide annotation of genes and noncoding RNAs of foxtail millet in response to simulated drought stress by deep sequencing. *Plant Molecular Biology*, *83*, 459–473. https://doi.org/10.1007/s11103-013-0104-6

Qin, T., Zhao, H., Cui, P., Albesher, N., & Xionga, L. (2017). A nucleus-localized long non-coding RNA enhances drought and salt stress tolerance. *Plant Physiology*, *175*, 1321–1336. https://doi.org/10.1104/pp.17.00574

Redillas, M. C. F. R., Jeong, J. S., Kim, Y. S., Jung, H., Bang, S. W., Choi, Y. D., Ha, S.-H., Reuzeau, C., & Kim, J.-K. (2012). The overexpression of *OsNAC9* alters the root architecture of rice plants enhancing drought resistance and grain yield under field conditions. *Plant Biotechnology Journal*, *10*, 792–805. https://doi.org/10.1111/j.1467-7652.2012.00697.x

Rinn, J. L., & Chang, H. Y. (2012). Genome regulation by long noncoding RNAs. *Annual Review of Biochemistry 81*, 145–66. doi: 10.1146/annurev-biochem-051410-092902

Rodriguez, A., Griffiths-Jones, S., Ashurst, J. L., & Bradley, A. (2004). Identification of mammalian microRNA host genes and transcription units. *Genome Research*, *14*, 1902–1910.

Röther, S., & Meister, G. (2011). Small RNAs derived from longer non-coding RNAs. *Biochimie*, *93*, 1905–1915.

Salzman, J. (2016). Circular RNA expression: Its potential regulation and function. *Trends in Genetics*, *32*, 309–316.

Sasaki, T., Suzaki, T., Soyano, T., Kojima, M., Sakakibara, H., & Kawaguchi, M. (2014). Shoot-derived cytokinins systemically regulate root nodulation. *Nature Communications*, *5*, 4983. https://doi.org/10.1038/ncomms5983

Šečić, E., Kogel, K.-H., & Ladera-Carmona, M. J. (2021). Biotic stress-associated microRNA families in plants. *Journal of Plant Physiology*, *263*, 153451. https://doi.org/10.1016/j.jplph.2021.153451

Seo Jeong, J., Kim, Y. S., Hun Baek, K., Jung, H., Ha, S.-H., Do Choi, Y., Kim, M., Reuzeau, C., & Kim, J.-K. (2010). Root-specific expression of *OsNAC10* improves drought tolerance and grain yield in rice under field drought conditions, *Plant Physiology*, *153*(1), 185–197. https://doi.org/10.1104/pp.110.154773

Song, X., Li, Y., Cao, X., & Qi, Y. (2019). MicroRNAs and their regulatory roles in plant–environment interactions. *Annual Review of Plant Biology*, *70*(1), 489–525. https://doi.org/10.1146/annurev-arplant-050718-100334

Song, X., Liu, G., Huang, Z., Duan, W., Tan, H., Li, Y., & Hou, X. (2016). Temperature expression patterns of genes and their coexpression with LncRNAs revealed by RNA-Seq in non-heading Chinese cabbage. *BMC Genomics*, *17*, 297. https://doi.org/10.1186/s12864-016-2625-2

Soto-Suarez, M., Baldrich, P., Weigel, D., Rubio-Somoza, I., & San Segundo, B. (2017). The *Arabidopsis* miR396 mediates pathogen-associated molecular pattern-triggered immune responses against fungal pathogens. *Scientific Reports*, *7*, 44898. https://doi.org/10.1038/srep44898

Sun, J., Cardoza, V., Mitchell, D. M., Bright, L., Oldroyd, G., & Harris, J. M. (2006). Crosstalk between jasmonic acid, ethylene and nod factor signaling allows integration of diverse inputs for regulation of nodulation. *The Plant Journal*, *46*, 961–970. https://doi.org/10.1111/j.1365-313X.2006.02751.x

van Zeijl, A., Op den Camp, R. H., Deinum, E. E., Charnikhova, T., Franssen, H., Op den Camp, H. J., Bouwmeester, H., Kohlen, W., Bisseling, T., & Geurts, R. (2015). Rhizobium Lipo-chitooligosaccharide signaling triggers accumulation of cytokinins in *Medicago truncatula* roots. *Molecular Plant*, *8*(8), 1213–1226. https://doi.org/10.1016/j.molp.2015.03.010

Waadt, R., Seller, C. A., Hsu, P. K., Takahashi, Y., Munemasa, S., & Schroeder, J. I. (2022). Plant hormone regulation of abiotic stress responses. *Nature Reviews. Molecular Cell Biology*, *23*(10), 680–694. https://doi.org/10.1038/s41580-022-00479-6

Wang, D., & Farhana, A. (2024). *Biochemistry, RNA structure*. StatPearls Publishing. [Updated 2023 Jul 29]. In: StatPearls [Internet].

Wang, X., Chang, X., Jing, Y., Zhao, J., Fang, Q., Sun, M., Zhang, Y., Li, W., & Li, Y. (2020). Identification and functional prediction of soybean CircRNAs involved in low-temperature responses. *Journal of Plant Physiology*, *250*, 153188. https://doi.org/10.1016/j.jplph.2020.153188

Wang, Y., Tyler, B. M., & Wang, Y. (2019). Defense and counter defense during plant-pathogenic oomycete infection. *Annual Review of Microbiology*, *73*, 667–696. https://doi.org/10.1146/annurev-micro-020518-120022

Xie, F. L., Wang, Q. L., Sun, R. R., & Zhang, B. (2015). Deep sequencing reveals important roles of microRNAs in response to drought and salinity stress in cotton. *Journal of Experimental Botany*, *66*, 789–804. https://doi.org/10.1093/jxb/eru437

Yadav, A., Mathan, J., Dubey, A. K., & Singh, A. (2024). The emerging role of non-coding RNAs (ncRNAs) in plant growth, development, and stress response signaling. *Non-Coding RNA*, *10*(1), 13. https://doi.org/10.3390/ncrna10010013

Yang, X., Liu, Y., Zhang, H., Wang, J., Zinta, G., Xie, S., Zhu, W., & Nie, W. F. (2020). Genome-wide identification of circular RNAs in response to low-temperature stress in tomato leaves. *Frontiers in Genetics*, *11*, 591806. https://doi.org/10.3389/fgene.2020.591806

Ye, C. Y., Chen, L., Liu, C., Zhu, Q. H., & Fan, L. (2015). Widespread noncoding circular RNAs in plants. *New Phytologist*, *208*, 88–95.

Yu, S., Huang, A., Li, J., Gao, L., Feng, Y., Pemberton, E., & Chen, C. (2018). OsNAC45 plays complex roles by mediating POD activity and the expression of development-related genes under various abiotic stresses in rice root. *Plant Growth Regulation*, *84*, 519–531.

Yu, X., Wang, H., Lu, Y., de Ruiter, M., Cariaso, M., Prins, M., van Tunen, A., & He, Y. (2012). Identification of conserved and novel microRNAs that are responsive to heat stress in *Brassica rapa*. *Journal of experimental Botany*, *63*(2), 1025–1038. https://doi.org/10.1093/jxb/err337

Yu, X., Yang, J., Li, X., Liu, X., Sun, C., Wu, F., & He, Y. (2013). Global analysis of cis-natural antisense transcripts and their heat-responsive nat-siRNAs in *Brassica rapa*. *BMC Plant Biology*, *13*, 208. https://doi.org/10.1186/1471-2229-13-208.

Zhang, C., Tang, G., Peng, X., Sun, F., Liu, S., & Xi, Y. (2018).Long non-coding RNAs of switchgrass (*Panicum virgatum* L.) in multiple dehydration stresses. *BMC Plant Biology*, *18*, 1–15. https://doi.org/10.1186/s12870-018-1288-3

Zhang, F., Luo, X., Zhou, Y., & Xie, J. (2016). Genome-wide identification of conserved microRNA and their response to drought stress in Dongxiang wild rice (Oryza rufipogon Griff.) *Biotechnology Letters*, *38*, 711–721. https://doi.org/10.1007/s10529-015-2012-0

Zhang, P., & Dai, M. (2022). CircRNA: A rising star in plant biology. *Journal of Genetics and Genomics = Yi chuan xue bao*, *49*(12), 1081–1092. https://doi.org/10.1016/j.jgg.2022.05.004

Zhang, W., Han, Z., Guo, Q., Liu, Y., Zheng, Y., Fangli, W., & Jin, W. (2014). Identification of maize long non-coding RNAs responsive to drought stress. *PLoS ONE*, *9*, e98958. https://doi.org/10.1371/journal.pone.0098958

Zhang, X., Zou, Z., Gong, P., Zhang, J., Ziaf, K., Li, H., Xiao, F., & Ye, Z. (2011). Over-expression of microRNA169 confers enhanced drought tolerance to tomato. *Biotechnology Letters*, *33*(2), 403–409. https://doi.org/10.1007/s10529-010-0436-0

Zhao, B., Liang, R., Ge, L., Li, W., Xiao, H., Lin, H., Ruan, K., & Jin, Y. (2007). Identification of drought-induced microRNAs in rice. *Biochemical and Biophysical Research Communications*, *354*, 585–590. https://doi.org/10.1016/j.bbrc.2007.01.022

Zhou, J., Xu, J., Zhang, L., Liu, S., Ma, Y., Wen, X., Hao, J., Li, Z., Ni, Y., Li, X., Zhou, F., Li, Q., Wang, F., Wang, X., Si, Y., Zhang, P., Liu, C., Bartolomei, M., Tang, F., Liu, B., Yu, J., & Lan, Y. (2019). Combined single-cell profiling of lncRNAs and functional screening reveals that H19 is pivotal for embryonic hematopoietic stem cell development. *Cell Stem Cell*, *24*(2), 285–298.e5. http://doi.org/10.1016/j.stem.2018.11.023.

Zuo, J., Wang, Q., Zhu, B., Luo, Y., & Gao, L. (2016). Deciphering the roles of circRNAs on chilling injury in tomato. *Biochemical and Biophysical Research Communications*, *479*(2), 132–138. https://doi.org/10.1016/j.bbrc.2016.07.032

12 Abiotic Stress Mitigation in Plants
The Critical Role of ncRNAs

Vibhuti Chandel, Priyanka Aggarwal, Vijay Chauhan and Chandresh Kumari

12.1 INTRODUCTION

Abiotic stressors are the main factors limiting agricultural productivity. These abiotic stresses unbalance the sustainability of agriculture, reduce potent yields for essential agricultural plants by more than 50% and restrict the genetic potential of plants, which results in a significant fall in crop output (Saharan et al., 2022). Abiotic stressors affect plant metabolism, which increases the synthesis of many secondary metabolites such as proline, phenol, polyamines and others. These compounds then either directly or indirectly increase abiotic stress tolerance by activating various stress response mechanisms (Bera et al., 2022). Plants can reduce abiotic stressors with the help of microorganisms' abundant metabolites (Munir et al., 2022). Drought is probably one of the most harmful abiotic conditions for plants and soil organisms. Since water and nutrients are essential for plant growth and survival, drought poses a serious threat to crops (Rai et al., 2023). Plants have evolved a number of stress response mechanisms to cope with stressful circumstances, such as tolerance, recovery and stress avoidance or escape. Plant cells receive stress signals and start a range of downstream processes to balance cellular RNA, proteins, metabolism and reactive oxygen species (ROS). These reactions include the production of different ROS and the buildup of osmotic regulating chemicals (Zhang et al., 2(Zhang et al.,2022) The advantages of next-generation sequencing technology and bioinformatics approaches are responsible for the discovery and description of several ncRNAs in plants, including miRNAs and lncRNAs (Wang et al., 2017). A number of ncRNA types, like microRNAs (miRNAs), small interfering RNAs (siRNAs), long non-coding RNAs (lncRNAs) and circular RNAs (circRNAs), have been identified as important actors in modifying gene expression networks as a result of developments in high-throughput sequencing and functional genomics (Figure 12.1 and Table 12.1 (Santosh et al., 2014).

Recent advancements in high-throughput sequencing technology and bioinformatics tools have fundamentally changed our understanding of ncRNA-mediated regulation in plants (Praveen et al., 2023). This chapter will delve into the diverse classes of ncRNAs, their biogenesis and their specific functions in regulating plant stress responses. We will explore how these molecular players contribute to the fine-tuning of gene expression, ultimately leading to enhanced stress tolerance. The possible uses of ncRNA-based techniques in crop improvement and the creation of stress-resistant plant types will also be covered, as these may become extremely important in handling the problems associated with global food security (Chand Jha et al., 2021). Despite their specific functions being unknown, most ncRNAs likely play a crucial role in chromosome dynamics, gene regulation, mRNA processing, splicing, translation, cell differentiation and development, among many other processes.

Three subgroups of ncRNAs have been identified based on their size (Table 12.1). Short non-coding RNAs (ncRNAs) are defined as those that are approximately 20–50 nucleotides (nt) long [12]. A total of 50–200 nt is the length of an intermediate or medium ncRNA, while >200 nt is the length of a long ncRNA (lncRNA). Short non-coding RNAs (ncRNAs) include transcription-initiating RNA (tiRNA), Piwi-interacting RNA (piRNA), small interfering RNA (siRNA) and microRNA (miRNA). Small nuclear RNAs (snRNAs), small nucleolar RNAs (snoRNAs), transcription start site-associated RNA (TSSARNA), transfer RNA (tRNA), ribosomal RNA (rRNA), promoter-associated small RNAs (PASRs), promoter upstream transcripts (PROMPTs) and small nuclear RNAs (snRNAs) are examples of medium-sized non-coding RNAs. Long-intergenic non-coding RNA (lincRNA), described as ultra-conserved RNAs (T-UCRs), and other long non-coding RNAs are examples of lncRNA. This chapter will discuss the structure, biosynthesis, mechanism of action and function of short- and medium-sized non-coding RNAs.

12.2 TYPES OF ncRNAs AND THEIR BIOGENESIS

12.2.1 MicroRNAs (miRNAs)

miRNAs are small, single-stranded ncRNAs that typically have a length of 20–24 nucleotides. Pre-miRNA, a hairpin-shaped loop structure with a 7-methylguanosine and poly(A) tail, is the result of microRNAs (miRNA) being produced by RNA Pol II. DiGeorge syndrome critical region 8 (DGCR8), sometimes referred to as 'Pasha' in invertebrates, and DROSHA form a microprocessor complex that breaks down this structure in the nucleus. When a debranching enzyme is present, pre-miRNAs can also resist DROSHA cleavage and emerge from the spliced-out introns, thereby controlling gene expression (Kryovrysanaki et al., 2021). When a debranching enzyme is present, pre-miRNAs can also regulate gene expression by passing DROSHA cleavage and emerging from the spliced-out introns (Figure 12.2).

DOI: 10.1201/9781003583257-12

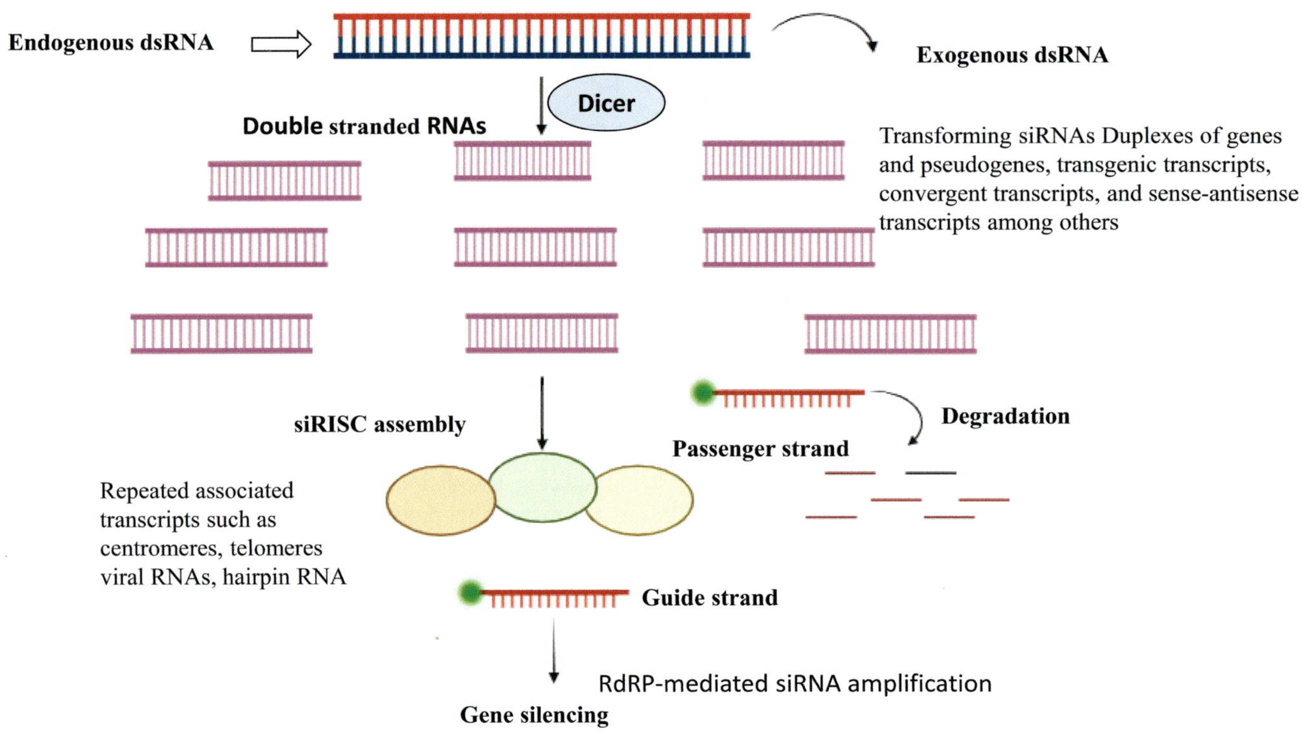

FIGURE 12.1 Classification of non-coding RNAs (ncRNA).

12.2.2 Small Interfering RNAs

The double-stranded RNA molecules known as siRNAs, which are usually 21–24 nucleotides long, are necessary for viral defence and gene silencing. The long double-stranded RNA precursors from which siRNAs are produced can be either endogenous or exogenous, in comparison to miRNAs (Kryovrysanaki et al., 2021). Dicer-like enzymes break down these precursors to create siRNAs, which join to Argonaute proteins to form the RISC complex (Figure 12.3).

The passenger strand created by the newly added RISC component is further broken down by the siRNA cleaver. The target mRNA is then directed by the guide strand to the active RISC and secondary siRNAs.

12.2.3 Long Non-Coding RNAs

Transcripts that are longer than 200 nucleotides and do not encode proteins are known as lncRNAs (Li et al., 2023). They are transcribed by RNA polymerase II and, like mRNAs, are susceptible to post-transcriptional changes such as polyadenylation, splicing and 5' capping. lncRNAs have a variety of modes of activity and can operate in the cytoplasm and nucleus. (Bhar & Roy, 2023).

12.3 ROLE OF ncRNAs IN ABIOTIC STRESS TOLERANCE

One of the most important abiotic factors impacting plant development and production globally is drought (Li et al.,

TABLE 12.1
Classification of ncRNAs According to Length

Name	Size (nt)	Functions	Examples
miRNAs	21–25	Controlling translation	miR-125a, miR LET7A1 and piRNAs, 24–33 DNA methylation, genome stability and transposon repression RASGRF1 and LI are the targets
tiRNAs	17–18	Regulation of transcription (unknown)	Related to the CAP1 genes
siRNAs	21–25	Protection against RNA interference and viral infection	–
piRNAs	24–33	Transposon suppression and genome stabilisation	Targeting the LINE, IAP and RASGRF1 components
snRNAs	70–350	Splicing and processing of mRNA	Structure-related RNAs U1, U2, U4, U5 and U6
PROMPTs	200	Activation of transcription	Linked to the genes rbm39 and EXT1
snoRNAs	70–300	Processing and modification of rRNA	SNORD
PASRs	22–200	Unknown	Protein-coding genes

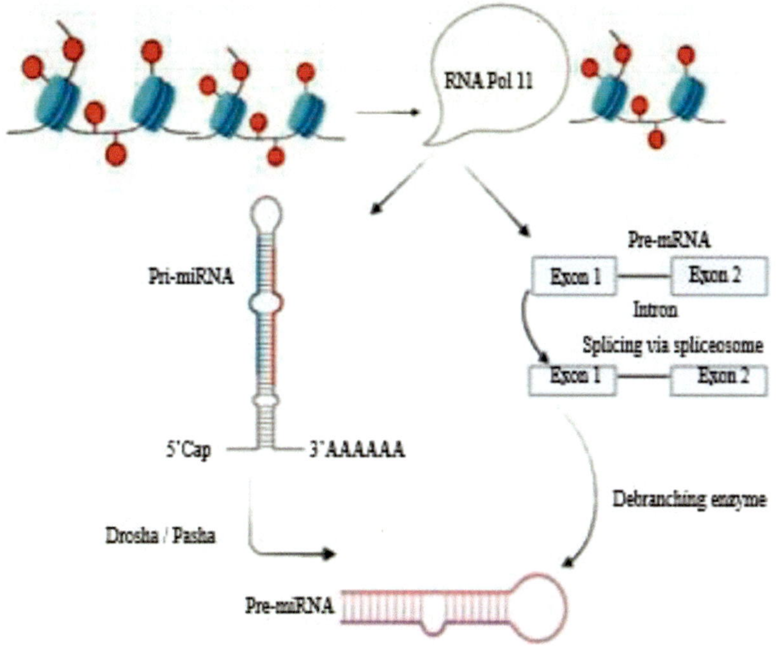

FIGURE 12.2 Silencing mechanism and miRNA biogenesis pathway (Tyagi et al., 2019).

2023). ncRNAs are necessary for controlling how plants react to water deficiency conditions. For example, it has been shown that miRNAs influence how different plant species react to drought stress. Salinity in the soil is another important abiotic stress that impacts plant yield and growth. miRNAs have been related to how different plant species react to salt stress (Islam et al., 2024). In tomatoes, for instance, miRNAs have been shown to be important in controlling reactions to salt stress (Bhar & Roy, 2023).

Plant growth and development are greatly impacted by both heat and cold stress. It has been shown that ncRNAs affect how plants react to severe temperatures. Heat or cold can activate miRNAs, showing their important functional role in temperature stress tolerance. These miRNAs protect the plant reproductive system, reduce ROS damage and activate heat shock proteins, transcription factors and phytohormones under temperature stress (Islam et al., 2024).

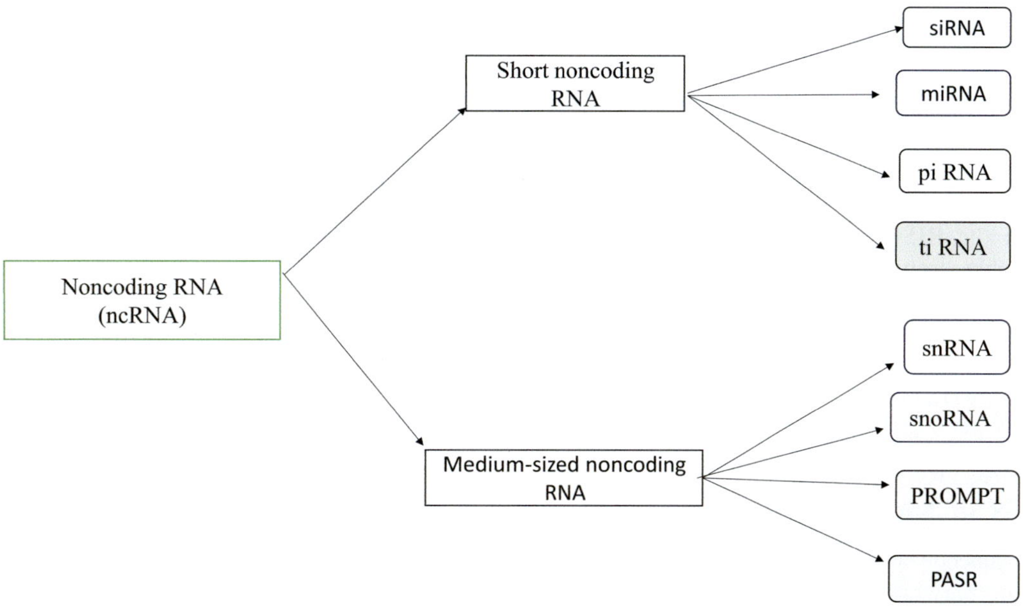

FIGURE 12.3 Double-stranded RNA, which can be either synthetically introduced or transcribed mRNA that is cleaved by the enzyme, is cut into slices, making it inactive. Another way that siRNAs can function is by unwinding and forming primers for RNA-dependent RNA polymers via Dicer, which is loaded into the RISC. The com-ases (RdRp) AGO2 is used to amplify the dsRNA.

Heavy metal contamination in soils poses a significant threat to plant growth and food safety. ncRNAs have been implicated in regulating plant responses to heavy metal stress. For instance, miRNAs in various plant species have been found to be involved in heavy metal stress responses, regulating genes crucial for metal uptake, transport and detoxification (Li et al., 2023).

12.4 ncRNAs' ROLE IN ABIOTIC STRESS TOLERANCE

Plants employ various defence mechanisms against bacterial pathogens, and ncRNAs play crucial roles in these responses (Kryovrysanaki et al., 2021). miRNAs have been found to regulate genes involved in plant immunity against bacterial pathogens (Bhar & Roy, 2023). lncRNAs have also been implicated in plant immunity against bacterial pathogens, often by interacting with other regulatory elements to promote the expression of defence-related genes (Kryovrysanaki et al., 2021). Plant responses to fungal diseases are significantly regulated by ncRNAs. It has been shown that miRNAs play a role in tomatoes' defence against a variety of fungal infections. It has been shown that miRNAs play a role in tomatoes' defence against a variety of fungal infections. Fungal disease resistance has also been related to circRNAs. Certain circRNAs, for instance, function as miRNA sponges in tomatoes, controlling the expression of genes related to resistance to fungal infections (Kryovrysanaki et al., 2021). Plants have evolved sophisticated RNA silencing mechanisms to combat viral infections, and ncRNAs play central roles in this process. siRNAs derived from viral genomes or double-stranded RNA intermediates guide the silencing of viral genes through the RNA-induced silencing complex (RISC). This mechanism serves as a primary antiviral defence in plants (Kryovrysanaki et al., 2021). In addition to siRNAs, other classes of ncRNAs contribute to antiviral immunity (Bhar & Roy, 2023).

ncRNAs are essential to the complex RNA silencing mechanisms that plants have developed to fight viral infections. Through the RISC, siRNAs produced from viral genomes or double-stranded RNA intermediates direct the silencing of viral genes. This system is the main defence against viruses in plants (Kryovrysanaki et al., 2021). Other kinds of ncRNAs, including siRNAs, also play a role in antiviral immunity (Li et al., 2023).

12.5 INTERACTION OF ncRNAs WITH STRESS SIGNALLING PATHWAYS

ncRNAs often function as integral components of complex stress signalling networks, interacting with various hormones and transcription factors to modulate plant stress responses. For example, the crosstalk between miRNAs and phytohormone signalling has been well-documented in both abiotic and biotic stress responses. miRNAs have the ability to control hormone biosynthesis-related genes, transport and signalling, thereby fine-tuning the plant's stress response (Bhar & Roy, 2023). lncRNAs also participate in hormone-mediated stress responses. The control of transcription factors or other regulatory factors in the stress response network is frequently a part of this interaction. Furthermore, ncRNAs can act as integrators of multiple stress signals (Islam et al., 2024). Abiotic and biotic stress responses are shown by certain ncRNAs, indicating their function in coordinating the plant's reaction to various environmental difficulties. This cross-regulation allows plants to fine-tune their responses to multiple, often concurrent, stresses in natural environments (Li et al., 2023).

12.6 ENGINEERING STRESS-TOLERANT PLANTS USING ncRNA-BASED APPROACHES

The growing understanding of ncRNA functions in plant stress responses has opened up new avenues for crop improvement. Several strategies have been employed to harness the potential of ncRNAs for developing stress-tolerant plants. Transgenic plants overexpressing stress-responsive miRNAs or lncRNAs have shown enhanced tolerance to various stresses. This method has been effective in increasing the tolerance of various crop species to temperature, salt and drought stress (Li et al., 2023). Silencing or downregulation of ncRNAs that negatively regulate stress tolerance genes can enhance plant stress resistance (Li et al., 2023). Several plant species have used this strategy to increase their resistance to a range of both abiotic and biotic stresses (Bhar & Roy, 2023). MiRNAs are molecules that are specifically designed to target genes and can be used to fine-tune stress responses (Bhar & Roy, 2023). This approach has been employed to enhance stress tolerance in various plant species by targeting stress-sensitive genes (Li et al., 2023).Genome-editing techniques can be used to modify ncRNA sequences or their regulatory elements, potentially altering their expression or target specificity. This approach offers precise manipulation of ncRNA function in stress responses (Islam et al., 2024). Synthetic RNA molecules designed to bind specific targets can be used to modulate stress signalling pathways or sequester stress-induced harmful molecules (Bhar & Roy, 2023).

12.7 CONCLUSION

Our knowledge of gene regulation and plant stress responses has completely changed as a result of the discovery of ncRNAs. These versatile substances are essential for regulating how plants react to biotic and abiotic stresses. The development of stress-tolerant crops that can flourish in harsh conditions is becoming more feasible as we continue to decipher the intricate networks of ncRNA-mediated regulation. Clarifying the exact mechanisms of ncRNA activity, investigating their functions in numerous plant species and under diverse combinations of stress and creating novel approaches to capitalise on their potential for crop improvement should be the main goals of future study. We may endeavour to develop robust and sustainable agricultural systems that can satisfy the rising global demand for food in the face of climate change and environmental issues by combining ncRNA-based techniques with other biotechnological technologies.

REFERENCES

Bera, K., Dutta, P., & Sadhukhan, S. (2022, Apr 1). Plant responses under abiotic stress and mitigation options towards agricultural sustainability. In *Plant stress: Challenges and management in the new decade* (pp. 3–28). Springer International Publishing.

Bhar, A., & Roy, A. (2023, Nov 23). Emphasizing the role of long non-coding RNAs (lncRNA), circular RNA (circRNA), and micropeptides (miPs) in plant biotic stress tolerance. *Plants*, *12*(23), 3951.

Chand Jha, U., Nayyar, H., Mantri, N., & Siddique, K. H. (2021, Jul 2). Non-coding RNAs in legumes: Their emerging roles in regulating biotic/abiotic stress responses and plant growth and development. *Cells*, *10*(7), 1674.

Islam, W., Adnan, M., Alomran, M. M., Qasim, M., Waheed, A., Alshaharni, M. O., & Zeng, F. (2024, Jan). Plant responses to temperature stress modulated by microRNAs. *Physiologia Plantarum*, *176*(1), e14126.

Kryovrysanaki, N., James, A., Tselika, M., Bardani, E., & Kalantidis, K. (2021, Nov 22). RNA silencing pathways in plant development and defense. *The International Journal of Developmental Biology*, *66*(1-2-3), 163–175.

Li, Q., Shen, H., Yuan, S., Dai, X., & Yang, C. (2023, Jan 11). miRNAs and lncRNAs in tomato: Roles in biotic and abiotic stress responses. *Frontiers in Plant Science*, *13*, 1094459.

Munir, N., Hanif, M., Abideen, Z., Sohail, M., El-Keblawy, A., Radicetti, E., Mancinelli, R., & Haider, G. (2022, Aug 30). Mechanisms and strategies of plant microbiome interactions to mitigate abiotic stresses. *Agronomy*, *12*(9), 2069.

Praveen, A., Dubey, S., Singh, S., & Sharma, V. K. (2023, Mar). Abiotic stress tolerance in plants: A fascinating action of defense mechanisms. *3 Biotech*, *13*(3), 102.

Rai, G. K., Khanday, D. M., Kumar, P., Magotra, I., Choudhary, S. M., Kosser, R., Kalunke, R., Giordano, M., Corrado, G., Rouphael, Y., & Pandey, S. (2023, Jun 14). Enhancing crop resilience to drought stress through CRISPR-Cas9 genome editing. *Plants*, *12*(12), 2306.

Saharan, B. S., Brar, B., Duhan, J. S., Kumar, R., Marwaha, S., Rajput, V. D., & Minkina, T. (2022, Oct 19). Molecular and physiological mechanisms to mitigate abiotic stress conditions in plants. *Life*, *12*(10), 1634.

Santosh, B., Varshney, A., & Yadava, P. K. (2015). Non-coding RNAs: Biological functions and applications. *Cell Biochemistry and Function*, *33*(1), 14–22.

Wang, J., Meng, X., Dobrovolskaya, O. B., Orlov, Y. L., & Chen, M. (2017, Oct). Non-coding RNAs and their roles in stress response in plants. *Genomics, Proteomics and Bioinformatics*, *15*(5), 301–12.

Zhang, Y., Xu, J., Li, R., Ge, Y., Li, Y., & Li., R. (2023). Plants' response to abiotic stress: Mechanisms and strategies. *International Journal of Molecular Sciences*, *24*(13), 10915.

13 Unlocking Plant Potential
The Role of ncRNAs in Abiotic Stress Management

Muhammad Anwar, Muhammad Asif Saleem, Fahad Hassan, Muhammad Farjad Ateeq, Muhammad Nauman and Muhammad Shoaib

13.1 INTRODUCTION

Francis Crick's 1958 discovery of the chain of actions in the passage of genomic information enclosed in the DNA to the functional universe of biological procedures through proteins, known as the central dogma, was the primary turning point in the molecular field. But as modern technologies along with reliable next-generation sequencing have emerged, massive global confederations like the Encyclopaedia of DNA Elements (ENCODE) and the Functional Annotation of the Mammalian Genome (FANTOM) have reported widespread transcription (that is, 80% of DNA transcribes into RNA, but only 1.5% is translated into protein) (Carninci et al., 2005; Hangauer et al., 2013; Katayama et al., 2005; Okazaki et al., 2002).

As a result, the RNA domain in cells is really divided into two different forms: a much shorter fragment of protein-coding RNA and non-coding RNA (ncRNA), which was formerly considered to be transformatory garbage as it is deficient in the capability to code proteins. According to the RNA world hypothesis, RNA was the first material to support life. Later, as DNA became more stable, RNA's function was replaced, leaving RNA to function as a messenger. However, it was eventually discovered that RNA is not only capable of storing genetic information but also possesses a wide range of diverse, albeit dormant, catalytic functions. As a result, despite its dethronement, RNA is the primarily valued candidate in epigenetics, diseases, and unidentified regulatory characteristics. It is thought that during evolution, RNA also changes alongside DNA and proteins (Robertson & Joyce, 2012).

At the moment, it is difficult to distinguish between junk RNA and functioning ncRNAs. Furthermore, current literature reveals that eukaryotes contain a much smaller proportion of protein-coding genes as compared to prokaryotes, which have 80–95% of their genomes coding proteins and only a small amount of ncRNA (varying even among closely related strains). This proteome's amount further declines in the organisms as the complexity increases in the traits, physiology, and development, for example from lower non-chordates to humans. This is because there are many intergenic sequences and introns, which are transcribed but translated equally without regard for differences in their sequence, although through regulatory variations, involving mainly splice variants (Frith et al., 2005; Mattick, 2001).

Cutting-edge technologies and other high-throughput sequencing have propelled the wave of ncRNA discovery, uncovering a vast ocean of molecules that go well beyond protein-coding genes (Wu et al., 2020). Among this varied cast, ncRNAs have taken centre stage as essential modulators of health and illness (Keshavarz Alikhani et al., 2022) ranging from neurological problems to malignancies that are linked to their dysregulation (Giordo et al., 2022; Zheng et al., 2022; Zhou et al., 2022). However, comprehending the intricate dances they conduct inside cells is necessary to fully realise their potential as therapeutic targets or diagnostic tools; this is an issue that has enormous potential for personalised medicine (Ceder, 2016; Van der Hauwaert et al., 2019).

13.2 TYPES OF ncRNA

13.2.1 MICRORNAs (MIRNAs)

It has been 30 years since the first comprehensions into microRNAs (miRNAs) in *Caenorhabditis elegans* development were gleaned from curiosity-driven research using this model organism (Lee et al., 1993), and even before the miRNAs had been identified in this species, in *Drosophila* regulatory and targeting principles of miRNAs were found through molecular genetic studies (Lai et al., 1998). In higher eukaryotes, miRNAs were identified as a large class of small RNAs (sRNAs) in 2001 (Lee & Ambros, 2001); this triggered new enormous research into the regulatory paradigm in biological sciences. In the subsequent decade, there were major strings of conserved molecular factors, mechanisms of crystallisation of miRNA biogenesis and function along with miRNA loci.

miRNAs are RNAs of ~22 nucleotides in length, derived from one or more hairpin-bearing long pri-miRNA transcripts. While non-coding transcripts or introns are the source for most miRNA hairpins, a small fraction overlaps protein-coding exon sequences. Almost all miRNAs generated in canonical ways are identified to reside in a pathway that involves two RNase III enzymes (Bartel, 2018). Next, nuclear microprocessor complex, comprises of one RNase III Drosha associated with its two identical copies accompanying

DOI: 10.1201/9781003583257-13

DGCR8 'crops' the hairpin bottom of a pri-miRNA transcript, yielding ~55–70 nucleotides long pri-miRNA hairpin. At least the most conserved miRNAs (those that are found in vertebrates) are produced using a perfectly monitored pathway indicated by the two RNase III proteins (Bartel, 2018).

Initially, the base region of the pri-miRNA hairpin is 'trimmed' by the nuclear microprocessor complex composed of a Drosha RNase III molecule and two DGCR8 proteins and yields a middle pre-miRNA hairpin of ~55–70 nucleotides in length. Taken out of nucleus into the cytoplasm is the pre-miRNA cleaved by the RNase III Dicer between the terminal loop to generate a miRNA duplex which binds with a member of Argonaute (Ago) protein family. After ablation of the microRNA, often called the passenger strand due to the strand selection process that depends on the nucleotide sequences and the overall architecture of the miRNA, the microRNA duplex, the single-stranded mature miRNA, directs the Ago complex for the repression of the target RNA species (Czech & Hannon, 2011). In addition, other non-canonical miRNA pathways also exist that are Drosha or Dicer independent (Yang & Lai, 2011). It is such knowledge which efficiently bypasses both RNase III enzymes to produce functional microRNAs by facilitating the design of synthetic substrates (Maurin et al., 2012; Shang et al., 2015).

The fundamental mechanisms of miRNA biogenesis and their functional roles were investigated using conventional methods such as cell biology, molecular genetics, and biochemistry together with state-of-the-art techniques, including deep sequencing and X-ray crystallography. These technological advancements and experimental methodologies such as cryo-electron microscopy (cryo-EM), high-throughput substrate assays, CRISPR screening, and single-molecule techniques have facilitated these radical approaches to understand the complexity of miRNA pathways (Shang et al., 2023).

13.2.2 Long non-Coding RNAs (lncRNAs)

Genomes are highly transcribed, leading to the synthesis of tens of thousands of long non-coding RNA (lncRNA) molecules, ncRNAs longer than 200 nucleotides are not transcribed into proteins. This broad definition encompasses a large and diverse population of transcripts that are driven from different biogenesis and genomic backgrounds. Human GENCODE statistics revealed that more than 16,000 lncRNA genes are present in human genome, while other lncRNA counts in humans are in tens of thousands (Fang et al., 2018; Uszczynska-Ratajczak et al., 2018).

These include lncRNAs by RNA polymerase II (Pol II) or other polymerases, or intergenic lncRNAs (lincRNAs), or sense or antisense lncRNA to other genes. At times these lncRNAs have 5′ capped with 7-methyl guanosine (m7G), are polyadenylated in 3′ end and undergo splicing similar to the mRNA. It was also observed that like eRNAs, enhancer and promoter regions are also transcribed (Wu et al., 2017).

There remains controversy regarding the number of functional lncRNAs. While most lncRNAs are currently untested for their functionality, leaving them as by-products of transcription, a large quantity of lncRNAs have been exposed to have functional characters in the cell. Many lncRNAs are expressed, and some function in key gene regulation mechanisms, but their expression is regulated. There are several lncRNAs that regulate the neighbouring genes' expression by changing their transcription, while also influencing other aspects of chromatin biology like DNA damage and repair response or DNA replication. Other lncRNAs are located distant from their site of transcription, and they have structural or regulatory functions related to mRNA maturation, stability and translation in addition to signalling pathways. Thus, lncRNAs are involved in a number of cellular processes that concern large issues of physiology and are implicated in many diseases. Strategies are being developed to target these functional lncRNAs as therapeutic targets with the potential-specific expression patterns of these functional lncRNAs as optimal disease biomarkers (Statello et al., 2021).

The vast majority of lncRNAs are even less conserved across species than the mRNA sequence of the proteome encoding. Majority of the mammalian lncRNA loci were initially assumed to have neutrally evolved (Pheasant & Mattick, 2007) based on the yardstick rate of divergence between human and mouse genomes of common 'ancient repeats', assuming these sequences are functional and maintain their unique distribution in the common ancestor of mammals. As data accumulate that transposable elements are commonly used as functional part of gene expression and structure for example as regulatory networks, promoters, exons and splice sites in protein-coding genes and lncRNAs (Faulkner et al., 2009; Fueyo et al., 2022; Kapusta et al., 2013; Kelley & Rinn, 2012) – they cannot be used as indicators of neutral evolution.

Like miRNAs, most lncRNAs have not been discovered in genetic screens, and their functions remain largely elusive. And there are two reasons for this: firstly, most of the previous genetic screening for mutations have been concerned with protein-coding mutations which more often have severe consequences that can be observed in the cases, while the regulatory mutations may have less dramatic impacts on quantitative traits. Secondly, causal mutations are often not distinguishable from hundreds of thousands of variants occurring in non-coding regions. Furthermore, most of these variations that are associated with human quantitative traits and complex disorders occur in non-coding regions that host several genes encoding lncRNAs (Bartonicek et al., 2017; de Goede et al., 2021).

13.2.3 Small RNAs (sRNAs)

sRNAs are endogenous small ncRNA molecules containing 18–20 nt which participate in gene expression regulation and hereditary stability. miRNAs have been categorised according to the total differences in plant sRNA biogenesis pathway and distinct functional styles and have been divided into two major groups, namely miRNAs and small interfering RNAs (siRNAs). The Dicer-like (DCL) proteins are involved in processing of miRNA and siRNA from miRNA gene (MIR) and dsRNA, respectively (Borges & Martienssen, 2015;, Ghildiyal & Zamore, 2009). With regard to siRNA in plants, the DCL2, 3 and 4 enzymes function to process a variety of endogenous and exogenous. For instance, we are learning about new forms of plant siRNA: heterochromatic siRNAs (hc-siRNAs), natural antisense siRNAs (nat-siRNAs), and phased secondary siRNAs (phasiRNAs) (Baulcombe, 2004; Bologna & Voinnet, 2014; Borges & Martienssen, 2015; Castel & Martienssen, 2013; Fei et al., 2013; Ghildiyal & Zamore, 2009).

Since it was reported that dsRNAs can cause gene silencing in *C. elegans* (Kamath et al.;2000), sRNAs have emerged as crucial factors that protect plant genomes and modulate their development and response to stresses. Basic plant sRNAs are small molecules with a size of less than 25 nt that originate from double-stranded RNA precursors or single-stranded hairpin RNA (Baulcombe, 2004; Fang & Qi, 2016; Song, Li, et al., 2019). NGS techniques significantly expand the possibilities of sRNA identification since it is possible to obtain a large amount of genetic sequences from a single plant sample. In combination with bioinformatics, computation has become an integral tool in the identification and quantification of sRNA and therefore understanding of sRNA world. This has led to improvement in the knowledge of types of sRNA and other intricate functions and relations to these molecules in gene regulation (Ivanova et al., 2022).

Functional roles have also been seen for some classes of sRNAs, although other sRNAs remain poorly characterised or unknown. Many biological processes are known or predicted to involve secondary sRNAs (phasiRNAs and their tasiRNA subsets). Yet, there are a lot of phasiRNA members PHAS loci in different plants, and how their precise practical roles have become unclear overall. These secondary sRNAs are triggered by miRNAs to play roles in many biological processes, including inhibition of tasiRNAs to target pentatricopeptide repeat-containing genes (PPR), plenty of which are related to abiotic stresses. For example, some tasiRNAs are observed to play part in plant thermotolerance through regulating heat stress (HS)-related TF (Li et al., 2014).

Experimental methods are not easy to identify the process of sRNA recognition. Secondly, sRNAs themselves lack the ability to encode proteins, as they are short, and single nucleotide mutations do not inhibit them easily. Furthermore, sRNAs are not translated, so their sequences cannot be gleaned by simple open reading frame pursuit. For this reason, a large number of sRNAs are identified and discovered using the combination of experimental methods and bioinformatics (Mraheil et al., 2010; Rath et al., 2017).

13.3 UNDERSTANDING ABIOTIC STRESS IN PLANTS

Stress can be defined as any external or internal factor that limits the plant's ability to convert energy into biomass, for example, by reducing the photosynthetic rate (Atafar et al., 2010). It can affect plants in different ways, including changes in cellular metabolism, gene expression, crop yields, growth rates, and overall stress responses.

Nevertheless, stress-tolerant plants can be exposed to a particular stress which induces resistance with time to that particular stress (Fahad et al., 2017). Abiotic and biotic stresses are the main types of stresses that plants face. Environmental factors that are placed on plants due to any chemical or physical obstruction constitute abiotic stress, while biotic stress includes diseases, insect pests, and other organisms that attack crop plants (Fich et al., 2016; Gimenez et al., 2018). In plants, some stresses do cause injury. However, these plants have numerous metabolic problems (Umar et al., 2021).

The negative impact of any environmental abiotic factor on a plant's growth and development in a given environment is termed abiotic stress (Zhang et al., 2023). The major detrimental elements affecting immobile plants development and growth are usually abiotic environmental factors (Zhang et al., 2022). In layman's terms, it's the effect (i.e., change) of any abiotic factor in a specific environment, on a plant, and this causes a variety of responses, ranging from the impact on biological processes (e.g., gene activation or cell metabolism) to growth and development (Zhang et al., 2023). Abiotic stress specifically includes drought, extreme temperature stress, metal stress and nutrient stress, salinity, flooding stress, and different stresses could lead to various responses (Zhang et al., 2020).

The primary environmental factors that restrict plant survival and influence their distribution in nature are drought, heat, and salinity. Deficiencies or excesses in abiotic environmental factors, such as water, salt, light, temperature and nutrients, can have drastic effects on plant growth and productivity, and even survival under abiotic stress. Plant growth is affected by stress due to the fact, by definition, that when plants are stressed, environmental conditions are suboptimal for growth processes, such as cell division and expansion. There is, however, not only a passive effect of the adverse, stressful environment, but it is also accompanied by a slower plant growth. Moreover, plants under stress go for slowing down their growth to adjust to the stress conditions (Skirycz et al., 2011; Zhang et al., 2022).

Significant yield losses are seen in different crops around the globe due to more frequent abiotic stress events (Mittler, 2006; Chaudhry and Sidhu, 2022). About two-thirds of the human food on Earth comes from these four crops: wheat, rice, maize and soybean, which are projected to experience yield reduction of 6.0%, 3.2%, 7.4%, and 3.1%, respectively, for each degree Celsius increase in global mean temperature (Zhao et al., 2017). Many would argue that these threats will further intensify under climate change and the stresses of increasing human population.

Recently, the global population reached 8 billion people (United Nations, 2023), and according to the latest UN projection, the world will have more than 8.5 billion people, likely 9.7 billion people by 2050 (United Nations, 2023). Climate extremes have an increasing poleward trend from a global perspective and a growing prevalence of heat extremes over cold ones in terms of temperature variability patterns.

13.3.1 Heat Stress

Global warming has emerged as a major concern and is not only an issue of increased temperature affecting development of plants but also decreased yield of plants inclusive of agricultural crops. HS is a condition whereby plant temperature rises above the optimal temperature that leads to harm dissection or plant growth and development (Kotak et al., 2007).

HS has evolved into the biggest challenge to crop production and therefore food security (Liliane & Charles, 2020). The high frequency of high temperature-mediated HS affects the availability of water acting as a threat to plant survival. Also, the escalation in temperatures is likely to increase and worsen the outbreak of diseases and epidemics affecting plants (Desaint et al., 2021).

Several mechanisms have, therefore, been developed by plants for escaping, avoiding or tolerating HS. These mechanisms include regulation of water balance, reinforcement of the antioxidant status, modulation of hormonal messages, membrane repair and equilibrium of ions. Moreover, the plants suffer from HS, and they have metabolic and genome alterations to acclimatise to the abiotic stress. For instance, chloroplasts respond positively to the occurrence of photosynthesis and temperature sensing, thereby promoting retrograde signalling for appropriate physiological changes (Zahra et al., 2023).

13.3.2 Drought Stress

Drought is defined as a state of affairs where rainfall is either absent or below the average threshold for such a long period of time that it causes a large hydrological disequilibrium. As a direct implication of water scarcity, plants, which are, by nature, stationary organisms, suffer from drought stress. Drought stress is proven to be increasing with a progressive reduction in water availability, while there is considerable increase in the demand for water. Clean water currently consumes 70% in the agricultural sector, and this figure is likely to grow even further in the future (Boretti & Rosa, 2019).

Another factor that affects yield loss is also the developmental stage of the plant that is infected by the drought. Of the different plant development stages, those affected by drought in the reproductive stage are comparatively more affected; hence, more effective plant development calendars will help to enhance yields (Wiegmann et al., 2019). Stress from drought at the reproductive phase reduces gains from filling the grains, the generation of flowers and seeds, the nutritional quality of seeds and plant life spans (Dufková et al., 2023; Farooq et al., 2017). When water shortage occurs during the vegetative period of the plants, some of the signs that are quite evident include diminished plant height, wilting of the leaves and change in the number of leaves as well as the area occupied by these leaves.

13.3.3 Salt Stress

Soil salinity is one of the most damaging abiotic factor that threaten crop yields globally (Sanwal et al., 2022). Almost all crop plants are glycophytes and are affected by salinity stress. For instance, it is now postulated that about half of the area of irrigated land is impaired by salinity (Gong et al., 2020).

Increased concentrations of dissolved salts in (i) water used for irrigation; (ii) nutrient solutions or (iii) soil solutions/extracts cause one of the most widespread abiotic stress disorders in both cultivated and wild plant species, known as 'salt stress', typically observed after a mid-term exposure of one to two weeks (Ondrasek et al., 2022; Song et al., 2017; Zhao et al., 2019).

However, longer durations of salinisation in permanent and non-reversible types of soil degradation such as desertification, crusting, dispersion of soil stable aggregates and structures, and swamping are viable possibilities under the geo-hydro-morphology and climate of a salinised land (Ondrasek et al., 2011, 2022).

13.3.4 Cold Stress

Cold stress is one among the important abiotic stresses that have a major impact on crop output as well as on the geographical location of many plant species (Zhu, 2016). Low-temperature stress harms many aspects of plants such as germination, growth and development rates, and reproduction (Moraes de Freitas et al., 2019). In general, two types of low-temperature stress are chilling and freezing which plants experience frequently. For plants, less means low temperatures, but it also means positive temperatures (0–15°C), though these can also vary with plant resistance and varieties. Cold temperature also has its effects depending on the air temperature and wind speed during the exposure. On the other hand, at freezing temperature (<0°C), the plant faces difficulty in resisting freezing stress (Thomashow, 2010).

With regard to frost tolerance, crops may be categorised into chilling-sensitive, chilling-tolerant or freezing-tolerant crops (Ding, Tie, et al., 2019). In chilling temperature conditions, plants and crops are at a disadvantage all through the vegetative stage, as well as reproductive stages. Chilling injury effects in sensitive plants include chlorosis, stiffness of seedlings, withering and the often fatal emergence of seedlings. Such phenotypes serve as tropism signalling factors for the separation of chilling-sensitive and tolerant varieties (Wei et al., 2021).

In particular, the examination of plant reaction to cold stress suggests that plants have developed complex strategies to sense the stress and respond to it. They try to detect changes in their physical and chemical environment such as the accessibility of water, the concentration of ions and temperature and then turn these changes into a biological signal using primary sensory processes (Lamers et al., 2020). Cold stress initiates a series of cell processes related to membrane changes such as membrane stiffening, lipid profiling, protoplasmic streaming, cytoskeletal changes, intracellular distribution of calcium ions and changes in the photosynthetic machinery and electron transport chain in plants (Tang et al., 2020). Thus, to tolerate cold stress, a plant will form different biomolecules such as proline, soluble sugars, antioxidants and cold affecting proteins, including anti-freezing proteins (AFPs), late embryogenesis protein (LEA) and cold shock protein (CSP) (Saleem et al., 2021).

13.4 ROLE OF ncRNAs IN ABIOTIC STRESS RESPONSES

Climate change has increased the interest of research in crop's response to abiotic and biotic stress factors, most specifically in abiotic stress factor because of their major impact on crop yield (Fracasso et al., 2016; Tan et al., 2020). A major focus of this research is on improving their resistance level in plants. It is studied that plant's responses to abiotic stresses include several types of responses which include missing production of messenger substances like reactive oxygen species, Ca^{2+} or phosphatidylinositol (Huang et al., 2012). Plants in response to these stresses also produced specific proteins which play a significant role

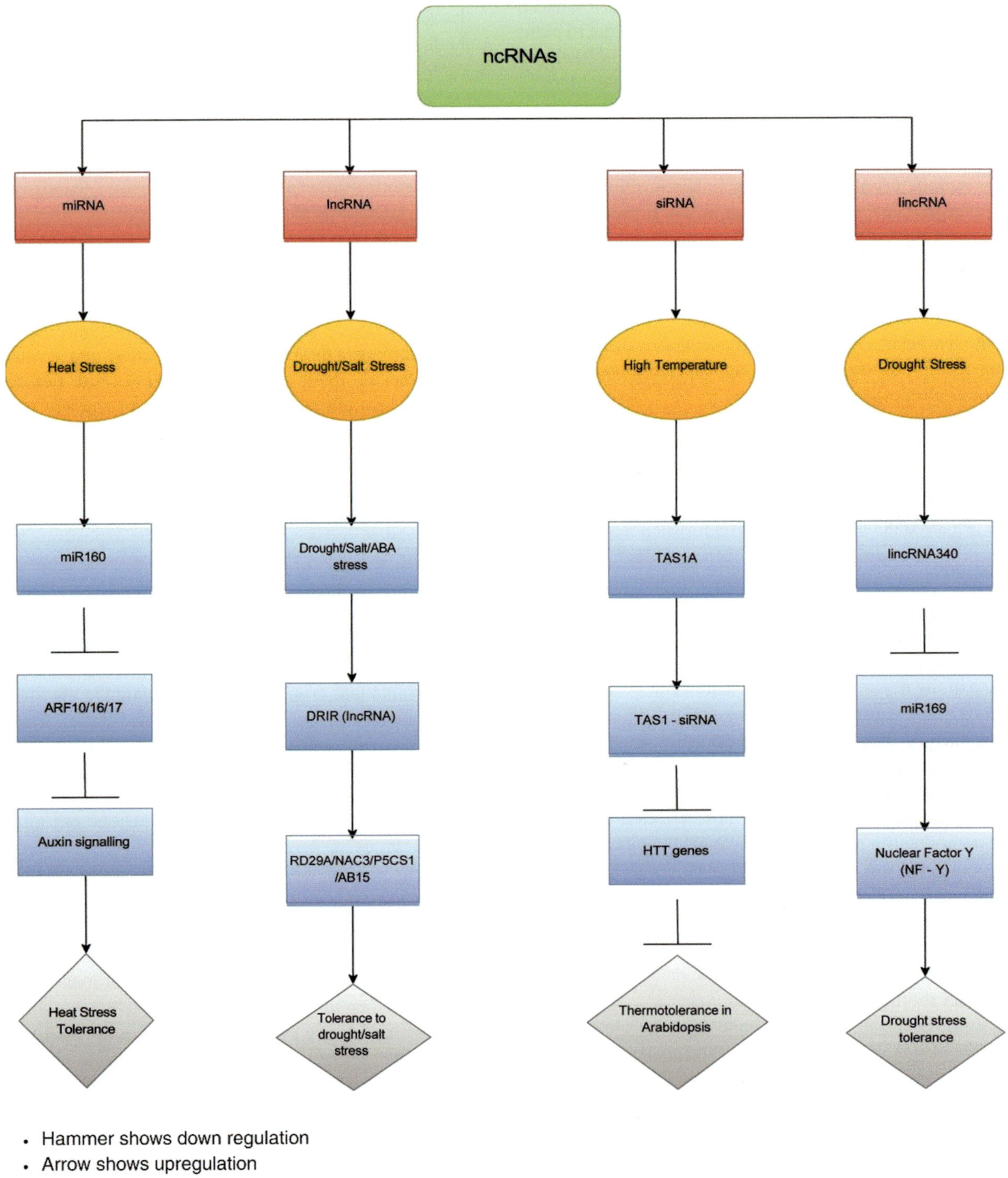

FIGURE 13.1 ncRNA response under abiotic stress.

in cell protection and modulation of other stress-specific genes (Ackah et al., 2024; Farooq et al., 2009). It also indicates that several ncRNAs play crucial roles in resistance development (Ackah et al., 2024; Farooq et al., 2009).

ncRNAs not only have role in plant growth and development but also regulate several metabolic pathways in stress tolerance. Activation of different regulatory RNA during abiotic stresses initiates crucial stress-responsive pathways through transcription (Bhogireddy et al., 2021). However, this response may vary in plant species depending on the nature of the stress. This response usually takes place through regulating targeted genes and their expression level (Bhogireddy et al., 2021; Ferdous et al., 2015; Megha et al., 2018; Song, Li, et al., 2019; Zhao et al., 2016).

In all the ncRNA types, one of the major studied classes is miRNAs, followed by siRNA and lncRNA (Figure 13.1) (Khraiwesh et al., 2012; Song et al., 2019; Yu et al., 2019). Researchers have conducted a detailed characterisation of this gene as regards different abiotic stresses (water deficit, heat, salt, cold, nutrient and pathogen) pointing at signal change. Some of the model crop plants which have been employed are Arabidopsis, rice, barley, wheat, and maize because they convey economic importance and since the genetic models exist (Barciszewska-Pacak et al., 2015; Bhogireddy et al., 2021; Ferdous et al., 2017; He et al., 2019; Hua et al., 2019; Mangrauthia et al., 2017; Sailaja et al., 2014).

13.5 microRNAs (miRNAs) MODULATED ABIOTIC STRESS RESPONSES

Different miRNAs have been mentioned regarding the genic upregulation controlling the diverse previous abiotic stresses (Liu et al., 2008; Zhou et al., 2012). They originate from RNA polymerase II transcribed pri-miRNAs and then upon processing targets mRNAs (Voinnet, 2009). Hence, it is assumed that they are controlled by cis-elements like protein-coding genes (Matsui et al., 2013).

Subsequent research on Arabidopsis has let the scientists identify over cold stress-responsive cis-regulatory elements, including DRE/LTRE-core (A/GCCGAC) (Yamaguchi-Shinozaki & Shinozaki, 1994), ABRE-core (ACGTGG/TC) (Mundy et al., 1990) and W-box (TTGAC) (Shang et al., 2010) in the promoter region of several cold-inducible miRNA genes (Matsui et al., 2013; Zhou et al., 2008).

It is also found that several miRNAs also play a maintenance role during phytohormone signalling in abiotic stress conditions. It is found that AGO1 homeostasis in ABA and abiotic stress responses in Arabidopsis is regulated by a miR168-mediated feedback regulatory loop as MIR168a-overexpressing plants show ABA hypersensitivity and drought tolerance. However, miR168a mutants have shown hypersensitivity during ABA and drought stress (Li et al., 2013; Matsui et al., 2013).

Further studies have revealed that Arabidopsis plants have shown hypersensitivity to ABA during expression of miR160-resistant ARF10 along with the development of abnormal leaf shape. However, the overexpression of miR160 also led the plants to hyposensitivity to ABA (Liu et al., 2008). Along with this several other miRNAs targeting auxin signalling have also been detected during abiotic stresses. For example, the expression of miR393 is seen upregulated during dehydration, salt and cold stresses and ABA because miR393 target gene, TR1, is directly involved in stress response (Sunkar & Zhu, 2004).

Moreover, universal conserved targets like miR160-ARF, miR159-MYB, miR156-SPL, miR167-ARF, miR164-NAC, miR394-LEAF Curling Responsiveness, miR319-TEOSINTE Branched/Cycloidea/Proliferating Cell Factors (TCP), and miR169-Nuclear Transcription Factor-Y (NFY) are known to play a vital regulatory part in various stress conditions to reduce the harmful effects (Bhogireddy et al., 2021).

13.5.1 Drought Stress

Several studies have indicated that not only protein-coding genes but many ncRNAs also play a significant role in stress tolerance (Joshi et al., 2016; Zhao et al., 2016).

miRNAs consist of 18–24 nucleotides and are a well-known type of ncRNAs that control gene expression regulation through messenger RNA degradation or post-transcriptional suppression (Ackah et al., 2024; Pasquinelli, 2012).

It is indicated that during drought conditions transcription factors in phytohormone regulation are directly targeted by miRNAs like auxin signalling pathways are controlled by the miR167-ARF module. Similarly, miR167 was seen upregulated in wheat, Arabidopsis, and maize, while negatively regulated in rice (Bhogireddy et al., 2021; Song, Li, et al., 2019).

Several miRNAs, including miR156, miR158, miR159, miR165, miR167, miR168, miR169, miR171, miR319, miR393, miR394 and miR396, were found regulated during drought stress in Arabidopsis. Upregulation of miR319, miR393, and miR397 in Arabidopsis during dehydration treatment has been reported earlier as well. A microarray platform was employed for the genome-wide analysis of drought stress response in rice seedlings at tillering to flowering stage.

This analysis led to the conclusion of 30 miRNA gene families that are distinctively expressed. These contain 17 miRNA gene families, which include miR156, miR159, miR168, miR170, miR171, miR172, miR319, miR396, miR397, miR408, miR529, miR896, miR1030, miR1035, miR1050, miR1088 and miR1126. In contrast, 14 miRNAs, including miR159, miR169, miR171, miR319, miR395, miR474, miR845, miR851, miR854, miR896, miR901, miR903, miR1026 and miR1125, showed increase in their expression level under drought stress (Gelaw & Sanan-Mishra, 2021).

Further studies have revealed that the decreased expression of miRNAs enhances drought tolerance in plants by regulating the opening of the guard cell. Like in the miR169-NFY module, the downregulated expression of miR169 in Arabidopsis, tomato and *Medicago* showed an increased expression of NF-Y during water-deficit conditions, resulting in enhanced drought tolerance (Megha et al., 2018).

Similarly, several other modules like miR156-SPL, miR160-ARF, miR164-NAC, miR159-MYB33, and miR172-AP2 have shown a similar response on genic regulation during drought stress (Bhogireddy et al., 2021; Song, Li, et al., 2019). Meanwhile, miR156 is known to be a major drought stress-responsive family in maize and rice due to its differential expression of isomiRs (Balyan et al., 2020).

In another report experimental analysis on the rice plant revealed the downregulation of miR1861d, miR319b, and miR164c and upregulation of miR408, miR172d, and miR166h under water stress condition. In rice, only pre-miRNA expression profiling was analysed, and the result showed that miR171f plays a role in root development and growth in rice in addition to drought stress. Another experiment on rice revealed that the miR169g and miR393 was upregulated weakly, but the expression of miR393 was stimulated quickly under drought. When rice plants were treated with *Piriformospora indica*, there was an increase in miR396 and, in turn, the expression of this target gene GRF, which slows down the rate of transpiration and thus facilitates plant tolerance of drought (Bhogireddy et al., 2021; Ma et al., 2022).

In *Vitis vinifera*, 12 exclusive and specie-specific candidate miRNAs were found regulated in drought conditions. Furthermore, 70 miRNAs were found conserved between control and drought-tolerant grapevines, and additional 28 miRNAs were detected in drought-tolerant grapevines only. Maize as a model organism for studying drought stress has provided sufficient information on the way differential regulation of miRNAs occurs. For instance, compared with the controls, levels of miR398 were higher in plants treated with polyethylene glycol and lower during soil water stress. Having integrated this knowledge downregulation of miR167 during drought leads to the upregulation of its target gene PLD (phospholipase D) which is helpful in regulation of stomata and ABA response (Ma et al., 2022).

Likewise upregulation of miR159 during drought resulted in suppression of transcription factors of GA-MYB, ARF, and HD-ZIP, promoting more adventitious as well as lateral roots. In addition, some of them have been proved, including that the expression of miR474 was regulated by drought stress to target proline dehydrogenase (PDH), and miR827 was also induced during drought stress to target NADP-binding and SPX (SYG1/Pho81/XPR) transcripts to prompt activation of drought stress signals (Gelaw & Sanan-Mishra, 2021).

For instance, miR169 has been reported to target several transcripts such as elongation factor 1-alpha (EF 1α) under drought conditions, and miR529, miR535 and miR156 have been involved in regulation of Squamosa promoter binding protein-like proteins transcripts (SPL) for controlling organ morphogenesis and development during stress conditions. Likewise, miR159 inhibits the transcription factor MYB33, and miR172 inhibits AP2 (Apetala 2) to control plant growth under drought stress conditions (Balyan et al., 2020; Gelaw & Sanan-Mishra, 2021).

13.5.2 Temperature Stress

Similarly, as in drought stress, miRNAs have an essential role in heat stress response as well (Mangrauthia et al., 2017; Ravichandran et al., 2019). Several investigations have shown that there is a diverse of miRNAs (miR156-SPL, miR172-AP2 and miR398-CSD) which plays a role in HS response in plants, including Populus, Brassica, and Arabidopsis. The miR398-CSD module is highly conserved across plant species and plays a key role in heat tolerance. When this module is overexpressed, it suppresses the genes CSD1, CSD2, and their copper chaperone (CCD). This suppression helps activate heat shock proteins (HSPs) and heat shock transcription factors (HSFs), which are essential for surviving heat stress. (Bhogireddy et al., 2021; Guan et al., 2013; Lu et al., 2013; Song, Li, et al., 2019; Zhao et al., 2016).

Besides HS, the roles of miRNAs as the key molecular regulators of genetic expression during low temperature stress in plants include miRNA394 and LCR module in which the overexpression of miR394 improves cold tolerance in *Arabidopsis thaliana*. In the same work it has also been pointed out that plants overexpressing miR394 contain higher levels of genes encoding DEHYDRATION-RESPONSIVE ELEMENT-BINDING FACTORS 1 (DREB1) OR C-REPEAT BINDING FACTORS (CBFs), which could aid in raising cold stress tolerance (Bhogireddy et al., 2021; Song et al., 2016).

Arabidopsis was the first plant species used to study plant miRNA regulation patterns under cold conditions (Liu et al., 2008; Tiwari et al., 2020; Zhou et al., 2008). Later, studies have reported the expression pattern of miRNAs under cold stress in several other plant species like poplar (Lu et al., 2008), rice (Barrera-Figueroa et al., 2012), sugarcane (Thiebaut et al., 2012), *Brachypodium* rice (Zhang et al., 2009), tomatoes (Cao et al., 2014), soybeans (Zhang et al., 2014), eggplant (Yang et al., 2017), grapes (Wang, Dai, et al., 2019), wheat (Tang et al., 2012), cotton (Wang et al., 2019) and astragalus (Huo et al., 2022).

However, miRNA expression patterns vary from species to species. For instance, different plants exhibit distinctive patterns of miR397 expression in reaction to cold; grapes have downregulated expression of miR397, while Arabidopsis has increased expression of miR397 (Huo et al., 2022; Sunkar & Zhu, 2004; Wang et al., 2019). Cold-induced miR402 upregulates seed germination in Arabidopsis and enhances seedling growth because it directly controls DNA methylation by targeting DEMETER-LIKE PROTEIN 3 (DML3) gene during cold stress (Huo et al., 2022).

In addition, miR394 and miR165/166 are also being induced in Arabidopsis after cold stress; however, they express differently because of the target genes LCR and homeodomain leucine zipper class-III transcription factor family (e.g., ATHB-8 and PHB). For example, overexpression of m5LCR reduces the freezing resistance and makes miR394 induce cold signals, while PHB induces apical meristem differentiation (Zhou et al., 2008).

Meanwhile, studies have revealed that the expression difference can be seen within same plants, but this pattern may not be too significant. In an experiment only seven miRNAs have been observed overlapped out of 107 differentially expressed (DE) miRNAs over the stress period of two days (Huo et al., 2022; Tiwari et al., 2021), previously reported as ten miRNAs for a stress period of 4°C for 24 hours (Liu et al. 2008). However, this number can vary depending on temperature, duration of cold stress and plant growth status (Huo et al., 2022).

It has also been seen in several plant species that the miRNA expression pattern differs during cold stress according to the growth and development stage of plant due to the presence of diverse cis-elements in their promoter regions (Huo et al., 2022). For example, in wheat L.15 (anther length = 1.5 mm) stage, the expression of taemiR167c observed deregulated, and at L3.0 (anther length = 3 mm) stage, the expression level observed upregulated after cold temperature stress (Tang et al., 2012). Similarly, Arabidopsis has shown upregulated expression levels of miR164 and miR159 after 1 hour of cold stress (Zhou et al., 2008).

Similar phenomenon has been studied in several other plant species like in Dongxiang common wild rice. The miR395d\k\w expression level was found upregulated 3 times, while the miR395e showed downregulated expression up to 12 times after 6 hours of cold stress. The actual signalling pathway is not yet discovered for rice, but in Arabidopsis main target gene for cold-induced miR408 was found to encode the blue copper protein, and it is also believed that there could be more than one target gene for some miRNAs, e.g., miR166/165. Therefore, more studies are needed to explain what actual factors are involved and how miRNAs respond to temperature stress at different developmental stages. (Huo et al., 2022; Yan et al., 2016)

Cold tolerance could also be improved by the overexpression of osa-miR528 in rice, Arabidopsis and pine. The Os06g06050 target gene, osa-miR528, can trigger the regulation of OsMYB30 and can incidentally block its expression. Meanwhile, OsMYB30 is known for being a negative regulator of low temperature stress tolerance since it can combine to the promoter region of the β-amylase gene, inhibiting BMY family gene expression. However, the upregulated expression of osa-miR528 results in reduced genetic regulation of OsMYB30 and enhanced regulation of BMY family genes. These genes help manage starch metabolism and increase the levels of fructose, sucrose and maltose. Therefore, upregulation of BMY gene expression results in maltose accumulation and increased cold resistance in plants (Huo et al., 2022).

13.5.3 Salinity Stress

Salinity has become a significant issue because of the rapid shift of uncultivated land to cultivated land. It affects crop development and growth and poses notable yield losses (Food and Agriculture Organization of the United Nations (FAO), 2015). Therefore, developing plants with salinity tolerance responses has become very important. However, like drought and temperature stress, ncRNAs play a notable role in in plants to salinity stress tolerance (Ismail & Horie, 2017).

The regulatory role of ncRNAs linked to various molecular mechanisms has been identified in plants during salt stress conditions. They can target several genes linked to plant development and growth like photosynthesis, uptake of toxic ions and salinity-responsive hormone signalling (Chen et al., 2019; Jha et al., 2021; Khandal et al., 2017; Sun et al., 2016; Wang et al., 2015).

Furthermore, it's been discovered that conserved miRNAs like the miR319-TCP module have shown significant salinity stress tolerance in bentgrass (*Agrostis stolonifera*) during overexpression of osa-miR319 (Zhou et al., 2013). Meanwhile, root tissue studies on cowpea have found 18 miRNAs (e.g., miR156/157, miR172, miR408, miR160, miR159, miR169) which induce tolerance under salinity stress (Paul et al., 2011). The qRT-PCR-based study further identified that expression levels of seven of them were upregulated (Jha et al., 2021; Long et al., 2015).

Evidence for the expression regulation of miRNAs in soybean through transcriptome analysis has shown that 22 identified miRNAs (like miR399a/b, miR1512b, miR156g and miR156j) had decreased expression levels and 14 identified miRNAs (like miR172f and miR390e) had increased levels of expression (Jha et al., 2021; Sun et al., 2016). Furthermore, 876 miRNAs related to alkalinity and salt stress were also found through genome-wide analysis in *Medicago truncatula* from which eight miRNAs (mtr-miR398a-3p, mtr-miR398a-5p, mtr-miR2628, mtr-miR171e-3p and four novel miRNAs) showed decreased expression regulation and 35 exhibited increased expression level (e.g., mtr-miR171, mtr-miR156 family and mtr-miR159a) under alkalinity and salinity stress (Cao et al., 2018).

In chickpeas, researchers have found 284 miRNAs under salinity stress from which miRNA397, Car-miR5507 and Car-novmiR2 have shown inversely correlated expression patterns for salt stress tolerance (Khandal et al., 2017). Similarly, a study on the Hassawi-3 genotype revealed that 284 of miRNAs out of total miRNAs found during genome sequence studies had upregulated expression, while 243 miRNAs had downregulated expression levels. However, the ILB4347 mutant population has shown somewhat differential expression for 298 miRNAs (upregulated) and 395 miRNAs (downregulated) under normal condition and salinity stress (Alzahrani et al., 2019; Jha et al., 2021).

Furthermore, sRNA sequencing of faba bean also identified 1220 salt-responsive miRNAs (Alzahrani et al., 2019). Furthermore, overexpression of miR156a showed decreased salt resistance in apple, while the target gene, MdSPL13, increases salt stress resistance. Furthermore, similar or various miRNA from this miRNA family can have diverse promoting or inhibitory effects in different plants under salt stress tolerance (Ma et al., 2022).

For example, the overexpression of Osa-miR393 in rice and Arabidopsis decreased salinity stress tolerance, and overexpression of mTIR1, miR393-resistant form, in Arabidopsis improved salinity tolerance. Similarly, overexpression of Osa-miR396c targeting GRF reduces alkalinity and salinity tolerance in Arabidopsis and rice but increases salt stress resistance in transgenic bentgrass through salinity hypersensitive (AsSOS-1)-mediated Na^{+} exclusion. Furthermore, in Arabidopsis, the overexpressed PeNAC070 targeted by miR164 increased sensitivity to salinity stress, as does overexpression of the miR169 target gene GmNFYA3. Meanwhile, another miRNA, miR414c and its targeted gene, GhFSD1 (iron superoxide dismutase-1), have also exhibited significant role in salt stress tolerance in cotton. However, the overexpression of miR414c can downgrade the expression pattern of its targeted gene GhFSD1 and increase sensitivity to salinity, controlling the regulation of reactive oxygen species metabolism (Ma et al., 2022).

13.6 SMALL INTERFERING RNAs (siRNAs) MODULATED ABIOTIC STRESS RESPONSES

Studies have reported several siRNAs that regulate the genic expression under abiotic stresses, enhancing plant stress tolerance and development. These are transcribed from the biological regulation of long double-strand RNAs (Mallory & Vaucheret, 2006). Further, it has been observed that abiotic stress adversely affects genome-wide DNA methylation which later causes activation of retrotransposons. Therefore, the transcriptional gene silencing (TGS)–related heterochromatic siRNAs are considered major transcription factors during abiotic stress (Ito et al., 2011).

Similarly, several siRNAs are believed to be driven from dsRNAs generated from miRNA-cleaved lncRNAs like trans-acting siRNAs (ta-siRNA) and natural antisense transcript siRNA (nat-siRNAs). These siRNAs are considered a special class in plants due to their expression in high abiotic stress environments (Matsui et al., 2013).

13.6.1 Drought Stress

Studies have revealed that rasiRNAs resulting from repetitive DNAs and transposable elements usually serve to suppress numerous abiotic stresses (e.g., drought) and retrotransposons subjective to their site of origin. Likewise, among rasiRNA and their targeted RNAs, complex feedback regulatory loops have been identified in Arabidopsis and maize. Meanwhile, tasiRNAs are supposed to result after miRNA-mediated transcriptive cleavage of TAS gene followed by phased cleavage of double-stranded RNA (Shengjun Li et al., 2016).

In Arabidopsis, TAS gene transcripts have shown differential regulatory patterns because the varying targeted miRNAs like TAS1 and TAS2 are triggered by miR173, TAS3 by miR390 and TAS4 by miR828. Further, through post-transcriptional silencing it has been found that they have a major regulatory function in the plant development. However, the regulatory patterns of some of the TAS genes (TAS1, TAS2 and TAS3) were found decreased under salt and drought stress. It has also been identified that some of the siRNAs play a crucial role in flower morphogenesis in plants. For instance, the tasiRNA-ARF (auxin response factor) is found triggered during flowering under salt and water stress conditions. Likewise, scientists have found two TAS3 gene counterparts in *Sorghum bicolor* that help regulate plant response to drought stress (Shengjun Li et al., 2016; Gelaw & Sanan-Mishra, 2021).

Furthermore, researchers have discovered a natural siRNA in *Craterostigma plantagineum* that is supposed to be triggered under drought condition and help plant induce tolerance. By using T-DNA tagging they found that by activating ABA-inducible genes *Craterostigma* gene enhances drought tolerance. The CDT-1 transcriptive levels in *C. plantagineum* were also found increased under drought and abscisic acid treatments. Remarkably, both sense and antisense siRNAs have been shown communicating to the CDT-1 locus, proposing that it produces sRNAs during stress targeting stress-responsive genes to improve tolerance (Khraiwesh et al., 2010).

Compared to the roles of miRNAs in drought tolerance, there are very less reports about the response of siRNAs under drought stress. During the study of wheat seedlings under drought stress, the regulatory patterns of four siRNAs were shown degraded. Meanwhile, drought-responsive siRNAs in maize were found derived from lncRNAs during drought situations (Zhang et al., 2014). Further, the expression levels of 24-nt siRNAs (linked with upregulated genes) were also observed decreased in these plant species under drought condition, suggesting their modulation by drought (Qi et al., 2013).

In barley, researchers have followed two steps to identify siRNAs. In the first step, they mapped all the obtained sRNA sequences through deep sequencing with the already identified barley genes. Later, in the second step forward, strand of various known factors e.g., genes and tRNAs was mapped with the 'antisense' reads. Using this working principle, a total of 914 differentially expressed siRNAs were discovered from which 172 siRNAs exhibited upregulation, whereas 642 were found decreased under drought stress. Interestingly, all tRNA-based siRNAs were observed increased during drought. Similarly, studies on wheat and barley have also revealed same number of differentially expressed siRNAs between normal and water-scarce environments (Shi BJ, unpublished data). This insight on cereal plants suggests that siRNAs regulated by drought stress may be conserved among them (Shi & Hussain, 2016).

13.6.2 Temperature Stress

The function of siRNAs is very prominent in reaction to temperature changes; however, the tasiRNAs are reported to have a significant function in the trade-off between plant development and their response to temperature stress. HS is known to be one of the main aspects for early flowering and reduced immunity in Arabidopsis. Studies have indicated that this negative regulation of plant development is mainly due to reduced SGS3 protein during HS which inhibits the tasiRNA biogenesis and relieves PTGS (Zhang et al., 2022; Zhong et al., 2013).

Researchers have suggested that effect of cold stress cannot be overlooked as it can affect the growth and development of a plant and its genomic integrity. As miRNAs, siRNAs are also major stress-induced factors which regulate the plant response under cold temperature stress. In a research, siRNAs were found controlling the regulation of the CBF-dependent pathway under low-temperature stress, leading to undetectable expression of CBF1 gene in mutant-type plants as of their wild-type plants. In a similar experiment in Arabidopsis, phasiRNA was also seen upregulated under low-temperature stress (Huo et al., 2022; Tiwari et al., 2020).

In the meantime, several researchers have demonstrated that there are approximately 38 trans-nat-siRNAs (resulting from the AT1G10522-AT5G53905 transcripts) and 104 cis-nat-siRNAs (resulting from AT3G05870-AT3G05880 transcripts) in Arabidopsis that are linked to stress response regulation during cold stress at different time intervals (from 3 hours to 2 days) (Huo et al., 2022 ;Tiwari et al., 2020).

Similarly, cold stress has a major impact on cucumbers (*Cucumis sativus* L.) and roses (*Rosa hybrida*). It affects the regulation of mCHH in cucumber by controlling temperature-dependent sex determination by demethylation and low-temperature-induced rise in rose petal number by upregulation (Iwasaki et al., 2019). Wheat seedlings have also shown differential expression of four nat-siRNAs under cold stress, and other abiotic stresses, e.g., nat-siRNA 005047_0654_1904.1, have shown upregulating expression under low-temperature stress and decreased expression levels under other stresses linked to abiotic factors (Huo et al., 2022).

Studies in wheat seedlings have found that abiotic stresses like salt, drought, heat or cold can alter the regulating expression of four nat-siRNAs. Meanwhile, miRNA target network module is also found triggered under low-temperature signalling based on the miRNAs and their target gene's expressional trends after low temperature stress. *Plant Signaling & Behavior* e2004035-7 005047_0654_1904.1 was found improved specifically under low-temperature stress and reduced under other abiotic stresses conditions. Recently, in Arabidopsis 38 trans-nat-siRNAs and 104 cis-nat-siRNAs were observed differentially regulated (up- or down-) at different time intervals (from 3 hours to 2 days) under cold stress conditions (Huo et al., 2022).

However, understanding the function of siRNAs in plant responses is still at initial stages as compared to miRNAs. Therefore, further studies on sources of siRNAs are needed to understand their production and target gene modulation. It has been observed that functions of siRNAs are somehow dependent on production and action of miRNAs, and it is estimated that epigenetic modifications of genes can be inherited by offspring domesticating plant evolution (Huo et al., 2022).

13.6.3 Salinity Stress

Like other ncRNAs, siRNAs play a major regulatory role in salt stress tolerance in plants, as initially shown by Sunkar and Zhu. For instance, the nat-siRNAs, a subclass, modulate double fertilisation in plants and increase salinity tolerance (Borsani et al., 2005; Ron et al., 2010; Zhang et al., 2022). Studies have revealed that nat-siRNAs (resulting from natural cis-antisense transcripts of P5CDH and SRO5 genes) in Arabidopsis have a major role in oxidative stress control and osmoprotection under salinity stress. However, these are genes expressed differentially, e.g., salt stress induces SRO5, while P5CDH is expressed consistently.

Studies have further identified that under high saline conditions, a 24-nt nat-siRNA is produced, related to the SRO5 mRNA, targeting P5CDH mRNA downregulation resulting in the production of 21-nt nat-siRNA population (Khraiwesh et al., 2010). Furthermore, the degradation of P5CDH also indicates the collection of proline, which improves the salt stress tolerance abilities in plants. However, the degraded P5CDH also stimulates the collection of ROS and a toxic metabolic intermediate P5C. Therefore, the SRO5-P5CDH nat-siRNAs, along with the SRO5 and P5CDH proteins, are important regulatory modules that regulate the production of ROS and the plant responses to saline conditions (Khraiwesh et al., 2010).

In *Triticum aestivum*, a pool of small ncRNAs has exhibited salinity stress responsiveness. During the studies of wheat seedlings, the expression of three siRNAs was found deferentially exhibited: siRNA 005047_0654_1904.1 found highly increased by low-temperature stress and decreased by drought, salinity and HS; siRNA002061_0636_3054.1 found highly decreased during drought, heat and salt stress; and siRNA007927_0100_2975.1 was observed downregulated by salinity, low temperature and drought stress (Khraiwesh et al., 2010; Matsui et al., 2013).

13.7 LONG NON-CODING RNAs (lncRNAs) MODULATED ABIOTIC STRESS RESPONSES

lncRNAs are the types that do not code protein but act as a signal to modulate the gene expression because of their close locality in the genome (Wang & Chang, 2011). Therefore, they have been identified as a major participant in plant response regulation to abiotic stress and growth and development (Chekanova, 2015; Jha et al., 2020; Tiwari & Lata, 2021; Yu et al., 2019; Zhang et al., 2022).

13.7.1 Drought Stress

lncRNAs have been known for their major contribution to drought tolerance and their part in plant development. For instance, in maize 1535 drought-sensitive lncRNAs were reported at various plant developmental stages (Pang et al., 2019; Zhang et al., 2014); lncRNAs found during the R1 stage (silking phase) in maize have been identified as a major contributor in drought tolerance (Jha et al., 2020); and 1832 lncRNAs in Arabidopsis have shown changing expression during drought, temperature, abscisic acid and salinity stress (Liu et al., 2012). Similarly, many different plants like wheat (*T. aestivum* L.) and rice (*Oryza sativa* L.) have shown evidence for lncRNAs (Ackah et al., 2024; Zhang et al., 2014).

Furthermore, researchers have reported that various other plant species have significant amounts of drought-responsive lncRNAs which improve their drought tolerance during stress conditions like 98 in rice (Chung et al., 2016), 504 in *Populus* spp. (Shuai et al., 2014), 19 in foxtail millet (Qi et al., 2013), 1597 in switchgrass (Zhang et al., 2018), and 185 in cassava (Li et al., 2017).

During drought stress conditions, lncRNAs are found to regulate the response by managing eTM-based complex mechanisms, antisense transcription-mediated modulation, chromatin modulation or transactional regulation of various drought-related genes (Li et al., 2017; Wang, Meng, et al., 2017; Yan et al., 2019). In foxtail millet, 584 lncRNAs (Qi et al., 2013) have remained their corresponding target genes veiled differential expression of two NAT lncRNAs and 17 lincRNAs under drought stress tolerance during the deep genomic sequencing. Meanwhile, a similar response in sorghum has been observed by one NAT lncRNA and 20 similar lincRNAs under drought conditions (Jha et al., 2020); however, only one drought-responsive lncRNA has shown sequence co-linearity with foxtail millet because of their low conserved genic nature (Qi et al., 2013).

One of the major known lncRNA modules is found in the potato which regulates the stomatal growth. During studies, the lncRNA StFLORE and central clock output transcription factor Cycling DOF Factor 1 (StCDF1) have been found to regulate plant growth and diurnal stomatal movement to reduce water loss during drought stress conditions. The upregulation of StFLORE, however, enhances drought stress tolerance in the potato but hinders vegetative reproduction (Ramirez Gonzales et al., 2021; Zhang et al., 2022).

Further genome studies using RT-qPCR of cassava for drought-responsive lncRNAs have revealed that three lincRNAs exhibit upregulation (lincRNA391, lincRNA101 and lincRNA356) and four lincRNAs exhibit downregulation (lincRNA392, lincRNA182, lincRNA64 and lincRNA350) during the period of water stress. However, strand-specific RNA-Seq data have found 40 lncRNAs exhibiting upregulatory expression under drought stress conditions (Jha et al., 2020). Studies on cassava have also revealed that the 11 lncRNAs work as target mimics for miRNAs, e.g., miR172, miR169, miR164 and miR156. Like lincRNA340, they have shown a similar mimic effect, exhibiting upregulated expression of Nuclear Factor Y (NF-Y) gene and decreased expression of target miR169 under water stress (Jha et al., 2020).

Rice genome studies have identified 98 drought-responsive lncRNAs, two of major importance, NAT Os02g0180800-01 (regulating cinnamoyl-CoA reductase gene) and lncRNAs NAT Os02g0250700-01 (regulating late embryogenesis abundant protein gene), because of the inversely correlated expression with their corresponding regulatory genes (Chung et al., 2016; Jha et al., 2020). Besides their role as eTM or NAT, lncRNAs can transcriptionally modulate various drought-responsive genes (Wang et al., 2019; Zhang et al., 2018).

In Arabidopsis, lncRNAs play a major part in drought tolerance. For example, the DROUGHT-INDUCED lncRNA (DRIR) containing a 755 nt-long lincRNA was identified in the nucleus which effectively regulates drought and salt stress genes, including aquaporin genes (NIP1, TIP4), annexin gene (ANNAT7), ABA-signalling genes (RD29B, P5CS1, RD29A and ABI5), fucosyltransferase 4 (FUT4) gene and transcription factor genes (NAC3, WARKY8) at the transcriptional level, but overexpressing seedlings have been found sensitive to ABA treatment activating signalling pathway (Jha et al., 2020; Qin et al., 2017; Wang & Chang, 2011).

13.7.2 Temperature Stress

Evidence of cold-induced lncRNA expression regulation has been found in many plants like rice (Yuan et al., 2011), alfalfa (Cui et al., 2019), banana and Arabidopsis (Tiwari et al., 2020). However, the expression regulation of lncRNAs is quite different from the other classes of ncRNAs, e.g., miRNAs and siRNAs. Studies have revealed that the majority of the lncRNAs found in Arabidopsis have expression regulating responses to cold stress, e.g., 46 DAS lncRNAs and 113 DE lncRNAs (Calixto et al., 2019; Huo et al., 2022). Furthermore, in Arabidopsis, the CBF1 gene expression was found deregulated by an lncRNA, SVALKA, to maximise cold tolerance (Kindgren et al., 2018).

Several lncRNAs have been studied for cold stress response and their role in modulating the vernalisation in plants. Among those lncRNAs, the two most common are COLDAIR and COOLAIR which modulate the plant vernalisation process to induce flowering by targeting the 3′ end of FLC mRNA and lowering its expression (Heo & Sung, 2011; Kim & Sung, 2017; Kim et al., 2017; Swiezewski et al., 2009; Zhang et al., 2022).

This usually happened during early cold stress while suppressing the FLC locus by COOLAIR generated through chromatin modification from the antisense orientation of the FLC gene (Csorba et al., 2014; Heo & Sung, 2011; Jha et al., 2020; Swiezewski et al., 2009). Furthermore, MAS (NAT lncRNA_2962), a cis-acting NAT lncRNA, has also been found triggered by MADs Affecting Flowering 4 (MAF4) a cold stress activating transcription of FLC family (Jha et al., 2020).

In recent studies about 46% of new lncRNAs were discovered in rice during the genomic sequencing of 7231 lncRNAs, from which 135 were identified as cold stress-responsive lncRNAs (29 induced and 106 repressed) (Yuan et al., 2018). However, because of less available studies on lncRNAs, their related molecular mechanism is not well known along with how tissue type and developmental stage affect their expression.

During the genome sequence analysis of grapevine, researchers have found several cold-responsive lncRNAs, including 284 upregulated, 182 downregulated and several differential stress-responsive genes like WRKY genes, late embryogenesis protein genes and CBF4 genes (Wang et al., 2019). Likewise, in cassava 318 drought and cold-responsive lncRNAs were found in genome-wide analysis, and lincRNA419, 207 and 234 were found increased during low-temperature conditions. Further, in cassava plants decreased expression levels were also observed for three miR164-target NAC genes (CUC2, NAM, ATAF1/2) because of the binding of lincRNA159 with conserved miR164 under low-temperature stress (Li et al., 2017).

Meanwhile, it has been reported that lncRNAs not only play a significant part in low-temperature stress but they also have a major contributing role in HS responses in plants. There are several pieces of evidence present about the molecular regularity of lncRNAs in plants under high-temperature stress like wheat studies that have found 54 heat-induced lncRNAs (Xin et al., 2011). However, qRT-PCR analysis has revealed that some of them are precursors of miRNAs and siRNAs that are closely linked to expression regulation of lncRNAs like upregulation of TahlnRNA27 and TalnRNA5, which corresponds to miR2010 and miR2004 during high-temperature stress (Jha et al., 2020). Additionally, a close interaction between lncRNAs and miR9748 was also found in cucumbers where it helps plants regulate thermo-tolerance through hormonal signalling like ethylene and indole-3-acetic acid (He et al., 2020; Zhang et al., 2022).

Like other ncRNAs, the expression of lncRNAs also varies according to plant tissues, developmental stages, and species because of their weak conservation nature. For example, TalnRNA5 in seed tissue showed an elevated expression level than other plant tissues under HS. Similar tissue-specific expression differences have also been noted in cassava (Ding, Wu, et al., 2019), maize (Huanca-Mamani et al., 2018; Pang et al., 2019), *Populus euphratica* and *Populus alba* var. pyramidalis under high-temperature stress (Jha et al., 2020).

In brassica, a similar increased and decreased expression level has been observed between differentially expressed genes (DEGs) and lncRNAs (e.g., TCONS_00048391 and TCONS_00010856). This differentially modulated expression change was found in ethylene, salicylic acid, auxin, and jasmonic acid hormone signalling pathways which encode several heat shock proteins during stress conditions (Wang et al., 2019). bra-miR164a has been observed upregulated and lncRNATCONS_00048391 downregulated under HS making target Bra030820 (NAC1) gene exhibit rendered heat tolerance (Li et al., 2017).

13.7.3 Salinity Stress

Several lncRNAs regulate salt tolerance in plants like cotton and Arabidopsis. However, their expression may vary depending on environmental conditions and neighbouring genes. In cotton, it has been observed that the expression level of lncRNA354 reduces during salinity stress due to the modulation of GhARF17/18 genes. However, cotton has seen an increase in plant height and salt tolerance upon silencing GhARF17/18 or lncRNA354. Meanwhile, higher expression of GhARF17/18 or lncRNA354 in Arabidopsis leads to reduced plant height and salinity resistance, but mostly the presence of lncRNAs results in expression upregulation of several genes (P5CS1, RD29B, FUT4, AtrbohB, RD29A, ANNAT7, and NAC3) involved in salinity stress (Qin et al., 2017; Zhang et al., 2021, 2022).

However, further soybean root transcriptome sequencing found 3030 lincRNAs and 275 lncNATs, 75% of which had shown upregulated expression under saline conditions (Chen et al., 2019). A similar study on groundnut revealed 1442 lncRNAs with differential expression varying with stress time. For example, studies on groundnut roots identified that after 12 hours of salinity stress, TCONS_00292946 lncRNA was found downregulated, while TCONS_00176941 and TCONS_00011551 were found upregulated, but after 24 hours of stress, TCONS_00292946 lncRNA was found upregulated (Tian et al., 2020; Jha et al., 2021).

Studies on *M. truncatula* found lncRNAs modulated during salinity and regulating stress tolerance by improving ROS-related stress responses. For example, TCONS_00116877 lncRNA in roots was found to induce the Medtr7g094600 gene responsible for producing glutathione peroxidase to reduce ROS-derived stress regulating salinity tolerance (Wang et al., 2015). Further studies have revealed several other lncRNAs involved in stress response regulation; in roots, TCONS_00020253 upregulates Na+/H+ exchanger Medtr1g081900 gene expression and TCONS_00046739 regulates cytochrome P450; in leaves, TCONS_00097188 upregulates photosynthetic activity by triggering Medtr6g006990 gene (Jha et al., 2020; Qin et al., 2017).

Genome-wide analysis of 1117 unique lncRNAs has identified 44 differentially expressed lincRNAs in cotton which on further validation revealed that lnc_883 and lnc_388lncRNA triggered by Gh_D03G0339MS channel protein gene had an upregulatory expression under salinity stress. During analysis, several lncRNAs acted as target mimics for miRNAs under salinity like lncRNAs lnc_253 and lnc_973 for ghr-156e and ghr-miR399 (Deng et al., 2018).

In Lluteño landrace of maize, 1710 salinity- and boron-induced lncRNAs were found in leaf and root tissues using transcriptome studies. However, during comparison, three native popular maize cultivars and the Lluteño landrace were studied alongside, and 955 conserved lncRNA transcripts (755 exclusive to Lluteño landrace) were found regulating expression under boron and salt stress conditions (Table 13.1) (Huanca-Mamani et al., 2018; Jha et al., 2020).

TABLE 13.1
Non-Coding RNAs and Their Targets in Response to Abiotic Stresses

Name of Plant	ncRNA Family	Type of ncRNA	Target	Overexpression/ Knockdown	Function	References
Rice	miR164b	miRNA	*OsNAC2*	Overexpression	Improves drought stress tolerance and yield	Jiang et al. (2019)
Wheat	TaMIR1119	miRNA	*TabHLH49, TaLZ*	Overexpression	Enhances drought tolerance	Shi et al. (2018)
Maize	zma-miR171b	miRNA	*GRAS*	Knockdown	Improves cadmium stress resistance	Gao et al. (2019)
Maize	miR319b-3p	miRNA	*ARF3, GAMYB*	Overexpression	Restrict shoot growth when exposed to arsenate stress	Ghosh et al. (2017)
Cotton	miRNVL5	miRNA	*GhCHR*	Overexpression	Hypersensitivity to salt stress	Gao et al. (2016)
Arabidopsis	TAS3	siRNA	*Nitrate transporter 2*	-	Triggers root development and N transport under low-nitrogen conditions	Fukuda et al. (2019)
Cotton	GhDAN1	lncRNA	-	Knockdown	Enhances drought tolerance	Tao et al. (2021)
Cotton	lnc_973, lnc_253	lncRNA	-	Overexpression	Enhances salinity stress tolerance	Deng et al. (2018)
Chinese cabbage	TCONS_00048391	lncRNA	-	Knockdown	Reduces heat tolerance	Wang et al. (2019)

13.8 CHALLENGES LINKED TO REGULATORY ROLE OF ncRNA RESEARCH AND APPLICATION

One of the possible applications of the available information about ncRNAs and their regulation could be to enhance global nutritional and food security. In order to apply ncRNA information to transpose important characteristics of different plants, several molecular genetic strategies have been employed. Several studies evidencing the impact of altering a single miRNA on multiple characteristics, including stress tolerance, biomass, and yield are available (Zhang et al., 2014).

For instance, in tomato plants, overexpression of the evolutionarily conserved miR156 exhibits a direct correlation with fruit sizes (Zhang et al., 2022). The miR156 targeted SPL genes were found to be positively regulated with the yield of rice (Wang et al., 2015; Jiao et al., 2019). In rice, it targets SPL13, SPL14 and SPL16, which are involved in the expression regulation of panicle and grain size (Wang et al., 2015; Tang & Chu, 2017; Jiao et al., 2019). Likewise, upregulation of miR397 exhibited a favourable relationship with yield and grain size by targeting its putative target gene LACCASE (LAC) in rice (Basso et al., 2019; Zhang et al., 2014).

Wang et al. (2016) found that RSV infection upregulated miR444 in rice by activating the RDR1-dependent antiviral RNA silencing pathway, which downregulated MAD23, MAD27a and MADS57. Further, artificial miRNAs (amiRNAs/amiRs) have also been considered as another potential for enhancing the crops by manipulating amiRNAs to control over the expression of the targeted gene (Rosa et al., 2018; Zhang et al. 2018). Of these, amiR159b intervention has been successfully applied in Arabidopsis to target three key genes governing seed oil metabolism –fatty acid elongase-1 (FAE1), fatty acid desaturase Δ12 (FAD2), and fatty acid ACP-B thioesterase (FATB) – to achieve high level record olein (Belide et al., 2012).

In this case, the constitutional character of miRNA/pre-miRNA sequence is critical for its activity, and changes in the single nucleotides/SNPs in pre-miRNA sequence affected the stability of their secondary structure. Specific differences in grain length and grain type corresponding to SNPs in the miR2923a pre-miRNA were also identified as potential agronomic traits (Wang et al., 2013). Of these ncRNA types, more can be done to enhance interesting agronomic traits. Likewise, the photoperiod-sensitive genetic male sterility 1 (Pmsl) locus produces an lncRNA, PMS1T, that is particularly regulated in youthful rice panicles (Bhogireddy et al., 2021).

In addition, investigations have revealed that site-specific nucleases, particularly the Clustered Regular Interspaced Short Palindromic Repeat Type II (CRISPR)/CRISPR-associated protein 9 (Cas9) system, are highly effective genome-editing tools for functional analyses in plants (Chen et al., 2016; Basso et al., 2019). In contrast to RNAi, CRISPR/Cas9 has capabilities to achieve gene knockout, which is complete in addition to lenient off-target effects (Basso et al., 2019; Bhogireddy et al., 2021).

However, due to the localisation of some of the ncRNA, the RNAi has difficulties in executing since it is already restricted in the cytoplasm where the RISC is present. The use of genome editing for crop improvement in plants using specific ncRNAs is still under development, albeit attempts to harness this system effectively (Basso et al., 2019). For example, Cas9 targeting of two soybean miRNAs, miR1514 and miR1509, was shown to target a short ncRNA by using the CRISPR/Cas9 technique (Jacobs et al., 2015).

In recent experiments, gene and genome (GO) and Kyoto Encyclopedia of Genes and Genomes (KEGG) enrichment analyses were used to identify the cellular locations, biological and molecular functions affected by stressful conditions, as well as the pathways by which lncRNAs are involved. Although studies of this type present a good foundation for the identification of new lncRNAs, they cannot conclusively prove the connection between lncRNAs and abiotic stresses. However, some other researchers were able to determine how lncRNAs are integral part of particular molecular mechanisms, but these still played a natural role (Imaduwage & Hewadikaram, 2024).

REFERENCES

Ackah, M., Jin, X., Zhang, Q., Amoako, F. K., Wang, L., Attaribo, T., …, & Zhao, W. (2024). Long noncoding RNA transcriptome analysis reveals novel lncRNAs in *Morus alba* 'Yu-711' response to drought stress. *The Plant Genome*, *17*(1), e20273.

Ahmed, W., Xia, Y., Li, R., Zhang, H., Siddique, K. H., & Guo, P. (2021). Identification and analysis of small interfering RNAs associated with heat stress in flowering

Alzahrani, S. M., Alaraidh, I. A., Khan, M. A., Migdadi, H. M., Alghamdi, S. S., & Alsahli, A. A. (2019). Identification and characterization of salt-responsive microRNAs in *Vicia faba* by high-throughput sequencing. *Genes*, *10*(3), 303.

Atafar, Z., Mesdaghinia, A., Nouri, J., Homaee, M., Yunesian, M., Ahmadimoghaddam, M., & Mahvi, A. H. (2010). Effect of fertilizer application on soil heavy metal concentration. *Environmental Monitoring and Assessment*, *160*, 83–89.

Balyan, S., Joseph, S. V., Jain, R., Mutum, R. D., & Raghuvanshi, S. (2020). Investigation into the miRNA/5′ isomiRNAs function and drought-mediated miRNA processing in rice. *Functional & Integrative Genomics*, *20*(4), 509–522. https://doi.org/10.1007/s10142-020-00731-2

Barciszewska-Pacak, M., Milanowska, K., Knop, K., Bielewicz, D., Nuc, P., Plewka, P., Pacak, A. M., Vazquez, F., Karlowski, W., Jarmolowski, A., & Szweykowska-Kulinska, Z. (2015). Arabidopsis microRNA expression regulation in a wide range of abiotic stress responses. *Frontiers in Plant Science*, *6*, 410. https://doi.org/10.3389/fpls.2015.00410

Barrera-Figueroa, B. E., Gao, L., Wu, Z., Zhou, X., Zhu, J., Jin, H., Liu, R., & Zhu, J.-K. (2012). High throughput sequencing reveals novel and abiotic stress-regulated microRNAs in the inflorescences of rice. *BMC Plant Biology*, *12*(1), 132. https://doi.org/10.1186/1471-2229-12-132

Bartel, D. P. (2018). Metazoan microRNAs. *Cell*, *173*(1), 20–51.

Bartonicek, N., Clark, M. B., Quek, X. C., Torpy, J. R., Pritchard, A. L., Maag, J. L. V., …, & Dinger, M. E. (2017). Intergenic disease-associated regions are abundant in novel transcripts. *Genome Biology*, *18*, 1–16.

Baulcombe, D. (2004). RNA silencing in plants. *Nature*, *431*(7006), 356–363.

Bhogireddy, S., Mangrauthia, S. K., Kumar, R., Pandey, A. K., Singh, S., Jain, A., …, & Kudapa, H. (2021). Regulatory noncoding RNAs: A new frontier in regulation of plant biology. *Functional & Integrative Genomics*, *21*(3), 313–330.

Bologna, N. G., & Voinnet, O. (2014). The diversity, biogenesis, and activities of endogenous silencing small RNAs in Arabidopsis. *Annual Review of Plant Biology*, *65*(1), 473–503.

Boretti, A., & Rosa, L. (2019). Reassessing the projections of the world water development report. *NPJ Clean Water*, *2*(1), 15.

Borges, F., & Martienssen, R. A. (2015). The expanding world of small RNAs in plants. *Nature Reviews Molecular Cell Biology*, *16*(12), 727–741.

Borsani, O., Zhu, J. H., Verslues, P. E., Sunkar, R., & Zhu, J. K. (2005). Endogenous siRNAs derived from a pair of natural cis-antisense transcripts regulate salt tolerance in *Arabidopsis*. *Cell*, *123*(6), 1279–1291. https://doi.org/10.1016/j.cell.2005.11.035

Calixto, C. P. G., Tzioutziou, N. A., James, A. B., Hornyik, C., Guo, W., Zhang, R., Nimmo, H. G., & Brown, J. W. S. (2019). Cold-dependent expression and alternative splicing of Arabidopsis long non-coding RNAs. *Frontiers in Plant Science*, *10*, 235.

Cao, C., Long, R., Zhang, T., Kang, J., Wang, Z., Wang, P., Sun, H., Yu, J., & Yang, Q. (2018). Genome-wide identification of microRNAs in response to salt/alkali stress in *Medicago truncatula* through high-throughput sequencing. *International Journal of Molecular Sciences*, *19*(12), 4076.

Cao, X., Wu, Z., Jiang, F., Zhou, R., & Yang, Z. (2014). Identification of chilling stress-responsive tomato microRNAs and their target genes by high-throughput sequencing and degradome analysis. *BMC Genomics*, *15*, 1130. https://doi.org/10.1186/1471-2164-15-1130

Carninci, P., Kasukawa, T., Katayama, S., Gough, J., Frith, M. C., Maeda, N., …, & Zhu, S. (2005). The transcriptional landscape of the mammalian genome. *Science*, *309*(5740), 1559–1563. [PMC][10.1126/science.1112014][16141072]

Castel, S. E., & Martienssen, R. A. (2013). RNA interference in the nucleus: Roles for small RNAs in transcription, epigenetics and beyond. *Nature Reviews Genetics*, *14*(2), 100–112.

Chaudhry, S., & Sidhu, G. P. S. (2022). Climate change regulated abiotic stress mechanisms in plants: A comprehensive review. *Plant Cell Reports*, *41*(1), 1–31.

Chekanova, J. A. (2015). Long non-coding RNAs and their functions in plants. *Current Opinion in Plant Biology*, *27*, 207–216. https://doi.org/10.1016/j.pbi.2015.08.003

Chen, L., Cheng, X., Cai, J., Zhan, L., Wu, X., Liu, Q., & Wu, X. (2016). Multiple virus resistance using artificial trans-acting siRNAs. *Journal of Virological Methods*, *228*, 16–20.

Chen, R., Li, M., Zhang, H., Duan, L., Sun, X., Jiang, Q., Zhang, H., & Hu, Z. (2019). Continuous salt stress-induced long non-coding RNAs and DNA methylation patterns in soybean roots. *BMC Genomics*, *20*, 1–12.

Chung, P. J., Jung, H., Jeong, D. H., Ha, S. H., Choi, Y. D., & Kim, J. K. (2016). Transcriptome profiling of drought responsive noncoding RNAs and their target genes in rice. *BMC Genomics*, *17*, 563.

Csorba, T., Questa, J. I., Sun, Q., & Dean, C. (2014). Antisense COOLAIR mediates the coordinated switching of chromatin states at FLC during vernalization. *Proceedings of the National Academy of Sciences of the United States of America*, *111*(46), 16160–16165.

Cui, G., Chai, H., Yin, H., Yang, M., Hu, G., Guo, M., & Zhang, P. (2019). Full-length transcriptome sequencing reveals the low-temperature tolerance mechanism of *Medicago falcata* roots. *BMC Plant Biology*, *19*, 575.

Czech, B., & Hannon, G. J. (2011). Small RNA sorting: Matchmaking for Argonautes. *Nature Reviews Genetics*, *12*(1), 19–31.

de Goede, O. M., Nachun, D. C., Ferraro, N. M., Gloudemans, M. J., Rao, A. S., Smail, C., …, & Li, Q. (2021). Population-scale tissue transcriptomics maps long non-coding RNAs to complex disease. *Cell*, *184*(10), 2633–2648.

Deng, F., Zhang, X., Wang, W., Yuan, R., & Shen, F. (2018). Identification of *Gossypium hirsutum* long non-coding RNAs (lncRNAs) under salt stress. *BMC Plant Biology*, *18*, 1–14. DOI: 10.1186/s12870-018-1238-0

Desaint, H., Aoun, N., Deslandes, L., Vailleau, F., Roux, F., & Berthomé, R. (2021). Fight hard or die trying: When plants face pathogens under heat stress. *New Phytologist*, *229*(2), 712–734.

Ding, Y., Shi, Y., & Yang, S. (2019). Advances and challenges in uncovering cold tolerance regulatory mechanisms in plants. *New Phytologist*, *222*(4), 1690–1704.

Ding, Z., Tie, W., Fu, L., Yan, Y., Liu, G., & Yan, W., et al. (2019). Strand-specific RNA-Seq based identification and functional prediction of drought-responsive lncRNAs in cassava. *BMC Genomics*, *20*, 214.

Ding, Z., Wu, C., Tie, W., Yan, Y., He, G., & Hu, W. (2019). Strand-specific RNA-Seq based identification and functional prediction of lncRNAs in response to melatonin and simulated drought stresses in cassava. *Plant Physiology and Biochemistry*, *140*, 96–104. https://doi.org/10.1016/j.plaphy.2019.05.008

Jacobs, T. B., LaFayette, P. R., Schmitz, R. J., & Parrott, W. A. (2015). Targeted genome modifications in soybean with CRISPR/Cas9. *BMC Biotechnology*, 15(1), 16.

Dufková, H., Berka, M., Psota, V., Brzobohatý, B., & Černý, M. (2023). Environmental impacts on barley grain composition and longevity. *Journal of Experimental Botany*, *74*(5), 1609–1628.

Fahad, S., Bajwa, A. A., Nazir, U., Anjum, S. A., Farooq, A., Zohaib, A., …, & Huang, J. (2017). Crop production under drought and heat stress: Plant responses and management options. *Frontiers in Plant Science*, *8*, 1147.

Fang, S., Zhang, L., Guo, J., Niu, Y., Wu, Y., Li, H., …, & Zhao, Y. (2018). NONCODEV5: A comprehensive annotation database for long non-coding RNAs. *Nucleic Acids Research*, *46*(D1), D308–D314.

Fang, X., & Qi, Y. (2016). RNAi in plants: An argonaute-centered view. *The Plant Cell*, *28*(2), 272–285.

Farooq, M., Gogoi, N., Barthakur, S., Baroowa, B., Bharadwaj, N., Alghamdi, S. S., & Siddique, K. H. (2017). Drought stress in grain legumes during reproduction and grain filling. *Journal of Agronomy and Crop Science*, *203*(2), 81–102.

Farooq, M., Wahid, A., Kobayashi, N., Fujita, D., & Basra, S. M. A. (2009). Plant drought stress: Effects, mechanisms, and management. *Agronomy for Sustainable Development*, *29*, 185–212. https://doi.org/10.1051/agro:2008021

Faulkner, G. J., Kimura, Y., Daub, C. O., Wani, S., Plessy, C., Irvine, K. M., …, & Carninci, P. (2009). The regulated retrotransposon transcriptome of mammalian cells. *Nature Genetics*, *41*(5), 563–571.

Fei, Q., Xia, R., & Meyers, B. C. (2013). Phased, secondary, small interfering RNAs in posttranscriptional regulatory networks. *The Plant Cell*, *25*(7), 2400–2415.

Ferdous, J., Hussain, S. S., & Shi, B. J. (2015). Role of microRNAs in plant drought tolerance. *Plant Biotechnology Journal*, *13*, 293–305. https://doi.org/10.1111/pbi.12318

Ferdous, J., Sanchez-Ferrero, J. C., Langridge, P., Milne, L., Chowdhury, J., Brien, C., & Tricker, P. J. (2017). Differential expression of microRNAs and potential targets under drought stress in barley. *Plant, Cell & Environment*, *40*, 11–24. https://doi.org/10.1111/pce.12764

Fich, E. A., Segerson, N. A., & Rose, J. K. (2016). The plant polyester cutin: Biosynthesis, structure, and biological roles. *Annual Review of Plant Biology*, *67*(1), 207–233.

Food and Agriculture Organization of the United Nations (FAO). (2015). *FAO land and plant nutrition management service*. FAO.

Fracasso, A., Trindade, L. M., & Amaducci, S. (2016). Drought stress tolerance strategies revealed by RNA-Seq in two sorghum genotypes with contrasting WUE. *BMC Plant Biology*, *16*, 115. https://doi.org/10.1186/s12870-016-0800-x

Frith, M. C., Pheasant, M., & Mattick, J. S. (2005). Genomics: The amazing complexity of the human transcriptome. *European Journal of Human Genetics*, *13*(8).

Fueyo, R., Judd, J., Feschotte, C., & Wysocka, J. (2022). Roles of transposable elements in the regulation of mammalian transcription. *Nature Reviews Molecular Cell Biology*, *23*(7), 481–497.

Fukuda, M., Nishida, S., Kakei, Y., Shimada, Y., & Fujiwara, T. (2019). Genome-wide analysis of long intergenic noncoding RNAs responding to low-nutrient conditions in *Arabidopsis thaliana*: Possible involvement of trans-acting siRNA3 in response to low nitrogen. *Plant Cell Physiology*, *60*, 1961–1973. https://doi.org/10.1093/pcp/pcz094

Gao, J., Luo, M., Peng, H., Chen, F., & Li, W. (2019). Characterization of cadmium-responsive MicroRNAs and their target genes in maize (Zea mays) roots. *BMC molecular biology*, 20(1), 14.

Gao, S., Yang, L., Zeng, H. Q., Zhou, Z. S., Yang, Z. M., Li, H., Sun, D., Xie, F., & Zhang, B. (2016). A cotton miRNA is involved in the regulation of plant response to salt stress. *Scientific Reports*, *6*, 1–14. https://doi.org/10.1038/srep20780

Gelaw, T. A., & Sanan-Mishra, N. (2021). Non-coding RNAs in response to drought stress. *International Journal of Molecular Sciences*, *22*(22), 12519. https://doi.org/10.3390/ijms222212519

Ghildiyal, M., & Zamore, P. D. (2009). Small silencing RNAs: An expanding universe. *Nature Reviews Genetics*, *10*(2), 94–108.

Ghosh, S., Singh, K., Shaw, A. K., Azahar, I., Adhikari, S., Ghosh, U., Basu, U., Roy, S., Saha, S., Sherpa, A. R., & Hossain, Z. (2017). Insights into the miRNA-mediated response of maize leaf to arsenate stress. *Environmental and Experimental Botany*, *137*, 96–109. https://doi.org/10.1016/j.envexpbot.2017.01.005

Gimenez, E., Salinas, M., & Manzano-Agugliaro, F. (2018). Worldwide research on plant defense against biotic stresses as improvement for sustainable agriculture. *Sustainability*, *10*(2), 391.

Giordo, R., Wehbe, Z., Posadino, A. M., Erre, G. L., Eid, A. H., Mangoni, A. A., & Pintus, G. (2022). Disease-associated regulation of non-coding RNAs by resveratrol: Molecular insights and therapeutic applications. *Frontiers in Cell and Developmental Biology*, *10*, 894305.

Gong, Z., Xiong, L., Shi, H., Yang, S., Herrera-Estrella, L. R., Xu, G., …, & Zhu, J. K. (2020). Plant abiotic stress response and nutrient use efficiency. *Science China Life Sciences*, *63*, 635–674.

Guan, Q., Lu, X., Zeng, H., Zhang, Y., & Zhu, J. (2013). Heat stress induction of miR398 triggers a regulatory loop that is critical for thermotolerance in *Arabidopsis*. *The Plant Journal*, *74*, 840–851. https://doi.org/10.1111/tpj.12169

Hangauer, M. J., Vaughn, I. W., & McManus, M. T. (2013). Pervasive transcription of the human genome produces thousands of previously unidentified long intergenic noncoding RNAs. *PLoS Genetics*, *9*(6), e1003569.

He, J., Jiang, Z., Gao, L., You, C., Ma, X., Wang, X., Xu, X., Mo, B., Chen, X., & Liu, L. (2019). Genome-wide transcript and small RNA profiling reveals transcriptomic responses to heat stress. *Plant Physiology*, *181*, 609–629. https://doi.org/10.1104/pp.19.00403

He, X., Guo, S., Wang, Y., Wang, L., Shu, S., & Sun, J. (2020). Systematic identification and analysis of heat-stress-responsive lncRNAs, circRNAs, and miRNAs with associated co-expression and ceRNA networks in cucumber (*Cucumis sativus* L.). *Physiologia Plantarum*, *168*, 736–754. https://doi.org/10.1111/ppl.12997

Heo, J. B., & Sung, S. (2011). Vernalization-mediated epigenetic silencing by a long intronic noncoding RNA. *Science*, *331*, 76–79. https://doi.org/10.1126/science.1197349

Hua, Y., Zhang, C., Shi, W., & Chen, H. (2019). High-throughput sequencing reveals microRNAs and their targets in response to drought stress in wheat (*Triticum aestivum* L.). *Biotechnology & Biotechnological Equipment*, *33*, 465–471. https://doi.org/10.1080/13102818.2019.1586586

Huanca-Mamani, W., Arias-Carrasco, R., Cárdenas-Ninasivincha, S., Rojas-Herrera, M., Sepúlveda-Hermosilla, G., Caris-Maldonado, J. C., et al. (2018). Long non-coding RNAs responsive to salt and boron stress in the hyper-arid Lluteño maize from the Atacama desert. *Genes*, *9*, 170.

Huang, G.-T., Ma, S.-L., Bai, L.-P., Zhang, L., Ma, H., Jia, P., Liu, J., Zhong, M., & Guo, Z. F. (2012). Signal transduction during cold, salt, and drought stresses in plants. *Molecular Biology Reports*, *39*, 969–987. https://doi.org/10.1007/s11033-011-0823-1

Huo, C., Zhang, B., & Wang, R. (2022). Research progress on plant noncoding RNAs in response to low-temperature stress. *Plant Signaling & Behavior*, *17*(1), 2004035.

Imaduwage, I. U. H., & Hewadikaram, M. (2024). Predicted roles of long non-coding RNAs in abiotic stress tolerance responses of plants. *Molecular Horticulture*, *4*(1), 20.

Ismail, A. M., & Horie, T. (2017). Genomics, physiology, and molecular breeding approaches for improving salt tolerance. *Annual Review of Plant Biology*, *68*, 19–19.30.

Ito, H., Gaubert, H., Bucher, E., Mirouze, M., Vaillant, I., & Paszkowski, J. (2011). An siRNA pathway prevents transgenerational retrotransposition in plants subjected to stress. *Nature*, *472*, 115–119.

Ivanova, Z., Minkov, G., Gisel, A., Yahubyan, G., Minkov, I., Toneva, V., & Baev, V. (2022). The multiverse of plant small RNAs: How can we explore it? *International Journal of Molecular Sciences*, *23*(7), 3979.

Iwasaki, M., Hyvärinen, L., Piskurewicz, U., & Lopez-Molina, L. (2019). Noncanonical RNA-directed DNA methylation participates in maternal and environmental control of seed dormancy. *eLife*, *8*, e37434. https://doi.org/10.7554/eLife.37434

Jha, C. U., Nayyar, H., Mantri, N., & Siddique, K. H. (2021). Non-coding RNAs in legumes: Their emerging roles in regulating biotic/abiotic stress responses and plant growth and development. *Cells*, *10*(7), 1674.

Jha, U. C., Nayyar, H., Jha, R., Khurshid, M., Zhou, M., Mantri, N., & Siddique, K. H. M. (2020). Long non-coding RNAs: Emerging players regulating plant abiotic stress response and adaptation. *BMC Plant Biology*, *20*, 1–20. https://doi.org/10.1186/s12870-020-02595-x

Jiang, D., Zhou, L., Chen, W., Ye, N., Xia, J., & Zhuang, C. (2019). Overexpression of a microRNA-targeted NAC transcription factor improves drought and salt tolerance in rice via ABA-mediated pathways. *Rice*, *12*, 1–11. DOI: 10.1186/s12284-019-0334-6

Jiao, F., Pahwa, K., Manning, M., Dichy, N., & Geuten, K. (2019). Cold-induced antisense transcription of FLOWERING LOCUS C in distant grasses. *Frontiers in Plant Science*, *10*, 72.

Joshi, R. K., Megha, S., Basu, U., Rahman, M. H., & Kav, N. N. V. (2016). Genome-wide identification and functional prediction of long noncoding RNAs responsive to *Sclerotinia sclerotiorum* infection in *Brassica napus*. *PLoS One*, *11*, e0158784. https://doi.org/10.1371/journal.pone.0158784

Kamath, Ravi S., et al. "Effectiveness of specific RNA-mediated interference through ingested double-stranded RNA in Caenorhabditis elegans." Genome biology 2.1 (2000): research0002-1.

Kapusta, A., Kronenberg, Z., Lynch, V. J., Zhuo, X., Ramsay, L., Bourque, G., …, & Feschotte, C. (2013). Transposable elements are major contributors to the origin, diversification, and regulation of vertebrate long noncoding RNAs. *PLoS Genetics*, *9*(4), e1003470.

Katayama, S., Tomaru, Y., Kasukawa, T., Waki, K., Nakanishi, M., Nakamura, M., …, & Wahlestedt, C. (2005). Antisense transcription in the mammalian transcriptome. *Science*, *309*(5740), 1564–1566.

Kelley, D., & Rinn, J. (2012). Transposable elements reveal a stem cell-specific class of long noncoding RNAs. *Genome Biology*, *13*, 1–14.

KeshavarzAlikhani, H., Pourhamzeh, M., Seydi, H., Shokoohian, B., Hossein-Khannazer, N., Jamshidi-Adegani, F., …, & Vosough, M. (2022). Regulatory non-coding RNAs in familial hypercholesterolemia, theranostic applications. *Frontiers in Cell and Developmental Biology*, *10*, 894800.

Khandal, H., Parween, S., Roy, R., Meena, M. K., & Chattopadhyay, D. (2017). MicroRNA profiling provides insights into post-transcriptional regulation of gene expression in chickpea root apex under salinity and water deficiency. *Scientific Reports*, *7*, 1–14.

Khraiwesh, B., Arif, M. A., Seumel, G. I., Ossowski, S., Weigel, D., Reski, R., & Frank, W. (2010). Transcriptional control of gene expression by microRNAs. *Cell*, *140*, 111–122.

Khraiwesh, B., Zhu, J. K., & Zhu, J. (2012). Role of miRNAs and siRNAs in biotic and abiotic stress responses of plants. *Biochimica et Biophysica Acta (BBA) - Gene Regulatory Mechanisms*, *1819*(2), 137–148. https://doi.org/10.1016/j.bbagrm.2011.11.002

Kindgren, P., Ard, R., Ivanov, M., & Marquardt, S. (2018). Transcriptional read-through of the long noncoding RNA SVALKA governs plant cold acclimation. *Nature Communications*, *9*, 4561. https://doi.org/10.1038/s41467-018-07010-6

Kotak, S., Larkindale, J., Lee, U., von Koskull-Döring, P., Vierling, E., & Scharf, K. D. (2007). Complexity of the heat stress response in plants. *Current Opinion in Plant Biology*, *10*(3), 310–316.

Lai, E. C., Burks, C., & Posakony, J. W. (1998). The K box, a conserved 3′ UTR sequence motif, negatively regulates accumulation of enhancer of split complex transcripts. *Development*, *125*(20), 4077–4088.

Lamers, J., Van Der Meer, T., & Testerink, C. (2020). How plants sense and respond to stressful environments. *Plant Physiology*, *182*(4), 1624–1635.

Lee, R. C., & Ambros, V. (2001). An extensive class of small RNAs in *Caenorhabditis elegans*. *Science*, *294*(5543), 862–864.

Lee, R. C., Feinbaum, R. L., & Ambros, V. (1993). The C. elegans heterochronic gene lin-4 encodes small RNAs with antisense complementarity to lin-14. *Cell*, *75*(5), 843–854.

Li, S., Castillo-González, C., Yu, B., & Zhang, X. (2017). The functions of plant small RNAs in development and in stress responses. *The Plant Journal*, *90*(4), 654–670. https://doi.org/10.1111/tpj.13314

Li, S., Liu, J., Liu, Z., Li, X., Wu, F., & He, Y. (2014). Heat-induced tas1 target1 mediates thermotolerance via heat stress transcription factor A1a–directed pathways in Arabidopsis. *The Plant Cell*, *26*(4), 1764–1780.

Li, W., Cui, X., Meng, Z., Huang, X., Xie, Q., Wu, H., Jin, H., Zhang, D., & Liang, W. (2013). Transcriptional regulation of Arabidopsis miR168a and ARGONAUTE1 homeostasis in abscisic acid and abiotic stress responses. *Plant Physiology*, *158*, 1279–1292.

Liliane, T. N., & Charles, M. S. (2020). Factors affecting yield of crops. *Agronomy-Climate Change & Food Security, 9.*

Liu, H. H., Tian, X., Li, Y. J., Wu, C. A., & Zheng, C. C. (2008). Microarray-based analysis of stress-regulated microRNAs in *Arabidopsis thaliana. RNA, 14*, 836–843. https://doi.org/10.1261/rna.895308

Liu, J., Jung, C., Xu, J., Wang, H., Deng, S., Bernad, L., Arenas-Huertero, C., & Chua, N.-H. (2012). Genome-wide analysis uncovers regulation of long intergenic noncoding RNAs in Arabidopsis. *The Plant Cell, 24*, 4333–4345. https://doi.org/10.1105/tpc.112.102855

Long, R. C., Li, M. N., Kang, J. M., Zhang, T. J., Sun, Y., & Yang, Q. C. (2015). Small RNA deep sequencing identifies novel and salt-stress-regulated microRNAs from roots of *Medicago sativa* and *Medicago truncatula. Physiologia Plantarum, 154*, 13–27.

Lu, S., Sun, Y. H., & Chiang, V. L. (2008). Stress-responsive microRNAs in Populus. *The Plant Journal: For Cell and Molecular Biology, 55*, 131–151. https://doi.org/10.1111/j.1365-313X.2008.03497.x

Lu, X., Guan, Q., & Zhu, J. (2013). Downregulation of CSD2 by a heat-inducible miR398 is required for thermotolerance in Arabidopsis. *Plant Signaling & Behavior, 8*, e24952. https://doi.org/10.4161/psb.24952

Ma, X., Zhao, F., & Zhou, B. (2022). The characters of non-coding RNAs and their biological roles in plant development and abiotic stress response. *International Journal of Molecular Sciences, 23*, 4124.

Mallory, A. C., & Vaucheret, H. (2006). Functions of microRNAs and related small RNAs in plants. *Nature Genetics, 38*, S31–S36.

Mangrauthia, S. K., Bhogireddy, S., Agarwal, S., Prasanth, V. V., Voleti, S. R., Neelamraju, S., & Subrahmanyam, D. (2017). Genome-wide changes in microRNA expression during short and prolonged heat stress and recovery in contrasting rice cultivars. *Journal of Experimental Botany, 68*, 2399–2412. https://doi.org/10.1093/jxb/erx111

Matsui, A., Nguyen, A. H., Nakaminami, K., & Seki, M. (2013). Arabidopsis non-coding RNA regulation in abiotic stress responses. *International Journal of Molecular Sciences, 14*(11), 22642–22654. https://doi.org/10.3390/ijms141122642

Mattick, J. S. (2001). Non-coding RNAs: The architects of eukaryotic complexity. *EMBO Reports, 2*(11), 986–991.

Maurin, T., Cazalla, D., Yang, J. S., Bortolamiol-Becet, D., & Lai, E. C. (2012). RNase III-independent microRNA biogenesis in mammalian cells. *RNA, 18*(12), 2166–2173.

Megha, S., Basu, U., & Kav, N. N. (2018). Regulation of low temperature stress in plants by microRNAs. *Plant, Cell & Environment, 41*, 1–15. https://doi.org/10.1111/pce.12956

Mittler, R. (2006). Abiotic stress, the field environment and stress combination. *Trends in Plant Science, 11*(1), 15–19.

Moraes de Freitas, G. P., Basu, S., Ramegowda, V., Thomas, J., Benitez, L. C., Braga, E. B., & Pereira, A. (2019). Physiological and transcriptional responses to low-temperature stress in rice genotypes at the reproductive stage. *Plant Signaling & Behavior, 14*(4), e1581557.

Mraheil, M. A., Billion, A., Kuenne, C., Pischimarov, J., Kreikemeyer, B., Engelmann, S., …, & Hain, T. (2010). Comparative genome-wide analysis of small RNAs of major Gram-positive pathogens: From identification to application. *Microbial Biotechnology, 3*(6), 658–676.

Mundy, J., Yamaguchi-Shinozaki, K., & Chua, N. H. (1990). Nuclear proteins bind conserved elements in the abscisic acid-responsive promoter of a rice rab gene. *Proceedings of the National Academy of Sciences of the United States of America, 87*, 1406–1410.

Ondrasek, G., Jelovica Badovinac, I., Peter, R., Petravić, M., Macan, J., & Rengel, Z. (2022). Humates and chlorides synergistically increase Cd phytoaccumulation in strawberry fruits, heightening health risk from Cd in human diet. *Exposure and Health, 14*(2), 393–410.

Pang, J., Zhang, X., Ma, X., & Zhao, J. (2019). Spatio-temporal transcriptional dynamics of maize long non-coding RNAs responsive to drought stress. *Genes, 10*, 138.

Pasquinelli, A. E. (2012). MicroRNAs and their targets: Recognition, regulation and an emerging reciprocal relationship. *Nature Reviews Genetics, 13*, 271–282. https://doi.org/10.1038/nrg3162

Paul, S., Kundu, A., & Pal, A. (2011). Identification and validation of conserved microRNAs along with their differential expression in roots of *Vigna unguiculata* grown under salt stress. *Plant Cell, Tissue and Organ Culture, 105*, 233–242.

Pheasant, M., & Mattick, J. S. (2007). Raising the estimate of functional human sequences. *Genome Research, 17*(9), 1245–1253.

Qi, X., Xie, S., Liu, Y., Yi, F., & Yu, J. (2013). Genome-wide annotation of genes and noncoding RNAs of foxtail millet in response to simulated drought stress by deep sequencing. *Plant Molecular Biology, 83*, 459–473.

Qin, T., Zhao, H., Cui, P., Albesher, N., & Xiong, L. (2017). A nucleus-localized long noncoding RNA enhances drought and salt stress tolerance. *Plant Physiology, 175*(3), 1321–1336. https://doi.org/10.1104/pp.17.00574

Ramirez Gonzales, L., Shi, L., Bergonzi, S. B., Oortwijn, M., Franco-Zorrilla, J. M., & Solano-Tavira, R., et al. (2021). Potato CYCLING DOF FACTOR 1 and its lncRNA counterpart StFLORE link tuber development and drought response. *Plant Journal, 105*, 855–869. https://doi.org/10.1111/tpj.15093

Rath, E. C., Pitman, S., Cho, K. H., & Bai, Y. (2017). Identification of streptococcal small RNAs that are putative targets of RNase III through bioinformatics analysis of RNA sequencing data. *BMC Bioinformatics, 18*, 111–120.

Ravichandran, S., Ragupathy, R., Edwards, T., Domaratzki, M., & Cloutier, S. (2019). MicroRNA-guided regulation of heat stress response in wheat. *BMC Genomics, 20*, 488. https://doi.org/10.1186/s12864-019-5799-6

Robertson, M. P., & Joyce, G. F. (2012). The origins of the RNA world. *Cold Spring Harbor Perspectives in Biology, 4*(5), a003608.

Ron, M., Saez, M. A., Williams, L. E., Fletcher, J. C., & McCormick, S. (2010). Proper regulation of a sperm-specific cis-nat-siRNA is essential for double fertilization in *Arabidopsis. Genes & Development, 24*, 1010–1021. https://doi.org/10.1101/gad.1882810

Sailaja, B., Anjum, N., Vishnu Prashanth, V., Sarla, N., Subrahmanyam, D., Voleti, S. R., Viraktamath, B. C., & Mangrauthia, S. K. (2014). Comparative study of susceptible and tolerant genotype reveals efficient recovery and root system contributes to heat stress tolerance in rice. *Plant Molecular Biology Reporter, 32*, 1228–1240. https://doi.org/10.1007/s11105-014-0728-y

Saleem, M., Fariduddin, Q., & Janda, T. (2021). Multifaceted role of salicylic acid in combating cold stress in plants: A review. *Journal of Plant Growth Regulation, 40*(2), 464–485.

Sanwal, S. K., Kumar, P., Kesh, H., Gupta, V. K., Kumar, A., Kumar, A., …, & Kumar, P. (2022). Salinity stress tolerance in potato cultivars: Evidence from physiological and biochemical traits. *Plants, 11*(14), 1842.

Shang, R., Lee, S., Senavirathne, G., & Lai, E. C. (2023). microRNAs in action: Biogenesis, function and regulation. *Nature Reviews Genetics, 24*(12), 816–833.

Shang, R., Zhang, F., Xu, B., Xi, H., Zhang, X., Wang, W., & Wu, L. (2015). Ribozyme-enhanced single-stranded Ago2-processed interfering RNA triggers efficient gene silencing with fewer off-target effects. *Nature Communications*, *6*(1), 8430.

Shang, Y., Yan, L., Liu, Z. Q., Zheng, C., Chao, M., Xin, Q., Wu, F. Q., Wang, X. F., Du, S. Y., Jiang, T., et al. (2010). The Mg-chelatase H subunit of *Arabidopsis* antagonizes a group of transcription repressors to relieve ABA-responsive genes of inhibition. *The Plant Cell*, *22*, 1909–1935.

Shi, B. J., & Hussain, S. S. (2016). miRNA/siRNA-based approaches to enhance drought tolerance of barley and wheat under drought stress. In *Water stress and crop plants: A sustainable approach* (pp. 248–260). CRC Press.

Shi, G. Q., Fu, J. Y., Rong, L. J., Zhang, P. Y., Guo, C. J., & Xiao, K. (2018). TaMIR1119, a miRNA family member of wheat (*Triticum aestivum*), is essential in the regulation of plant drought tolerance. *Journal of Integrative Agriculture*, *17*, 2369–2378. DOI: 10.1016/S2095-3119(17)61879-3

Shuai, P., Liang, D., Tang, S., Zhang, Z., Ye, C. Y., & Su, Y., et al. (2014). Genome-wide identification and functional prediction of novel and drought-responsive lincRNAs in *Populus trichocarpa*. *Journal of Experimental Botany*, *65*, 4975–4983.

Skirycz, A., Vandenbroucke, K., Clauw, P., Maleux, K., De Meyer, B., Dhondt, S., …, & Inzé, D. (2011). Survival and growth of Arabidopsis plants given limited water are not equal. *Nature Biotechnology*, *29*(3), 21.

Song, T., Xu, H., Sun, N., Jiang, L., Tian, P., Yong, Y., …, & Cui, G. (2017). Metabolomic analysis of alfalfa (*Medicago sativa* L.) root-symbiotic rhizobia responses under alkali stress. *Frontiers in Plant Science*, *8*, 1208.

Song, X., Li, Y., Cao, X., & Qi, Y. (2019). MicroRNAs and their regulatory roles in plant–environment interactions. *Annual Review of Plant Biology*, *70*(1), 489–525.

Song, X. M., Liu, G. F., Huang, Z. N., Duan, W. K., Tan, H. W., & Li, Y., et al. (2016). Temperature expression patterns of genes and their co-expression with LncRNAs revealed by RNA-Seq in non-heading Chinese cabbage. *BMC Genomics*, *17*, 297.

Statello, L., Guo, C. J., Chen, L. L., & Huarte, M. (2021). Gene regulation by long non-coding RNAs and its biological functions. *Nature Reviews Molecular Cell Biology*, *22*(2), 96–118.

Sun, Z., Wang, Y., Mou, F., Tian, Y., Chen, L., Zhang, S., Jiang, Q., & Li, X. (2016). Genome-wide small RNA analysis of soybean reveals auxin-responsive microRNAs that are differentially expressed in response to salt stress in root apex. *Frontiers in Plant Science*, *6*, 1273.

Sunkar, R., & Zhu, J. K. (2004). Novel and stress-regulated microRNAs and other small RNAs from *Arabidopsis*. *The Plant Cell*, *16*, 2001–2019. https://doi.org/10.1105/tpc.104.022830

Swiezewski, S., Liu, F., Magusin, A., & Dean, C. (2009). Cold-induced silencing by long antisense transcripts of an *Arabidopsis* Polycomb target. *Nature*, *462*, 799–802. https://doi.org/10.1038/nature08618

Tan, X., Li, S., Hu, L., & Zhang, C. (2020). Genome-wide analysis of long noncoding RNAs (lncRNAs) in two contrasting rapeseed (*Brassica napus* L.) genotypes subjected to drought stress and re-watering. *BMC Plant Biology*, *20*, 81. https://doi.org/10.1186/s12870-020-2286-9

Tang, J., & Chu, C. (2017). MicroRNAs in crop improvement: Fine-tuners for complex traits. *Nature Plants*, *3*, 17077. https://doi.org/10.1038/nplants.2017.77

Tang, K., Zhao, L., Ren, Y., Yang, S., Zhu, J. K., & Zhao, C. (2020). The transcription factor ICE1 functions in cold stress response by binding to the promoters of CBF and COR genes. *Journal of Integrative Plant Biology*, *62*(3), 258–263.

Tang, Z., Zhang, L., Xu, C., Yuan, S., Zhang, F., Zheng, Y., & Zhao, C. (2012). Uncovering small RNA-mediated responses to cold stress in a wheat thermosensitive genic male-sterile line by deep sequencing. *Plant Physiology*, *159*(2), 721–738. https://doi.org/10.1104/pp.112.196048

Tao, X., Li, M., Zhao, T., Feng, S., Zhang, H., Wang, L., Han, J., Gao, M., Lu, K., Chen, Q., Zhou, B., & Guan, X. (2021). Neofunctionalization of a polyploidization-activated cotton long intergenic non-coding RNA DAN1 during drought stress regulation. *Plant Physiology*, *186*, 2152–2168. https://doi.org/10.1093/plphys/kiab077

Thiebaut, F., Grativol, C., Carnavale-Bottino, M., Rojas, C. A., Tanurdzic, M., Farinelli, L., Martienssen, R. A., Hemerly, A. S., & Ferreira, P. C. G. (2012). Computational identification and analysis of novel sugarcane microRNAs. *BMC Genomics*, *13*(1), 290. https://doi.org/10.1186/1471-2164-13-290

Thomashow, M. F. (2010). Molecular basis of plant cold acclimation: Insights gained from studying the CBF cold response pathway. *Plant Physiology*, *154*(2), 571–577.

Tian, H., Guo, F., Zhang, Z., Ding, H., Meng, J., Li, X., Peng, Z., & Wan, S. (2020). Discovery, identification, and functional characterization of long noncoding RNAs in *Arachis hypogaea* L. *BMC Plant Biology*, *20*, 1–16.

Tiwari, B., Habermann, K., Arif, M. A., Weil, H. L., Garcia-Molina, A., Kleine, T., Mühlhaus, T., & Frank, W. (2020). Identification of small RNAs during cold acclimation in *Arabidopsis thaliana*. *BMC Plant Biology*, *20*(1), 298. https://doi.org/10.1186/s12870-020-02511-3

Tiwari, S., & Lata, C. (2021). Recent advancements in long noncoding RNA-mediated stress responses in rice. In S. K. Upadhyay (Ed.), *Long noncoding RNAs in plants* (pp. 63–74). Academic Press.

Umar, O. B., Ranti, L. A., Abdulbaki, A. S., Bola, A. L., Abdulhamid, A. K., Biola, M. R., & Victor, K. O. (2021). *Stresses in plants: Biotic and abiotic. Current trends in wheat research*, 1–8.

United Nations. 2023. Available online: https://www.un.org (accessed on 29 November 2022).

Uszczynska-Ratajczak, B., Lagarde, J., Frankish, A., Guigó, R., & Johnson, R. (2018). Towards a complete map of the human long non-coding RNA transcriptome. *Nature Reviews Genetics*, *19*(9), 535–548.

Van der Hauwaert, C., Glowacki, F., Pottier, N., & Cauffiez, C. (2019). Non-Coding RNAs as new therapeutic targets in the context of renal fibrosis. *International Journal of Molecular Sciences*, *20*(8), 1977. RNA and the Reproductive System, 155–170.

Voinnet, O. (2009). Origin, biogenesis, and activity of plant microRNAs. *Cell*, *136*, 669–687.

Wang, A., Hu, J., Gao, C., Chen, G., Wang, B., Lin, C., Song, L., Ding, Y., & Zhou, G. (2019). Genome-wide analysis of long non-coding RNAs unveils the regulatory roles in the heat tolerance of Chinese cabbage (*Brassica rapa* ssp. chinensis). *Scientific Reports*, *9*, 1–14. 10.1038/s41438-021-00484-4

Wang, G., & Reinke, V. (2008). A C. elegans Piwi, PRG-1, regulates 21U-RNAs during spermatogenesis. *Current Biology*, *18*(12), 861–867.

Wang, J., Yu, W., Yang, Y., Li, X., Chen, T., Liu, T., Ma, N., Yang, X., Liu, R., & Zhang, B. (2015). Genome-wide analysis of tomato long noncoding RNAs and identification as endogenous target mimic for microRNA in response to TYLCV infection. *Scientific Reports*, *5*, 16946. https://doi.org/10.1038/srep16946

Wang, K. C., & Chang, H. Y. (2011). Molecular mechanisms of long noncoding RNAs. *Molecular Cell*, *43*, 904–914.

Wang, P., Dai, L., Ai, J., Wang, Y., & Ren, F. (2019). Identification and functional prediction of cold-related long noncoding RNA (lncRNA) in grapevine. *Scientific Reports*, *9*, 6638.

Wang, P., Yang, Y., Shi, H., Wang, Y., & Ren, F. (2019). Small RNA and degradome deep sequencing reveal respective roles of cold-related microRNAs across Chinese wild grapevine and cultivated grapevine. *BMC Genomics*, *20*, 740. https://doi.org/10.1186/s12864-019-6111-5

Wang, Q., Liu, N., Yang, X., Tu, L., & Zhang, X. (2016). Small RNA-mediated responses to low- and high-temperature stresses in cotton. *Scientific Reports*, *6*, 35558. https://doi.org/10.1038/srep35558

Wei, X., Ke, H., Wen, A., Gao, B., Shi, J., & Feng, Y. (2021). Structural basis of microRNA processing by Dicer-like 1. *Nature Plants*, *7*, 1389–1396. https://doi.org/10.1038/s41477-021-01000-1

Wiegmann, M., Maurer, A., Pham, A., March, T. J., Al-Abdallat, A., Thomas, W. T., …, & Pillen, K. (2019). Barley yield formation under abiotic stress depends on the interplay between flowering time genes and environmental cues. *Scientific Reports*, *9*(1), 6397.

Wu, H., Yang, L., & Chen, L. L. (2017). The diversity of long noncoding RNAs and their generation. *Trends in Genetics*, *33*(8), 540–552.

Wu, P., Mo, Y., Peng, M., Tang, T., Zhong, Y., Deng, X., …, & Xiong, W. (2020). Emerging role of tumor-related functional peptides encoded by lncRNA and circRNA. *Molecular Cancer*, *19*(1), 14.

Xin, M., Wang, Y., Yao, Y., Song, N., Hu, Z., Qin, D., et al. (2011). Identification and characterization of wheat long non-protein coding RNAs responsive to powdery mildew infection and heat stress by using microarray analysis and SBS sequencing. *BMC Plant Biology*, *11*, 61.

Yamaguchi-Shinozaki, K., & Shinozaki, K. (1994). A novel cis-acting element in an Arabidopsis gene is involved in responsiveness. *Plant Cell*, *6*, 251–264.

Yan, J., Zhao, C., Zhou, J., Yang, Y., Wang, P., Zhu, X., Tang, G., Bressan, R. A., & Zhu, J. K. (2016). The miR165/166 mediated regulatory module plays critical roles in ABA homeostasis and response in *Arabidopsis thaliana*. *PLoS Genetics*, *12*, e1006416.

Yan, Q., Wu, F., Yan, Z., Li, J., Ma, T., Zhang, Y., et al. (2019). Differential co-expression networks of long non-coding RNAs and mRNAs in *Cleistogenes songorica* under water stress and during recovery. *BMC Plant Biology*, *19*, 23. doi: 10.1186/s12870-018-1626-5

Yang, J. S., & Lai, E. C. (2011). Alternative miRNA biogenesis pathways and the interpretation of core miRNA pathway mutants. *Molecular Cell*, *43*(6), 892–903.

Yang, X., Liu, F., Zhang, Y., Wang, L., & Cheng, Y. F. (2017). Cold-responsive miRNAs and their target genes in the wild eggplant species *Solanum aculeatissimum*. *BMC Genomics*, *18*, 1000. https://doi.org/10.1186/s12864-017-4341-y

Yu, Y., Zhang, Y., Chen, X., & Chen, Y. (2019). Plant noncoding RNAs: Hidden players in development and stress responses. *Annual Review of Cell and Developmental Biology*, *35*, 407–431. https://doi.org/10.1146/annurev-cellbio-100818-125218

Yuan, J. H., Yang, F., Chen, B. F., Lu, Z., Huo, X. S., Zhou, W. P., Wang, F., & Sun, S. H. (2011). The histone deacetylase 4/SP1/microRNA-200a regulatory network contributes to aberrant histone acetylation in hepatocellular carcinoma. *Hepatology*, *54*, 2025–2035.

Zahra, N., Hafeez, M. B., Ghaffar, A., Kausar, A., Al Zeidi, M., Siddique, K. H., & Farooq, M. (2023). Plant photosynthesis under heat stress: Effects and management. *Environmental and Experimental Botany*, *206*, 105178.

Zhang, C., Dong, Y., Ren, Y., Wang, S., & Yang, M. (2023). Conjoint analysis of genome-wide lncRNA and mRNA expression during the salicylic acid response in *Populus × euramericana*. *Plants*, *12*(6), 1377. https://doi.org/10.3390/plants12061377

Zhang, C., Gu, K., Gu, D., Zhang, S., & Wu, J. (2021). Quantifying the effect of low-temperature events on the grain quality formation of wheat. *Journal of Cereal Science*, *100*, 103257.

Zhang, C., Tang, G., Peng, X., Sun, F., Liu, S., & Xi, Y. (2018). Long non-coding RNAs of switchgrass (*Panicum virgatum* L.) in multiple dehydration stresses. *BMC Plant Biology*, *18*, 79.

Zhang, H., Liu, S., Ren, T., Niu, M., Liu, X., Liu, C., …, & Xia, X. (2023). Crucial abiotic stress regulatory network of NF-Y transcription factor in plants. *International Journal of Molecular Sciences*, *24*(5), 4426.

Zhang, H., Zhao, Y., & Zhu, J. K. (2020). Thriving under stress: How plants balance growth and the stress response. *Developmental Cell*, *55*(5), 529–543.

Zhang, H., Zhu, J., Gong, Z., & Zhu, J. K. (2022). Abiotic stress responses in plants. *Nature Reviews Genetics*, *23*(2), 104–119.

Zhang, J., Xu, Y., Huan, Q., & Chong, K. (2009). Deep sequencing of *Brachypodium* small RNAs at the global genome level identifies microRNAs involved in cold stress response. *BMC Genomics*, *10*, 449. https://doi.org/10.1186/1471-2164-10-449

Zhang, L., Meng, X., Pan, C., Qu, F., Gan, W., Xiang, Z., …, & Li, D. (2020). piR-31470 epigenetically suppresses the expression of glutathione S-transferase pi 1 in prostate cancer via DNA methylation. *Cellular Signalling*, *67*, 109501.

Zhang, Q., Zhu, Y., Cao, X., Tan, W., Yu, J., Lu, Y., …, & Li, E. (2023). The epigenetic regulatory mechanism of PIWI/piRNAs in human cancers. *Molecular Cancer*, *22*(1), 45.

Zhang, S., Wang, Y., Li, K., Zou, Y., Chen, L., & Li, X. (2014). Identification of cold-responsive miRNAs and their target genes in nitrogen-fixing nodules of soybean. *International Journal of Molecular Sciences*, *15*, 13596–13614. https://doi.org/10.3390/ijms150813596

Zhang, W., Han, Z., Guo, Q., Liu, Y., Zheng, Y., & Wu, F., et al. (2014). Identification of maize long non-coding RNAs responsive to drought stress. *PLoS One*, *9*, e98958.

Zhang, Y., Xu, J., Li, R., Ge, Y., Li, Y., & Li, R. (2023). Plants' response to abiotic stress: Mechanisms and strategies. *International Journal of Molecular Sciences*, *24*(13), 10915.

Zhang, Y., Zhou, Y., Zhu, W., Liu, J., & Cheng, F. (2022). Non-coding RNAs fine-tune the balance between plant growth and abiotic stress tolerance. *Frontiers in Plant Science*, *13*, 965745.

Zhao, C., Liu, B., Piao, S., Wang, X., Lobell, D. B., Huang, Y., …, & Asseng, S. (2017). Temperature increase reduces global yields of major crops in four independent estimates. *Proceedings of the National Academy of Sciences*, *114*(35), 9326–9331.

Zhao, G., Zhao, Y., Lou, W., Su, J., Wei, S., Yang, X., …, & Shen, W. (2019). Nitrate reductase-dependent nitric oxide is crucial for multi-walled carbon nanotube-induced plant tolerance against salinity. *Nanoscale*, *11*(21), 10511–10523.

Zhao, J., He, Q., Chen, G., Wang, L., & Jin, B. (2016). Regulation of noncoding RNAs in heat stress responses of plants. *Frontiers in Plant Science*, *7*, 1213. https://doi.org/10.3389/fpls.2016.01213

Zheng, C., Ye, M., Sang, M., & Wu, R. (2019). A regulatory network for miR156-SPL module in *Arabidopsis thaliana*. *International Journal of Molecular Sciences*, *20*, 6166. https://doi.org/10.3390/ijms20246166

Zheng, Q., Reid, G., Eccles, M. R., & Stayner, C. (2022). Non-coding RNAs as potential biomarkers and therapeutic targets in polycystic kidney disease. *Frontiers in Physiology*, *13*, 1006427.

Zhong, S. H., Liu, J. Z., Jin, H., Lin, L., Li, Q., Chen, Y., et al. (2013). Warm temperatures induce transgenerational epigenetic release of RNA silencing by inhibiting siRNA biogenesis in Arabidopsis. *Proceedings of the National Academy of Sciences of the United States of America*, *110*, 9171–9176. https://doi.org/10.1073/pnas.1219655110

Zhou, M., Li, D., Li, Z., Hu, Q., Yang, C., Zhu, L., & Luo, H. (2013). Constitutive expression of a miR319 gene alters plant development and enhances salt and drought tolerance in transgenic creeping bentgrass. *Plant Physiology*, *161*, 1375–1391. https://doi.org/10.1104/pp.112.208702

Zhou, X., Ao, X., Jia, Z., Li, Y., Kuang, S., Du, C., …, & Liu, Y. (2022). Non-coding RNA in cancer drug resistance: Underlying mechanisms and clinical applications. *Frontiers in Oncology*, *12*, 951864.

Zhou, X., Wang, G., Sutoh, K., Zhu, J. K., & Zhang, W. (2008). Identification of cold-inducible microRNAs in plants by transcriptome analysis. *Biochimica et Biophysica Acta*, *1779*, 780–788. https://doi.org/10.1016/j.bbagrm.2008.04.005

Zhou, Z. S., Huang, S. Q., & Yang, Z. M. (2008). Bioinformatic identification and expression analysis of new microRNAs from *Medicago truncatula*. *Biochemical and Biophysical Research Communications*, *374*, 538–542.

Zhou, Z. S., Zeng, H. Q., Liu, Z. P., & Yang, Z. M. (2012). Genome-wide identification of *Medicago truncatula* microRNAs and their targets reveals their differential regulation by heavy metal. *Plant, Cell & Environment*, *35*, 86–99.

Zhu, J. K. (2016). Abiotic stress signaling and responses in plants. *Cell*, *167*(2), 313–324.

14 The Silent Guardians
Understanding ncRNAs in Plant Abiotic Stress Management

Shumaila Ijaz, Javed Iqbal, Banzeer Ahsan Abbasi, Naila Ijaz, Zakir Ullah, Rooma Waqar, Ayush Madan, Tabassum Yaseen, Ghulam Murtaza, Rashid Iqbal, Tayyaba Shahzadi and Seweta Srivastava

14.1 INTRODUCTION

Abiotic stresses are responsible for 50–70% of crop yield reductions worldwide (Ganguly et al., 2020). It is widely acknowledged that agriculture worldwide is susceptible to alterations in the climate. Effective mitigation techniques must be implemented as soon as feasible to reduce the adverse effects of climate change (Anekwe et al., 2024). Raising awareness of agriculture's adaptability requires a deep comprehension of the factors and motivations underlying the suggestion of successful climate-smart farming practices (Taishykov et al., 2024). The current state of global climate change is a critical concern because of its multifaceted infiltration capacity, which has presented significant obstacles to the food production agroecosystem and related socio-economic sectors (Amedie, 2013). Unexpected changes in the environment and global warming make it difficult to meet objectives for global food security. Additionally, the world's arable land area is contracting by around 1.2–1.5 billion hectares (Ahmed et al., 2021). A key tactic to address climate change is the growth of the agriculture sector with improved climatic resilience. Abiotic stressors, which are growing more common, have a higher detrimental effect on plants as a result of climate change, as may be observed on a vast geographic scale. A total of 96.5% of the world's arable land is reportedly stressed in some way (Apostolova, 2023; Imran et al., 2021). The atmospheric enrichment of greenhouse gases caused rainfall patterns to change and the average global surface temperature to rise by 0.85°C between the years 1880 and 2012 (Awanthi & Navaratne, 2018). Since a 1°C increase in temperature increases a plant's water demand by 4–5%, it is expected that crop cultivation in drought-affected areas will become more difficult (Sachdev et al., 2021).

Raising the temperature may cause chemical contaminants to bio-transform into more dangerous or bioactive forms, which could aggravate environmental irritability and interfere with plant homeostasis (Mansinhos et al., 2024). A decrease in precipitation and an increase in temperature promote the volatilisation of xenobiotic compounds as persistent organic pollutants, which exacerbates air pollution. However, excessive precipitation increases the amount of air pollutants deposited on land and strengthens the process of pollutants and nutrients from the soil leaking into groundwater, leading to nutrient imbalance, salinity, aquifer contamination and soil pollution (Behera et al., 2024; Di Micco, 2024). Variations in temperature and precipitation, nutritional imbalances, waterlogging and other factors are examples of temporary and competitive plant growth limits under normal circumstances. Nevertheless, major weather events and changes in typical weather patterns cause stress to prolong and worsen, significantly beyond plant function, beyond the point of recovery.

Crops are grown in agro-environments that are continuously changing and typically unsuitable for their opulent development. The global research community faces significant challenges due to abiotic factors of the agro-ecosystem, including the presence of salt (beyond threshold limit), the lack of phyto-available water, toxic elements like Cd, As, Cr and Al, the restrictions of essential phyto-nutrients and unpredictable agro-climatic fluctuations. The global yield gap is widening due to the multifaceted challenges given by abiotic stress, either by itself or in conjunction with biotic variables, hence threatening food security. The researchers' top priority at the moment is to demonstrate how crops (i) recognise stressor-mediated signals and (ii) adapt to harsh environments. It is well recognised that abiotic stressors can disrupt the phytohormonal system and increase the production of TFs, which in turn trigger the activation of stress-reactive genes. ROS, or reactive oxygen species, frequently serve as essential molecules and are crucial in triggering downstream metabolic machinery throughout the whole process, i.e., stress response. One of the well-known tactics is plant/crop breeding, which usually depends on finding a specific crop that is appropriate for the intended agroecosystem. The breeding process acclimatises the varieties that are being tested in a particular agro-environment where climate change is significant. Advanced remote-sensing and GIS-based technologies and methods for crop performance monitoring might aid in the creation of appropriate agronomic practices.

 DOI: 10.1201/9781003583257-14

Novel crops may become the most important crops soon due to climate change and the sustainability of agricultural systems. However, one of the most important factors is the ability of current crops to adapt to the new agro-ecosystem.

'Omics-based' methods may not only offer a platform for the creation of genome-edited varieties with desired features, but they also reduce the period due to the advancements in genetic engineering and biotechnological interventions.

Unquestionably, one of the most important advantages of these genome editing technologies is their capacity to add complicated features that are otherwise impossible to incorporate using conventional upgrading approaches. Above all, plant biotechnology is essential to crop genome editing, which aims to advance both mankind and science at the same time. Furthermore, wild relatives of a variety of crops have been searched for their potential as trustworthy sources of significant features (osmoprotection, tolerance to diverse abiotic stressors through secondary metabolite synthesis). In addition to the genetic viewpoint, the metabolome profile showed promise in depicting the crops' reactions to abiotic stressors (Figure 14.1). Non-coding RNAs (ncRNAs) have become a vital bioactive element in recent decades, contributing significantly to both genomic and phenotypic diversity. In eukaryotes, including plants, only around 2% of the 90% of transcribed RNAs get translated into proteins. Once thought to have no function in gene functioning, residual parts have now been found to primarily function as ncRNA transcripts that are crucial for regulating several metabolic and developmental processes, including stress responses. Over the past ten years, a significant amount of work has been done to understand how plants respond to the elements of the agro-environment. When a plant is subjected to one or more stressors, some stress-induced entities (genes, proteins etc.) and their intricate web of connections have been discovered. It has been noted that the exact regulation and expression of genes at different levels control stress-induced reactions (Shiade et al., 2024).

14.2 BACKGROUND, TYPES AND PRIMARY FEATURES OF ncRNAs

Many intricate mechanisms that crops typically use at the molecular level to maintain their 'cellular homeostasis' when exposed to adverse agro-environmental factors have been successfully identified by the global research community (Zhang et al., 2019). According to their ability to encode proteins, RNAs can be classified

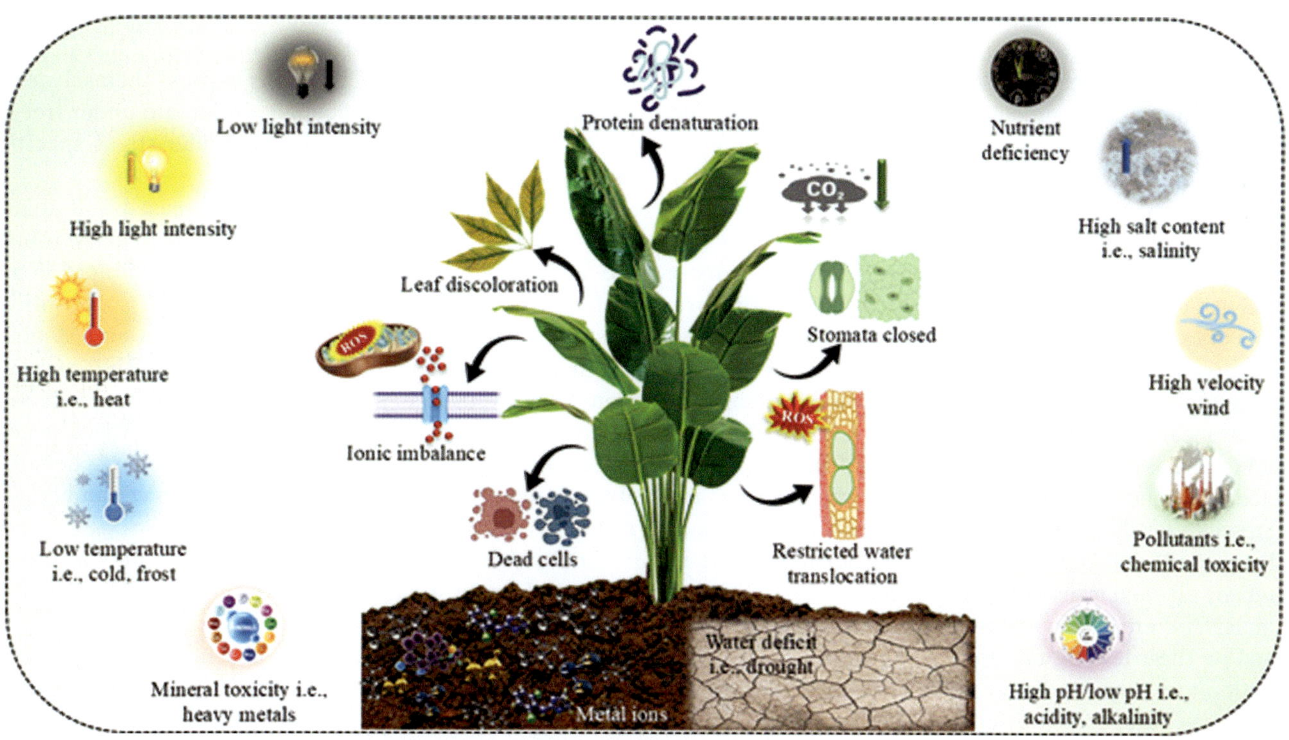

FIGURE 14.1 The visual representation of the typical reactions of plants to a range of abiotic stressors, including intense and low light levels, harsh heat and cold, water shortages (drought) and soil elements like salt and nutritional deficiencies. Specific physiological and biochemical reactions are triggered by each stressor: ionic imbalances can cause cell death, leaf browning signals stress and stomatal closure minimises water loss. Additionally emphasised are the effects of pollution and the toxicity of heavy metals. All things considered, this figure highlights the intricacy of plant adaptations to harsh abiotic stress, demonstrating their adaptability and survival strategies.

as either non-coding or coding (Zhang et al., 2022). The discovery of 21–24-nt-long miRNAs, one of the first regulatory ncRNAs to be identified, and their controlling mechanisms of tension-induced gene expression/silencing and diverse developmental processes have significantly advanced our understanding of plant gene regulation. These findings have opened the door for the identification of numerous new ncRNAs, including long ncRNAs (lncRNAs), and their roles in regulating a broad range of biological functions, such as plant development and stress-induced responses. A growing number of miRNAs in plants have been identified as a result of the quick advancement of instruments and techniques (Zhang, 2015). Because of their tissue and stage specificity, variable subcellular localisation and spatiotemporal expression, ncRNAs are essential components of a vast eukaryotic RNA-regulated network. According to their biogenesis and modes of action, ncRNAs may be divided into two main classes: housekeeping ncRNAs and regulatory ncRNAs. According to Ponting et al., the former group includes constitutively expressed ribosomal, transfer, small nuclear and small nucleolar RNAs. The term 'sRNA' usually refers to regulatory ncRNAs with a length of less than 200 nucleotides, whereas 'lncRNA' refers to those with a length of more than 200 nucleotides.

The parallels between mRNAs and lncRNAs are many. The majority of them depend on RNA polymerase II for alternative splicing, transcription, polyadenylation and 5′-UTR capping. There are a few slight variations between mRNAs and lncRNAs, though. The length of mRNA transcripts is greater than that of lncRNAs. The latter have been less conserved across species, have fewer exons, a lower level of expression and higher variability in their natural expression during evolution. Furthermore, unlike mRNAs, which are produced in the nucleus and transported to the cytosol for protein translation, most lncRNAs are present in the cytosol or inner compartments of cells or in the nucleus, where they are connected to chromatin.

The function of lncRNAs is often revealed by their subcellular localisation. LncRNAs control a number of important biological processes in both plants and animals. These include epigenetic structural chromatin remodelling, the expression levels of neighbouring protein-coding genes (Zhang et al., 2014) and the control of blooming time in plants by cis- and trans-regulation of genes.

ncRNAs are often categorised into various types based on their length, chromosomal location, biological function and sequence/structure conservation. They are sometimes categorised as either multi-exonic (spliced), which indicates that they have numerous exons, or mono-exonic (unspliced). ncRNAs are further classified into the following groups based on their length: medium-sized ncRNAs (31–200 nt) > short ncRNAs or sRNAs (18–30 nt) > long ncRNAs (200+ nt). CircRNAs are another new family of ncRNAs that were described by Ye et al.

Only a few plant lncRNAs have had their biological roles fully examined and validated. A total of 6480 lncRNAs (stress-induced or organ-specific expression) were identified by Liu et al. (2012) as being implicated among the 200 transcriptome datasets of *Arabidopsis thaliana*. Later in 2014, Wang et al. identified 37,238 long non-coding NATs (lncNATs) using the same model plant (*A. thaliana*). More than 26,066 of these were linked to mRNAs that were discovered (Wang et al., 2014). Polyadenylated lncRNAs were found in *A. thaliana* seedlings subjected to a variety of stressors with the use of RNA-sequencing methods. When exposed to drought stress, 125 stress-responsive lncRNAs were shown to be involved in other agricultural plants, such as wheat, whereas 584 lncRNAs were present.

Polyadenylated lncRNAs were found in *A. thaliana* seedlings subjected to a variety of stressors with the use of RNA-sequencing methods. When exposed to drought stress, 125 stress-responsive lncRNAs were shown to be involved in other crop plants, such as wheat, whereas millet showed 584 lncRNAs (Qi et al., 2013). Moreover, additional lncRNAs were discovered in *Brassica*, *Populus trichocarpa* and *Prunus persica*, among other fruits and crops, using RNA sequencing methods (Shuai et al., 2014). Numerous lncRNAs that are produced by stress have been identified, and their functions in various plants have been investigated. For example, salt and phyto-available water limitation in *Arabidopsis* leads to increased expression of DRIR (an lncRNA). It has already been demonstrated that CircRNAs express themselves in rice during nutritional deficiency, in tomato under freezing stress and in wheat and barley under stress-specific circumstances (Qin et al., 2017;).

Amor et al. (2009) found that increased salt tolerance and root formation (primary and secondary roots) are associated with increased expression of npc536 in *Arabidopsis*, a salinity stress-reactive lncRNA. According to studies like Di et al., Zhu et al. and Nejat and Mantri, TFs may successfully alter the pattern of lncRNA expression when exposed to stressors of both biotic and abiotic origin. The expression pattern of stress resistance genes is consistently associated with reduced fitness, demonstrating the innate connection between growth and stress responses.

LncRNAs are becoming balancing elements in crops cultivated under a variety of agro-environmental conditions, as evidenced by their role in plant development both in the absence of stressors and in stress responses. Without a doubt, lncRNAs might be good targets for researchers working to rationalise yield stability while developing stress tolerance across a range of crops.

14.2.1 MicroRNAs (miRNAs)

Small non-coding RNA molecules called microRNAs (miRNAs), which are usually 20–25 nucleotides long, are crucial for controlling gene expression in plants and impacting a range of biological functions, including growth, development and stress reactions. According to Luo et al. (2024) and Li et al. (2024), they are essential for plant defence mechanisms against biotic pressures like pathogen assaults and abiotic stresses like heavy metal toxicity and the effects of climate change. Plant immunity depends on hormone signalling pathways and systemic acquired resistance, both of which are impacted by miRNAs' modulation of target mRNA expression (Li et al., 2024; Luo et al., 2024). Furthermore, they let plants communicate across great distances. Research indicates that some miRNAs react to environmental stresses, such as excess nickel, and go from leaves to roots to control stress reactions (Rafiq et al., 2025).

14.2.2 Small Interfering RNAs (siRNAs)

A class of short ncRNAs known as small interfering RNAs (siRNAs) is derived from a variety of double-stranded RNAs. They are around 22 nucleotides in length and are caused by the activity of DCL (DICER-LIKE) proteins. RNA polymerase produces dsRNAs in an RNA-dependent manner, which serves as the starting point for the biogenesis of siRNAs. Certain AGO proteins handle them after synthesis to create distinct siRNA classes (Axtell, 2013). The structure of siRNA is homoduplex, whereas that of miRNA is heteroduplex. While siRNAs are often created from transposons or heterochromatic areas, miRNAs are derived from particular genes. The participation of different AGO proteins is the second key distinction between the miRNA and siRNA pathways. AGO2 and other AGO proteins are particularly used by siRNAs; however, any AGO protein between AGO1 and AGO10 can be used in miRNA-mediated gene silencing. siRNAs are essential for DNA methylation and the suppression of several mRNA targets, and they are mostly conserved across animals (Jha et al., 2021). Transposons or repetitive DNA are the source of TE-derived siRNAs; repeat-associated siRNAs (ra-siRNAs); lnc-single-stranded RNAs are the source of trans-acting siRNAs (ta-siRNAs); intergenic/repetitive genomic regions are the source of heterochromatic siRNAs (hc-siRNAs); and repeats are the source of long small interfering siRNAs (lsiRNAs). Furthermore, phasiRNAs are a different secondary form of siRNAs that are regularly spaced and generated from PHAS loci (Axtell, 2013; Borges & Martienssen, 2015; Liu et al., 2020). The location of siRNA genesis and the activity of RDRs, which subsequently change ssRNA into dsRNA, determine the origin of siRNAs. The majority of siRNAs participate in direct DNA methylation by RNA-directed DNA methylation (RdDM) and/or transcriptional gene silencing (TGS) by histone methylation (Yu et al., 2019).

14.2.3 Long Non-Coding RNAs (lncRNA)

RNA transcripts longer than 200 nt that cannot code are known as lncRNAs. Because of their modest expression in comparison to mRNAs and sequence conservation, early research questioned the function of lncRNAs in plants. Over time, chemicals that were previously believed to be merely 'transcriptional noise' have become important regulators of gene expression in a variety of eukaryotic organisms, ranging from single-celled eukaryotes like yeast to plants and humans (Van Werven et al., 2012). Although RNA Pol II is responsible for the transcription of lncRNAs in nature, reports have also shown the participation of two other RNA Pols, Pol IV and V. RNA Pol IV-assisted lncRNA transcription frequently acts as a precursor to siRNA production. According to Liu et al. (2012), in plants, lncRNAs are linked to functional transcriptome components because they are controlled throughout certain developmental stages and environmental changes. LncRNAs can originate from intergenic, coding and intronic regions oriented in both sense and antisense directions. Based on their genesis from different genomic areas, NATs (natural antisense transcripts), circRNAs (circular lncRNAs), incRNAs (intronic ncRNAs) and lincRNAs (long intergenic ncRNAs) are the main types. NATs are produced from the complementary strand of the gene in question (Mattick & Rinn, 2015). The majority of lncRNAs serve as guides, sponges, backbones or decoy molecules. Furthermore, they can act as regulators of alternative splicing and eTMs for different miRNAs (Jha et al., 2021).

14.3 ROLE OF miRNAs IN ABIOTIC STRESS RESPONSE

One of the main strategies used by plants to withstand a variety of stressors and carry on with their growth and development is RNA interference (RNAi) induced gene silence. It is now recognised that essential reactions to environmental changes require miRNA-mediated gene silencing (Jha et al., 2021). These include the auxin signalling system, which downregulates the target genes in response to ABA, drought, heat, nutritional deprivation and salt. Zhang (2015) talked about the dynamics of miRNA gene expression in tissue, stress and genotype. This indicates that both species-specific and conserved miRNAs function differently depending on the genotype and abiotic stress. MiR156 was the first miRNA found in plants (Axtell, 2013).

Additionally, research shows that exposure to various environmental stressors induces miR156. Stief et al. (2014) demonstrated that miR156 plays a role in stress memory and that plants with overexpressed miR156 were more tolerant of heat stress. Furthermore, several conditions, including heat, dehydration and a lack of adequate water, can produce drought stress in plants. According to Song et al. (2020), conserved miRNAs such as miR169, miR408, miR396, miR171, miR164, miR398, miR399 and miR2118 are important in the body's reaction to drought stress. It is now simpler and faster to identify miRNA targets with high diversity and rapid divergence because of the latest advancements in NGS technology.

Short Tandem Target Mimic (STTM), a large-scale knockdown technique of miRNA, has been extensively employed in the past ten years to interfere with the function of the miRNA family in plants (Zhang et al., 2017). One work used rice (*Oryza sativa*) to characterise a xylem-localised miRNA called miR166 by inducing its knockdown. Because of their decreased transpiration rate, the STTM166 plants showed improved resistance to drought stress. Reduced bulliform cells in rice leaves, leaf rolling and modified xylem vessels in the stem were also noted, in addition to drought stress resistance. OsHB4 was identified as one of the primary targets of miR166 in rice by molecular analysis, which also demonstrated that the miR166-OsHB4 module contributes to rice's ability to withstand drought stress, roll its leaves and generate xylem. In a different study, rice pri-miR393a was overexpressed in bentgrass plants to better understand the role of miR393 in creeping bentgrass (*Agrostis stolonifera*) (Zhao et al., 2019). This led to increased resistance to heat stress and drought, as well as increased potassium absorption and heat shock protein expression.

Similarly, it is known that during various phases of plant growth, the miR156/SPL module contributes to the definition of the plant architecture (Ma et al., 2021). Its role in stress tolerance hasn't been well studied, though. A recent study found that the miR156/SPL13 module controls salt stress by causing apple (*Malus domestica*) to produce MdWRKY100 (Ma et al., 2021). Both MdmiR156a and MdSPL3 overexpression strains were produced for functional characterisation. Overexpression of MdmiR156a (OE) reduced apple plants' ability to withstand salt stress. On the other hand, MdSPL13's OE improved its resistance to salt stress. Subsequent research showed that MdSPL13 imparts salt stress resistance in apples by targeting the MdWRKY100 promoter. One such salt-responsive miRNA, miR172, positively regulates monocots' ability to withstand salt stress (Cheng et al., 2021). MIR172 loss-of-function mutant lines and OE of miR172 were produced for functional validation. Remarkably, miR172a and miR172b helped to demonstrate resistance to salt stress, but miR172c and miR173d did not, according to an examination of promoter-GUS fusion lines of miR172 transcript variations. This demonstrates how the members of the miR172 family define the salt stress response with functional differentiation and specificity. Furthermore, miRNAs have an impact on essential transcriptional networks in plants, which are at the core of intricate gene regulatory networks and guarantee the correct operation of hormone pathways and stress signalling. Plants that overexpress miRNA exhibit either tolerance or vulnerability, depending on the target gene that controls various environmental responses. Many of these findings show promise for using precision miRNA-biotechnology to increase the productivity of plants growing under unfavourable circumstances.

14.4 THE FUNCTION OF siRNAs IN THE BODY'S REACTION TO ABIOTIC STRESS

The roles of several siRNAs in response to abiotic stress are described in fewer, but more recent, publications (Figures 14.1–2). PhasiRNA and trans-acting siRNA (tasiRNA) are known to target complementary mRNA for destruction due to their roles in development and stress response. Additionally, their biogenesis and functions overlap (Liu et al., 2020). According to Yu et al., phasiRNAs in plants are often produced from protein-coding genes, such as lncRNAs from monocots' PHAS loci and dicots' PENTATRICOPEPTIDE REPEAT (PPR) and NUCLEOTIDE-BINDING LEUCINE-RICH REPEAT (NB-LRR) genes. In one study, thermotolerance was linked to two tasiRNA-targeting genes, HTT1 and HTT2 (HEAT-INDUCED TAS1 TARGET) (Li et al., 2014). *Arabidopsis* became less thermotolerant as a result of TAS1a overexpression, whose tasiRNA target was the HTT genes. This led to an increase in the accumulation of TAS1-siRNA, which in turn reduced the expression of HTT genes. In poplar (*P. trichocarpa*), phasiRNAs triggered by several miRNAs, including miR482, miR828 and miR6445, were found to be susceptible to drought stress (Shuai et al., 2016).

In Jerusalem artichokes (*Helianthus tuberosus*), plants treated with 100-mM NaCl exhibited enhanced cleavage of TAS3 due to miR390 expression, according to recent research that included MIR390, miR390, TAS3a/b/c, tasiARF-1/2/3 (tasiRNA influencing ARFs) and ARF1/2/3 (Wen et al., 2020). The large production of tasi-ARFs, which enhanced the cleavage of ARF3/4 and ultimately made Jerusalem artichokes resistant to salt stress, was generated by a complex regulatory module. Interestingly, tasiRNA is an essential component of miRNA-tasiRNA modules, which have been connected to trade-offs between plant development and growth. One study found that siRNA-based RNAi downregulated the cytokinin oxidase (OsCKX2) unique to the floral meristem of rice plants (Joshi et al., 2018).

Transgenic plants with OsCKX2 knockdown demonstrated salt stress resistance at 200-mM NaCl. Surprisingly, in addition to resistance, panicle branching and grain filling were unchanged even under salty conditions. This study shows how OsCKX2 adversely affects inflorescence meristem activity, which affects plant productivity and tolerance to salt stress. Furthermore, productivity losses are often caused by male sterility in crops caused by extreme temperature stress. Additionally, DNA methylation is a process that characterises stress reactions, growth and development (Ma et al., 2018). The molecular function of DNA methylation in heat-induced male sterility, however, remains unknown. Heat stress alters the methylation of the genome's DNA, per a study on cotton (*Gossypium hirsutum*) anthers (Ma et al., 2018). Changes in CHH methylation were only seen in heat-sensitive cotton lines; tolerant lines did not. The hypermethylation investigation demonstrated the involvement of hc-siRNA in this DNA methylation, which changed the target genes' methylation pattern. The other indehiscence and microspore abortion that resulted in male sterility in the heat-sensitive cotton lines may be connected to this.

14.5 ROLE OF LNCRNAS IN ABIOTIC STRESS RESPONSE

The genome-wide discovery and functional characterisation of several lncRNAs that respond to the environment and describe how they aid in the regulation of abiotic stress are the subject of numerous papers nowadays. For instance, several plants, including *Arabidopsis*, rice, maize, poplar and foxtail millet (Qi et al., 2013), have been reported to contain some lncRNAs that respond to drought. In another study, the overexpression of a soybean lncRNA77580 was used to evaluate the lines' resilience to abiotic stress; as a result, overexpression lines were more resilient to drought stress than the wild-type soybean (Chen et al., 2023).

At the seedling stage, however, there was an increased vulnerability to salt stress. When under water deficiency stress during the reproductive stage, the overexpression lines produced more seeds per plant, increasing their output. Because they co-express with mRNAs, lncRNAs are essential for tomatoes' ability to withstand drought stress (Eom et al., 2019). While sly-miR169c was downregulated during drought stress, lncRNA535 was discovered to be upregulated as a target. Furthermore, during drought stress, lncRNA-tomato-503 was downregulated while sly-miR396b was upregulated. According to research, lncRNAs serve as transcriptional regulators of genes linked to the drought stress response that are involved in transporter activity, signalling and tolerance. Co-expression research revealed that 10 TFs and about 21 temperature and drought-sensitive lncRNAs co-express each other. In cassava (*Manihot esculenta*), RNA-Seq data analysis under various drought and cold conditions identified 682 lncRNAs. Sixteen of them were shown to be putative TMs of certain known miRNAs. Additionally, 42 lnc-NATs that might produce NAT-siRNAs were found following a comparison with small RNA-Seq data. Overexpression of lncRNA-973, which is produced by salt stress, enhanced cotton's ability to withstand salt stress (Zhang et al., 2019a, 2019b). In addition to harming environmental agricultural production, heavy metal stress presents a serious health concern to humans since it may lead to the ingestion of edible plant products grown in heavy metal-contaminated soil. Chen et al. (2018) discovered that 75 lncRNAs were downregulated and 69 were upregulated when rice was exposed to cadmium (Cd) stress. Through pathway analysis, it was demonstrated that under Cd stress, secondary metabolites such as phenylpropanoids, phenylalanine and genes involved in photosynthesis were significantly altered. To find lncRNAs in maize, researchers analysed the PAC Bio fl-cDNA and total RNA databases (Lv et al., 2019). For evaluation, the proportion of lncRNAs associated with abiotic stress responses such as heat, cold, drought and salt stress was employed (Figure 14.2).

Gaining an understanding of these systems is essential to creating plans to increase plants' resistance to abiotic

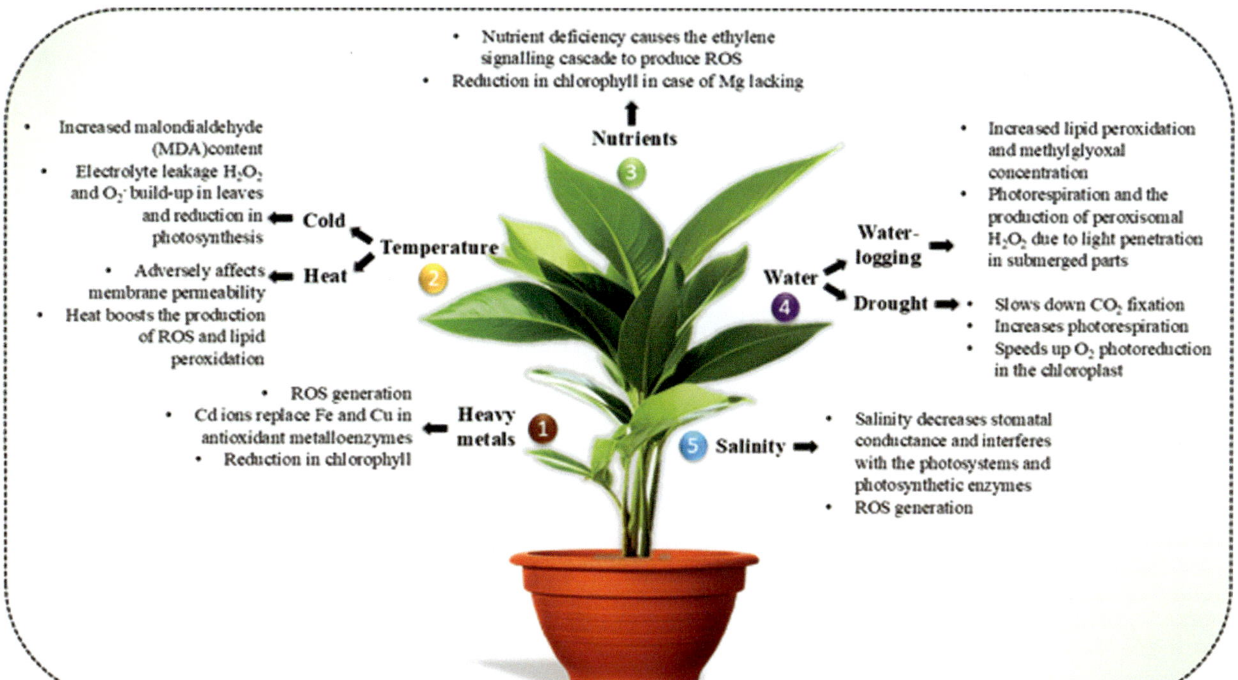

FIGURE 14.2 This figure illustrates the several abiotic stressors that plants encounter, emphasising their unique effects and the resulting production of reactive oxygen species (ROS). High temperatures, salt, dryness and heavy metals are examples of stressors that cause physiological and biochemical reactions. ROS generation is a frequent response that frequently leads to oxidative stress, which damages proteins, lipids and DNA and impairs cellular function. The growth, development and production of plants may be impacted by these detrimental impacts.

stress, which will eventually promote food security and agricultural sustainability.

It subsequently emerged how a lncRNA contributes to the generation of ethylene, which causes light-induced anthocyanin development in apples (Yu et al., 2022). The discovered lncRNA MdLNC610, which was upstream of the ethylene biosynthesis gene 1-aminocyclopropane-1-carboxylate oxygenase (ACO), or MdACO1, most likely had an effect on anthocyanin accumulation following high light treatment. The use of reporter gene assay and overexpression of MdACO1 and MdLNC610 in apple fruit and calli revealed the advantageous role of MdACO1 expression and ethylene production in encouraging the synthesis of anthocyanins at high light levels in apple fruits. Nutrient deficiencies are another common problem that leads to significant physiological and biochemical changes in plants. Low phosphorus (P) levels can lower agricultural productivity since P is essential to the basic structure and metabolism of plants. Low-phosphate contrasting soybean cultivars were studied to identify circRNAs (Lv et al., 2020). A total of 120 of the 371 newly discovered circRNAs exhibited differential expression. Increased circRNA expression in P-tolerant cultivars and GO enrichment analysis showed a plausible correlation between the expression patterns of P stress-inducible circRNAs and their host genes in soybeans. The aforementioned lncRNA results will serve as the foundation for future studies on the regulatory roles of several novel lncRNAs in response to abiotic stress and their regulation.

14.6 ROLE OF ncRNAs IN VARIOUS ABIOTIC STRESS

Numerous abiotic stressors, such as heat, drought, salt, floods, low temperatures and others, can affect plants since they are sessile organisms. With reported effects ranging from 50 to 70% globally, these stressors have the potential to significantly impair plant development and overall production (Hazra et al., 2023). On a wide geographic scale, abiotic stresses particularly those associated with global warming have a significant and widespread detrimental effect on plants. Global agricultural and plant ecosystems face serious problems due to the increasing frequency of these stressors, which are made worse by variables like climate change. Devoted research, the creation of crop types resistant to stress and the application of sustainable agricultural techniques are necessary to manage this situation successfully. These methods seek to strengthen plants' resistance to environmental shocks while lessening the detrimental effects of abiotic stressors on plant output. Non-coding RNAs (ncRNAs) are crucial for plants to adapt to a range of abiotic stresses, such as drought, salt and extreme temperatures (Figure 14.3). Unlike protein-coding RNAs, non-coding RNAs (ncRNAs) regulate gene

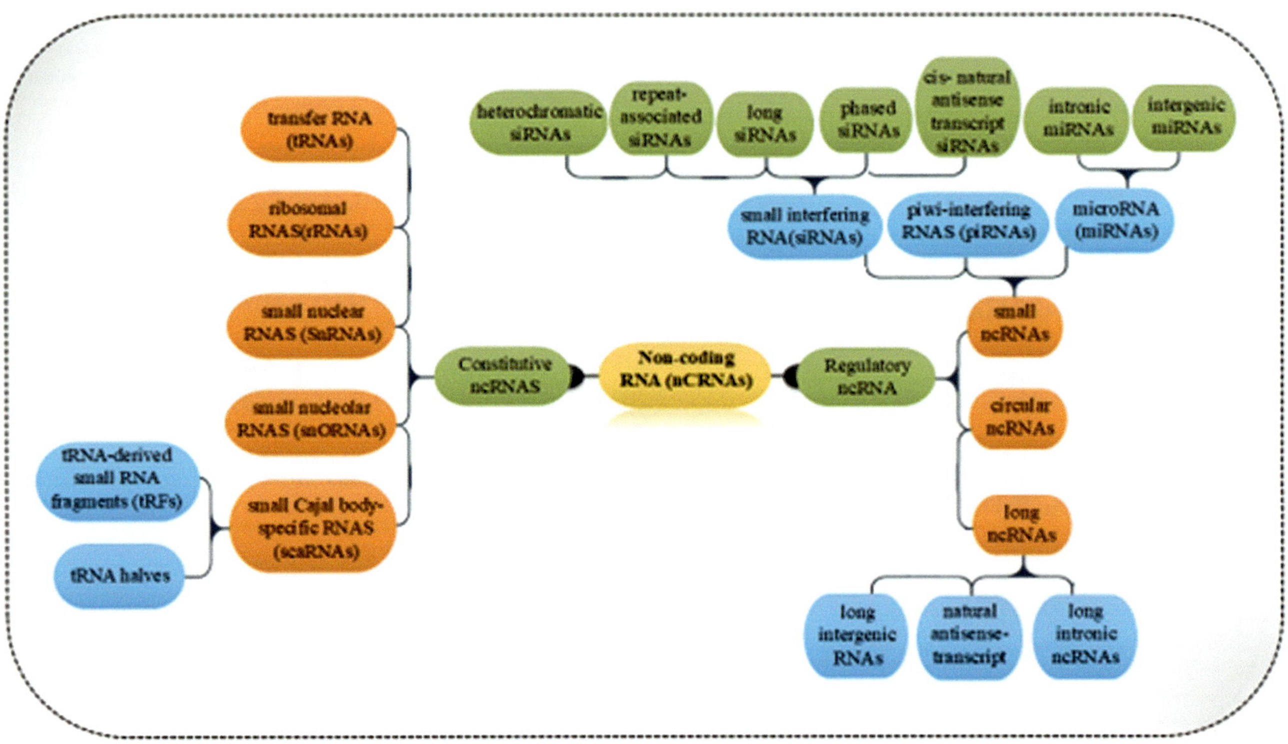

FIGURE 14.3 The classification of different non-coding RNA (ncRNA) types within plant systems is shown in this illustration. Transfer RNAs (tRNAs), small nuclear RNAs (snRNAs) and ribosomal RNAs (rRNAs) are examples of constitutive ncRNAs. On the contrary, regulatory ncRNAs include small interfering RNAs (siRNAs), microRNAs (miRNAs) and other small and long ncRNAs. Different subclasses, including phased siRNAs, long intergenic ncRNAs and heterochromatic siRNAs, are highlighted in this figure, indicating how these ncRNAs affect gene regulation and stress responses.

expression, chromatin dynamics and cellular activities through RNAi and transcriptional and post-transcriptional activity regulation. By altering how plants respond to environmental stressors, ncRNAs aid in the growth, development and resilience of plants in challenging conditions. Their involvement expands our understanding of the molecular processes that underlie stress tolerance and paves the way for biotechnological applications aimed at improving agricultural performance and sustainability under climate change.

14.6.1 Drought Stress

A significant agricultural concern is drought stress, which is why there is an urgent global push to develop drought-tolerant crops. Plants under drought stress suffer from osmotic imbalance, membrane system deterioration, decreased respiration and decreased photosynthetic rates, among other negative effects. These reactions reduce crop production and quality in addition to impeding plant development and metabolism throughout its life cycle. The majority of ncRNAs target genes related to calcium signalling, starch and sucrose production, ethylene and ABA synthesis or signalling and several other metabolic activities. Three different tissues (ear, tassel and leaf) from four developmental phases in both well-watered and drought-stricken maize were shown to contain a large number of lncRNAs (Pang et al., 2019). MSTRG.2834.1, 1 of the 3,488 high-confidence discovered lncRNAs, reacted to drought stress exclusively in the ear, showing the most changes in expression at the R1 stage and the smallest changes at the V12, V14 and V18 stages. The lncRNAs MSTRG.6838.1 and vpp4 may be promising cis-acting pairs with the potential for co-regulation, as evidenced by the substantial correlation between their expressions across different tissues and developmental stages and their downregulation during drought stress.

Drought-responsive lncRNA (DRIR) was discovered, indicating its critical function in controlling plant resistance to salt and drought stressors (Dong et al., 2022). Recently, Zhang et al. (2019) carried out extensive profiling and discovered 1354 high-confidence circRNAs in *Arabidopsis* and 2174 in maize. Through back-splicing between exons, these circRNAs, circCPK and circCPK, were created, causing alleles in the maize panel to be inserted and deleted. The insertion alleles significantly increased the survival rate and circCPK expression levels of maize plants compared to those with the deletion alleles in both normal and drought circumstances. This finding raises the possibility that circRNA expression and plant drought tolerance are related. Additionally, the GORK (guard cell outward-rectifying K+-channel) gene hosts circGORK, which the researchers examined in another study. According to Becker et al. (2003), GORK encodes a potassium output channel that is essential for controlling ABA signalling and water stress responses. In transgenic *Arabidopsis* plants, the ABA response was adversely affected by circGORK overexpression. Transgenic seedlings in particular had a much lower germination rate on ABA plates than control seeds, indicating that circGORK may alter ABA signalling pathways and affect responses to water stress (Yadav et al., 2024).

14.6.2 Salt Stress

A significant obstacle to plant growth and development, salt stress affects over 33% of arable land. In plants, it raises intracellular and ionic osmotic pressures, which leads to osmotic stress, oxidative damage and enhanced toxicity. According to Hasanuzzaman and Fujita (2022), the majority of cultivated cultivars of rice have an electrical conductivity threshold of 3 dSm^{-1}, making them susceptible to salt stress. lncRNA DRIR was shown to have a function in drought stress and to be a positive regulator of the plant response to salt stress. Although DRIR expression was low under normal circumstances, drought, salt stress and ABA therapy all markedly increased it. The effect of DRIR on gene expression in response to stress was examined using a transcriptome analysis of the whole genome. The findings showed that both the DRIR mutant and overexpressing plants had altered expression of genes related to ABA signalling, water transport and other stress-relieving mechanisms. The finding sheds important light on how DRIR regulates the intricate web of chemical reactions that plants go through in response to salt and drought stress (Yadav et al., 2024).

14.6.3 Waterlogging/Flooding/Hypoxia Stress

Waterlogging occurs when the soil is either entirely saturated or close to saturation most of the time, whereas flooding occurs when water temporarily submerges land. According to Fukao et al. (2019), both processes interfere with cellular energy generation, impede photosynthesis and upset ionic equilibrium, among other physiological functions in plants. Hypoxia, a condition that commonly occurs in plants' natural environments, is a lower oxygen concentration than ideal circumstances that exacerbates these effects. Plant growth and development are hindered by the limited supply of oxygen in the root zone, which has a negative effect on the plant's total metabolism (Aslam et al., 2023; Panda et al., 2022; Yu et al., 2020).

A sucrose non-fermenting-related protein kinase (SnRK1) has shown that flooding is also linked to decreased energy metabolism (Renziehausen et al., 2025). By controlling SQUAMOSA-promoter binding protein-like (SPL) transcription factors, overexpression of miR156 has been demonstrated to increase abiotic stress tolerance in a variety of plants (Agrawal et al., 2025; Kumari et al., 2025). The function and potential mechanism of action of the miR156/SPL module in alfalfa flooding tolerance were further investigated by Feyissa et al. (2021). Numerous genetic constructs were employed, such as wild-type, empty vector plants, SPL6RNAi, SPL13_RNAi, miR156 overexpression (miR156OE) and alfalfa cultivars that were flood-tolerant

(AAC-Trueman) and flood-sensitive (AC-Caribou). According to the findings, when flooding occurs in alfalfa, the miR156/SPL module regulates physiological responses. In particular, SnRK1 expression rose in an ABA-dependent way in miR156OE alfalfa plants. In plants, SnRK1 is a crucial regulator of energy and stress signals. The results emphasise the complex regulatory mechanisms involved in plant stress responses and imply that the miR156/SPL module aids in flooding tolerance by controlling physiological processes and SnRK1 expression (Feyissa et al., 2021; Jeyakumar et al., 2020; Yadav et al., 2024).

14.6.4 Heat and Cold Stress

Both high and low temperatures have an effect on plant growth, development and some physiological processes (freezing at <0°C and chilling at 0–15°C). Many temperature-responsive regulatory genes, including C-repeat/dehydration-responsive element binding factors (CBF/DREB1s) and heat shock factors, bind to the promoters of cold-regulated genes (COR) and heat stress response genes, respectively, to increase plant resistance to cold stress and heat stress (Wang et al., 2022). There are several siRNAs, circRNAs and miRNA-targets among the variety of ncRNAs that respond to heat and cold stress. The following are notable examples of miRNAs and their targets associated with heat and cold stress: miR156-SPL (Tang et al., 2025); miR159-GAMYB; miR160-ARF; miR164-NAC; miR165/166-homeodomain leucine zipper class-III transcription factor primarily PHBULOSA and ATHB-8; miR167-ARF6; miR168-AGO1; miR171-GRAS gene family; miR172-APETALA2; miR319-TCP; miR393-auxin receptors with F-box domains; TIR1/AFB2, miR396-GRF, miR397-laccase gene; lignin-degrading enzyme, miR402-DEMETER-LIKE PROTEIN3; a 5-methylcytosine DNA glycosidase (Arshad & Hannoufa, 2022; Tsai et al., 2023). In the maize leaves at the V3 stage under heat stress, genome-wide transcriptome studies have also shown conserved miRNA (miR166) and novel miRNAs (miR168, miR444 and miR528) (Kong et al., 2025). Certain miRNAs, including IbmiR162, IbmiR164, IbmiR171 and IbmiR397, have been linked to physiological alterations in sweet potato seedlings, namely oxidative stress under heat and chilling stress (Yadav et al., 2024; Yu et al., 2020).

Chip-based microarray and strand-specific sequencing techniques were also used to identify some lncRNAs for both heat and cold stress, with excellent specificity in tissue types and developmental stages (Lin et al., 2025). For example, during heat stress, only the target gene – rather than the lncRNA itself – was expressed if lnc-173 targets the SUCROSE SYNTHASE4 gene (Lin et al., 2025; Song et al., 2020). The sensitivity and adaptation of eukaryotes to heat stress are significantly influenced by heat shock transcription factors (HSFs). Heat stress was discovered to produce asHSFB2a (natural antisense transcript of HSFB2a) in *Arabidopsis*, which in turn adversely regulates HSFB2a expression (Zhang et al., 2022). Pear has recently been shown to contain Heat-induced Long Intergenic noncoding RNA 1 (HILinc1) (Jampala et al., 2021; Yadav et al., 2024). Multiprotein bridging factor 1c is upregulated during heat response as a result of the target of this lncRNA binding with HSF to increase its transcriptional activity.

Plants that experience cold stress go through a process called vernalisation, which stops them from flowering during winter vegetative development and encourages them to do so during the reproductive phase when conditions are right. FLC is in charge of this vernalisation regulation mechanism. Through chromatin remodelling, several lncRNAs, including COLDAIR, COLDWRAP and COOLAIR, inhibit FLC expression (Jampala et al., 2021). Additionally, via regulating the transcription of the CBF1 gene, the lncRNAs SVALKA, cryptic antisense CBF1 (asCBF1) and CIR (CBF intergenic lncRNA) control cold tolerance (Jha et al., 2023). While the cis-acting NAT lncRNA_2962 stimulates transcription of cold-responsive MADS AFFECTING FLOWERING 4, the COLD-INDUCED lncRNA1 (CIL) contributes to the low-temperature stress response by influencing the reactive oxygen route or osmotic adjustment substance. By controlling the expression of many stress-related genes throughout the embryonic stage, these lncRNAs have an impact on how plants react to cold.

14.7 COMBINED STRESSES

Unexpected climatic circumstances frequently expose plants to several stressors at once, which has a substantial effect on their development, growth and output. Numerous miRNAs have been discovered that, in situations of combined stress, target both hormone-signalling genes and transcription factors. For example, it has been discovered that miR156 improves the ability of rice and *Arabidopsis* plants to withstand cold and heat stress, respectively (Mmbando, 2025). In contrast, miR156 downregulated the SPL gene in alfalfa, increasing plant resistance to drought, heat and salt stress (Peng et al., 2025). Another noteworthy example is miR393, which forms the miR393-TIR1/AFBs module by targeting the auxin receptors with F-box domains, TIR1 and AFBs. According to Palani et al. (2025), this module, which is largely conserved in many plant species, connects auxin response to transcriptional control of plants under drought and salt stress. Through ABA-mediated and CBF-dependent cold acclimation pathways, the LEAF CURLING RESPONSIVENESS GENE, an F-box protein (SKP1-Cullin/CDC53-F-box), controls responses to salt stress and drought (Yadav et al., 2024). miR394 targets this gene. Furthermore, miR408 improves plant resistance to oxidative stress, salt and cold by targeting the genes for laccase and blue copper protein (Balyan et al., 2023; Yadav et al., 2024).

A new long non-coding RNA (lncRNA) called DRIR was discovered by Qin et al. (Qin et al., 2017) in the context of hormone-mediated responses to combined stress. It controls responses to salt and drought stress by controlling the production of ABA signalling genes, stomatal closure and proline accumulation. Furthermore, in soybeans, lncRNA77580 regulates the expression of many transcription factors in response to drought and salt stress (Chen et al., 2023). Moreover, about 467 circRNAs were discovered in tomatoes in response to heat, drought and combination stressors (Yadav et al., 2024) utilising high-throughput sequencing techniques. Out of them, 156 circRNAs were common across all stress circumstances. CircRNAs 197 and 29 were found to have high expression levels in three different stress conditions and are thought to be crucial for photosynthesis and other biological activities.

14.8 CONCLUSION

Plant growth and reactions to abiotic stress are two biological processes that are impacted by ncRNAs, which are essential for controlling mRNA translation. While phasi- and circ-RNAs need more investigation, the finding of certain ncRNA types, such as miRNAs, has been given priority. The discovery and characterisation of ncRNAs are aided by advancements in sequencing technology and bioinformatics; yet, difficulties persist because of the paucity of research on some plant species and the requirement for more effective small RNA sequencing methods. Plants utilise complex regulatory networks formed by miRNA–siRNA–lncRNA–target gene interactions to manage stress responses. However, improvements in stress tolerance often come at the cost of reduced growth and yield, highlighting the need for better understanding of these trade-offs. Future efforts may include using CRISPR/Cas9 technology to modify ncRNAs related to abiotic stress response, potentially leading to the development of climate-resilient crops. Key research questions involve the regulation of ncRNA homeostasis, their relationship with regulatory proteins and the feasibility of ncRNA-assisted breeding programmes.

REFERENCES

Ache, P., Becker, D., Ivashikina, N., Dietrich, P., Roelfsema, M. R. G., & Hedrich, R. (2000). GORK, a delayed outward rectifier expressed in guard cells of *Arabidopsis thaliana*, is a K+-selective, K+-sensing ion channel. *FEBS Letters*, *486*, 93–98.

Agrawal, R., Singh, A., & Thakur, J. K. (2025). Epigenetic memory in plants for stress response and adaptation. In *Exogenous priming and engineering of plant metabolic and regulatory genes* (pp. 243–251). Academic Press.

Ahmed, Z., Guha, G. S., Shew, A. M., & Alam, G. M. (2021). Climate change risk perceptions and agricultural adaptation strategies in vulnerable riverine char islands of Bangladesh. *Land Use Policy*, *103*, 105295.

Amedie, F. A. (2013). *Impacts of climate change on plant growth, ecosystem services, biodiversity, and potential adaptation measures* [MasterThesis. Program Study of Biological and Enviromental Science, University of Gothenberg, Sweden].

Amor, B. B., Wirth, S., Merchan, F., Laporte, P., d'Aubenton-Carafa, Y., Hirsch, J., … & Crespi, M. (2009). Novel long non-protein coding RNAs involved in Arabidopsis differentiation and stress responses. *Genome Research*, *19*(1), 57–69.

Anekwe, I. M. S., Zhou, H., Mkhize, M. M., & Akpasi, S. O. (2024). Addressing climate change challenges in South Africa: A study in KwaZulu Natal province. In *Climate crisis: Adaptive approaches and sustainability* (pp. 475–496).

Apostolova, E. L. (2023). Molecular mechanisms of plant defense against abiotic stress. *International Journal of Molecular Sciences*, *24*(12), 10339.

Arikit, S., Xia, R., Kakrana, A., Huang, K., Zhai, J., Yan, Z., … & Meyers, B. C. (2014). An atlas of soybean small RNAs identifies phased siRNAs from hundreds of coding genes. *The Plant Cell*, *26*(12), 4584–4601.

Arshad, M., & Hannoufa, A. (2022). Alfalfa transcriptome profiling provides insight into MiR156-mediated molecular mechanisms of heat stress tolerance. *Genome*, *65*, 315–330.

Aslam, A., Mahmood, A., Ur-Rehman, H., Li, C., Liang, X., Shao, J., Negm, S., Moustafa, M., Aamer, M., & Hassan, M. U. (2023). Plant adaptation to flooding stress under changing climate conditions: Ongoing breakthroughs and future challenges. *Plants*, *12*, 3824.

Awanthi, M. G. G., & Navaratne, C. M. (2018). Carbon footprint of an organization: A tool for monitoring impacts on global warming. *Procedia Engineering*, *212*, 729–735.

Axtell, M. J. (2013). Classification and comparison of small RNAs from plants. *Annual Review of Plant Biology*, *64*(1), 137–159.

Balyan, S., Kansal, S., Jajo, R., Behere, P. R., Chatterjee, R., & Raghuvanshi, S. (2023). Delineating the tissue-mediated drought stress governed tuning of conserved miR408 and its targets in rice. *Functional & Integrative Genomics*, *23*(2), 187.

Becker, D., Hoth, S., Ache, P., Wenkel, S., Roelfsema, M. R. G., Meyerhoff, O., Hartung, W., & Hedrich, R. (2003). Regulation of the ABA-sensitive Arabidopsis potassium channel Gene GORK in response to water stress. *FEBS Letters*, *554*, 119–126.

Behera, J. K., Jena, A. K., Bhattacharya, M., & Behera, B. (2024). India's current situation with regard to the effects of water pollution on agricultural productivity and public health. In *Soil, water pollution and mitigation strategies: A spatial approach* (pp. 431–464). Springer Nature Switzerland.

Bhatia, G., Singh, A., Verma, D., Sharma, S., & Singh, K. (2020). Genome-wide investigation of regulatory roles of lncRNAs in response to heat and drought stress in *Brassica juncea* (Indian mustard). *Environmental and Experimental Botany*, *171*, 103922.

Borges, F., & Martienssen, R. A. (2015). The expanding world of small RNAs in plants. *Nature Reviews Molecular Cell Biology*, *16*(12), 727–741.

Bouyer, D., Kramdi, A., Kassam, M., Heese, M., Schnittger, A., Roudier, F., & Colot, V. (2017). DNA methylation dynamics during early plant life. *Genome Biology*, *18*(1), 12.

Calarco, J. P., Borges, F., Donoghue, M. T., Van Ex, F., Jullien, P. E., Lopes, T., … & Martienssen, R. A. (2012). Reprogramming of DNA methylation in pollen guides epigenetic inheritance via small RNA. *Cell*, *151*(1), 194–205.

Campalans, A., Kondorosi, A., & Crespi, M. (2004). Enod40, a short open reading frame–containing mRNA, induces cytoplasmic localization of a nuclear RNA binding protein in *Medicago truncatula*. *The Plant Cell*, *16*(4), 1047–1059.

Cao, L., Lu, X., Wang, G., Zhang, P., Fu, J., Wang, Z., … & Wang, T. (2021). Transcriptional regulatory networks in response to drought stress and rewatering in maize (*Zea mays* L.). *Molecular Genetics and Genomics*, 1–17.

Chen, L., Shi, S., Jiang, N., Khanzada, H., Wassan, G. M., Zhu, C., ... & Bian, J. (2018). Genome-wide analysis of long noncoding RNAs affecting roots development at an early stage in the rice response to cadmium stress. *BMC Genomics*, *19*, 1–10.

Chen, L., Zhu, Q. H., & Kaufmann, K. (2020). Long non-coding RNAs in plants: Emerging modulators of gene activity in development and stress responses. *Planta*, *252*(5), 92.

Chen, X. (2010). Small RNAs–secrets and surprises of the genome. *The Plant Journal*, *61*(6), 941–958.

Chen, X., Jiang, X., Niu, F., Sun, X., Hu, Z., Gao, F., ... & Jiang, Q. (2023). Overexpression of lncRNA77580 regulates drought and salinity stress responses in soybean. *Plants*, *12*(1), 181.

Cheng, X., He, Q., Tang, S., Wang, H., Zhang, X., Lv, M., ... & Liu, J. (2021). The miR172/IDS1 signaling module confers salt tolerance through maintaining ROS homeostasis in cereal crops. *New Phytologist*, *230*(3), 1017–1033.

Chung, P. J., Jung, H., Jeong, D. H., Ha, S. H., Choi, Y. D., & Kim, J. K. (2016). Transcriptome profiling of drought responsive noncoding RNAs and their target genes in rice. *BMC Genomics*, *17*, 1–12.

Conn, V. M., Hugouvieux, V., Nayak, A., Conos, S. A., Capovilla, G., Cildir, G., ... & Conn, S. J. (2017). A circRNA from SEPALLATA3 regulates splicing of its cognate mRNA through R-loop formation. *Nature Plants*, *3*(5), 1–5.

Csorba, T., Questa, J. I., Sun, Q., & Dean, C. (2014). Antisense COOLAIR mediates the coordinated switching of chromatin states at FLC during vernalization. *Proceedings of the National Academy of Sciences*, *111*(45), 16160–16165.

Di Micco, D. (2024). *Exploring climate change impacts and groundwater contamination dynamics in a west-Mediterranean coastal basin* [Doctoral dissertation, Politecnico di Torino].

Ding, B., Xia, R., Lin, Q., Gurung, V., Sagawa, J. M., Stanley, L. E., ... & Yuan, Y. W. (2020). Developmental genetics of corolla tube formation: Role of the tasiRNA-ARF pathway and a conceptual model. *Plant Cell*, *32*(11), 3452–3468.

Ding, J., Lu, Q., Ouyang, Y., Mao, H., Zhang, P., Yao, J., ... & Zhang, Q. (2012). A long noncoding RNA regulates photoperiod-sensitive male sterility, an essential component of hybrid rice. *Proceedings of the National Academy of Sciences*, *109*(7), 2654–2659.

Dong, S.-M., Xiao, L., Li, Z.-B., Shen, J., Yan, H.-B., Li, S.-X., Liao, W.-B., & Peng, M. (2022). A novel long non-coding RNA, DIR, increases drought tolerance in cassava by modifying stress-related gene expression. *Journal of Integrative Agriculture*, *21*, 2588–2602.

Eom, S. H., Lee, H. J., Lee, J. H., Wi, S. H., Kim, S. K., & Hyun, T. K. (2019). Identification and functional prediction of drought-responsive long non-coding RNA in tomato. *Agronomy*, *9*(10), 629.

Feyissa, B. A., Amyot, L., Nasrollahi, V., Papadopoulos, Y., Kohalmi, S. E., & Hannoufa, A. (2021). Involvement of the MiR156/SPL module in flooding response in *Medicago sativa*. *Science Reports*, *11*, 3243.

Fukao, T., Barrera-Figueroa, B. E., Juntawong, P., & Peña-Castro, J. M. (2019). Submergence and waterlogging stress in plants: A review highlighting research opportunities and understudied aspects. *Frontiers in Plant Science*, *10*, 340.

Ganguly, P., Roy, D., Das, T., Kundu, A., Cartieaux, F., Ghosh, Z., & DasGupta, M. (2021). The natural antisense transcript DONE40 derived from the lncRNA ENOD40 locus interacts with SET domain protein ASHR3 during inception of symbiosis in *Arachis hypogaea*. *Molecular Plant-Microbe Interactions*, *34*(9), 1057–1070.

Ganguly, S., Saha, S., Vangaru, S., Purkayastha, S., Das, D., Saha, A. K., ... & Bhattacharyya, S. (2020). Identification and analysis of low light tolerant rice genotypes in field conditions and their SSR-based diversity in various abiotic stress tolerant lines. *Journal of Genetics*, *99*, 1–9.

Geisler, S., & Coller, J. (2013). RNA in unexpected places: Long non-coding RNA functions in diverse cellular contexts. *Nature Reviews Molecular Cell Biology*, *14*(11), 699–712.

Hasanuzzaman, M., & Fujita, M. (2022). Plant responses and tolerance to salt stress: Physiological and molecular interventions. *International Journal of Molecular Sciences*, *23*, 15740.

Hazra, S., Moulick, D., Mukherjee, A., Sahib, S., Chowardhara, B., Majumdar, A., Upadhyay, M. K., Yadav, P., Roy, P., Santra, S. C. et al. (2023). Evaluation of efficacy of non-coding RNA in abiotic stress Nakaya management of field crops: Current status and future prospective. *Plant Physiology and Biochemistry*, *203*, 107940.

He, J., Jiang, Z., Gao, L., You, C., Ma, X., Wang, X., Xu, X., Mo, B., Chen, X., & Liu, L. (2019). Genome-wide transcript and small RNA profiling reveals transcriptomic responses to heat stress. *Plant Physiology*, *181*, 609–629.

Imran, Q. M., Falak, N., Hussain, A., Mun, B. G., & Yun, B. W. (2021). Abiotic stress in plants; stress perception to molecular response and role of biotechnological tools in stress resistance. *Agronomy*, *11*(8), 1579.

Jampala, P., Garhewal, A., & Lodha, M. (2021). Functions of long non-coding RNA in *Arabidopsis thaliana*. *Plant Signaling & Behavior*, *16*, 1925440.

Jeyakumar, J. M. J., Ali, A., Wang, W. M., & Thiruvengadam, M. (2020). Characterizing the role of the MiR156-SPL network in plant development and stress response. *Plants*, *9*, 1206.

Jha, U. C., Nayyar, H., Roychowdhury, R., Prasad, P. V. V., Parida, S. K., & Siddique, K. H. M. (2023). Non-coding RNAs (NcRNAs) in plant: Master regulators for adapting to extreme temperature conditions. *Plant Physiology and Biochemistry*, *205*, 108164.

Jha, V., Narjala, A., Basu, D., TN, S., Pachamuthu, K., Chenna, S., ... & Shivaprasad, P. V. (2021). Essential role of γ-clade RNA-dependent RNA polymerases in rice development and yield-related traits is linked to their atypical polymerase activities regulating specific genomic regions. *New Phytologist*, *232*(4), 1674–1691.

Joshi, R., Sahoo, K. K., Tripathi, A. K., Kumar, R., Gupta, B. K., Pareek, A., & Singla-Pareek, S. L. (2018). Knockdown of an inflorescence meristem-specific cytokinin oxidase–OsCKX2 in rice reduces yield penalty under salinity stress condition. *Plant, Cell & Environment*, *41*(5), 936–946.

Kim, D. H., Xi, Y., & Sung, S. (2017). Modular function of long noncoding RNA, COLDAIR, in the vernalization response. *PLoS Genetics*, *13*(7), e1006939.

Kong, X., Peng, K., Shan, Y., Yun, Z., Dalmay, T., Duan, X., ... & Zhu, H. (2025). Transcriptional regulation of miR528-PPO module by miR156 targeted SPLs orchestrates chilling response in banana. *Molecular Horticulture*, *5*(1), 2.

Kumari, C., Malik, I., Shekhawat, N., Sharma, M., Singh, R., & Gochhwal, N. (2025). MicroRNAs in dicot plant like Arabidopsis. In *MicroRNA advances and application in plant biology* (pp. 279–292). Academic Press.

Li, S., Liu, J., Liu, Z., Li, X., Wu, F., & He, Y. (2014). Heat-induced tas1 target1 mediates thermotolerance via heat stress transcription factor A1a–directed pathways in Arabidopsis. *The Plant Cell*, *26*(4), 1764–1780.

Li, S., Yu, X., Lei, N., Cheng, Z., Zhao, P., He, Y., … & Peng, M. (2017). Genome-wide identification and functional prediction of cold and/or drought-responsive lncRNAs in cassava. *Scientific Reports*, *7*(1), 45981.

Li, Z., Yang, J., Cai, X., Zeng, X., Zou, J. J., & Xing, W. (2024). Withdrawn: A review on the function of mirnas in plant responses to stresses. *Plant Stress*, 100347.

Lin, K., Yi, Z., Lv, S., Zhang, B., Guo, Z., & Li, Y. (2025). Uncovering the key lncRNAs in regulating cadmium accumulation and translocation in sweet sorghum. *Planta*, *261*(1), 1–15.

Liu, J., Jung, C., Xu, J., Wang, H., Deng, S., Bernad, L., … & Chua, N. H. (2012). Genome-wide analysis uncovers regulation of long intergenic noncoding RNAs in Arabidopsis. *The Plant Cell*, *24*(11), 4333–4345.

Liu, Y., Teng, C., Xia, R., & Meyers, B. C. (2020). PhasiRNAs in plants: Their biogenesis, genic sources, and roles in stress responses, development, and reproduction. *Plant Cell*, *32*(10), 3059–3080.

Luo, C., Bashir, N. H., Li, Z., Liu, C., Shi, Y., & Chu, H. (2024). Plant microRNAs regulate the defense response against pathogens. *Frontiers in Microbiology*, *15*, 1434798.

Lv, L., Yu, K., Lü, H., Zhang, X., Liu, X., Sun, C., … & Zhang, D. (2020). Transcriptome-wide identification of novel circular RNAs in soybean in response to low-phosphorus stress. *PLoS ONE*, *15*(1), e0227243.

Lv, Y., Hu, F., Zhou, Y., Wu, F., & Gaut, B. S. (2019). Maize transposable elements contribute to long non-coding RNAs that are regulatory hubs for abiotic stress response. *BMC Genomics*, *20*, 1–17.

Ma, Y., Min, L., Wang, M., Wang, C., Zhao, Y., Li, Y., … & Zhang, X. (2018). Disrupted genome methylation in response to high temperature has distinct affects on microspore abortion and anther indehiscence. *The Plant Cell*, *30*(7), 1387–1403.

Ma, Y., Xue, H., Zhang, F., Jiang, Q., Yang, S., Yue, P., … & Zhang, Z. (2021). The miR156/SPL module regulates apple salt stress tolerance by activating MdWRKY100 expression. *Plant Biotechnology Journal*, *19*(2), 311–323.

Mansinhos, I., Gonçalves, S., & Romano, A. (2024). How climate change-related abiotic factors affect the production of industrial valuable compounds in Lamiaceae plant species: A review. *Frontiers in Plant Science*, *15*, 1370810.

Mao, Y., Xu, J., Wang, Q., Li, G., Tang, X., Liu, T., … & Lu, Y. (2021). A natural antisense transcript acts as a negative regulator for the maize drought stress response gene ZmNAC48. *Journal of Experimental Botany*, *72*(7), 2790–2806.

Markulin, L., Škiljaica, A., Tokić, M., Jagić, M., Vuk, T., Bauer, N., & Leljak Levanić, D. (2021). Taking the wheel–De novo DNA methylation as a driving force of plant embryonic development. *Frontiers in Plant Science*, *12*, 764999.

Martínez, G., Panda, K., Köhler, C., & Slotkin, R. K. (2016). Silencing in sperm cells is directed by RNA movement from the surrounding nurse cell. *Nature Plants*, *2*(4), 1–8.

Mattick, J. S., & Rinn, J. L. (2015). Discovery and annotation of long noncoding RNAs. *Nature Structural & Molecular Biology*, *22*(1), 5–7.

Mmbando, G. S. (2025). The current relationship between ultraviolet-B radiation stress and drought stress in rice: An avenue for developing multiple resistant crops. *Journal of Crop Health*, *77*(1), 1–13.

Ning, L. H., Du, W. K., Song, H. N., Shao, H. B., Qi, W. C., Sheteiwy, M. S. A., & Yu, D. Y. (2019). Identification of responsive miRNAs involved in combination stresses of phosphate starvation and salt stress in soybean root. *Environmental and Experimental Botany*, *167*, 103823.

Palani, T., Selvakumar, D., Nathan, B., Shanmugam, V., Duraisamy, K., & Mannu, J. (2025). Deciphering the impact of microRNAs in plant biology: A review of computational insights and experimental validation. *Molecular Biology Reports*, *52*(1), 1–13.

Panda, A. K., Rawal, H. C., Jain, P., Mishra, V., Nishad, J., Chowrasia, S., Sarkar, A. K., Sen, P., Naik, S. K., & Mondal, T. K. (2022). Identification and analysis of MiRNAs-LncRNAs-MRNAs modules involved in stem-elongation of deepwater rice (*Oryza sativa* L.). *Physiologia Plantarum*, *174*, e13736.

Pang, J., Zhang, X., Ma, X., & Zhao, J. (2019). Spatio-temporal transcriptional dynamics of maize long non-coding RNAs responsive to drought stress. *Genes*, *10*, 138.

Parmar, S., Gharat, S. A., Tagirasa, R., Chandra, T., Behera, L., Dash, S. K., & Shaw, B. P. (2020). Identification and expression analysis of miRNAs and elucidation of their role in salt tolerance in rice varieties susceptible and tolerant to salinity. *PLoS ONE*, *15*(4), e0230958.

Peng, W., Cai, W., Pan, J., Su, X., & Dou, L. (2025). Molecular mechanisms of alfalfa response to abiotic stresses. *Plants*, *14*(3), 487.

Qi, X., Xie, S., Liu, Y., Yi, F., & Yu, J. (2013). Genome-wide annotation of genes and noncoding RNAs of foxtail millet in response to simulated drought stress by deep sequencing. *Plant Molecular Biology*, *83*, 459–473.

Qin, T., Zhao, H., Cui, P., Albesher, N., & Xiong, L. (2017). A nucleus-localized long non-coding RNA enhances drought and salt stress tolerance. *Plant Physiology*, *175*(3), 1321–1336.

Rafiq, S., Wagay, N. A., Hadi, A., Ishrath, A., & Kaloo, Z. A. (2025). miRNAs and plant development. In *miRNAomics and stress management in plants* (pp. 36–54). CRC Press.

Ravichandran, S., Ragupathy, R., Edwards, T., Domaratzki, M., & Cloutier, S. (2019). MicroRNA-guided regulation of heat stress response in wheat. *BMC Genomics*, *20*, 1–16.

Ravichandran Song, X., Li, Y., Cao, X., & Qi, Y. (2019). MicroRNAs and their regulatory roles in plant–environment interactions. *Annual Review of Plant Biology*, *70*(1), 489.

Renziehausen, T., Chaudhury, R., Hartman, S., Mustroph, A., & Schmidt-Schippers, R. R. (2025). A mechanistic integration of hypoxia signaling with energy, redox, and hormonal cues. *Plant Physiology*, *197*(1), kiae596.

Rohrig, H., Schmidt, J., Miklashevichs, E., Schell, J., & John, M. (2002). Soybean ENOD40 encodes two peptides that bind to sucrose synthase. *Proceedings of the National Academy of Sciences*, *99*(4), 1915–1920.

Sachdev, S., Ansari, S. A., Ansari, M. I., Fujita, M., & Hasanuzzaman, M. (2021). Abiotic stress and reactive oxygen species: Generation, signaling, and defense mechanisms. *Antioxidants*, *10*(2), 277.

Shiade, S. R. G., Zand-Silakhoor, A., Fathi, A., Rahimi, R., Minkina, T., Rajput, V. D., … & Chaudhary, T. (2024). Plant metabolites and signaling pathways in response to biotic and abiotic stresses: Exploring bio stimulant applications. *Plant Stress*, 100454.

Shuai, P., Liang, D., Tang, S., Zhang, Z., Ye, C. Y., Su, Y., … & Yin, W. (2014). Genome-wide identification and functional prediction of novel and drought-responsive lincRNAs in *Populus trichocarpa*. *Journal of Experimental Botany*, *65*(17), 4975–4983.

Shuai, P., Su, Y., Liang, D., Zhang, Z., Xia, X., & Yin, W. (2016). Identification of phasiRNAs and their drought-responsiveness in *Populus trichocarpa*. *FEBS Letters*, *590*(20), 3616–3627.

Song, Y., Chen, P., Liu, P., Bu, C., & Zhang, D. (2020). High-temperature-responsive poplar LncRNAs modulate target gene expression via RNA interference and act as RNA scaffolds to enhance heat tolerance. *International Journal of Molecular Sciences*, *21*, 808.

Stief, A., Altmann, S., Hoffmann, K., Pant, B. D., Scheible, W. R., & Bäurle, I. (2014). Arabidopsis miR156 regulates tolerance to recurring environmental stress through SPL transcription factors. *The Plant Cell*, *26*(4), 1792–1807.

Sun, X., Zheng, H. X., Li, S., Gao, Y., Dang, Y., Chen, Z., … & Sui, N. (2023). MicroRNAs balance growth and salt stress responses in sweet sorghum. *The Plant Journal*, *113*(4), 677–697.

Taishykov, Z., Tolysbayeva, M., Zhumanazarov, K., Ibraimova, S., & Mizambekova, Z. (2024). Management of innovation processes in agriculture. *World Development Perspectives*, *33*, 100566.

Tang, X., Feng, X., Xu, Y., Yang, B., Wang, Y., Zhou, Y., … & Lu, Y. (2025). CircZmMED16 delays plant flowering by negatively regulating starch content through its binding to ZmAPS1. *Journal of Integrative Plant Biology*.

Tsai, W. A., Sung, P. H., Kuo, Y. W., Chen, M. C., Jeng, S. T., & Lin, J. S. (2023). Involvement of MicroRNA164 in responses to heat stress in Arabidopsis. *Plant Science*, *329*, 111598.

Van Werven, F. J., Neuert, G., Hendrick, N., Lardenois, A., Buratowski, S., van Oudenaarden, A., … & Amon, A. (2012). Transcription of two long noncoding RNAs mediates mating-type control of gametogenesis in budding yeast. *Cell*, *150*(6), 1170–1181.

Wang, H., Lu, S., Guan, X., Jiang, Y., Wang, B., Hua, J., & Zou, B. (2022). Dehydration-responsive element binding protein 1C, 1E, and 1G promote stress tolerance to chilling, heat, drought, and salt in rice. *Frontiers in Plant Science*, *13*, 851731.

Wang, M., Zhao, W., Gao, L., & Zhao, L. (2018). Genome-wide profiling of long non-coding RNAs from tomato and a comparison with mRNAs associated with the regulation of fruit ripening. *BMC Plant Biology*, *18*, 1–10.

Wang, Y., Fan, X., Lin, F., He, G., Terzaghi, W., Zhu, D., & Deng, X. W. (2014). Arabidopsis noncoding RNA mediates control of photomorphogenesis by red light. *Proceedings of the National Academy of Sciences*, *111*(28), 10359–10364.

Wen, F. L., Yue, Y., He, T. F., Gao, X. M., Zhou, Z. S., & Long, X. H. (2020). Identification of miR390-TAS3-ARF pathway in response to salt stress in *Helianthus tuberosus* L. *Gene*, *738*, 144460.

Wunderlich, M., Groß-Hardt, R., & Schöffl, F. (2014). Heat shock factor HSFB2a involved in gametophyte development of *Arabidopsis thaliana* and its expression is controlled by a heat-inducible long non-coding antisense RNA. *Plant Molecular Biology*, *85*, 541–550.

Yadav, A., Mathan, J., Dubey, A. K., & Singh, A. (2024). The emerging role of non-coding RNAs (ncRNAs) in plant growth, development, and stress response signaling. *Non-Coding RNA*, *10*(1), 13.

Yang, H., Cui, Y., Feng, Y., Hu, Y., Liu, L., & Duan, L. (2023). Long non-coding RNAs of plants in response to abiotic stresses and their regulating roles in promoting environmental adaption. *Cells*, *12*, 729.

Yu, F., Tan, Z., Fang, T., Tang, K., Liang, K., & Qiu, F. (2020). A comprehensive transcriptomics analysis reveals long non-coding RNA to be involved in the key metabolic pathway in response to waterlogging stress in maize. *Genes*, *11*, 267.

Yu, J., Qiu, K., Sun, W., Yang, T., Wu, T., Song, T., … & Tian, J. (2022). A long noncoding RNA functions in high-light-induced anthocyanin accumulation in apple by activating ethylene synthesis. *Plant Physiology*, *189*(1), 66–83.

Yu, J., Su, D., Yang, D., Dong, T., Tang, Z., Li, H., Han, Y., Li, Z., & Zhang, B. (2020). Chilling and heat stress-induced physiological changes and MicroRNA-related mechanism in sweetpotato (*Ipomoea batatas* L.). *Frontiers in Plant Science*, *11*, 687.

Yu, Y., Zhou, Y., Zhang, Y., & Chen, Y. (2018). Grass phasiRNAs and male fertility. *Science China Life Sciences*, *61*, 148–154.

Zhai, J., Jeong, D. H., De Paoli, E., Park, S., Rosen, B. D., Li, Y., … & Meyers, B. C. (2011). MicroRNAs as master regulators of the plant NB-LRR defense gene family via the production of phased, trans-acting siRNAs. *Genes & Development*, *25*(23), 2540–2553.

Zhang, B. (2015). MicroRNA: A new target for improving plant tolerance to abiotic stress. *Journal of Experimental Botany*, *66*(7), 1749–1761.

Zhang, H., Zhang, J., Yan, J., Gou, F., Mao, Y., Tang, G., … & Zhu, J. K. (2017). Short tandem target mimic rice lines uncover functions of miRNAs in regulating important agronomic traits. *Proceedings of the National Academy of Sciences*, *114*(20), 5277–5282.

Zhang, J., Zhang, H., Srivastava, A. K., Pan, Y., Bai, J., Fang, J., … & Zhu, J. K. (2018). Knockdown of rice microRNA166 confers drought resistance by causing leaf rolling and altering stem xylem development. *Plant Physiology*, *176*(3), 2082–2094.

Zhang, P., Fan, Y., Sun, X., Chen, L., Terzaghi, W., Bucher, E., Li, L., & Dai, M. (2019). A large-scale circular RNA profiling reveals universal molecular mechanisms responsive to drought stress in maize and Arabidopsis. *Plant Journal*, *98*, 697–713.

Zhang, W., Han, Z., Guo, Q., Liu, Y., Zheng, Y., Wu, F., & Jin, W. (2014). Identification of maize long non-coding RNAs responsive to drought stress. *PLoS ONE*, *9*(6), e98958.

Zhang, X., Dong, J., Deng, F., Wang, W., Cheng, Y., Song, L., … & Shen, F. (2019). The long non-coding RNA lncRNA973 is involved in cotton response to salt stress. *BMC Plant Biology*, *19*(1), 16.

Zhang, Y., Wang, S., Li, W., Wang, S., Hao, L., Xu, C., Yu, Y., Xiang, L., Li, T., & Jiang, F. (2022). A long noncoding RNA HILinc1 enhances pear thermotolerance by stabilizing PbHILT1 transcripts through complementary base pairing. *Communications Biology*, *5*, 1134.

Zhao, J., Yuan, S., Zhou, M., Yuan, N., Li, Z., Hu, Q., … & Luo, H. (2019). Transgenic creeping bentgrass overexpressing Osa-miR393a exhibits altered plant development and improved multiple stress tolerance. *Plant Biotechnology Journal*, *17*(1), 233–251.

Zhao, X., Li, J., Lian, B., Gu, H., Li, Y., & Qi, Y. (2018). Global identification of arabidopsis LncRNAs reveals the regulation of MAF4 by a natural antisense RNA. *Nature Communications*, *9*, 5056.

Zhao, Y., Liu, J., Li, Y., Wen, H., Teotia, S., Zhang, X., … & Peng, T. (2022). osa-miR528 promotes seedling growth by enhancing nitrate uptake under nitrogen deficiency in rice. *Environmental and Experimental Botany*, *202*, 105040.

15 Decoding Plant Resilience
The Role of ncRNA in Abiotic Stress Tolerance

Talib Ali, Anosha Asghar, Zakir Ullah, Javed Iqbal, Banzeer Ahsan Abbasi, Shumaila Ijaz, Rooma Waqar, Sadia Zafar and Tariq Mahmood

15.1 INTRODUCTION

Plant growth and productivity along with survival are extensively limited by non-biological environmental factors known as abiotic stressors (Mosa et al., 2017). The combination of drought stressors with salinity issues along with extreme temperature conditions, heavy metals toxicity and dietary imbalances disrupts up to 50% of global agricultural yields (Partheeban et al., 2017). A rise in abiotic pressures is projected to happen in the coming decades because of climate change, which will increase temperature levels as well as modify precipitation patterns and degrade soils, thus becoming a major issue for agricultural production (Thakur et al., 2024). Nearly all agricultural areas experience the frequent occurrence of abiotic stress, which affects their crops. Drought proves to be the leading concern because it impacts more than 60% of cultivated land area, resulting in small plants and death of reproductive stages and subpar crop outcomes (Gupta et al., 2022). Soil salinisation affects irrigated agricultural systems extensively because it causes root functional problems by disrupting water and ion absorption processes. Semi-arid and dry lands experience dual effects of water limitation alongside growing soil salinity because these areas are highly reactive to both conditions (Mishra et al., 2023). The growth of plants along with agricultural farm productivity faces substantial interference from two main abiotic stresses, including excessive heat and metal toxicity. The combination of heat stress causes three primary effects on plants: it intensifies water loss while damaging photosynthesis and affecting reproductive growth to deliver less than desirable grain content and yield reductions (Ullah et al., 2022a). Plant germination reduces when subjected to cold stress, and the stress simultaneously affects membrane fluidity and disorganises cellular homeostasis as well as enzyme activity (Zhao et al., 2017). Plant stress severity increases when heavy metals like cadmium and lead appear together with nutrient deficiencies because they create toxic conditions that disrupt metabolic functions (Vasilachi et al., 2023).

Plants exhibit both physiological and biochemical modifications because of abiotic stress. Drought stress induces water deficits, stomatal closure and reduced transpiration, decreasing photosynthetic efficiency and biomass accumulation. Reactive oxygen species (ROS) production leads to protein and lipid destruction and damages DNA structures (Anjum et al., 2017). The excessive Na^+ and Cl^- ions generate osmotic and ionic imbalances in plants under salinity stress, which disrupt cell processes leading to death (El Sabagh et al., 2021). Membrane destabilisation along with protein denaturation occurs when plants experience heat stress, but cold stress impacts membrane fluidity and causes difficulties in nutrient transport as well as signal transduction activities (Jahed et al., 2023; Sato et al., 2024).

Greater than 60% growth of worldwide food production will be necessary by 2050 to fulfil demand because staple crops like wheat, rice, maize and soybeans face threats from abiotic stress. Agricultural regions with wheat cultivation will suffer yield reductions reaching 30% as the middle of this century develops (O'Leary et al., 2018). Rice production stands as the primary crop affected by water salinity since it impacts 20% of the overall irrigated terrain (Fahad et al., 2019). The extreme temperatures in tropical and subtropical areas represent a threat that puts yields in danger (Praveen & Sharma, 2019). Plants fight off stress by generating deeper roots while adjusting osmotic pressure with substances such as proline and glycine betaine and using antioxidants that remove ROS (Gowtham et al., 2022). The hormone signal pathway and photosynthetic processes are controlled by genes which respond to stress (Jogawat et al., 2021).

Plant stress responses receive major regulation through non-coding RNAs (ncRNAs). Stress-related regulation of gene expression depends on both microRNAs (miRNAs) and long non-coding RNAs (lncRNAs), which target superoxide dismutase genes to regulate oxidative stress and modify chromatin remodelling to start transcriptional activation (Ijaz, Iqbal, et al., 2024). The three methods for abiotic stress mitigation include traditional breeding and biotechnology alongside sustainable practices. There are exciting prospects in developing resilient ncRNA-enhanced crops that gain support from investments in policy and infrastructure systems (Asghar et al., 2024; Ijaz, Liu, et al., 2024; Zenda et al., 2021). Research into molecular stress reactions helps build robust agricultural systems which secure food production, according to Asghar et al. (2024).

The cellular signals detected by plants activate both signalling systems and transcriptional control networks. Plants

DOI: 10.1201/9781003583257-15

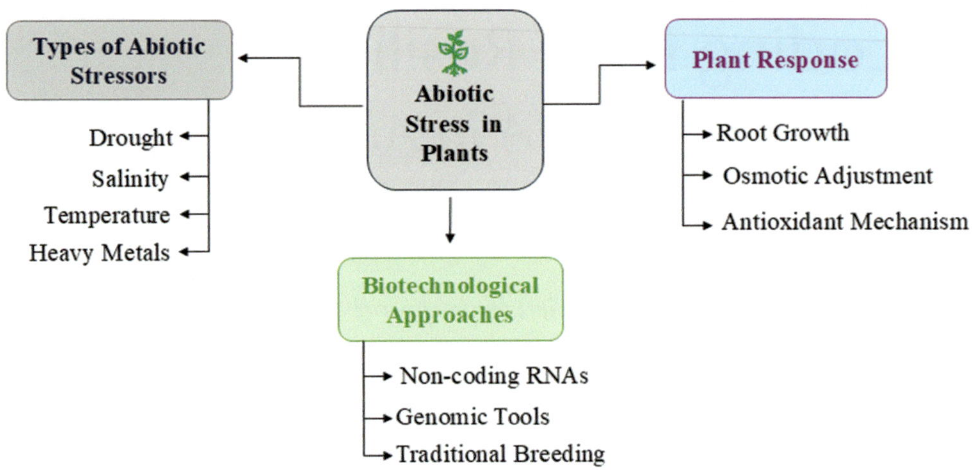

FIGURE 15.1 Abiotic stresses in plants: challenges and responses.

under drought conditions trigger abscisic acid (ABA)-dependent signalling that causes stomatal closure while performing osmotic adjustment (Muhammad Aslam et al., 2022). Salinity tolerance occurs through two mechanisms of sodium exclusion along with vacuolar compartmentalisation, according to Chakraborty et al. (2018). Stress-created gene expression patterns become decodable using genomics, transcriptomics and metabolomics methods, which reveal transcription factors (TFs) and small RNA regulators (Sari et al., 2022). Plants use antioxidants to protect themselves against ROS (Hasanuzzaman et al., 2020). The stress response system in plants gets additional control through DNA methylation and histone acetylation modifications in epigenetic regulation (Banerjee & Roychoudhury, 2017).

CRISPR/Cas9 genomic tools allow researchers to perform exact modifications of genes that increase stress survival abilities. The tolerance to drought and salt stress increases when TFs NAC and WRKY are overexpressed (Erpen et al., 2018). Stress-related gene expression gets controlled through non-coding RNAs which include both miRNAs and siRNAs and further include lncRNAs and circular RNAs (circRNAs) that affect chromatin and transcription (Yadav, Mathan, et al., 2024). Through the combination of genomics data analysis with bioinformatics techniques, researchers can develop better stress-resistant plant cultivars (Sahoo et al., 2020). Resourceful crop cultivation through molecular techniques enables scientists to create stress-resistant plants because of food security concerns (Figure 15.1) (Ahlawat et al., 2024).

15.2 ABIOTIC STRESSES IN PLANTS: A GLOBAL CHALLENGE

The agricultural world faces severe threats from various abiotic factors that combine drought stress with salinity issues, high temperatures, toxic metal contamination and problems with nutrient levels. Plants suffer adverse effects on their development and agricultural output under these environmental conditions, which result in substantial worldwide agricultural losses. The stressors from climate change intensify through modifications of rainfall patterns and enhanced temperatures, as well as degradation of soil conditions (Furtak & Wolińska, 2023). Soil drought is regarded as one of the most damaging environmental issues because it impacts approximately 60% of usable lands and causes growth reduction together with fruit failure alongside reduced crop value (Dietz et al., 2021). The irrigation process creates soil salinity that degrades root functions and halts ion absorption and endangers staple crops, especially rice and wheat (Majeed & Muhammad, 2019). Plant physiological processes suffer substantial damage from both heat and cold stress temperatures because heat exposure causes fast water loss, poor pollen quality and disrupted photosynthesis and exposes plants to cold-related enzyme damage and cellular breakdown (Muhammad et al., 2021). Heavy metal contamination and nutrient deficiencies create disturbances in key metabolic pathways that cause further reductions in plant resilience and crop yield (Vasilachi et al., 2023). Earth's food security depends strictly on solving abiotic stress challenges because scientists and agricultural practitioners need to improve their plant management techniques and sustainable farming methods.

15.3 TYPES AND PHYSIOLOGICAL IMPACTS OF ABIOTIC STRESSES

Abiotic stress affects plants in multiple ways, including disruptions in water balance, nutrient uptake and photosynthesis. Flooding stress occurs when excess water in the soil limits oxygen availability to plant roots, resulting in hypoxia or anoxia. Oxygen deprivation impairs cellular respiration, resulting in energy deficiencies and the accumulation of harmful by-products such as ethanol (Habibi et al., 2023). Flood-tolerant plants, such as rice, exhibit adaptations such as aerenchyma development, which promotes oxygen transfer to submerged tissues, and the activation of anaerobic metabolic pathways (Bridgemohan et al., 2020). The presence of dangerous quantities of heavy metals, such as cadmium, lead and mercury, in soils adversely affects plant growth.

These metals affect enzyme functioning, create ROS and interfere with nutrient intake. Plants adopt methods including chelation through phytochelatins, compartmentalisation in vacuoles and antioxidant defences to ameliorate heavy metal toxicity (Ghuge et al., 2023). Plant physiological processes suffer substantial damage from both heat and cold stress temperatures because heat exposure causes fast water loss, poor pollen quality and disrupted photosynthesis and exposes plants to cold-related enzyme damage and cellular breakdown (Muhammad et al., 2021). Heavy metal pollution as well as nutrient imbalances harm metabolic pathways in plants, which weakens their resilience and causes lower crop productivity (Vasilachi et al., 2023). Earth's food security depends strictly on solving abiotic stress challenges because scientists and agricultural practitioners need to improve their plant management techniques and sustainable farming methods. Heat stress alters proteins and disorganises membranes, as well as speeding up normal cell ageing, whereas cold stress creates icy conditions and reduces enzyme functioning, which decreases photosynthetic ability. Plants utilise protective mechanisms involving heat shock proteins (HSPs) and antifreeze proteins together with lipid modifications of cellular membranes to combat these challenges (Choudhary et al., 2024; Ullah et al., 2022b). UV radiation stress results in DNA, along with protein and cellular structure damage, which reduces photosynthetic efficiency for plants. Under radiation stress conditions, plants create flavonoids as protective metabolites while boosting their antioxidant activity to limit oxidative damage to tissues (Ferreyra et al., 2021). These abiotic stress conditions create detrimental effects on plant survival together with growth and yield reduction which demands additional research for crop variety development.

15.4 MOLECULAR RESPONSES TO ABIOTIC STRESSES

Specific organelles and membranes contain stress-detection mechanisms that launch elaborate signal pathways to interpret abiotic stress signals from the environment. Secondary messenger signals that activate from stress perception lead to the activation of calcium ions (Ca^{2+}) and ROS, as well as phytohormone ABA, to control stress-responsive gene expression (Ravi et al., 2023; Rawat et al., 2021). The regulation of stress response genes depends heavily on the TFs DREB along with WRKY and MYB. Stress-induced genes produce protective proteins described as HSPs as well as late embryogenesis abundant (LEA) proteins, which guard against cellular deterioration in harsh environmental conditions, according to Wu et al. (2022). Superoxide dismutase (SOD) and catalase existing as antioxidant enzymes work to reduce ROS excess and fight oxidative stress (Gusti et al., 2021). The body of a plant regulates salt tolerance through ion transporters while removing toxic sodium ions to maintain equilibrium (Joshi et al., 2022). Plants create osmolytes, such as proline, glycine betaine and sugars, in response to water scarcity in order to stabilise proteins and membranes and preserve osmotic equilibrium (Dikilitas et al., 2020). By generating signalling molecules like salicylic acid and jasmonates, which alter defensive mechanisms for protection, plants engage their preventive defence systems (Meraj et al., 2020). Because epigenetic changes that employ DNA methylation and histone acetylation regulate gene expression to create stress memory, plants in subsequent generations get more tolerant (Sharma et al., 2022). By combining ABA pathways with ROS and calcium systems, plants can manage several environmental difficulties simultaneously and improve their chances of surviving environmental changes (Mohanta et al., 2018). Researchers can use genetic techniques and technologies to improve the sophisticated, flexible mechanisms found in extensive stress control networks in plants (Kollist et al., 2019). According to Sharma et al. (2022), the epigenetic mechanism of DNA methylation and histone acetylation creates stress memories that allow for the survival of multiple generations. Because of their intricate signalling system between ABA pathways with ROS and calcium systems, which enables them to manage many stress components simultaneously, plants increase their chances of surviving when their environments change (Mohanta et al., 2018). Studies in genetic methods and biotechnology benefit from plant stress response regulation molecular networks due to their advanced adaptive strategies (Kollist et al., 2019). *lncRNAs*: Emerging as molecular scaffolds and decoys, lncRNAs interact with transcriptional machinery and chromatin remodellers, playing roles in stress signal transduction (Nadhan et al., 2022). *circRNAs*: These operate as sponges for miRNAs, preventing them from targeting mRNAs and so controlling gene expression under stress (Lavenniah et al., 2020). ncRNAs regulate important pathways in response to drought, salinity, temperature extremes and oxidative stress: in drought stress, miRNAs such as miR398 and miR396 perform roles in osmotic adjustment and root architecture change, improving water uptake (Ahmad et al., 2024; Ijaz et al., 2024; Islam et al., 2022). The ionic balance of salinity-tolerant plants becomes regulated by lncRNAs that control the expression of sodium compartmentation-related genes HKT and NHX (Bouzroud et al., 2023). Under heat stress conditions, both miR159- and SVALKA-type lncRNAs protect cell structures by managing hormone communication pathways according to Praveen et al. (2023). The epigenetic regulation actions of ncRNAs use DNA methylation along with histone changes to remodel chromatin and make genes more accessible, during which plants develop memories of past stress events to better manage recurring Stressful challenges (Junaid et al., 2024). The production of drought- and salinity-tolerant crops relies on introducing elevated levels of stress-responsive microRNA and short-interference RNAs to plants. The modification of *Arabidopsis* with miR319 resulted in improved resistance to drought conditions as described by Singroha et al. (2021). Scientists are developing synthetic lncRNAs along with circRNAs as molecular switches to control stress-responsive pathways with high precision, according to Jha and his research team (2020). The potential implementation of ncRNAs faces ongoing challenges because of their intricate regulatory networks intertwined with the possible unwanted effects. The combination of modern sequencing technology with computational modelling will help researchers understand ncRNA functions better, which will lead to their maximum agricultural application (Satam et al., 2023). An evolutionary breakthrough for plant science is delivered through ncRNAs' involvement in plant stress management. Scientists studying the regulatory mechanism and power of ncRNAs will create new techniques to improve crop durability, which supports sustainable agriculture against climate change, according to Shah et al. (2021).

15.5 GLOBAL IMPLICATIONS OF ABIOTIC STRESS FOR FOOD SECURITY

Food security encounters significant threats because abiotic factors decrease crop outputs while impacting agricultural farming results. Drought leads to 60% of all yield reductions in rainfed agricultural systems, ranking it among the most destructive stressors (Ahmad et al., 2022, 2024; Meza et al., 2021). Climate change intensifies already challenging food problems because it causes more frequent and intense dry seasons, salt content increases and temperature extremes in sub-Saharan Africa and South Asia as well as other impacted regions (Kidane 2024). Abiotic stress generates a severe economic effect which both diminishes farmer profits and weakens national economic systems. Plant failure caused by stress factors drives food prices upward and diminishes exporting capabilities, which leads to dependent food import systems in these nations (Shiferaw et al., 2013). Abiotic stress environments directly affect the nutritional composition of strategic food cereals. Plants under stress develop toxic substances while losing important nutrients, including proteins, vitamins and minerals, which results in decrease in nutritional value in fundamental food products (Figure 15.2) (Haider et al., 2021, 2022; Proietti et al., 2015).

15.6 ADAPTATION AND MITIGATION STRATEGIES FOR ABIOTIC STRESS

Multiple solutions are needed to handle abiotic stress through genetic techniques integrated with technical improvements and agricultural practices. The creation of stress-tolerant crop varieties needs biological development with genetic engineering and conventional breeding because environmental stress needs management. Genomic technology advances through the implementation of CRISPR/Cas9 methods now allow researchers to modify genes precisely, which increases crop resistance to drought, salinity and heat conditions (Proietti et al., 2015). The productivity of crops can be preserved by adopting improved irrigation methods together with proper soil management practices and the implementation of warning systems (Adeyemi et al., 2017). The necessary steps for enhancing agricultural resilience in vulnerable areas require both international collaboration and supportive policies which maintain sustainable food production amid unpredictable climate conditions. Research and policy collaboration between scientists and policymakers will create farming systems that resist environmental challenges while protecting world food security.

15.7 NON-CODING RNAs: CLASSIFICATION AND FUNCTIONS

Sequences of RNA without protein-coding potential function are essential regulatory molecules through which they regulate gene expression while maintaining cellular homeostasis. The plant scientific community studies these regulatory molecules closely because they provide stress defence mechanisms and beneficial effects on improved crop production systems (Santosh et al., 2015; Shomali et al., 2022). Plants achieve environmental stress resistance through ncRNA

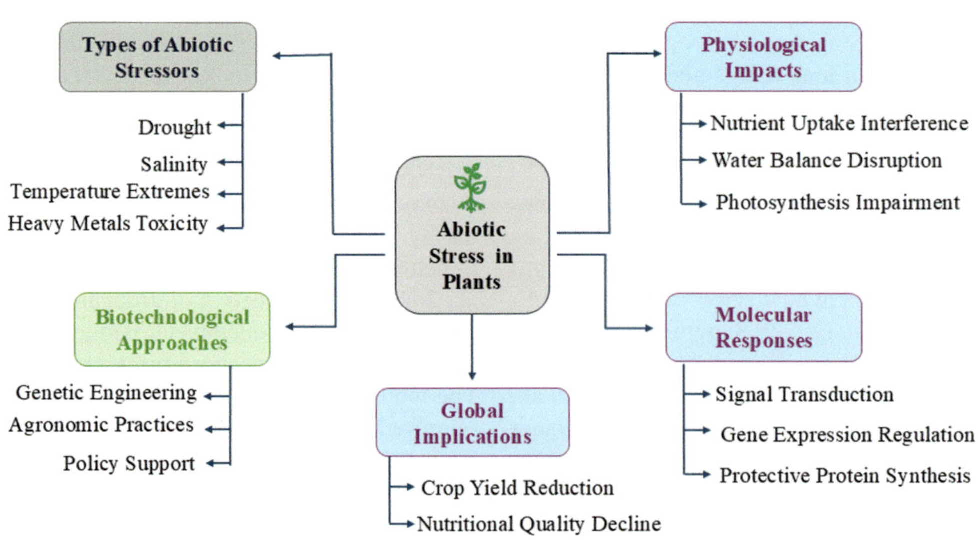

FIGURE 15.2 Abiotic stresses in plants; challenges and solutions.

regulation of gene regulatory pathways, which makes these biomolecules promising subjects for genetic modification and agricultural development.

15.7.1 Types of Non-Coding RNAs

The different types of NcRNAs receive classification through assessments of size and functionality. MicroRNA (miRNA) together with small interfering RNA (siRNA) represent short ncRNA types that measure 20–24 nucleotides and operate to control gene expression at the post-transcriptional phase (Holley & Topkara, 2011). The regulatory potential of non-coding RNAs (NcRNAs) increases with their length as lncRNAs span more than 200 nucleotides while the regulatory power of circRNAs rises from their covalently closed-loop structure which also extends their stability (Zhang Wang, et al., 2019). Multi-faceted regulatory networks made up of ncRNAs operate singularly and collectively to control plant responses during development and stress situations.

15.7.2 MicroRNAs (miRNAs)

Endogenous ncRNAs called microRNAs (miRNAs) perform gene silencing functions by directing messenger RNAs (mRNAs) to destruction or translational inhibition. After transcription as the primary products pri-miRNAs, they undergo processing to produce precursor miRNAs pre-miRNAs which eventually result in cleaved mature miRNAs through Dicer enzyme activity (Starega-Roslan et al., 2011). During RNA-induced silencing complex (RISC) activity, miRNAs function as essential molecular regulators for plant development, together with stress response and hormone signalling. Oxidative stress tolerance receives a boost from miR398 because this miRNA directs its action at copper/zinc superoxide dismutase genes (Li et al., 2022). Plants use miRNAs as regulatory drivers to enhance their ability to overcome abiotic challenges while simultaneously generating benefits for agricultural improvement.

15.8 SHORT INTERFERING RNAs (siRNAs)

Each 21–25-nucleotide short interfering RNA (siRNA) serves as a component of RNA interference (RNAi) that protects against viruses and maintains genomic integrity (Bhan et al., 2017). The development of short interfering RNAs (siRNAs) originates from double-stranded RNA (dsRNA) that results from transposons, viruses or external environmental sources, according to Sanan-Mishra et al. (2021). The Dicer enzyme processes siRNAs until they join RISC to perform sequence-specific mRNA degradation in accord with Jaskiewicz and Filipowicz (2008). Through siRNA regulation, plants maintain silencing of transposable elements and viruses as well as epigenetic gene regulation, which results in increased stress tolerance to abiotic threats (Wang & Chekanova, 2016).

15.8.1 Long Non-Coding RNAs (lncRNAs)

By regulating many levels of gene expression, from transcription to epigenetics and post-transcription, long non-coding RNAs (lncRNAs) influence vital activities like immunological responses and stress adaptation in addition to flowering (Zhang, Wang, et al., 2019). The processing mechanism of long non-coding RNAs stands apart from that of miRNAs and siRNAs since they skip mRNA breakdown activity. The multiple functions executed by lncRNAs involve being molecular scaffolds and chromatin modifiers and acting as decoy elements. The regulatory functions of lncRNAs happen by drawing chromatin-modifying complexes towards genes for accessibility and also by stopping miRNA binding for mRNA stability (Sebastian-delaCruz et al., 2021).

15.8.2 Circular RNAs (circRNAs)

Stable circRNA structures called circRNAs maintain their shape into closed-loop molecular structures, which defend them from degradation by exonucleases (Geng et al., 2020). The production of circRNAs happens through a back-splicing process, which links a 5′ splice site sequence to a 3′ splice site sequence (Zhang et al., 2016). The function of circRNAs consists of capturing miRNAs and blocking their access to target mRNAs. Some circRNAs accomplish transcription and RNA splicing control by establishing interactions with RNA-binding proteins. Plants require additional research into circRNAs since scientists have observed these RNA molecules to participate in stress response and developmental regulation as well as hormonal signalling pathways (Zang et al., 2020).

15.9 FUNCTIONAL INTERACTIONS AND MOLECULAR MECHANISMS OF ncRNAs

Complex regulatory networks formed by ncRNAs achieve precise genetic expression control through stressful environmental conditions and changes in circumstances. Long non-coding RNAs function as decoys which protect miRNAs from binding their target mRNAs in plants (Di Martino et al., 2021). CircRNAs control the activity of miRNAs by acting as sponge molecules. miRNA-mediated mechanisms(MMO), indicates gene silencing methods of siRNAs parallel to those of miRNAs, which together regulate gene networks precisely (Moazed, 2009). Through these interactive systems, plants obtain an adaptable mechanism that enables effective control over stress-related gene networks (Nejat & Mantri, 2018). Different classes of ncRNAs show precise coordination because these molecules play a vital role in enhancing plant resilience as well as maximising agricultural productivity (Bashir & Husaini, 2025).

The biogenesis pathway of NcRNAs starts with RNA polymerase transcription, which then leads to capping modifications, splicing processes and polyadenylation steps. The mechanism of small ncRNA production starts with double-stranded RNA precursors, which get processed through specialised enzymes, whereas long ncRNAs retain an mRNA-like format but do not carry coding sequences (Gomes et al., 2013). The stability of circular RNAs increases due to their exonuclease-resistant structures, which stem from back-splicing events (Liu et al., 2022).

15.10 GENERAL ROLES OF ncRNAs IN PLANT GROWTH AND DEVELOPMENT

Different stages of plant growth including seed germination through flowering as well as root architecture and stress responses differ from each other under ncRNA regulation (Fahad et al., 2024). Genetic regulation happens on multiple stages through ncRNAs, so plants can connect an assortment of internal cues together with environmental signals to develop optimised physiological responses. During root development, miRNAs manage TFs responsible for lateral root establishment and nutrient uptake, which enables plants to enhance their defence against soil stimuli (Prabhakaran, 2024). The proper leaf structure and function depend on the influence of ncRNAs, which independently control leaf morphogenesis while directing vascular differentiation and chlorophyll biosynthesis (Ma, Zhao, et al., 2022).

Flowering triggers an interaction between ncRNA molecules, which enables their regulation of signalling pathways that control the floral transition timetable. LncRNAs together with miRNAs enable plants to modify flowering time while processing photoperiod and temperature signals that drive their reproductive success (Datta & Paul, 2019; Traubenik et al., 2024). Plant responses to environmental stress factors depend heavily on the function of NcRNAs. Stress-responsive ncRNAs work under drought, salinity and extreme temperature conditions to manage osmotic equilibrium and water maintenance and antioxidant defence for better tolerance (Gelaw & Sanan-Mishra, 2021). Specific miRNAs control genes involved in stress resistance through their management of TF activities, thus allowing plants to survive harsh conditions (Pagano et al., 2021). The immune response management capabilities of ncRNA include precise control of pathogen defence genes together with maintenance of growth-defence balance and defence against biotic threats (Figure 15.3) (Ning et al., 2017). Research teams can create sustainable agricultural systems by using ncRNA-based approaches to advance genetic engineering technologies and biotechnology, as explained in Zhang et al. (2022).

15.11 ncRNA-MEDIATED REGULATION OF ABIOTIC STRESS RESPONSES

The existence of plants is endangered, together with production loss caused by four major abiotic stress elements that include drought and salinity stress, extreme temperature stress and oxidative stress. Plants develop drought resistance due to the regulatory capacity of ncRNA elements, which manage stomatal movements as well as osmotic pressure and water preservation mechanisms (Yadav et al., 2024). Under salinity conditions, certain ncRNAs function as regulators of ion transporters to stop sodium accumulation, thus preserving ionic equilibrium (Balasubramaniam et al., 2023). Temperature variations cause specific ncRNAs which activate HSPs together with antifreeze proteins to guarantee stable proteins, intact membranes and balanced metabolism (Balasubramaniam et al., 2023). The accumulation of ROS occurs under oxidative stress situations, yet specific ncRNAs control antioxidant enzyme activities to reduce ROS damage (Fuschi et al., 2019). Stress memory in plants develops due to ncRNAs, which perform epigenetic changes that enhance plant responsiveness to recurring stress situations (Marasca et al., 2018). Research into ncRNA regulatory functions demonstrates their practical value for crop biotechnology since genetically modified levels of ncRNA expression could enhance plant adaptability to stress, along with improving sustainable farming under climate adjustments (Rajput et al., 2021). Generation of ROS detoxifying enzymes through miR528 regulation

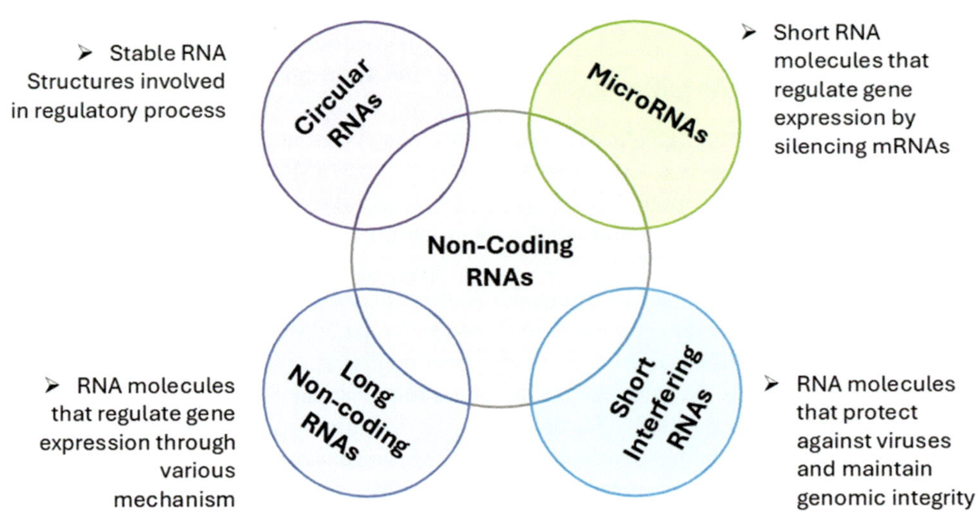

FIGURE 15.3 Overview of types of ncRNA and their functions.

protects plants from oxidative damage under high-salinity conditions (Ghosh et al., 2022).

15.12 miRNAs, siRNAs AND lncRNAs IN STRESS RESPONSE

During abiotic stress conditions, miRNAs act as major regulatory elements which direct their targets towards stress-responsive genes and TFs. Plants experience developmental problems, and their cellular mechanisms become disordered because of both high and low temperature stress conditions. The stress-responsive gene response to temperature adaptation relies on specific miRNAs (Pagano et al., 2021). The SPL TFs important for blooming time and thermotolerance come under regulative control from MiR156 and miR319, which targets the TCP TFs, which manage the processes linked to heat stress. The NF-YA TFs receive miR169 regulation, which enables proper cellular functioning under cold stress conditions (Xu et al., 2014). Oxidative stress occurs primarily through other abiotic conditions to create harmful ROS accumulations that harm cells. The function of miR395 and miR398 on sulphur metabolism and antioxidant enzyme production leads to enhanced stress tolerance (Yang et al., 2021). MiR164 targets the NAC TFs that regulate programmed cell death, hence reducing the tissue damage brought on by oxidative stress. Research on the role of miRNA in stress tolerance provides important insights for agricultural innovation. According to Basso et al. (2019), genetic engineering that alters miRNA levels gives plants increased environmental resilience. According to Xu et al. (2019), researchers have created stress-resistant plant breeds that can sustain agricultural operations in altered climate circumstances by combining novel CRISPR/Cas9 gene editing technologies with miRNA-based interventions. According to Li et al. (2022), miR398 regulates the genes for the antioxidant enzyme superoxide dismutase (SOD) during drought circumstances to increase plant resilience to water deficit stress. As per Yadav et al. (2024), miR166 promotes root xylem differentiation under drought. When auxin-root development signals are disrupted by naturally occurring salt stress, miRNAs alter their expression, improving resistance to salt (Fahad et al., 2025).

When applied using RNA interference (RNAi) techniques, small interfering RNAs (siRNAs) aid in producing stress-induced gene silence. Before being converted into 21–24 nucleotide fragments that integrate into the RISC for complementary mRNA targeting and degradation, the dsRNA precursors enter the system and become substrates for Dicer-like (DCL) enzymes, starting the pathway (Fukudome & Fukuhara, 2017). In order to increase their capacity for adaptation to changes in their environment, plants that are undergoing gene silencing turn on genes that respond to stress (Sihag et al., 2024). Drought siRNAs regulate stomatal processes while maximising the effectiveness of water utilisation, improving plant performance in water-restricted environments (Pradhan et al., 2015). When plants are under salt stress, siRNAs regulate the expression of sodium transporters, protecting them against ion toxicity (Tanna et al., 2024). siRNA control, which includes genes for HSPs, cold-responsive elements and ROS detoxification pathways, provides protection for plants exposed to oxidative damage or temperature stress (Circu & Aw, 2010; Megha et al., 2018). In order to maximise the dynamics of their stress response, plants use RNA regulatory factors, such as miRNAs and lncRNAs, to activate a complex regulatory system (Ratti et al., 2020). According to Younis and colleagues (2014), the agricultural industry uses siRNA-mediated gene silencing to create RNAi-based transgenic crops that increase resilience to diseases, drought and salinity (Figure 15.4). Through advancements in

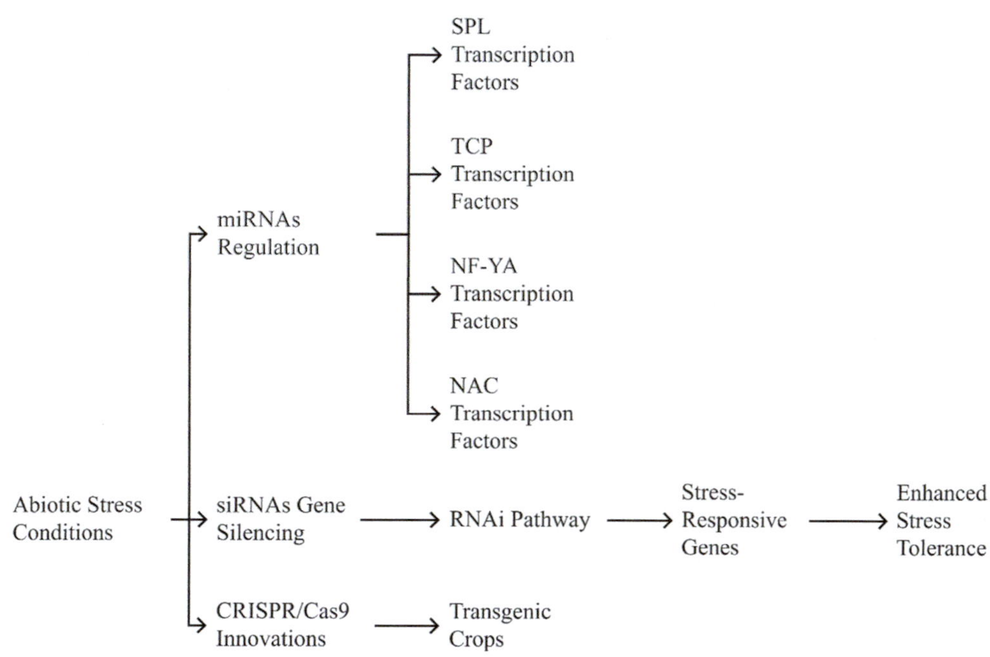

FIGURE 15.4 Role of miRNAs and siRNAs in stress response.

the CRISPR/Cas genome editor, siRNA-based systems can be precisely regulated, opening up new possibilities for sustainable agriculture (Mujtaba et al., 2021).

15.13 MOLECULAR NETWORKS OF ncRNAs IN ABIOTIC STRESS TOLERANCE

In order to regulate TFs and form ribonucleoprotein (RNP) complexes that dictate stress tolerance mechanisms, plants employ non-coding RNAs (ncRNAs). Plants can develop stress tolerance through the regulation of gene expression at various levels through interactions with ncRNAs (Ma et al., 2022;).

15.13.1 ncRNA-Mediated Regulation of Transcription Factors and Proteins

MicroRNAs' (miRNAs') regulation mechanism consists of either blocking translation processes or destroying their target mRNAs. The expression of miR159 rises during drought in order to lower the levels of MYB TFs, which regulate stress responses mediated by ABA. In order to precisely manage ABA signalling and optimise resource allocation, miR159 expression rises in response to drought exposure, which causes MYB suppression (Deng et al., 2024; Millar et al., 2019). Because of its evolutionary significance, stress causes miR398 to be downregulated, which allows for higher quantities of CSD protein to fight oxidative stress (Zhang et al., 2022). Through the determination of essential physiological adaptations and energy conservation, each interaction between a TF and non-coding RNA (ncRNA) aids plants in adhering to particular stress situations (Ragupathy et al., 2016; Shriram et al., 2016; Singh et al., 2025). According to Nejat and Mantri (2018), long non-coding RNAs (lncRNAs) regulate TFs by preventing their destruction, regulating their activity and guiding epigenetic changes. CBF1, a TF involved in the thermal stress response, is stabilised by SVALKA lncRNA. SVALKA improves cold resistance in plants by stabilising CBF1 and preventing its degradation, which sustains the expression of genes responsive to cold (Kiger & Schroeder, 2024; Lindemose et al., 2013). Because SVALKA is an lncRNA that alters chromatin states to make transcription machinery more accessible, it supports the cold stress adaptation mechanism (Kiger & Schroeder, 2024). Direct binding to TFs, co-regulator recruitment and environmental feedback mechanisms that sustain gene expression are all ways that lncRNAs impact TFs (Dykes & Emanueli, 2017; Long et al., 2017; Valadkhan & Valencia-Hipólito, 2015). According to Białoskórska et al. (2024), plants exploit these particular interactions as a biological mechanism to increase their capacity for stress tolerance while retaining flexible regulatory flexibility in a variety of environmental situations.

15.13.2 ncRNA–Protein Complex Formation and Stress Adaptation

Non-coding RNAs, in particular lncRNAs, frequently interact with proteins to form RNP complexes that regulate the stability, location and activity of the proteins. Plant stress response mechanisms heavily depend on these essential complexes that enable specific gene regulation, according to Bhatti et al. (2021). The development of RNP complexes proceeds through fundamental processes involving TF binding for maintenance as well as serving to assemble protein complexes in genomic areas and functioning as directional signals that guide proteins to specific stress-related genomic elements (Ahmad et al., 2021; Dykes & Emanueli, 2017; Sauvageau, 2019).

The DREB (dehydration-responsive element binding) TFs show vital response to drought and cold stress with activation help from lncRNA interactions. The binding of lncRNAs to DREB TFs under stressed conditions guides the proteins to activate key genes RD29A and COR genes through their promoters, which leads to an improvement in plant stress tolerance (Ritonga & Chen, 2020). Through lncRNA–DREB complex interactions, plants experience more accessible stress-responsive promoters which the transcriptional machinery can easily reach.

Three biological consequences result from stress-related complexes between lncRNAs and DREBs as they allow plants to improve their stress resilience by enhancing targeted gene activation, achieve dynamic regulatory flexibility and executing precise post-transcriptional control of stress-responsive mRNAs (Agarwal et al., 2017; Dhingra et al., 2023; Iqbal et al., 2024). The mechanisms demonstrate that ncRNAs support plant survival through adapting to harsh conditions while ensuring effective stress response against drought, cold and oxidative conditions (Altaf et al., 2023; Herman et al., 2022). Plants' ability to withstand abiotic stress is mostly dependent on complex molecular networks that comprise ncRNAs, TFs and proteins. Because lncRNAs and TFs have a regulatory interaction with miRNAs, these transcripts can regulate gene expression in stressful plant environments, allowing for proper physiological adaptation. Protein–ncRNA combinations result in enhanced transcriptional activation characteristics, chromatin remodelling capabilities and stress signal transmission mechanisms. Researchers can create more effective crop development tactics by studying these molecular mechanisms and creating genetically modified plants that are more resilient to stress. More efficient ways to use these molecules for sustainable agriculture will be established with the aid of research on ncRNA regulatory systems.

15.14 BIOTECHNOLOGICAL APPLICATIONS OF ncRNAs IN IMPROVING ABIOTIC STRESS TOLERANCE

According to Ahmed et al. (2024), the use of non-coding RNAs (ncRNAs) in genetic engineering is a crucial way to boost plant resistance against abiotic stressors. Molecular biology and bioinformatics have created exact techniques to modify miRNAs, lncRNAs and siRNAs so plants can become more resilient (Bhar & Roy, 2023).

The modification of miRNAs produces successful outcomes that enable stress adaptation. The overexpression of miR398 in transgenic plants results in better oxidative stress tolerance because it controls copper/zinc superoxide dismutase (CSD) genes, which promote ROS scavenging when plants face drought or salinity stress (Li et al., 2022; Zhang et al., 2022). Placing miR156 under reduced expression to target SQUAMOSA PROMOTER BINDING PROTEIN-LIKE (SPL) genes results in increased root growth together with better drought protection (Chaudhary et al., 2021; Wang et al., 2025). Short tandem target mimicry (STTM) provides researchers with a precise method to sequester miR319 and other vital miRNAs that enhance heat and salt stress tolerance (Othman et al., 2023; Pentimone & Ciancio, 2021).

Scientists have started to uncover how lncRNAs function as regulatory elements in mitigating plant stress responses. When SVALKA lncRNA concentrations are elevated in the cell, this RNA molecule maintains CBF1 protein stability, thereby boosting cold tolerance (Lindemose et al., 2013). Artificial lncRNA molecules were developed to replicate or destroy stress-related molecular connections, which facilitate exact gene regulation (Jathar et al., 2017; Zafar et al., 2020). The utilisation of siRNA-based methods allows for RNA-directed DNA methylation (RdDM) to activate or suppress stress-responsive genes, thus building stress memory according to studies conducted by Akhter et al. (2021) and Gallusci et al. (2023). Through their conjunction, CRISPR-Cas technology and siRNA pathways enable specific modifications of plant epigenetics that improve stress responses without disrupting plant growth (Jain, 2015; Jogam et al., 2022).

Synthetic biology develops modular regulatory systems through artificial miRNA integration with transcriptional activators, along with repressors, according to Deng et al. (2022). Plants benefit from targeted CRISPR-Cas9 genome editing of ncRNA regulatory elements for better stress response optimisation without effects on agronomic traits (Rajput et al., 2021; Zafar et al., 2020).

RNA-based biostimulants present themselves as a modern agricultural technology that allows precise regulation of plant genes to enhance stress tolerance (Ali et al., 2024). The biological agents from dsRNA and sRNAs along with synthetic RNA molecules affect stress-responsive pathways directly in plants while leaving their genomic DNA intact (Pérez-Pizá et al., 2024; Scarpin, 2024).

The main operational principle of RNA-based biostimulants includes RNA interference (RNAi) that allows dsRNA molecules obtained from outside sources to become siRNAs, which destroy or stop target mRNAs that control stress responses in plants (Das & Sherif, 2020; Hung & Slotkin, 2021). Scientists applied RNA biostimulants to control ethylene biosynthesis genes, which protected plants from drought-induced senescence while improving photosynthesis along with water-use efficiency, according to Hasanuzzaman et al. (2021). Biostimulants through antioxidant enzyme expression enhancement of superoxide dismutase and catalase enable plants to fight against ROS-induced stress in the presence of heat and salinity pressure (Ma et al., 2022).

Research into RNA delivery systems succeeded in increasing the stability and performance levels of RNA biostimulants. The application of layered double hydroxides (LDHs) and carbon nanotubes and liposomes for RNA delivery ensures both environmental protection and controlled substance release according to Jing et al. (2022) and Ali Zaidi et al. (2023). The procedure of applying RNA through plant leaves has demonstrated efficient delivery to cell systems alongside stress protection (Carbonell, 2023; Cui et al., 2024).

SiRNAs, which regulate DNA methylation and histone changes at stress-responsive genomic regions, enable stress memory development in crops by RNA-based biostimulant treatments (Kinoshita & Seki, 2014). Applying RNA biostimulants to maize and soybean crops has been shown to enhance drought resistance and yield sustainability across a number of growing years (Van Oosten et al., 2017). Despite their potential, RNA-based biostimulants are not yet commercialised due to high production costs and complicated regulatory requirements (Chandler et al., 2020; De Schutter et al., 2022). Researchers have started to reduce the cost of RNA synthesis in order to combine these technologies with smart sensors and drone-assisted spraying for more useful applications in large-scale agriculture (Chandile & Patil, 2025).

According to Xu et al. (2019), the development of climate-resilient crops depends on the critical roles played by non-coding RNAs (ncRNAs). In order for plants to adjust to environmental factors such as drought, salt exposure and temperature fluctuations, the molecules regulate stress-responsive pathways that operate at transcriptional, post-transcriptional and epigenetic stages (Yang et al., 2023). According to successful models that direct future agricultural growth, researchers can effectively use ncRNAs for crop resilience improvement through genetic engineering and molecular breeding.

15.15 FUTURE DIRECTIONS IN ncRNA RESEARCH FOR ABIOTIC STRESS

Scientists use multi-omics integration to explore new ncRNAs in plants under abiotic stress conditions through the combination of genomics with transcriptomics to proteomics then metabolomics and epigenomics (Dar et al., 2023; Vaschetto, 2024). Through these methods scientists can discover ncRNA regulatory pathways which aid in creating crops resistant to climate change (Wang et al., 2023).

Genomics, together with transcriptomics, functions as the initial approach to detect ncRNA transcripts, yet proteomics reveals interprotein connections between ncRNAs, and metabolomics identifies outcomes (Aydın, 2021; Lin et al., 2023). Transcriptomic along with degradomic studies established miR156 as a TF regulator in *Oryza sativa*, which leads to drought tolerance through improved root growth and lower water consumption (Archana et al., 2024). Research into ncRNA becomes more powerful through multi-omics integration because it builds regulatory networks utilising systems biology methods (Correa-Aguila et al., 2022). Research that combines sRNAome and transcriptome analysis of tetraploid wheat (*Triticum turgidum*) identified 690 miRNAs during grain development under drought and heat stress conditions (Liu et al., 2020). ASRmiRNA, together with other machine learning models, helps analyse multiple omics datasets to improve the identification of stress-responsive ncRNAs, according to Meher et al. (2022).

The integration and computational analysis and cost-related barriers limit the execution of multi-omics research according to Misra et al. (2019) and Gallo (2024). Future study will maximise the usefulness of ncRNA-based crop enhancements by applying deep learning and AI-driven analysis tools to improve ncRNA prediction accuracy (Yang et al., 2021).

The steady functionality is necessary for ncRNA field applications since it dictates how effective they are at resiliency to stress. Since these events reduce their longevity, environmental elements such as UV radiation, temperature fluctuations and microbial interactions are necessary for the successful function of ncRNAs (Asghar et al., 2024; Yu et al., 2019).

In addition to controlling their release profiles, the nanoparticle-based delivery systems, such as liposomes and LDHs, improve the stability of ncRNA (Das & Sherif, 2020; Jasinski et al., 2017). Under stress, the delivery mechanisms allow plants to efficiently absorb ncRNA and transfer it throughout the body for regulatory reasons (Touzdjian Pinheiro Kohlrausch Távora et al., 2022). Multiple stress analyses are necessary to optimise practical applications since ncRNA functionality is dependent on environmental conditions (Ding et al., 2020; Jin et al., 2024).

Plants regulate stress memory through epigenetic changes that are guided by siRNAs to improve tolerance and 'stress remembrance' for subsequent generations (Ramakrishnan et al., 2022). Research by Matsui et al. (2019) has demonstrated that the use of extrinsic siRNA in maize and soybean agriculture produces stress tolerance that is inherited by subsequent generations.

The three primary barriers to ncRNA field deployment are their ability to remain active in the environment, handle wide areas and be cost-effective (Nitnavare et al., 2021; Rinaldi et al., 2024). Academic researchers create biodegradable delivery systems while utilising drone-based nanoribonucleic acid spraying to achieve better applicability and minimise environmental harm (Jin et al., 2025).

The delivery of ncRNAs into the field requires successful protection against the damaging effects of enzymes as well as UV radiation and temperature variations (Davalos et al., 2019). Nanoparticles, specifically lipid-based carriers (LNPs), protect ncRNAs through their delivery mechanism and facilitate cellular uptake along with controlled substance release (Aldosari et al., 2021; Dasgupta & Chatterjee, 2021). RNAi-based biostimulants for drought and salinity tolerance have achieved effective delivery through LDHs, according to Venu et al. (2024) and Hu, Gu, et al. (2022).

The biodegradable delivery system of chitosan and cellulose derivatives represents biopolymer-based carriers that serve as alternatives for ncRNA delivery (Srivastava et al., 2024; Subhan & Torchilin, 2023). The carriers strengthen delivery penetration and stability, which has led to successful applications of miRNA delivery for salinity tolerance improvement (Fiol et al., 2021). CRISPR-Cas systems, when combined with ncRNA delivery, achieve targeted precision, according to Mujtaba et al. (2021). According to Sivakumar et al. (2019), aptamer-conjugated nanoparticles and RNA-binding proteins will be further developed to improve delivery selectivity and operational efficiency. These new developments increase the likelihood that ncRNA technology will be applied in agricultural biotechnology more quickly.

Artificial intelligence's (AI's) capacity to do comprehensive data analysis and create predictive modelling pathways gives ncRNA research transformative power (Pachchigar et al., 2025). Researchers can better detect ncRNAs, forecast their effects and develop regulatory models to improve functional research with machine learning programmes (Wang et al., 2024). AI-powered tools like DARIO and LP-HCLUS use deep learning for ncRNA annotation and function prediction (Fasold et al., 2011). To guide genetic engineering efforts, researchers can use AI to develop network-based predictive models for TF and protein and metabolite association with ncRNAs (Morishita & Nakamura, 2024).

REFERENCES

Adeyemi, O., Grove, I., Peets, S., & Norton, T. (2017). Advanced monitoring and management systems for improving sustainability in precision irrigation. *Sustainability*, *9*(3), 353.

Agarwal, P. K., Gupta, K., Lopato, S., & Agarwal, P. (2017). Dehydration responsive element binding transcription factors and their applications for the engineering of stress tolerance. *Journal of Experimental Botany*, *68*(9), 2135–2148.

Ahlawat, U., Naruka, A., Changdeo, W. B., Rehsawla, R., Sansanwal, R., Mishra, R., & Singh, V. K. (2024). A review of cutting-edge biotechnological solutions for next-generation farming. *Journal of Experimental Agriculture International*, *46*(7), 1187–1204.

Ahmad, P., Bensaoud, C., Mekki, I., Rehman, M. U., & Kotsyfakis, M. (2021). Long non-coding RNAs and their potential roles in the vector–host–pathogen triad. *Life*, *11*(1), 56.

Ahmad, Y., Haakim, Z., Iqbal, J., Abbasi, B. A., Mahmood, T., Iqbal, R., & Murtaza, G. (2024). Crosstalk between ncRNAs and stress signaling pathways unmasking the regulatory networks. In *ncRNAs: Mediated regulation: The unarguable biological function to environmental constrains* (pp. 265–282). Springer Nature Switzerland.

Ahmad, Y., Haider, S., Iqbal, J., Abbasi, B. A., Yaseen, T., & Mahmood, T. (2022). The mechanisms of genome editing technologies in crop plants. In *Principles and practices of OMICS and genome editing for crop improvement* (pp. 295–313). Springer International Publishing.

Ahmad, Y., Haider, S., Iqbal, J., Naseer, S., Attia, K. A., Mohammed, A. A., & Mahmood, T. (2024). In-silico analysis and transformation of OsMYB48 transcription factor driven by CaMV35S promoter in model plant – *Nicotiana tabacum* L. conferring abiotic stress tolerance. *GM Crops & Food*, *15*(1), 130–149.

Ahmed, H. G. M. D., Fatima, T., & Khaliq, N. (2024). Non-coding RNA-mediated regulation of genes expression in response to abiotic stresses. In *ncRNAs: Mediated regulation: The unarguable biological function to environmental constrains* (pp. 235–264).

Akhter, Z., Bi, Z., Ali, K., Sun, C., Fiaz, S., Haider, F. U., & Bai, J. (2021). In response to abiotic stress, DNA methylation confers epigenetic changes in plants. *Plants*, *10*(6), 1096.

Aldosari, B. N., Alfagih, I. M., & Almurshedi, A. S. (2021). Lipid nanoparticles as delivery systems for RNA-based vaccines. *Pharmaceutics*, *13*(2), 206.

Ali, S., Akhtar, M. S., Siraj, M., & Zaman, W. (2024). Molecular communication of microbial plant biostimulants in the rhizosphere under abiotic stress conditions. *International Journal of Molecular Sciences*, *25*(22), 12424.

Ali Zaidi, S. S., Fatima, F., Ali Zaidi, S. A., Zhou, D., Deng, W., & Liu, S. (2023). Engineering siRNA therapeutics: Challenges and strategies. *Journal of Nanobiotechnology*, *21*(1), 381.

Altaf, M. A., Mandal, S., Behera, B., Mangal, V., Naz, S., Kumar, R., … & Aftab, T. (2023). Salinity stress tolerance in solanaceous crops: Current understanding and its prospects in genome editing. *Journal of Plant Growth Regulation*, *42*(7), 4020–4036.

Anjum, S. A., Ashraf, U., Zohaib, A., Tanveer, M., Naeem, M., Ali, I., … & Nazir, U. (2017). *Growth and development responses of crop plants under drought stress: A review.*

Archana, P., Ramamoorthy, P., Swaminathan, M., Muthurajan, R., Alagarsamy, S., & Kathiresan, P. (2024). *Multi-omics approaches for abiotic stress tolerance in rice (Oryza sativa L.).* Plant Science Today.

Asghar, A., Ullah, Z., Ali, T., Iqbal, J., Abbasi, B. A., Ijaz, S., & Mahmood, T. (2024). Introduction to nc-RNA and their crucial role in biological regulation under environmental stress. In *ncRNAs: Mediated regulation: The unarguable biological function to environmental constrains* (pp. 1–46). Springer Nature Switzerland.

Aydın, B. (2021). *Systems biomedicine oriented multi-omics data integration to unveil molecular signatures and repositioned therapeutics in pituitary neuroendocrine tumors* [Master's thesis, Marmara Universitesi (Turkey)].

Balasubramaniam, T., Shen, G., Esmaeili, N., & Zhang, H. (2023). Plants' response mechanisms to salinity stress. *Plants*, *12*(12), 2253.

Banerjee, A., & Roychoudhury, A. (2017). Epigenetic regulation during salinity and drought stress in plants: Histone modifications and DNA methylation. *Plant Gene*, *11*, 199–204.

Bashir, T., & Husaini, A. M. (2025). Non-coding RNAs and their role in plants: Prospective omics-tools for improving growth, development and stress tolerance in field crops. *Molecular Biology Reports*, *52*(1), 1–27.

Basso, M. F., Ferreira, P. C. G., Kobayashi, A. K., Harmon, F. G., Nepomuceno, A. L., Molinari, H. B. C., & Grossi-de-Sa, M. F. (2019). Micro RNA s and new biotechnological tools for its modulation and improving stress tolerance in plants. *Plant Biotechnology Journal*, *17*(8), 1482–1500.

Bhan, A., Deb, P., Soleimani, M., & Mandal, S. S. (2017). The short and medium stories of noncoding RNAs: microRNA and siRNA. In *Gene regulation, epigenetics and hormone signaling* (pp. 137–168).

Bhar, A., & Roy, A. (2023). Emphasizing the role of long non-coding RNAs (lncRNA), circular RNA (circRNA), and micropeptides (miPs) in plant biotic stress tolerance. *Plants*, *12*(23), 3951.

Bhatti, G. K., Khullar, N., Sidhu, I. S., Navik, U. S., Reddy, A. P., Reddy, P. H., & Bhatti, J. S. (2021). Emerging role of non-coding RNA in health and disease. *Metabolic Brain Disease*, *36*, 1119–1134.

Białoskórska, M., Rucińska, A., & Boczkowska, M. (2024). Molecular mechanisms underlying freezing tolerance in plants: Implications for cryopreservation. *International Journal of Molecular Sciences*, *25*(18), 10110.

Bouzroud, S., Henkrar, F., Fahr, M., & Smouni, A. (2023). Salt stress responses and alleviation strategies in legumes: A review of the current knowledge. *3 Biotech*, *13*(8), 287.

Bridgemohan, P., Mohammed, M., Bridgemohan, R. S., & Mohammed, Z. (2020). Ecophysiology of anaerobiosis stress due to flood and waterlogging in rice. *Journal of Horticulture and Postharvest Research*, *3*(Special Issue-Abiotic and Biotic Stresses), 113–128.

Carbonell, A. (2023). *Next-generation artificial small RNA-based RNAi for crop improvement.*

Chakraborty, K., Basak, N., Bhaduri, D., Ray, S., Vijayan, J., Chattopadhyay, K., & Sarkar, R. K. (2018). Ionic basis of salt tolerance in plants: Nutrient homeostasis and oxidative stress tolerance. In *Plant nutrients and abiotic stress tolerance* (pp. 325–362).

Chandile, P. G., & Patil, K. S. (2025). Use of biopesticides and entomopathogens. *Advances in Applied Entomology*, 121.

Chandler, M., Panigaj, M., Rolband, L. A., & Afonin, K. A. (2020). Challenges in optimizing RNA nanostructures for large-scale production and controlled therapeutic properties. *Nanomedicine*, *15*(13), 1331–1340.

Chaudhary, S., Grover, A., & Sharma, P. (2021). MicroRNAs: Potential targets for developing stress-tolerant crops. *Life*, *11*.

Choudhary, P., Bhatt, S., & Chatterjee, S. (2024). From freezing to functioning: Cellular strategies of cold-adapted bacteria for surviving in extreme environments. *Archives of Microbiology*, *206*(7), 329.

Circu, M. L., & Aw, T. Y. (2010). Reactive oxygen species, cellular redox systems, and apoptosis. *Free Radical Biology and Medicine*, *48*(6), 749–762.

Correa-Aguila, R., Alonso-Pupo, N., & Hernández-Rodríguez, E. W. (2022). Multi-omics data integration approaches for precision oncology. *Molecular Omics*, *18*(6), 469–479.

Cui, Y., Wang, K., & Zhang, C. (2024). Carbon nanomaterials for plant priming through mechanostimulation: Emphasizing the role of shape. *ACS Nano*, *18*(16), 10829–10839.

Dar, M. A., Arafah, A., Bhat, K. A., Khan, A., Khan, M. S., Ali, A., … & Rehman, M. U. (2023). Multiomics technologies: Role in disease biomarker discoveries and therapeutics. *Briefings in Functional Genomics*, *22*(2), 76–96.

Das, P. R., & Sherif, S. M. (2020). Application of exogenous dsRNAs-induced RNAi in agriculture: Challenges and triumphs. *Frontiers in Plant Science*, *11*, 946.

Dasgupta, I., & Chatterjee, A. (2021). Recent advances in miRNA delivery systems. *Methods and Protocols*, *4*(1), 10.

Datta, R., & Paul, S. (2019). Long non-coding RNAs: Fine-tuning the developmental responses in plants. *Journal of Biosciences*, *44*(4), 77.

Deng, F., Zeng, F., Shen, Q., Abbas, A., Cheng, J., Jiang, W., … & Chen, Z. H. (2022). Molecular evolution and functional modification of plant miRNAs with CRISPR. *Trends in Plant Science*, *27*(9), 890–907.

Deng, Y., Yang, J., Wei, J., Liu, S., Yang, L., Wang, X., … & Si, H. (2024). Stu-miR159a negatively regulates anthocyanin-specific MYB transcription factor to mediate drought stress tolerance in potato. *Horticultural Plant Journal*, 11(5), 1917–1929.

De Schutter, K., Taning, C. N. T., Van Daele, L., Van Damme, E. J., Dubruel, P., & Smagghe, G. (2022). RNAi-based biocontrol products: Market status, regulatory aspects, and risk assessment. *Frontiers in Insect Science*, *1*, 818037.

Dhingra, Y., Gupta, S., Gupta, V., Agarwal, M., & Katiyar-Agarwal, S. (2023). The emerging role of epitranscriptome in shaping stress responses in plants. *Plant Cell Reports*, *42*(10), 1531–1555.

Dietz, K. J., Zörb, C., & Geilfus, C. M. (2021). Drought and crop yield. *Plant Biology*, *23*(6), 881–893.

Dikilitas, M., Simsek, E., & Roychoudhury, A. (2020). Role of proline and glycine betaine in overcoming abiotic stresses. In *Protective chemical agents in the amelioration of plant abiotic stress: Biochemical and molecular perspectives* (pp. 1–23).

Di Martino, M. T., Riillo, C., Scionti, F., Grillone, K., Polerà, N., Caracciolo, D., … & Tassone, P. (2021). miRNAs and lncRNAs as novel therapeutic targets to improve cancer immunotherapy. *Cancers*, *13*(7), 1587.

Ding, Y., Huang, L., Jiang, Q., & Zhu, C. (2020). MicroRNAs as important regulators of heat stress responses in plants. *Journal of Agricultural and Food Chemistry*, *68*(41), 11320–11326.

Dykes, I. M., & Emanueli, C. (2017). Transcriptional and post-transcriptional gene regulation by long non-coding RNA. *Genomics, Proteomics & Bioinformatics*, *15*(3), 177–186.

El Sabagh, A., Islam, M. S., Skalicky, M., Ali Raza, M., Singh, K., Anwar Hossain, M., … & Arshad, A. (2021). Salinity stress in wheat (*Triticum aestivum* L.) in the changing climate: Adaptation and management strategies. *Frontiers in Agronomy*, *3*, 661932.

Erpen, L., Devi, H. S., Grosser, J. W., & Dutt, M. (2018). Potential use of the DREB/ERF, MYB, NAC and WRKY transcription factors to improve abiotic and biotic stress in transgenic plants. *Plant Cell, Tissue and Organ Culture (PCTOC)*, *132*, 1–25.

Fahad, M., Tariq, L., Li, W., & Wu, L. (2025). MicroRNA gatekeepers: Orchestrating rhizospheric dynamics. *Journal of Integrative Plant Biology*, 67(3), 845–876.

Fahad, M., Tariq, L., Muhammad, S., & Wu, L. (2024). Underground communication: Long noncoding RNA signaling in the plant rhizosphere. *Plant Communications*, 5(7), 1009.

Fahad, S., Adnan, M., Noor, M., Arif, M., Alam, M., Khan, I. A., … & Wang, D. (2019). Major constraints for global rice production. In *Advances in rice research for abiotic stress tolerance* (pp. 1–22). Woodhead Publishing.

Farooq, M., Gogoi, N., Barthakur, S., Baroowa, B., Bharadwaj, N., Alghamdi, S. S., & Siddique, K. H. (2017). Drought stress in grain legumes during reproduction and grain filling. *Journal of Agronomy and Crop Science*, *203*(2), 81–102.

Farooqi, M., Nawaz, G., Wani, S., Choudhary, J., Rana, M., Sah, R., Afzal, M., Zahra, Z., Ganie, S., Razzaq, A., Reyes, V., Mahmoud, E., Elansary, H., El-Abedin, T., & Siddique, K. (2022). Recent developments in multi-omics and breeding strategies for abiotic stress tolerance in maize (*Zea mays* L.). *Frontiers in Plant Science*, *13*.

Fasold, M., Langenberger, D., Binder, H., Stadler, P. F., & Hoffmann, S. (2011). DARIO: A ncRNA detection and analysis tool for next-generation sequencing experiments. *Nucleic Acids Research*, *39*(suppl_2), W112–W117.

Feng, C., Zhang, X., Du, B., Xiao, Y., Wang, Y., Sun, Y., … & Li, T. H. (2023). MicroRNA156ab regulates apple plant growth and drought tolerance by targeting transcription factor MsSPL13. *Plant Physiology*, *192*(3), 1836–1857.

Ferreyra, M. L. F., Serra, P., & Casati, P. (2021). Recent advances on the roles of flavonoids as plant protective molecules after UV and high light exposure. *Physiologia Plantarum*, *173*(3), 736–749.

Fiol, D. F., Terrile, M. C., Frik, J., Mesas, F. A., Álvarez, V. A., & Casalongué, C. A. (2021). Nanotechnology in plants: Recent advances and challenges. *Journal of Chemical Technology & Biotechnology*, *96*(8), 2095–2108.

Fletcher, S. J., Reeves, P. T., Hoang, B. T., & Mitter, N. (2020). A perspective on RNAi-based biopesticides. *Frontiers in Plant Science*, *11*, 51.

Fukudome, A., & Fukuhara, T. (2017). Plant dicer-like proteins: Double-stranded RNA-cleaving enzymes for small RNA biogenesis. *Journal of Plant Research*, *130*, 33–44.

Furtak, K., & Wolińska, A. (2023). The impact of extreme weather events as a consequence of climate change on the soil moisture and on the quality of the soil environment and agriculture – A review. *Catena*, *231*, 107378.

Fuschi, P., Maimone, B., Gaetano, C., & Martelli, F. (2019). Noncoding RNAs in the vascular system response to oxidative stress. *Antioxidants & Redox Signaling*, *30*(7), 992–1010.

Gallo, E. (2024). The rise of big data: Deep sequencing-driven computational methods are transforming the landscape of synthetic antibody design. *Journal of Biomedical Science*, *31*(1), 29.

Gallusci, P., Agius, D. R., Moschou, P. N., Dobránszki, J., Kaiserli, E., & Martinelli, F. (2023). Deep inside the epigenetic memories of stressed plants. *Trends in Plant Science*, *28*(2), 142–153.

Gelaw, T. A., & Sanan-Mishra, N. (2021). Non-coding RNAs in response to drought stress. *International Journal of Molecular Sciences*, *22*(22), 12519.

Geng, X., Jia, Y., Zhang, Y., Shi, L., Li, Q., Zang, A., & Wang, H. (2020). Circular RNA: Biogenesis, degradation, functions and potential roles in mediating resistance to anticarcinogens. *Epigenomics*, *12*(3), 267–283.

Ghosh, S., Adhikari, S., Adhikari, A., & Hossain, Z. (2022). Contribution of plant miRNAome studies towards understanding heavy metal stress responses: Current status and future perspectives. *Environmental and Experimental Botany*, *194*, 104705.

Ghuge, S. A., Nikalje, G. C., Kadam, U. S., Suprasanna, P., & Hong, J. C. (2023). Comprehensive mechanisms of heavy metal toxicity in plants, detoxification, and remediation. *Journal of Hazardous Materials*, *450*, 131039.

Gill, S. S., Khan, N. A., Agarwala, N., Singh, K., Sunkar, R., & Tuteja, N. (2024). ncRNAs in plant development and stress responses. *Plant Physiology and Biochemistry*, *214*, 108950.

Gomes, A. Q., Nolasco, S., & Soares, H. (2013). Non-coding RNAs: Multi-tasking molecules in the cell. *International Journal of Molecular Sciences*, *14*(8), 16010–16039.

Gottesman, S. (2010). Roles of mRNA stability, translational regulation, and small RNAs in stress response regulation. In *Bacterial stress responses* (pp. 59–73).

Gowtham, H. G., Singh, S. B., Shilpa, N., Aiyaz, M., Nataraj, K., Udayashankar, A. C., … & Sayyed, R. Z. (2022). Insight into recent progress and perspectives in improvement of antioxidant machinery upon PGPR augmentation in plants under drought stress: A review. *Antioxidants*, *11*(9), 1763.

Grossniklaus, U., & Paro, R. (2014). Transcriptional silencing by polycomb-group proteins. *Cold Spring Harbor Perspectives in Biology*, *6*(11), a019331.

Gupta, A., Mishra, R., Rai, S., Bano, A., Pathak, N., Fujita, M., … & Hasanuzzaman, M. (2022). Mechanistic insights of plant growth promoting bacteria mediated drought and salt stress tolerance in plants for sustainable agriculture. *International Journal of Molecular Sciences*, *23*(7), 3741.

Gupta, O. P., Sharma, P., Gupta, R. K., & Sharma, I. (2014). MicroRNA mediated regulation of metal toxicity in plants: Present status and future perspectives. *Plant Molecular Biology*, *84*(1), 1–18.

Gusti, A. M., Qusti, S. Y., Alshammari, E. M., Toraih, E. A., & Fawzy, M. S. (2021). Antioxidants-related superoxide dismutase (SOD), catalase (CAT), glutathione peroxidase (GPX), glutathione-S-transferase (GST), and nitric oxide synthase (NOS) gene variants analysis in an obese population: A preliminary case-control study. *Antioxidants*, *10*(4), 595.

Habibi, F., Liu, T., Shahid, M. A., Schaffer, B., & Sarkhosh, A. (2023). Physiological, biochemical, and molecular responses of fruit trees to root zone hypoxia. *Environmental and Experimental Botany*, *206*, 105179.

Haider, S., Iqbal, J., Naseer, S., Yaseen, T., Shaukat, M., Bibi, H., & Mahmood, T. (2021). Molecular mechanisms of plant tolerance to heat stress: Current landscape and future perspectives. *Plant Cell Reports*, 1–25.

Haider, S., Raza, A., Iqbal, J., Shaukat, M., & Mahmood, T. (2022). Analyzing the regulatory role of heat shock transcription factors in plant heat stress tolerance: A brief appraisal. *Molecular Biology Reports*, *49*(6), 5771–5785.

Hall, C., Dawson, T. P., Macdiarmid, J. I., Matthews, R. B., & Smith, P. (2017). The impact of population growth and climate change on food security in Africa: Looking ahead to 2050. *International Journal of Agricultural Sustainability*, *15*(2), 124–135.

Hasanuzzaman, M., Bhuyan, M. B., Zulfiqar, F., Raza, A., Mohsin, S. M., Mahmud, J. A., … & Fotopoulos, V. (2020). Reactive oxygen species and antioxidant defense in plants under abiotic stress: Revisiting the crucial role of a universal defense regulator. *Antioxidants*, *9*(8), 681.

Hasanuzzaman, M., Parvin, K., Bardhan, K., Nahar, K., Anee, T. I., Masud, A. A. C., & Fotopoulos, V. (2021). Biostimulants for the regulation of reactive oxygen species metabolism in plants under abiotic stress. *Cells*, *10*(10), 2537.

Herman, A. B., Tsitsipatis, D., & Gorospe, M. (2022). Integrated lncRNA function upon genomic and epigenomic regulation. *Molecular Cell*, *82*(12), 2252–2266.

Holley, C. L., & Topkara, V. K. (2011). An introduction to small non-coding RNAs: miRNA and snoRNA. *Cardiovascular Drugs and Therapy*, *25*, 151–159.

Hu, T., Gu, Z., Williams, G. R., Strimaite, M., Zha, J., Zhou, Z., … & Liang, R. (2022). Layered double hydroxide-based nanomaterials for biomedical applications. *Chemical Society Reviews*, *51*(14), 6126–6176.

Hua, X. (2023). *Systematic analysis of small RNAs and long non-coding RNAs in the context of abscisic acid-dependent abiotic stress responses in arabidopsis* [Doctoral dissertation, University of Pennsylvania].

Hueso, M., Mallén, A., Suñé-Pou, M., Aran, J. M., Suñé-Negre, J. M., & Navarro, E. (2021). ncRNAs in therapeutics: Challenges and limitations in nucleic acid-based drug delivery. *International Journal of Molecular Sciences*, *22*(21), 11596.

Hung, Y. H., & Slotkin, R. K. (2021). The initiation of RNA interference (RNAi) in plants. *Current Opinion in Plant Biology*, *61*, 102014.

Huntzinger, E., & Izaurralde, E. (2011). Gene silencing by microRNAs: Contributions of translational repression and mRNA decay. *Nature Reviews Genetics*, *12*(2), 99–110.

Hussain, S., Shaukat, M., Ashraf, M., Zhu, C., Jin, Q., & Zhang, J. (2019). Salinity stress in arid and semi-arid climates: Effects and management in field crops. In *Climate change and agriculture*. IntechOpen.

Hussain, S. S., Ali, A., Abbas, M., Sun, Y., Li, Y., Li, Q., & Ragauskas, A. J. (2024). Harnessing miRNA156: A molecular Toolkit for reshaping plant development and achieving ideal architecture. *Plant Physiology and Biochemistry*, *215*, 109071.

Ijaz, S., Iqbal, J., Abbasi, B. A., Yaseen, T., Ullah, Z., Alqahtani, T., & Mahmood, T. (2024). nc-RNA: A key player in stress response pathways, a molecular perspective. In *ncRNAs: Mediated regulation: The unarguable biological function to environmental constrains* (pp. 139–163). Springer Nature Switzerland.

Ijaz, S., Iqbal, J., Abbasi, B. A., Ullah, Z., Yaseen, T., Munir, F., & Mahmood, T. (2024). From single nc-RNAs to networks: Understanding the complexity of environmental adaption. In *ncRNAs: Mediated regulation: The unarguable biological function to environmental constrains* (pp. 283–310). Springer Nature Switzerland.

Iqbal, B., Ahmad, N., Li, G., Jalal, A., Khan, A. R., Zheng, X., & Du, D. (2024). Unlocking plant resilience: Advanced epigenetic strategies against heavy metal and metalloid stress. *Plant Science*, 112265.

Isaacs, F. J., Dwyer, D. J., & Collins, J. J. (2006). RNA synthetic biology. *Nature Biotechnology*, *24*(5), 545–554.

Islam, W., Waheed, A., Naveed, H., & Zeng, F. (2022). MicroRNAs mediated plant responses to salt stress. *Cells*, *11*(18), 2806.

Iyengar, B. R., Choudhary, A., Sarangdhar, M. A., Venkatesh, K. V., Gadgil, C. J., & Pillai, B. (2014). Non-coding RNA interact to regulate neuronal development and function. *Frontiers in Cellular Neuroscience*, *8*, 47.

Jahed, K. R., Saini, A. K., & Sherif, S. M. (2023). Coping with the cold: Unveiling cryoprotectants, molecular signaling pathways, and strategies for cold stress resilience. *Frontiers in Plant Science*, *14*, 1246093.

Jain, M. (2015). Function genomics of abiotic stress tolerance in plants: A CRISPR approach. *Frontiers in Plant Science*, *6*.

Jasinski, D., Haque, F., Binzel, D. W., & Guo, P. (2017). Advancement of the emerging field of RNA nanotechnology. *ACS Nano*, *11*(2), 1142–1164.

Jaskiewicz, L., & Filipowicz, W. (2008). Role of Dicer in posttranscriptional RNA silencing. In *RNA interference* (pp. 77–97).

Jathar, S., Kumar, V., Srivastava, J., & Tripathi, V. (2017). Technological developments in lncRNA biology. In *Long non coding RNA biology* (pp. 283–323). Springer Nature.

Javed, M., Sinha, A., & Shukla, L. I. (2019). Evaluation of mature miR398 family, expression analysis and the post-transcriptional regulation evidence in gamma-irradiated and nitrogen-stressed *Medicago sativa* seedlings. *International Journal of Radiation Biology*, *95*(5), 585–596.

°°Jha, U. C., Nayyar, H., Jha, R., Khurshid, M., Zhou, M., Mantri, N., & Siddique, K. H. (2020). Long non-coding RNAs: Emerging players regulating plant abiotic stress response and adaptation. *BMC Plant Biology*, *20*, 1–20.

Jin, D., Wei, Y., & Zhang, K. (2025). The research on soil water utilization mechanism in drought-resistant crops based on big data. *Advances in Resources Research*, *5*(1), 329–349.

Jin, X., Wang, Z., Li, X., Ai, Q., Wong, D. C. J., Zhang, F., … & Si, H. (2024). Current perspectives of lncRNAs in abiotic and biotic stress tolerance in plants. *Frontiers in Plant Science*, *14*, 1334620.

Jing, G., Yang, L., Wang, H., Niu, J., Li, Y., & Wang, S. (2022). Interference of layered double hydroxide nanoparticles with pathways for biomedical applications. *Advanced Drug Delivery Reviews*, *188*, 114451.

Jogam, P., Sandhya, D., Alok, A., Peddaboina, V., Allini, V. R., & Zhang, B. (2022). A review on CRISPR/Cas-based epigenetic regulation in plants. *International Journal of Biological Macromolecules*, *219*, 1261–1271.

Jogawat, A., Yadav, B., Chhaya, Lakra, N., Singh, A. K., & Narayan, O. P. (2021). Crosstalk between phytohormones and secondary metabolites in the drought stress tolerance of crop plants: A review. *Physiologia Plantarum*, *172*(2), 1106–1132.

Joshi, S., Nath, J., Singh, A. K., Pareek, A., & Joshi, R. (2022). Ion transporters and their regulatory signal transduction mechanisms for salinity tolerance in plants. *Physiologia Plantarum*, *174*(3), e13702.

Junaid, M. D., Chaudhry, U. K., Şanlı, B. A., Gökçe, A. F., & Öztürk, Z. N. (2024). A review of the potential involvement of small RNAs in transgenerational abiotic stress memory in plants. *Functional & Integrative Genomics*, *24*(2), 74.

Kaikkonen, M. U., Lam, M. T., & Glass, C. K. (2011). Non-coding RNAs as regulators of gene expression and epigenetics. *Cardiovascular Research*, *90*(3), 430–440.

Kajla, M., Roy, A., Singh, I. K., & Singh, A. (2023). Regulation of the regulators: Transcription factors controlling biosynthesis of plant secondary metabolites during biotic stresses and their regulation by miRNAs. *Frontiers in Plant Science*, *14*, 1126567.

Kapoore, R. V., Wood, E. E., & Llewellyn, C. A. (2021). Algae biostimulants: A critical look at microalgal biostimulants for sustainable agricultural practices. *Biotechnology Advances*, *49*, 107754.

Karthik, S., Mohan, S., Magesh, I., Bharathy, A., Kolipaka, R., Ganesamoorthi, S., … & Selvamurugan, N. (2024). Chitosan nanocarriers for non-coding RNA therapeutics: A review. *International Journal of Biological Macromolecules*, 263(Part 1), 130361.

Kesawat, M. S., Satheesh, N., Kherawat, B. S., Kumar, A., Kim, H. U., Chung, S. M., & Kumar, M. (2023). Regulation of reactive oxygen species during salt stress in plants and their crosstalk with other signaling molecules—Current perspectives and future directions. *Plants*, *12*(4), 864.

Kidane, G. (2024). *Dryland agriculture and climate change adaptation in sub-Saharan Africa: A case of policies, technologies, and strategies in Ethiopia*.

Kiger, N. M., & Schroeder, S. J. (2024). SVALKA: A long non-coding cis-natural antisense RNA that plays a role in the regulation of the cold response of *Arabidopsis thaliana*. *Non-Coding RNA*, *10*(6), 59.

Kinoshita, T., & Seki, M. (2014). Epigenetic memory for stress response and adaptation in plants. *Plant and Cell Physiology*, *55*(11), 1859–1863.

Kiran, B. R., Prasad, M. N. V., & Mohan, S. V. (2023). Farm to fork: Sustainable agrifood systems. In *Sustainable and circular management of resources and waste towards a green deal* (pp. 25–38). Elsevier.

Kollist, H., Zandalinas, S. I., Sengupta, S., Nuhkat, M., Kangasjärvi, J., & Mittler, R. (2019). Rapid responses to abiotic stress: priming the landscape for the signal transduction network. *Trends in Plant Science*, *24*(1), 25–37.

Kopecká, R., Kameniarová, M., Černý, M., Brzobohatý, B., & Novák, J. (2023). Abiotic stress in crop production. *International Journal of Molecular Sciences*, *24*(7), 6603.

Kotowska-Zimmer, A., Pewinska, M., & Olejniczak, M. (2021). Artificial miRNAs as therapeutic tools: Challenges and opportunities. *Wiley Interdisciplinary Reviews: RNA*, *12*(4), e1640.

Kumar, R. (2014). Role of microRNAs in biotic and abiotic stress responses in crop plants. *Applied Biochemistry and Biotechnology*, *174*, 93–115.

Kumar, S., Gonzalez, E. A., Rameshwar, P., & Etchegaray, J. P. (2020). Non-coding RNAs as mediators of epigenetic changes in malignancies. *Cancers*, *12*(12), 3657.

Lakhotia, S. C. (2012). Long non-coding RNAs coordinate cellular responses to stress. *Wiley Interdisciplinary Reviews: RNA*, *3*(6), 779–796.

Latham, J. R., & Wilson, A. K. (2015). *Off-target effects of plant transgenic RNAi: Three mechanisms lead to distinct toxicological and environmental hazards*.

Lavenniah, A., Luu, T. D. A., Li, Y. P., Lim, T. B., Jiang, J., Ackers-Johnson, M., & Foo, R. S. Y. (2020). Engineered circular RNA sponges act as miRNA inhibitors to attenuate pressure overload-induced cardiac hypertrophy. *Molecular Therapy*, *28*(6), 1506–1517.

Lelandais-Briere, C., Moreau, J., Hartmann, C., & Crespi, M. (2016). Noncoding RNAs, emerging regulators in root endosymbioses. *Molecular Plant-Microbe Interactions*, *29*(3), 170–180.

Leng, X. P., Wang, P., Zhu, X., Li, X., Zheng, T., Shangguan, L., & Fang, J. (2017). Ectopic expression of CSD1 and CSD2 targeting genes of miR398 in grapevine is associated with oxidative stress tolerance. *Functional & Integrative Genomics*, *17*, 697–710.

Li, D., Zhang, J., & Li, J. (2020). Role of miRNA sponges in hepatocellular carcinoma. *Clinica Chimica Acta*, *500*, 10–19.

Li, J., Song, Q., Zuo, Z. F., & Liu, L. (2022). MicroRNA398: A master regulator of plant development and stress responses. *International Journal of Molecular Sciences*, *23*(18), 10803.

Li, Q., Shen, H., Yuan, S., Dai, X., & Yang, C. (2023). miRNAs and lncRNAs in tomato: Roles in biotic and abiotic stress responses. *Frontiers in Plant Science*, *13*, 1094459.

Li, Q., Wang, Y., Sun, Z., Li, H., & Liu, H. (2024). The biosynthesis process of small RNA and its pivotal roles in plant development. *International Journal of Molecular Sciences*, *25*(14), 7680.

Li, X., Qi, H., Cui, W., Wang, Z., Fu, X., Li, T., … & Yu, T. (2022). Recent advances in targeted delivery of non-coding RNA-based therapeutics for atherosclerosis. *Molecular Therapy*, *30*(10), 3118–3132.

Li, Z., & Rana, T. M. (2012). Molecular mechanisms of RNA-triggered gene silencing machineries. *Accounts of Chemical Research*, *45*(7), 1122–1131.

Lin, K. Y., Wu, S. Y., Hsu, Y. H., & Lin, N. S. (2022). MiR398-regulated antioxidants contribute to Bamboo mosaic virus accumulation and symptom manifestation. *Plant Physiology*, *188*(1), 593–607.

Lin, Y., Tan, H., Yu, G., Zhan, M., & Xu, B. (2023). Molecular mechanisms of noncoding RNA in the occurrence of castration-resistant prostate cancer. *International Journal of Molecular Sciences*, *24*(2), 1305.

Lindemose, S., O'Shea, C., Jensen, M., & Skriver, K. (2013). Structure, function and networks of transcription factors involved in abiotic stress responses. *International Journal of Molecular Sciences*, *14*, 5842–5878.

Liu, H., Able, A., & Able, J. (2020). Multi-omics analysis of small RNA, transcriptome, and degradome in *T. turgidum*—Regulatory networks of grain development and abiotic stress response. *International Journal of Molecular Sciences*, *21*.

Liu, X., Zhang, Y., Zhou, S., Dain, L., Mei, L., & Zhu, G. (2022). Circular RNA: An emerging frontier in RNA therapeutic targets, RNA therapeutics, and mRNA vaccines. *Journal of Controlled Release*, *348*, 84–94.

Long, Y., Wang, X., Youmans, D. T., & Cech, T. R. (2017). How do lncRNAs regulate transcription? *Science Advances*, *3*(9), eaao2110.

Ma, X., Zhao, F., & Zhou, B. (2022). The characters of non-coding RNAs and their biological roles in plant development and abiotic stress response. *International Journal of Molecular Sciences*, *23*(8), 4124.

Ma, Y., Freitas, H., & Dias, M. C. (2022). Strategies and prospects for biostimulants to alleviate abiotic stress in plants. *Frontiers in Plant Science*, *13*, 1024243.

Mahajan, S., & Tuteja, N. (2005). Cold, salinity and drought stresses: An overview. *Archives of Biochemistry and Biophysics*, *444*(2), 139–158.

Mahmoud, M. M., Sanad, E. F., & Hamdy, N. M. (2021). MicroRNAs' role in the environment-related non-communicable diseases and link to multidrug resistance, regulation, or alteration. *Environmental Science and Pollution Research*, *28*(28), 36984–37000.

Majeed, A., & Muhammad, Z. (2019). Salinity: A major agricultural problem—causes, impacts on crop productivity and management strategies. In *Plant abiotic stress tolerance: Agronomic, molecular and biotechnological approaches* (pp. 83–99).

Marasca, F., Bodega, B., & Orlando, V. (2018). How polycomb-mediated cell memory deals with a changing environment: Variations in PcG complexes and proteins assortment convey plasticity to epigenetic regulation as a response to environment. *Bioessays*, *40*(4), 1700137.

Mates, J. M. (2000). Effects of antioxidant enzymes in the molecular control of reactive oxygen species toxicology. *Toxicology*, *153*(1-3), 83–104.

Megha, S., Basu, U., & Kav, N. N. (2018). Regulation of low temperature stress in plants by microRNAs. *Plant, Cell & Environment*, *41*(1), 1–15.

Meher, P., Begam, S., Sahu, T., Gupta, A., Kumar, A., Kumar, U., Rao, A., Singh, K., & Dhankher, O. (2022). ASRmiRNA: Abiotic stress-responsive miRNA prediction in plants by using machine learning algorithms with pseudo K-tuple nucleotide compositional features. *International Journal of Molecular Sciences*, *23*.

Meraj, T. A., Fu, J., Raza, M. A., Zhu, C., Shen, Q., Xu, D., & Wang, Q. (2020). Transcriptional factors regulate plant stress responses through mediating secondary metabolism. *Genes*, *11*(4), 346.

Meza, I., Rezaei, E. E., Siebert, S., Ghazaryan, G., Nouri, H., Dubovyk, O., … & Hagenlocher, M. (2021). Drought risk for agricultural systems in South Africa: Drivers, spatial patterns, and implications for drought risk management. *Science of the Total Environment*, *799*, 149505.

Millar, A. A., Lohe, A., & Wong, G. (2019). Biology and function of miR159 in plants. *Plants*, *8*(8), 255.

Miller, A. (2014). Delivering the promise of small ncRNA therapeutics. *Therapeutic delivery*, *5*(5), 569–589.

Mishra, A. K., Das, R., George Kerry, R., Biswal, B., Sinha, T., Sharma, S., … & Kumar, M. (2023). Promising management strategies to improve crop sustainability and to amend soil salinity. *Frontiers in Environmental Science*, *10*, 962581.

Mishra, E., & Thakur, M. K. (2024). Recovery of memory decline during aging-role of epigenetics. *Annals of the National Academy of Medical Sciences (India)*, *60*(2), 101–112.

Mishra, K., & Kanduri, C. (2019). Understanding long noncoding RNA and chromatin interactions: What we know so far. *Non-Coding RNA*, *5*(4), 54.

Misra, B. B., Langefeld, C., Olivier, M., & Cox, L. A. (2019). Integrated omics: Tools, advances and future approaches. *Journal of Molecular Endocrinology*, *62*(1), R21–R45.

Moazed, D. (2009). Small RNAs in transcriptional gene silencing and genome defence. *Nature*, *457*(7228), 413–420.

Mohanta, T. K., Bashir, T., Hashem, A., Abd_Allah, E. F., Khan, A. L., & Al-Harrasi, A. S. (2018). Early events in plant abiotic stress signaling interplay between calcium, reactive oxygen species and phytohormones. *Journal of Plant Growth Regulation*, *37*(4), 1033–1049.

Moreau, P. R. (2024). *Exploring the non-coding RNA landscape in atherosclerosis: Cellular and stimulus-specific profiling of key regulatory elements* [Doctoral dissertation, Itä-Suomen yliopisto].

Morishita, E. C., & Nakamura, S. (2024). Recent applications of artificial intelligence in RNA-targeted small molecule drug discovery. *Expert Opinion on Drug Discovery*, *19*(4), 415–431.

Mosa, K. A., Ismail, A., Helmy, M., Mosa, K. A., Ismail, A., & Helmy, M. (2017). Introduction to plant stresses. In *Plant stress tolerance: An integrated omics approach (pp.* 1–19*)*.

Muhammad, I., Shalmani, A., Ali, M., Yang, Q. H., Ahmad, H., & Li, F. B. (2021). Mechanisms regulating the dynamics of photosynthesis under abiotic stresses. *Frontiers in Plant Science*, *11*, 615942.

Muhammad Aslam, M., Waseem, M., Jakada, B. H., Okal, E. J., Lei, Z., Saqib, H. S. A., … & Zhang, Q. (2022). Mechanisms of abscisic acid-mediated drought stress responses in plants. *International Journal of Molecular Sciences*, *23*(3), 1084.

Mujtaba, M., Wang, D., Carvalho, L. B., Oliveira, J. L., Espirito Santo Pereira, A. D., Sharif, R., … & Fraceto, L. F. (2021). Nanocarrier-mediated delivery of miRNA, RNAi, and CRISPR-Cas for plant protection: current trends and future directions. *ACS Agricultural Science & Technology*, *1*(5), 417–435.

Mustafa, G., Akhtar, M. S., & Abdullah, R. (2019). Global concern for salinity on various agro-ecosystems. *Salt Stress, Microbes, and Plant Interactions: Causes and Solution: Volume*, *1*, 1–19.

Nadarajah, K., & Rahman, N. (2021). The role of non-coding RNA in Rice immunity. *Agronomy, 12*(1), https://doi.10.3390/agronomy12010039.

Nadeem, M., Li, J., Yahya, M., Sher, A., Ma, C., Wang, X., & Qiu, L. (2019). Research progress and perspective on drought stress in legumes: A review. *International Journal of Molecular Sciences*, *20*(10), 2541.

Nadhan, R., Isidoro, C., Song, Y. S., & Dhanasekaran, D. N. (2022). Signaling by LncRNAs: Structure, cellular homeostasis, and disease pathology. *Cells*, *11*(16), 2517.

Nejat, N., & Mantri, N. (2018). Emerging roles of long non-coding RNAs in plant response to biotic and abiotic stresses. *Critical Reviews in Biotechnology*, *38*(1), 93–105.

Nephali, L., Piater, L. A., Dubery, I. A., Patterson, V., Huyser, J., Burgess, K., & Tugizimana, F. (2020). Biostimulants for plant growth and mitigation of abiotic stresses: A metabolomics perspective. *Metabolites*, *10*(12), 505.

Nesshöver, C., Assmuth, T., Irvine, K. N., Rusch, G. M., Waylen, K. A., Delbaere, B., … & Wittmer, H. (2017). The science, policy and practice of nature-based solutions: An interdisciplinary perspective. *Science of the Total Environment*, *579*, 1215–1227.

Nguyen, N. H., Vu, N. T., & Cheong, J. J. (2022). Transcriptional stress memory and transgenerational inheritance of drought tolerance in plants. *International Journal of Molecular Sciences*, *23*(21), 12918.

Ning, Y., Liu, W., & Wang, G. L. (2017). Balancing immunity and yield in crop plants. *Trends in Plant Science*, *22*(12), 1069–1079.

Nitnavare, R. B., Bhattacharya, J., Singh, S., Kour, A., Hawkesford, M. J., & Arora, N. (2021). Next generation dsRNA-based insect control: Success so far and challenges. *Frontiers in Plant Science*, *12*, 673576.

Niu, M., Lin, Y., & Zou, Q. (2021). sgRNACNN: Identifying sgRNA on-target activity in four crops using ensembles of convolutional neural networks. *Plant Molecular Biology*, *105*, 483–495.

O'Leary, G. J., Aggarwal, P. K., Calderini, D. F., Connor, D. J., Craufurd, P., Eigenbrode, S. D., … & Hatfield, J. L. (2018). Challenges and responses to ongoing and projected climate change for dryland cereal production systems throughout the world. *Agronomy*, *8*(4), 34.

Othman, S. M. I. S., Mustaffa, A. F., Che-Othman, M. H., Samad, A. F. A., Goh, H. H., Zainal, Z., & Ismail, I. (2023). Overview of repressive miRNA regulation by short tandem target mimic (STTM): Applications and impact on plant biology. *Plants*, *12*, 669.

Pagano, L., Rossi, R., Paesano, L., Marmiroli, N., & Marmiroli, M. (2021). miRNA regulation and stress adaptation in plants. *Environmental and Experimental Botany*, *184*, 104369.

Pandita, D., & Wani, S. H. (2019). MicroRNA as a tool for mitigating abiotic stress in rice (*Oryza sativa* L.). In *Recent approaches in omics for plant resilience to climate change* (pp. 109–133).

Partheeban, C., Chandrasekhar, C. N., Jeyakumar, P., Ravikesavan, R., & Gnanam, R. (2017). Effect of PEG induced drought stress on seed germination and seedling characters of maize (*Zea mays* L.) genotypes. *International Journal of Current Microbiology and Applied Sciences*, *6*(5), 1095–1104.

Pentimone, I., & Ciancio, A. (2021). miRNA-based approaches for sustainable control of diseases. *CABI Reviews*.

Pérez-Pizá, M. C., Sautua, F. J., Szparaga, A., Bohata, A., Kocira, S., & Carmona, M. A. (2024). New tools for the management of fungal pathogens in extensive cropping systems for friendly environments. *Critical Reviews in Plant Sciences*, *43*(2), 63–93.

Pervaiz, T., Amjid, M. W., El-Kereamy, A., Niu, S. H., & Wu, H. X. (2021). MicroRNA and cDNA-microarray as potential targets against abiotic stress response in plants: Advances and prospects. *Agronomy*, *12*(1), 11.

Piergentili, R., Basile, G., Nocella, C., Carnevale, R., Marinelli, E., Patrone, R., & Zaami, S. (2022). Using ncRNAs as tools in cancer diagnosis and treatment—The way towards personalized medicine to improve patients' health. *International Journal of Molecular Sciences*, *23*(16), 9353.

Piergentili, R., Gullo, G., Basile, G., Gulia, C., Porrello, A., Cucinella, G., … & Zaami, S. (2023). Circulating miRNAs as a tool for early diagnosis of endometrial cancer-implications for the fertility-sparing process: Clinical, biological, and legal aspects. *International Journal of Molecular Sciences*, *24*(14), 11356.

Pirasteh-Anosheh, H., Saed-Moucheshi, A., Pakniyat, H., & Pessarakli, M. (2016). Stomatal responses to drought stress. *Water Stress and Crop Plants: A Sustainable Approach*, *1*, 24–40.

Prabhakaran, S. (2024). RNA in plant nutrient availability. *Noncoding RNA: The Dark Matter of the Genome*, *165*.

Pradhan, A., Naik, N., & Sahoo, K. K. (2015). RNAi mediated drought and salinity stress tolerance in plants. *American Journal of Plant Sciences*, *6*(12), 1990–2008.

Praveen, A., Dubey, S., Singh, S., & Sharma, V. K. (2023). Abiotic stress tolerance in plants: A fascinating action of defense mechanisms. *3 Biotech*, *13*(3), 102.

Praveen, B., & Sharma, P. (2019). A review of literature on climate change and its impacts on agriculture productivity. *Journal of Public Affairs*, *19*(4), e1960.

Proietti, I., Frazzoli, C., & Mantovani, A. (2015). Exploiting nutritional value of staple foods in the world's semi-arid areas: Risks, benefits, challenges and opportunities of sorghum. *Healthcare*, *3*(2), 172–193. Multidisciplinary Digital Publishing Institute.

Qi, H. Y., Zhang, D. D., Liu, B., Chen, J. Y., Han, D., & Wang, D. (2024). Leveraging RNA interference technology for selective and sustainable crop protection. *Frontiers in Plant Science*, *15*, 1502015.

Ragupathy, R., Ravichandran, S., Mahdi, M. S. R., Huang, D., Reimer, E., Domaratzki, M., & Cloutier, S. (2016). Deep sequencing of wheat sRNA transcriptome reveals distinct temporal expression pattern of miRNAs in response to heat, light and UV. *Scientific Reports*, *6*(1), 39373.

Rajput, M., Choudhary, K., Kumar, M., Vivekanand, V., Chawade, A., Ortiz, R., & Pareek, N. (2021). RNA interference and CRISPR/Cas gene editing for crop improvement: Paradigm shift towards sustainable agriculture. *Plants*, *10*(9), 1914.

Ramakrishnan, M., Zhang, Z., Mullasseri, S., Kalendar, R., Ahmad, Z., Sharma, A., ... & Wei, Q. (2022). Epigenetic stress memory: A new approach to study cold and heat stress responses in plants. *Frontiers in Plant Science*, *13*, 1075279.

Ramesh, S. V. (2013). Non-coding RNAs in crop genetic modification: Considerations and predictable environmental risk assessments (ERA). *Molecular Biotechnology*, *55*, 87–100.

Ranjan, A., Sinha, R., Singla-Pareek, S. L., Pareek, A., & Singh, A. K. (2022). Shaping the root system architecture in plants for adaptation to drought stress. *Physiologia Plantarum*, *174*(2), e13651.

Rao, S., Balyan, S., Jha, S., Bansal, C., Das, J. R., Gupta, A., & Mathur, S. (2020). Orchestration of microRNAs and transcription factors in the regulation of plant abiotic stress response. *Plant stress biology: Strategies and trends* (pp. 251–277).

Ratti, M., Lampis, A., Ghidini, M., Salati, M., Mirchev, M. B., Valeri, N., & Hahne, J. C. (2020). MicroRNAs (miRNAs) and long non-coding RNAs (lncRNAs) as new tools for cancer therapy: First steps from bench to bedside. *Targeted Oncology*, *15*, 261–278.

Ravi, B., Foyer, C. H., & Pandey, G. K. (2023). The integration of reactive oxygen species (ROS) and calcium signalling in abiotic stress responses. *Plant, Cell & Environment*, *46*(7), 1985–2006.

Rawal, H. C., Ali, S., & Mondal, T. K. (2023). Role of non-coding RNAs against salinity stress in *Oryza* species: Strategies and challenges in analyzing miRNAs, tRFs and circRNAs. *International Journal of Biological Macromolecules*, *242*, 125172.

Rawat, N., Singla-Pareek, S. L., & Pareek, A. (2021). Membrane dynamics during individual and combined abiotic stresses in plants and tools to study the same. *Physiologia Plantarum*, *171*(4), 653–676.

Ren, L., Zhang, D., Chen, G. Q., Reed, B. M., Shen, X. H., & Chen, H. Y. (2015). Transcriptomic profiling revealed the regulatory mechanism of *Arabidopsis* seedlings response to oxidative stress from cryopreservation. *Plant Cell Reports*, *34*, 2161–2178.

Rinaldi, S., Moroni, E., Rozza, R., & Magistrato, A. (2024). Frontiers and challenges of computing ncRNAs biogenesis, function and modulation. *Journal of Chemical Theory and Computation*, *20*(3), 993–1018.

Ritonga, F. N., & Chen, S. (2020). Physiological and molecular mechanism involved in cold stress tolerance in plants. *Plants*, *9*(5), 560.

Roychowdhury, R., Das, S. P., Gupta, A., Parihar, P., Chandrasekhar, K., Sarker, U., … & Sudhakar, C. (2023). Multi-omics pipeline and omics-integration approach to decipher plant's abiotic stress tolerance responses. *Genes*, *14*(6), 1281.

Sahoo, J. P., Behera, L., Sharma, S. S., Praveena, J., Nayak, S. K., & Samal, K. C. (2020). Omics studies and systems biology perspective towards abiotic stress response in plants. *American Journal of Plant Sciences*, *11*(12), 2172.

Sanan-Mishra, N., Abdul Kader Jailani, A., Mandal, B., & Mukherjee, S. K. (2021). Secondary siRNAs in plants: Biosynthesis, various functions, and applications in virology. *Frontiers in Plant Science*, *12*, 610283.

Santosh, B., Varshney, A., & Yadava, P. K. (2015). Non-coding RNAs: biological functions and applications. *Cell Biochemistry and Function*, *33*(1), 14–22.

Sari, B., Isik, M., Eylem, C. C., Bektas, C., Okesola, B. O., Karakaya, E., … & Derkus, B. (2022). Omics technologies for high-throughput-screening of cell–biomaterial interactions. *Molecular Omics*, *18*(7), 591–615.

Satam, H., Joshi, K., Mangrolia, U., Waghoo, S., Zaidi, G., Rawool, S., … & Malonia, S. K. (2023). Next-generation sequencing technology: Current trends and advancements. *Biology*, *12*(7), 997.

Sato, H., Mizoi, J., Shinozaki, K., & Yamaguchi-Shinozaki, K. (2024). Complex plant responses to drought and heat stress under climate change. *The Plant Journal*, *117*(6), 1873–1892.

Sauvageau, M. (2019). Diverging RNPs: Toward understanding lncRNA-protein interactions and functions. *The biology of mRNA: Structure and function* (pp. 285–312).

Scandalios, J. G. (2005). Oxidative stress: molecular perception and transduction of signals triggering antioxidant gene defenses. *Brazilian Journal of Medical and Biological Research*, *38*, 995–1014.

Scarpin, D. (2024). *Innovative technologies enabling plants to increase resilience as a pivotal feature in the conservation of agro-ecosystems.*

Scheideler, M., Vidakovic, I., & Prassl, R. (2020). Lipid nanocarriers for microRNA delivery. *Chemistry and Physics of Lipids*, *226*, 104837.

Sebastian-delaCruz, M., Gonzalez-Moro, I., Olazagoitia-Garmendia, A., Castellanos-Rubio, A., & Santin, I. (2021). The role of lncRNAs in gene expression regulation through mRNA stabilization. *Non-Coding RNA*, *7*(1), 3.

Shah, A., Nazari, M., Antar, M., Msimbira, L. A., Naamala, J., Lyu, D., … & Smith, D. L. (2021). PGPR in agriculture: A sustainable approach to increasing climate change resilience. *Frontiers in Sustainable Food Systems*, *5*, 667546.

Sharma, M., Kumar, P., Verma, V., Sharma, R., Bhargava, B., & Irfan, M. (2022). Understanding plant stress memory response for abiotic stress resilience: Molecular insights and prospects. *Plant Physiology and Biochemistry*, *179*, 10–24.

Shelake, R. M., Kadam, U. S., Kumar, R., Pramanik, D., Singh, A. K., & Kim, J. Y. (2022). Engineering drought and salinity tolerance traits in crops through CRISPR-mediated genome editing: Targets, tools, challenges, and perspectives. *Plant Communications*, *3*(6).

Shi, M., Wang, C., Wang, P., Yun, F., Liu, Z., Ye, F., … & Liao, W. (2023). Role of methylation in vernalization and photoperiod pathway: A potential flowering regulator? *Horticulture Research*, *10*(10), uhad174.

Shiferaw, B., Smale, M., Braun, H. J., Duveiller, E., Reynolds, M., & Muricho, G. (2013). Crops that feed the world 10. Past successes and future challenges to the role played by wheat in global food security. *Food Security*, *5*, 291–317.

Shomali, A., Das, S., Arif, N., Sarraf, M., Zahra, N., Yadav, V., … & Hasanuzzaman, M. (2022). Diverse physiological roles of flavonoids in plant environmental stress responses and tolerance. *Plants*, *11*(22), 3158.

Shriram, V., Kumar, V., Devarumath, R. M., Khare, T. S., & Wani, S. H. (2016). MicroRNAs as potential targets for abiotic stress tolerance in plants. *Frontiers in Plant Science*, *7*, 817.

Shukla, L. I., Chinnusamy, V., & Sunkar, R. (2008). The role of microRNAs and other endogenous small RNAs in plant stress responses. *Biochimica et Biophysica Acta (BBA)-Gene Regulatory Mechanisms*, *1779*(11), 743–748.

Shukla, P., Mantin, E., Adil, M., Bajpai, S., Critchley, A., & Prithiviraj, B. (2019). Ascophyllum nodosum-based biostimulants: Sustainable applications in agriculture for the stimulation of plant growth, stress tolerance, and disease management. *Frontiers in Plant Science*, *10*.

Sihag, N., Singh, T., Sheoran, S., Singh, O., Malik, R., Kumar, L., & Singh, J. (2024). Role of RNA interference in drought stress management: Physiological, biochemical and molecular approach. *Crop and Pasture Science*, *75*(1), NULL-NULL.

Singh, A., Sutradhar, M., Prasad, S. S., Ramteke, P. W., Rajeshwar, J. R., & Vagvolgyi, C. (2025). *Plant novel transcription factors and their roles in gene regulation.*

Singroha, G., Sharma, P., & Sunkur, R. (2021). Current status of microRNA-mediated regulation of drought stress responses in cereals. *Physiologia Plantarum*, *172*(3), 1808–1821.

Sivakumar, P., Kim, S., Kang, H. C., & Shim, M. S. (2019). Targeted siRNA delivery using aptamer-siRNA chimeras and aptamer-conjugated nanoparticles. *Wiley Interdisciplinary Reviews: Nanomedicine and Nanobiotechnology*, *11*(3), e1543.

Sonjaya, Y., Noch, M. Y., & Sutisna, E. (2024). The role of appropriate technology in sustainable development design. *Advances in Community Services Research*, *2*(1), 24–36.

Srijyothi, L., Ponne, S., Prathama, T., Ashok, C., & Baluchamy, S. (2018). Roles of non-coding RNAs in transcriptional regulation. *Transcriptional and Post-Transcriptional Regulation*, *55*.

Srivastava, G. K., Martinez-Rodriguez, S., Md Fadilah, N. I., Looi Qi Hao, D., Markey, G., Shukla, P., … & Panetsos, F. (2024). Progress in wound-healing products based on natural compounds, stem cells, and microRNA-based biopolymers in the European, USA, and Asian markets: Opportunities, barriers, and regulatory issues. *Polymers*, *16*(9), 1280.

Starega-Roslan, J., Koscianska, E., Kozlowski, P., & Krzyzosiak, W. J. (2011). The role of the precursor structure in the biogenesis of microRNA. *Cellular and Molecular Life Sciences*, *68*, 2859–2871.

Subhan, M. A., & Torchilin, V. P. (2023). Biopolymer-based nanosystems for siRNA drug delivery to solid tumors including breast cancer. *Pharmaceutics*, *15*(1), 153.

Sun, M., Shen, Y., Chen, Y., Wang, Y., Cai, X., Yang, J., Jia, B., Dong, W., Chen, X., & Sun, X. (2022). Osa-miR1320 targets the ERF transcription factor OsERF096 to regulate cold tolerance via JA-mediated signaling. *Plant Physiology*, *189*(4), 2500–2516.

Swida-Barteczka, A., & Szweykowska-Kulinska, Z. (2019). Micromanagement of developmental and stress-induced senescence: the emerging role of microRNAs. *Genes*, *10*(3), 210.

Tafech, A., Bassett, T., Sparanese, D., & Lee, C. H. (2006). Destroying RNA as a therapeutic approach. *Current Medicinal Chemistry*, *13*(8), 863–881.

Tang, J., Gu, X., Liu, J., & He, Z. (2021). Roles of small RNAs in crop disease resistance. *Stress Biology*, *1*(1), 6.

Tanna, S., Doshi, G., & Godad, A. (2024). siRNA as potential therapeutic strategy for hypertension. *European Journal of Pharmacology*, *969*, 176467.

Thakur, Y., Sharma, A., & Sharma, V. K. (2024). A review on the relationship of climate variability and extremes with crop production. *International Journal of Environment and Climate Change*, *14*(4), 513–529.

Tian, Z., Wang, J. W., Li, J., & Han, B. (2021). Designing future crops: challenges and strategies for sustainable agriculture. *The Plant Journal*, *105*(5), 1165–1178.

Touzdjian Pinheiro Kohlrausch Távora, F., de Assis dos Santos Diniz, F., de Moraes Rêgo-Machado, C., Chagas Freitas, N., Barbosa Monteiro Arraes, F., Chumbinho de Andrade, E., … & Correa Molinari, H. B. (2022). CRISPR/Cas-and topical RNAi-based technologies for crop management and improvement: Reviewing the risk assessment and challenges towards a more sustainable agriculture. *Frontiers in Bioengineering and Biotechnology*, *10*, 913728.

Traubenik, S., Charon, C., & Blein, T. (2024). From environmental responses to adaptation: The roles of plant lncRNAs. *Plant Physiology*, *195*(1), 232–244.

Turner, A. W., Wong, D., Khan, M. D., Dreisbach, C. N., Palmore, M., & Miller, C. L. (2019). Multi-omics approaches to study long non-coding RNA function in atherosclerosis. *Frontiers in Cardiovascular Medicine*, *6*, 9.

Ullah, A., Nadeem, F., Nawaz, A., Siddique, K. H., & Farooq, M. (2022a). Heat stress effects on the reproductive physiology and yield of wheat. *Journal of Agronomy and Crop Science*, *208*(1), 1–17.

Ullah, I., Kamel, E. A., Shah, S. T., Basit, A., Mohamed, H. I., & Sajid, M. (2022b). Application of RNAi technology: A novel approach to navigate abiotic stresses. *Molecular Biology Reports*, *49*(11), 10975–10993.

Valadkhan, S., & Valencia-Hipólito, A. (2015). lncRNAs in stress response. *Current Topics in Microbiology and Immunology*, *394*, 203–36.

Valencia-Sanchez, M. A., Liu, J., Hannon, G. J., & Parker, R. (2006). Control of translation and mRNA degradation by miRNAs and siRNAs. *Genes & Development*, *20*(5), 515–524.

Van Oosten, M., Pepe, O., De Pascale, S., Silletti, S., & Maggio, A. (2017). The role of biostimulants and bioeffectors as alleviators of abiotic stress in crop plants. *Chemical and Biological Technologies in Agriculture*, *4*.

Vaschetto, L. M. (2024). DNA methylation, histone modifications, and non-coding RNA pathways. In *Epigenetics in crop improvement: Safeguarding food security in an ever-changing climate* (pp. 15–27). Springer Nature Switzerland.

Vasilachi, I. C., Stoleru, V., & Gavrilescu, M. (2023). Analysis of heavy metal impacts on cereal crop growth and development in contaminated soils. *Agriculture*, *13*(10), 1983.

Vatanparast, M., Merkel, L., & Amari, K. (2024). Exogenous application of dsRNA in plant protection: efficiency, safety concerns and risk assessment. *International Journal of Molecular Sciences*, *25*(12), 6530.

Venu, E., Ramya, A., Babu, P. L., Srinivas, B., Kumar, S., Reddy, N. K., … & Manik, S. (2024). Exogenous dsRNA-mediated RNAi: Mechanisms, applications, delivery methods and challenges in the induction of viral disease resistance in plants. *Viruses*, *17*(1), 49.

Vidanagamachchi, S. M., & Waidyarathna, K. M. G. T. R. (2024). Opportunities, challenges and future perspectives of using bioinformatics and artificial intelligence techniques on tropical disease identification using omics data. *Frontiers in Digital Health*, *6*, 1471200.

Villalobos-López, M. A., Arroyo-Becerra, A., Quintero-Jiménez, A., & Iturriaga, G. (2022). Biotechnological advances to improve abiotic stress tolerance in crops. *International Journal of Molecular Sciences*, *23*(19), 12053.

Waititu, J., Zhang, C., Liu, J., & Wang, H. (2020). Plant non-coding RNAs: Origin, biogenesis, mode of action and their roles in abiotic stress. *International Journal of Molecular Sciences*, *21*.

Wang, D., Jin, X., & Li, S. (2024, October). *Advancing disease diagnosis and biomarker discovery: The role of machine learning in ncRNA analysis*. In Proceedings of the 2024 International Conference on Smart Healthcare and Wearable Intelligent Devices (pp. 29–35).

Wang, H. L. V., & Chekanova, J. A. (2016). Small RNAs: Essential regulators of gene expression and defenses against environmental stresses in plants. *Wiley Interdisciplinary Reviews: RNA*, *7*(3), 356–381.

Wang, M., Jiang, S., Zhou, L., Yu, F., Ding, H., Li, P., … & Wang, K. (2019). Potential mechanisms of action of curcumin for cancer prevention: focus on cellular signaling pathways and miRNAs. *International Journal of Biological Sciences*, *15*(6), 1200.

Wang, M., Li, R., & Zhao, Q. (2023). Multi-omics techniques in genetic studies and breeding of forest plants. *Forests*, *14*(6), 1196.

Wang, S., Hu, W., Zhang, X., Liu, Y., & Liu, F. (2025). Identification and Characterization of SQUAMOSA Promoter Binding Protein-like Transcription Factor Family Members in Zanthoxylum bungeanum and Their Expression Profiles in Response to Abiotic Stresses. *Plants*, *14*(4), 520.

Wang, Z., Li, Y., Kong, D., Ahmad, A., Banerjee, S., & Sarkar, F. H. (2010). Cross-talk between miRNA and Notch signaling pathways in tumor development and progression. *Cancer Letters*, *292*(2), 141–148.

Wani, S., Kumar, V., Khare, T., Tripathi, P., Shah, T., Ramakrishna, C., Aglawe, S., & Mangrauthia, S. (2020). miRNA applications for engineering abiotic stress tolerance in plants. *Biologia*, *75*, 1063–1081.

Willow, J., & Veromann, E. (2022). Integrating RNAi technology in smallholder farming: *Accelerating Sustainable Development Goals*, *6*.

Wu, F., & Rodricks, J. V. (2022). Risk assessment and regulatory decision-making in environmental health. In *Risk assessment for environmental health* (pp. 39–55). CRC Press.

Wu, J., Gao, T., Hu, J., Zhao, L., Yu, C., & Ma, F. (2022). Research advances in function and regulation mechanisms of plant small heat shock proteins (sHSPs) under environmental stresses. *Science of the Total Environment*, *825*, 154054.

Xie, X., He, Z., Chen, N., Tang, Z., Wang, Q., & Cai, Y. (2019). The roles of environmental factors in regulation of oxidative stress in plant. *BioMed Research International*, *2019*(1), 9732325.

Xu, J., Hou, Q. M., Khare, T., Verma, S. K., & Kumar, V. (2019). Exploring miRNAs for developing climate-resilient crops: A perspective review. *Science of the Total Environment*, *653*, 91–104.

Xu, M., Li, H., Liu, Z., Wang, X., Xu, P., Dai, S., Cao, X., & Cui, X. (2021). The soybean CBL-interacting protein kinase, GmCIPK2, positively regulates drought tolerance and ABA signaling. *Plant Physiology and Biochemistry: PPB*, *167*, 980–989.

Xu, M. Y., Zhang, L., Li, W. W., Hu, X. L., Wang, M. B., Fan, Y. L., … & Wang, L. (2014). Stress-induced early flowering is mediated by miR169 in *Arabidopsis thaliana*. *Journal of Experimental Botany*, *65*(1), 89–101.

Yadav, A., Kumar, S., Verma, R., Narayan, S., Gupta, U., Lata, C., … & Sanyal, I. (2024). Overexpression of miR166 in response to root rhizobacteria enhances drought adaptive efficacy by targeting HD-ZIP III family genes in chickpea. *Journal of Soil Science and Plant Nutrition*, *24*(3), 6024–6039.

Yadav, A., Mathan, J., Dubey, A. K., & Singh, A. (2024). The emerging role of non-coding RNAs (ncRNAs) in plant growth, development, and stress response signaling. *Non-Coding RNA*, *10*(1), 13.

Yadav, R., Chauhan, S., Sharma, H., Tiwari, R., & Singh, C. P. (2025). RNAi-based biopesticide: A powerful tool for crop protection. In *Non-coding RNAs for crop improvement: Concepts and applications* (pp. 225–243). Springer Nature Singapore.

Yan, G., Hua, Y., Jin, H., Huang, Q., Zhou, G., Xu, Y., … & Zhu, Z. (2023). Sly-miR398 participates in cadmium stress acclimation by regulating antioxidant system and cadmium transport in tomato (*Solanum lycopersicum*). *International Journal of Molecular Sciences*, *24*(3), 1953.

Yang, H., Cui, Y., Feng, Y., Hu, Y., Liu, L., & Duan, L. (2023). Long non-coding RNAs of plants in response to abiotic stresses and their regulating roles in promoting environmental adaption. *Cells*, *12*(5), 729.

Yang, Y., Saand, M., Huang, L., Abdelaal, W., Zhang, J., Wu, Y., Li, J., Sirohi, M., & Wang, F. (2021). Applications of multi-omics technologies for crop improvement. *Frontiers in Plant Science*, *12*.

Yoo, C. Y., Pence, H. E., Hasegawa, P. M., & Mickelbart, M. V. (2009). Regulation of transpiration to improve crop water use. *Critical Reviews in Plant Science*, *28*(6), 410–431.

Younis, A., Siddique, M. I., Kim, C. K., & Lim, K. B. (2014). RNA interference (RNAi) induced gene silencing: a promising approach of hi-tech plant breeding. *International Journal Of Biological Sciences*, *10*(10), 1150.

Yu, Y., Zhang, Y., Chen, X., & Chen, Y. (2019). Plant noncoding RNAs: Hidden players in development and stress responses. *Annual Review of Cell and Developmental Biology*, *35*(1), 407–431.

Zafar, S., Zaidi, S., Gaba, Y., Singla-Pareek, S., Dhankher, O., Li, X., Mansoor, S., & Pareek, A. (2020). Engineering abiotic stress tolerance via CRISPR-Cas mediated genome editing. *Journal of Experimental Botany*, *71*(2), 470–479.

Zang, J., Lu, D., & Xu, A. (2020). The interaction of circRNAs and RNA binding proteins: An important part of circRNA maintenance and function. *Journal of Neuroscience Research*, *98*(1), 87–97.

Zenda, T., Liu, S., Dong, A., & Duan, H. (2021). Advances in cereal crop genomics for resilience under climate change. *Life*, *11*(6), 502.

Zenda, T., Liu, S., Dong, A., Li, J., Wang, Y., Liu, X., … & Duan, H. (2021). Omics-facilitated crop improvement for climate resilience and superior nutritive value. *Frontiers in Plant Science*, *12*, 774994.

Zhang, B. (2015). MicroRNA: a new target for improving plant tolerance to abiotic stress. *Journal of Experimental Botany*, *66*(7), 1749–1761.

Zhang, D., Zhang, Y., Sun, L., Dai, J., & Dong, H. (2023). Mitigating salinity stress and improving cotton productivity with agronomic practices. *Agronomy*, *13*(10), 2486.

Zhang, G., Luo, Y., Dai, X., & Dai, Z. (2023). Benchmarking deep learning methods for predicting CRISPR/Cas9 sgRNA on- and off-target activities. *Briefings in Bioinformatics*, *24*(6), bbad333.

Zhang, H., Zhu, J., Gong, Z., & Zhu, J. K. (2022). Abiotic stress responses in plants. *Nature Reviews Genetics*, *23*(2), 104–119.

Zhang, L., Lin, T., Zhu, G., Wu, B., Zhang, C., & Zhu, H. (2023). LncRNAs exert indispensable roles in orchestrating the interaction among diverse noncoding RNAs and enrich the regulatory network of plant growth and its adaptive environmental stress response. *Horticulture Research*, *10*(12), uhad234.

Zhang, P., Wu, W., Chen, Q., & Chen, M. (2019). Non-coding RNAs and their integrated networks. *Journal of Integrative Bioinformatics*, *16*(3), 20190027.

Zhang, X., Wang, W., Zhu, W., Dong, J., Cheng, Y., Yin, Z., & Shen, F. (2019). Mechanisms and functions of long non-coding RNAs at multiple regulatory levels. *International Journal of Molecular Sciences*, *20*(22), 5573.

Zhang, X. O., Dong, R., Zhang, Y., Zhang, J. L., Luo, Z., Zhang, J., & Yang, L. (2016). Diverse alternative back-splicing and alternative splicing landscape of circular RNAs. *Genome Research*, *26*(9), 1277–1287.

Zhang, Y., Zhou, Y., Zhu, W., Liu, J., & Cheng, F. (2022). Non-coding RNAs fine-tune the balance between plant growth and abiotic stress tolerance. *Frontiers in Plant Science*, *13*, 965745.

Zhao, C., Wang, P., Si, T., Hsu, C. C., Wang, L., Zayed, O., … & Zhu, J. K. (2017). MAP kinase cascades regulate the cold response by modulating ICE1 protein stability. *Developmental Cell*, *43*(5), 618–629.

Zhao, J., Lu, Z., Wang, L., & Jin, B. (2020). Plant responses to heat stress: physiology, transcription, noncoding RNAs, and epigenetics. *International Journal of Molecular Sciences*, *22*(1), 117.

Zheng, Y., Hivrale, V., Zhang, X., Valliyodan, B., Lelandais-Brière, C., Farmer, A. D., … & Sunkar, R. (2016). Small RNA profiles in soybean primary root tips under water deficit. *BMC Systems Biology*, *10*, 1–10.

Zhong, L., Zhen, M., Sun, J., & Zhao, Q. (2021). Recent advances on the machine learning methods in predicting ncRNA-protein interactions. *Molecular Genetics and Genomics*, *296*(2), 243–258.

Zhou, B., Zheng, B., & Wu, W. (2023). The ncRNAs involved in the regulation of abiotic stress-induced anthocyanin biosynthesis in plants. *Antioxidants*, *13*(1), 55.

Zhou, W. Y., Cai, Z. R., Liu, J., Wang, D. S., Ju, H. Q., & Xu, R. H. (2020). Circular RNA: Metabolism, functions and interactions with proteins. *Molecular Cancer*, *19*, 1–19.

Zuo, Z. F., He, W., Li, J., Mo, B., & Liu, L. (2021). Small RNAs: the essential regulators in plant thermotolerance. *Frontiers in Plant Science*, *12*, 726762.

16 Non-coding RNA
The Unsung Heroes of Plant Stress Tolerance

Javed Iqbal, Shumaila Ijaz, Naila Ijaz, Banzeer Ahsan Abbasi, Faiza Munir, Zakir Ullah, Muhammad Anas and Rahmat Ali

16.1 INTRODUCTION TO NON-CODING RNAs IN PLANTS

Being immobile, plants have to develop avoidance and/or tolerance mechanisms to thrive amid constantly changing environmental conditions. Chilli plants are sensitive to abiotic stressors, such as cold, heat, hypoxia, drought, nutrient shortage and salinity, which can disrupt plant development and production, resulting in a major yield loss (Ahmad et al., 2022; Ahmad et al., 202; Hirayama & Shinozaki, 2010). Moreover, soil quality, global warming, the increasing global population and loss of available land for food production due to rapid urbanisation are predicted to exacerbate food security issues (Mittler, 2006). Recent years have been witnesses of a great surge in the significance of the non-coding RNAs (ncRNAs) as bioactive molecules that add to the genomic and phenotypic diversity. With up to 90% of the eukaryotic genome transcribed into RNAs, it becomes important to mention that only 2% of transcribed RNAs are eventually translated into proteins (Pauli et al., 2011; Rai et al., 2019). Because they are not conserved and do not have protein-coding potential, the remaining transcriptome, ncRNA transcripts, were previously considered as transcriptional noise (Ariel et al., 2015; Cech & Steitz, 2014; Pauli et al., 2011). However, experimental validation and high-throughput sequencing analysis revealed that ncRNAs have common transcriptional, post-transcriptional and epigenetic gene regulatory roles (Peschansky & Wahlestedt, 2014; Ponjavic et al., 2007). Thus, the discovery of ncRNAs in plants has been largely reliant on the adoption of efficient next-generation sequencing (NGS) technologies (Eddy, 2001; Shimoni et al., 2007) combined with homology-based and/or experimental methods.

Since their discovery, ncRNAs have been categorised into multiple classes, which are based on biogenesis, origin and activity mechanism. ncRNA transcripts can be classified into two major groups – regulatory and housekeeping ncRNAs. Housekeeping ncRNAs have two major functions: cellular and ribosomal. This includes ribosomal RNAs (rRNAs), small nuclear RNAs (snRNAs), transfer RNAs (tRNAs) and small nucleolar RNAs (snoRNAs). They are essential housekeeping ncRNAs involved in protein synthesis and maintenance of cellular structural integrity. Without these molecules, plant health would decline because ribosomes and the entire cellular apparatus would fail to function. In contrast, the regulatory ncRNAs include long non-coding RNAs (lncRNAs), piwi-interacting RNAs (piRNAs), short interfering RNAs (siRNAs) and microRNAs (miRNAs) (Morris & Mattick, 2014). Although they cannot be translated into proteins, these regulatory ncRNAs are transcribed from DNA (Brosnan & Voinnet, 2009; Peschansky & Wahlestedt, 2014). Unlike housekeeping ncRNAs, regulatory ncRNAs are involved in important biological processes by primarily modulating gene expression. They are crucial for tuning gene regulatory networks, since they either silence target genes or promote the degradation of mRNAs.

They do, however, accomplish multiple critical functions in governing transcription and post-transcription of plant growth, development and abiotic stress responses. They control cell differential, developmental transitions and responses to external stimulation. To adapt and survive against adversities such as salt and drought, some miRNAs and lncRNAs were revealed as the abnoxious regulators of the signalling pathways involved in homeostasis. Moreover, the multifaceted interactions of these ncRNAs with other molecular players such as transcription factors and signal transduction elements point to the sophistication of plant regulatory networks. This intricate control highlights the significant roles that ncRNAs play in supporting plant health and productivity, particularly in the context of climate change and decreasing agricultural viability.

In recent decades, substantial progress has been achieved in uncovering the complex and dynamic mechanisms that deliver plant responses to environmental stress. To date, much focus has been put on identifying stress-responsive genes and their corresponding protein interaction networks. Further investigation unveiled an intricate regulation of gene expression, mediated by multiple mechanisms at different stages, including transcriptional and post-transcriptional regulators, which became evident as key components of the plant stress response (Ali et al., 2022; D'Ario et al., 2017). Underlying resilience to severe environments, ncRNAs are confirmed to contribute significantly to complex functions that improve resistance to environmental stressors alongside their auxiliary role in regulating gene expression. However, many studies focused on the regulatory mechanisms of coding proteins; only a few recent studies discovered the essential functions of ncRNAs in transcriptional and post-transcriptional regulation of gene expression levels in leaf developmental and environmental stress responses in plants (Campalans et al., 2004; Sunkar & Zhu, 2004).

DOI: 10.1201/9781003583257-16

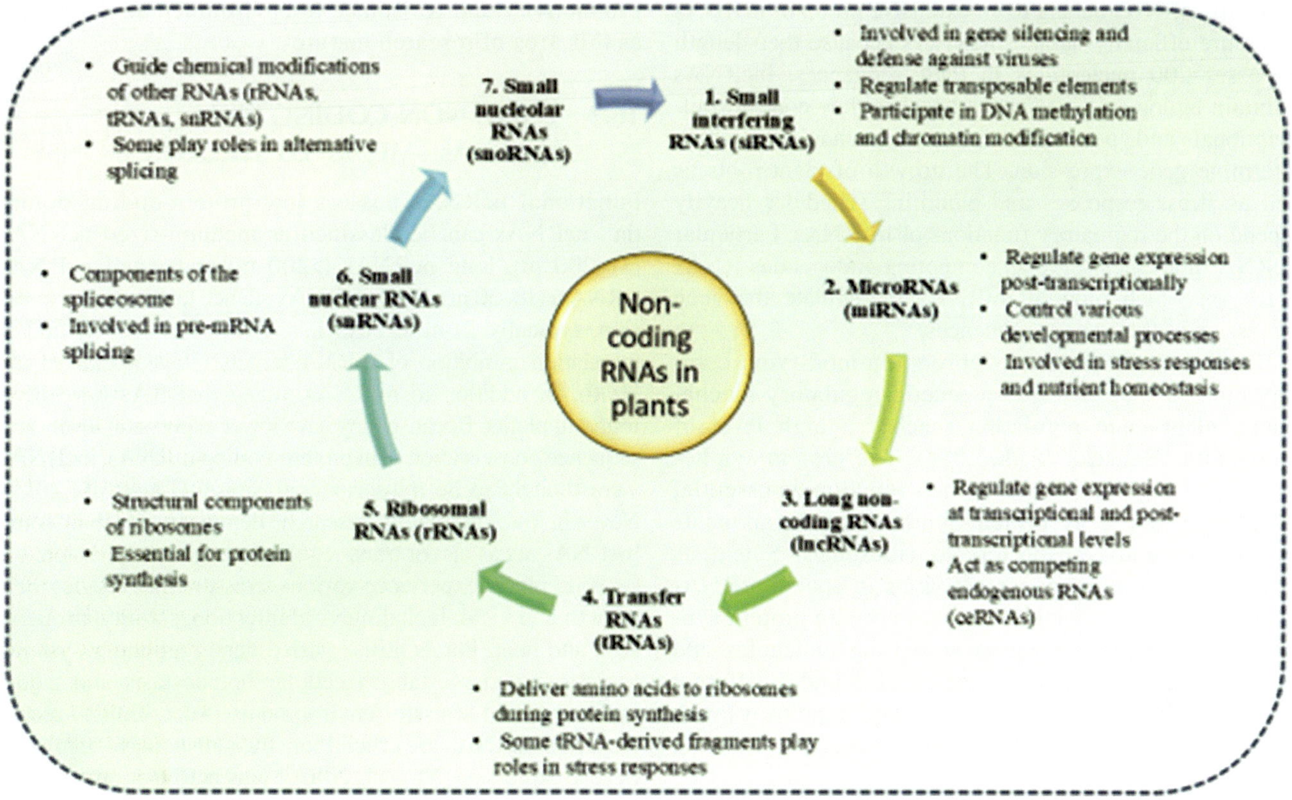

FIGURE 16.1 Overview of the roles of non-coding RNAs (ncRNAs) in plants. This figure categorises different types of ncRNAs, including small interfering RNAs (siRNAs), microRNAs (miRNAs), long non-coding RNAs (lncRNAs), transfer RNAs (tRNAs), ribosomal RNAs (rRNAs), small nuclear RNAs (snRNAs) and small nucleolar RNAs (snoRNAs).

Explaining these processes can lead to new approaches in increasing crop stress tolerance, sustainable agriculture and food security (Figure 16.1). As unfolded in this chapter, 'Non-coding RNA: The Unsung Heroes of Plant Stress Tolerance', ncRNAs are involved in modulating plant response to a wide range of abiotic stressors. In this chapter, we discuss how these ncRNAs, once thought to be no more than transcriptional noise, have developed into important gene expression regulators at various levels, ranging from transcription to post-transcriptional processes. The obtained data emphasise the necessity of developing plants with higher tolerance to environmental stresses which include salt and temperature alongside drought conditions. The study of ncRNA relationships to stress response mechanisms enables their development as useful targets to boost crop resistance against stress conditions.

16.2 NON-CODING RNA TYPES AND THEIR ROLES IN PLANTS

Plants utilise ncRNAs in addition to their multiple classes that play distinct functions within the plant system. A set of RNA molecules which cannot produce proteins known as ncRNAs perform multiple cellular functions in plant organisms. Multiple classifications exist for RNAs based on their dimensional characteristics and cellular positions as well as functional effects, according to Song et al. (2021). siRNAs represent established ncRNA molecules that consist of double-stranded RNA with 20–25 nucleotide length. RNAi processes that use siRNAs are essential for gene silencing and simultaneously perform vital functions in controlling transposable elements and protecting plants from viruses as well as modifying DNA and altering chromosome structure (George et al., 2024). The compounds achieve genomic stability and gene control through their ability to inhibit mRNA translation while also inducing its degradation (Deogharia & Gurha, 2022). The supporting group of this mechanism consists of siRNAs, but the essential components are miRNAs which function as 21–24 nucleotide molecules that control gene expression after transcription in small ncRNA families (Santosh et al., 2015).

Once miRNA identifies target mRNAs in the cell, it performs two main functions by blocking translation while also destroying target mRNAs. Plant life implements control systems for leaf growth as well as flower development and root formation while also having systems that determine their response to environmental threats, including salt stress and drought conditions and pathogen attacks. The complex system featuring miRNAs and siRNAs in plant development response to stress is depicted in Figure 16.1, as explained by Ahmad et al. (2024).

Small ncRNAs belong to an extensive group of ncRNAs which are officially named lncRNAs because their length surpasses 200 nucleotides in their sequences. lncRNAs maintain biological importance because they control transcriptional and post-transcriptional mechanisms which determine gene expression. The growth of plant roots as well as stress responses and blooming schedules heavily depend on the regulatory functions of lncRNAs. Particular lncRNA molecules serve as competing endogenous RNAs (ceRNAs) which bind to miRNAs to regulate the gene expression of their target sequences.

Because lncRNAs act as chromatin-modifying complex binders that provide epigenetic regulatory mechanisms, plant gene regulation reaches a high level of complexity (Benacka et al., 2024). The protein synthesis process requires tRNAs which function as essential ncRNAs to serve as protein-synthesising components and amino acid transporters to ribosomes. Scientists found plant stress response activities in specific tRNA-derived fragments despite their main role in protein synthesis. The fragments act as signalling molecules and regulatory elements for gene expression under different stress situations by connecting translation pathways with stress response mechanisms (Haider et al., 2021; Li et al., 2017; Pecoraro et al., 2022).

rRNAs work as ribosome components to enable protein synthesis and exist as ncRNAs within the cellular structure. The ribosomes' catalytic core contains these molecules which assist ribosomes with decoding mRNA and peptide bond formation for translation activities. The protein synthetic efficiency depends on the stability alongside the functional ability of rRNAs as cellular ribosome components (Cui et al., 2024; Jiao et al., 2023). The spliceosome requires short nuclear RNAs (snRNAs) to function normally as a protein-RNA complex in pre-mRNA splicing processes. The recognition of splice sites along with pre-mRNA intron excision depends on these responsible molecules (Reddy, 2001). Plants need alternative splicing processes to build protein diversity while managing gene expression patterns that respond to environmental and developmental signals. The capacity of snRNAs for splicing tasks proves that ncRNAs play a critical role in shaping transcriptomic outcomes as described by Lee and Rio (2015) and Sahebi et al. (2016).

The three RNA types, rRNAs and tRNAs as well as snRNAs, undergo chemical processing by snoRNAs while located in the nucleolus for short periods of time. RNA modification stands as crucial because it enables effective work along with constant splicing and translation processes. Modulating alternate splicing and stress responses in plants have also been attributed to some snoRNAs (Decatur & Fournier, 2003; Jouravleva & Zamore, 2025; Lee et al., 2023). These clusters of ncRNAs demonstrate their role in regulating biological activities such as stress response, plant growth and general cellular function. Further understanding of ncRNAs will lead to opportunities for enhancing agricultural productivity and resistance to environmental challenges as this area of research matures.

16.3 LONG NON-CODING RNAs (lncRNAs)–MEDIATED STRESS RESPONSE

Functional ncRNAs possess low protein-coding potential. ncRNAs can be classified as medium-sized ncRNAs (31–200 nt), long ncRNAs (>200 nt) and small ncRNAs (sRNAs) (18–30 nt). Plant miRNAs silence target gene expression, typically 21-nt sRNAs that direct degradation and/or translation inhibition of mRNA targets (Rajagopalan et al., 2006). In addition to miRNAs, many lncRNAs have been found in plants. Because they have lower expression levels and sequence conservation than protein-coding mRNAs, lncRNAs were thought to be transcriptional 'noise' (Liu et al., 2015). Nevertheless, mounting research demonstrates that some lncRNAs act as cis- or trans-regulators of gene expression.

Wild plants experience various stressors that restrict their growth and yield, including viral infection, salinity, drought, cold and heat. Plants utilise varied gene regulatory systems to restore and reestablish cellular homeostasis and adjust and survive in stressful environments. According to recent research, ncRNAs are crucial for stress-responsive regulation of gene expression (Xu et al., 2016). Many ncRNAs, particularly miRNAs and lncRNAs, have been identified and characterised in plants by NGS technology and bioinformatics tools. In the following sections, we review the current state of knowledge regarding plant miRNAs and lncRNAs, including their biogenesis, resources and bioinformatics tools (Figure 16.2).

Identification of lncRNAs is conducted by two significant stages. Firstly, tiling microarray or high-throughput sequencing data was used to identify lncRNA transcript units. Secondly, codon statistics and alignment with known protein-coding sequences were used to assess the coding potential of such lncRNA transcript units. The post-chemical processing of rRNAs along with tRNAs and snRNAs by snoRNAs in brief nucleolus periods produces these RNA types. RNA requires chemical modifications prior to work efficiently for both translation processes and steady splicing. Multiple programmes exist for measuring coding potential, including PhyloCSF created by Lin et al. (2011), CPC by Wang et al. (2017), CPAT by Wang et al. (2013) and CNCI by Wang et al. (2017). The codon substitution frequency (CSF) score allows PhyloCSF to distinguish between protein-coding and non-protein-coding transcript units. The basic concept of CPC and CNCI systems depends on support vector machines (SVM), whereas CPAT uses sequence features within a logistic regression model as its foundational approach. The identification of lncRNAs from transcriptome sequencing data utilises two independent standalone programmes named iSeeRNA (Sun et al., 2013) together with DeepLNC (Tripathi et al., 2016). The deep neural network-classification model powers the functioning of DeepLNC. The iSeeRNA programme employs SVM technology to detect lncRNAs but differs from the DeepLNC model. Multiple investigations showed lncRNAs carry out different vital biological activities. The nucleic acid thermodynamics-based programme lncRNATargets (Hu et al., 2016) serves as a tool for

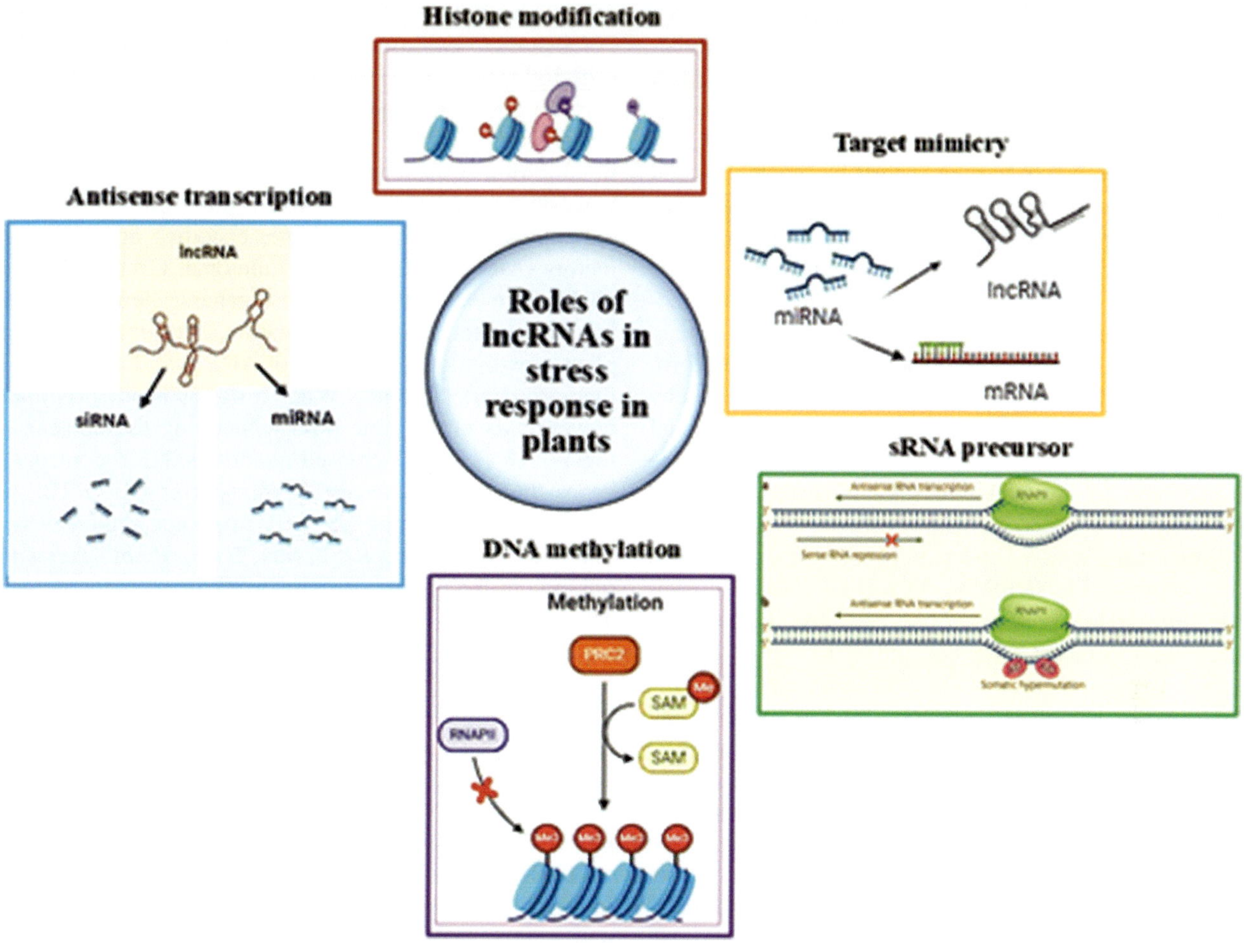

FIGURE 16.2 Long non-coding RNAs (lncRNAs) play a part in plant stress responses through mechanisms like target mimicry, DNA methylation, antisense transcription, histone modification and sRNA precursor involvement.

predicting the interactions which occur between lncRNAs and mRNAs or between lncRNAs and DNAs. ceRNAs are detectable with spongeScan (Furio-Tari et al., 2016) to predict interactions between lncRNA and miRNA. Besides, an lncRNA list can be accessed using RegRNA (Huang et al., 2006) and TF2LncRNA (Jiang et al., 2014) to determine the regulatory RNA motifs/functional sites and common transcription factors. Recent research has unambiguously shown that lncRNAs may be involved in regulatory functions in many developmental events other than mere transcriptional noise. Functions of lncRNAs at the biological level in plants have been concluded and discussed in various articles (Liu et al., 2015; Rajagopalan et al., 2006; Xu et al., 2016). Moreover, growing evidence demonstrates that lncRNAs are involved in stress response. According to Xuan et al. (2015), 1060 stress-related lncRNAs were harvested and recorded in 43 plant species under 17 abiotic or biotic stresses in the PLNlncRbase database. Five mechanisms have been described by which plant lncRNAs perform their functions in stress response. Firstly, some plant lncRNAs have the ability to function as ceRNAs, i.e., that are specifically targeted by specific miRNAs through target mimicry, preventing miRNAs from binding to their targets.

For instance, in Arabidopsis, IPS1, whose expression is a product of phosphate (Pi) deficiency, functions as a miR-399 target mimic, protecting the miR-399 target gene encoding PHOSPHATE2 (PHO2), which binds PHO1 and controls Pi homeostasis (Franco-Zorrilla et al., 2007). Furthermore, recently, lncRNAs have been reported to compete with genes to carry out significant roles in rice under Pi deficiency by creating a ceRNA network (Xu et al., 2016). Similarly, in tomato TYLCV infection, many lncRNAs are tomato miRNA target mimics (Wang et al., 2015). Secondly, some lncRNAs in plants can be precursors of sRNAs, generating sRNAs, such as siRNAs and miRNAs. Thirteen lncRNAs have been identified in a recent study as precursors for 96 miRNAs for *Brassica napus* resistance against *Sclerotinia sclerotiorum* infection (Joshi et al., 2016). A total of 14 lncRNAs harbouring miRNA precursor sequences were found in N deficit population by Chen et al. (2016); 9 lncRNAs acted as precursors of 11 reported and 5 of 14 new miRNAs. Three lncRNAs played precursor roles of miR-2004 and miR-2066 in another study on wheat, and 16 lncRNAs were found as precursors of 97 siRNAs targeted by powdery mildew infection (Xin et al., 2011). Thirdly, plant antisense lncRNAs respond to stress by interacting with sense mRNAs.

Our recent study found 828 lncRNAs participating in NAT pairings in *Arabidopsis thaliana*. Conversely, our previous work built the PlantNATsDB, which includes plant natural antisense transcripts (NATs) in 70 plant genomes (Chen et al., 2012). A number of gene expression regulatory mechanisms, including transcription, genomic imprinting, RNA stability, chromatin modification, transcription interference and alternative splicing, have been implicated with asRNAs (Wight & Werner, 2013). Various reports have demonstrated the involvement of antisense lncRNAs in stress responses. For example, five antisense lncRNAs were identified in poplar under N deficiency (Chen et al., 2016), and the same has been found in drought-stressed maize, stripe rust pathogen-infected wheat and light-stressed Arabidopsis as well (Zhang et al., 2014). Antisense lncRNAs have the potential to form double-stranded RNA duplexes with sense mRNAs, thereby influencing gene expression on the complementary strand (Wang et al., 2014). According to Wang et al. (2016), some plant lncRNAs are involved in lncRNA-mediated chromatin modifications (lncR2Epi) regulatory networks.

In Arabidopsis stressed under cold conditions, it is noticed that expression of FLOWERING LOCUS C (FLC) is downregulated through expression of two lncRNAs induced during cold stress, namely COOLAIR and COLDAIR, using the lncR2Epi pathway (Csorba et al., 2014). FC and COLDAIR are MADS-box transcription factors that repress flowering in cold conditions (Michaels and Amasino, 1999). In FLC, however, COOLAIR can mediate a decrease in H3K36me3 or H3K4me2 [92]. Conversely, COLDAIR promotes accumulation of H3K27me3 at FLC by directly binding to the polycomb repressive complex 2 (PRC2) (Csorba et al., 2014). Finally, RNA-directed DNA methylation (RdDM) plays a significant role in the stress response pathway through lncRNAs.

The RdDM pathway involves the de novo cytosine methylation by the action of 24-nt siRNAs and lncRNAs. RNAPIV-dependent transcripts produce siRNAs that initiate the RdDM complex. The siRNAs assemble a silencing complex by binding to Argonaute 4 (AGO4) protein. AGO4 is subsequently recruited by RNAPII-dependent lncRNAs that are AGO4-associated siRNAs. DNA methyltransferase domains reorganising methyltransferase 2 (DRM2) are ultimately delivered by the AGO4–siRNA–lncRNA complex, leading to de novo methylation of cytosine in all three contexts. Abiotic and biotic stress tolerance are thought to be significantly influenced by the RdDM pathway. Stress tolerance was highly enhanced by suppressing major RdDM pathway components. For example, under wild-type and SlAGO4A-overexpressing plants, SlAGO4A downregulation, an essential component of the RdDM pathway in tomato, greatly increased salt and drought stress tolerance (Haider et al., 2022; Huang et al., 2016). Conditions are much the same for DNA methyltransferase 1 (MET1) (Steward et al., 2000).

16.4 SMALL INTERFERING RNAs (siRNAs)

During the last few years, several review articles have stressed the important roles of miRNAs in plant stress responses as post-transcriptional regulators of genes (Kumar et al., 2025). Similar to protein-coding genes, stress can trigger up- or downregulation of miRNA expression. For instance, expression of miR-160 is induced, whereas miR-169 expression is repressed in Arabidopsis under nitrogen (N) deficiency (Ijaz et al., 2024; Koroban et al., 2016). Under stress, the altered expression of these miRNAs results in inhibition of plant growth and development. N deficiency, however, triggers the expression of miR-160, which reduces the expression of its target genes encoding auxin-response factors (ARFs). ARF17 is a regulator of Gretchen Hagen3 (GH3)–like early auxin-response genes, whereas ARF16 is involved in root cap cell formation. Through suppressing plant growth, downregulation of ARF16 and ARF17 would increase stress tolerance. When miR-169 is downregulated, however, its target gene family encoding the nuclear factor Y (NFY) is in increased quantities. NFY controls the genes that encode the nitrate transporters AtNRT2.1 and AtNRT1.1 by binding to their promoter regions (Singh et al., 2023; Singroha & Sharma, 2023). Plant species also possess miRNAs that are involved in stress adaptation. The response to maize N deficiency, for example, includes a range of miRNAs associated with Arabidopsis N deficiency, including miR-160, miR-169, miR-171, miR-395, miR-397, miR-398, miR-399, miR-408 and miR-827 (Ijaz et al., 2024). Their expression profiles were the same (miR-160 upregulated, while the others downregulated). Two crop species, however, were discovered to have opposite miRNA expression profiles when under stress. Although miR-169 expression was downregulated in Arabidopsis but upregulated in rice, miR-156, miR-319 and miR-396 were upregulated in Arabidopsis but suppressed in rice in the case of drought stress (Figure 16.3) (Sabana et al., 2020; Wang et al., 2017). Several miRNAs tend to be involved with various different stressors. On the other hand, various different stressors have a tendency to induce or suppress the generation of some miRNAs. For instance, at least seven stressors have been reported to induce Arabidopsis to produce miR-393. On the contrary, biotic stress, salt, cold and ABA treatment repressed the expression of miR-398. Nevertheless, there exist miRNAs that respond oppositely towards specific stressors. For instance, in Arabidopsis, miR-169 expression is repressed in drought, heat and ABA treatment but stimulated by salt, cold and UVB irradiation (Basso et al., 2019; Olejniczak et al., 2018). The stress-dependent abnormal expression of miRNA occurs under stressful conditions. That is, a particular stressor defines the pattern of miRNA expression. Furthermore, it has been demonstrated that the abnormal expression of miRNAs induced by stress is tissue-dependent. Eldem et al. investigated the miRNA expression patterns in drought-stressed root and leaf tissues. They revealed that peach root cells have more aberrant production of miRNAs than in the leaf tissue, indicating susceptibility to dehydration for the roots (Ludwig et al., 2016; Wang et al., 2017). Moreover, several investigations have also indicated that expression levels of particular miRNAs may vary depending on the organs when under stress. Although the levels of miR-156, miR-171 and miR-408 were elevated in leaves but not in roots, the

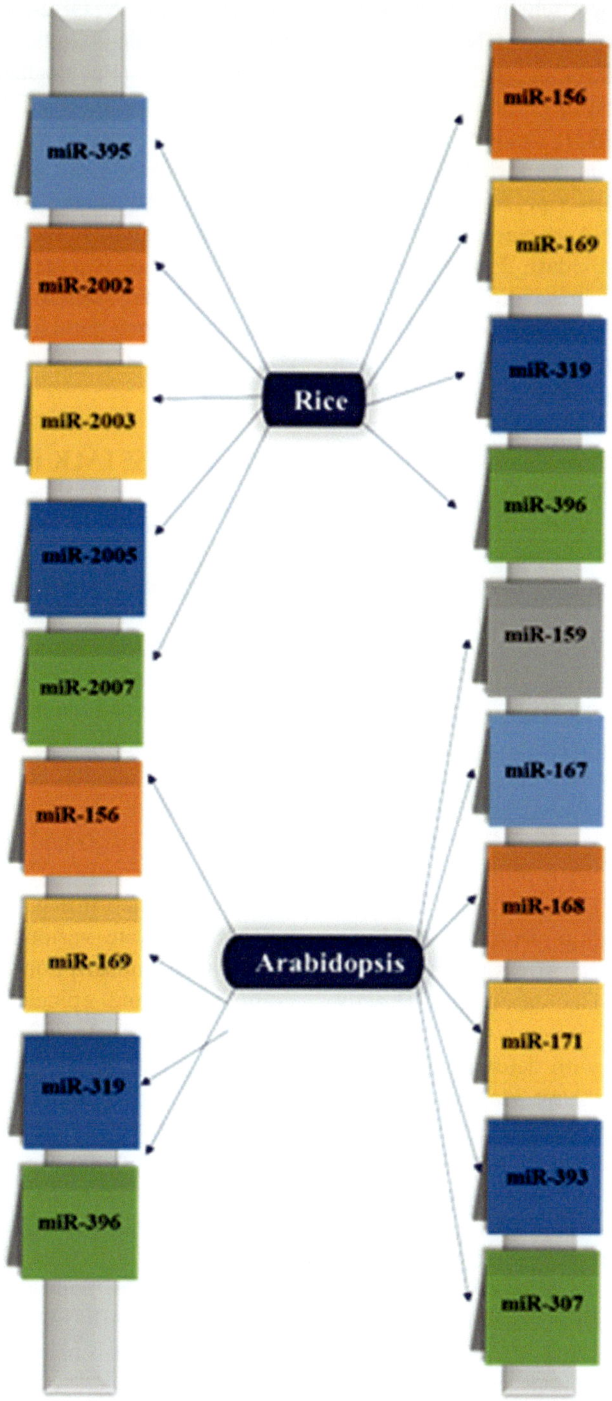

FIGURE 16.3 This figure illustrates the roles of miRNAs in gene regulation and adaptation between monocot and dicot plants. It shows the conserved (e.g., miR-156, miR-319) and species-specific miRNAs in rice and Arabidopsis. While shared miRNAs imply evolutionary conservation, unique miRNAs (rice: miR-2002/2005; Arabidopsis: miR-159/167) may regulate distinct traits.

expression level of miR-166 was elevated in leaves but reduced in roots. The siRNAs are responsible for various processes involved in the development of plants, including the role in environmental stress regulation and resistance to pathogens. siRNAs function through silencing specific genes, allowing plants to control growth and developmental processes and respond appropriately to environmental stresses and virus infections. Additionally, by silencing transposons, siRNAs defend the plant genome from the uncontrolled mobilisation of genetic elements (Figure 16.3).

Thus, siRNAs promote plant adaptation by augmenting genetic stability and survival. The following are the primary mechanisms through which siRNAs operate at the plant level. Triggered by siRNA, the so-called environmental memory or plant immune memory, which is transmitted epigenetically, is thought to be formed by siRNAs that arise as a reaction to environmental stresses and preserve the information gained during the lifetime of an organism in its offspring (Kryovrysanaki et al., 2021; Sanan-Mishra et al., 2021). Environmental adaptation is supported by siRNAs by degradation or silencing of transcripts for abiotic stress tolerance. *Hordeum vulgare* siRNAs have been described to target HvCKX2.1 (cytokinin oxidase 2.1) to modulate plant responsiveness to drought stress during grain fill. Cytokinin oxidase is responsible for cytokinin breakdown and control over germination timing, and higher levels of DNA methylation within the HvCKX2.1 promoter area and higher levels of 24 nt siRNA production under drought stress during the immature grain period affect germination rate and emergence of shoots through the accumulation of cytokinin ribosides (Surdonja et al., 2017; Yadav et al., 2024). It was determined that TAS3-tasiRNAs play a role in the stress adaptability of *Bruguiera gymnorrhiza*. In *H. vulgare* and *Triticum aestivum*, 24 nt phasiRNAs accumulate prior to meiosis initiation, and AGO6 proteins are supposed to bind them. Under environmental stress, the phasiRNAs ensure male fertility (Liu et al., 2020; Wen et al., 2016). The nat-siRNAs are involved in Arabidopsis osmoprotection against salt and oxidative stress, which corresponded to the processing of nat-siRNAs derived from SRO5 and P5CDH gene transcripts. During salt stress, a 24 nt nat-siRNA complementary to the SRO5 mRNA is produced; this produces a 24 nt nat-siRNA, which cleaves the P5CDH transcript, which results in accumulation of proline and hence increases the plants' tolerance to salt stress (Borsani et al., 2005; Chinnusamy et al., 2007). Decreased P5CDH activity results in the accumulation of poisonous P5C, which is detoxified by the SRO5 protein with detoxifying activity in the mitochondria. Researchers have identified that the nat-siRNA SRO5-P5CDH is involved in a regulatory cycle that contributes to generating ROS and salt stress response (Borsani et al., 2005; Chinnusamy et al., 2007). The major function of siRNAs is to silence viral RNAs.

However, siRNAs are crucial in preventing cellular infection, as indicated in the recent discovery of cross-species RNA interference (RNAi). Current research shows that endogenous plant siRNAs have a unique mode of action, particularly in suppressing fungal infections through pathogen gene silencing (Motamedi et al., 2024; Zeng et al., 2019).

This research demonstrates that siRNA tools function as random 'shotgun' weapons to effect effective gene treatment of target pathogen genes (Diallo & Provost, 2020). Scientists have not yet comprehended the full mechanism of RNAi in different animal species. The fungus cells receive plant-derived siRNAs that use gene silencing to disable the disease-causing genetic sets in the microorganism. The application of this method enables plants to protect themselves better from fungal assaults, resulting in reduced transmission of illnesses and infections (Mann et al., 2023; Mapuranga et al., 2022). The identification of the first siRNA from plant sources regarding defence against biotic stress occurred with ATGB2-nat-siRNA. Evidence shows that siRNA molecules handle gene expression linked to pathogen resistance (Puchta-Jasińska et al., 2025; Ruiz-Ferrer & Voinnet, 2009). Plant biotic stress tolerance research reached its most vital discovery through the identification of nat-siRNA/ATGB2. Plants utilise the essential regulatory mechanism which siRNA provides for their defensive mechanisms against ETI-induced resistance. Research on plant inoculation resistance has advanced because scientists now modify nat-siRNA and ATGB2 to enhance protective abilities against pathogenic attacks (Bandameedi et al., 2022; Pandey et al., 2012; Shukla et al., 2008).

16.5 CROSSTALK BETWEEN NON-CODING RNAs AND OTHER STRESS RESPONSE PATHWAYS

Plants maintain sophisticated protective systems which defend them from biological hazards and environmental threats, including pathogens as well as excessive heat stress and dry conditions and salt contamination. ncRNAs constitute essential elements of plant stress responses because they control multiple stress response regulatory processes. Plants enable adaptation to adverse situations by uniting stress response components with ncRNA, according to Nawaz et al. (2023) and Umer et al. (2021). The three primary categories of ncRNAs which achieve stress response in plants include miRNAs, lncRNAs and siRNAs. These ncRNAs activate various stress response regulatory processes and transcription factors and hormone signalling components and epigenetic transformations (Jha et al., 2020; Mattick et al., 2023).

As regulatory RNAs of 20–24 nucleotide length, the main function of miRNAs is to direct messenger RNA (mRNA) to translational silencing or degradation pathways. During stress responses miRNAs mainly target components of transcription factors and stress-responsive genes as well as hormone signalling pathway elements (Catalanotto et al., 2016; Valinezhad Orang et al., 2014). For instance, by regulating the major phosphate uptake and translocation player PHO2, miR-399 plays a role in maintaining phosphate homeostasis. PHO2 is suppressed, and phosphate uptake is enhanced with increased expression of miR-399 under phosphate starvation (Li et al., 2016; Yang et al., 2024). Notably, hormone signalling pathways and miRNAs communicate. One of the key hormones in abiotic stress responses, abscisic acid (ABA), controls stress tolerance through targeting multiple miRNAs. For example, miR-160 modulates auxin-response factors (ARFs) that affect ABA sensitivity, and miR-159 targets MYB transcription factors participating in ABA signalling. This kind of interaction between hormone signalling and miRNAs offers precise control over plant stress responses (Figure 16.4) (Singh et al., 2021; Zhang et al., 2022).

Plants activate complex protective mechanisms to fight pathogen intrusions and three major types of environmental

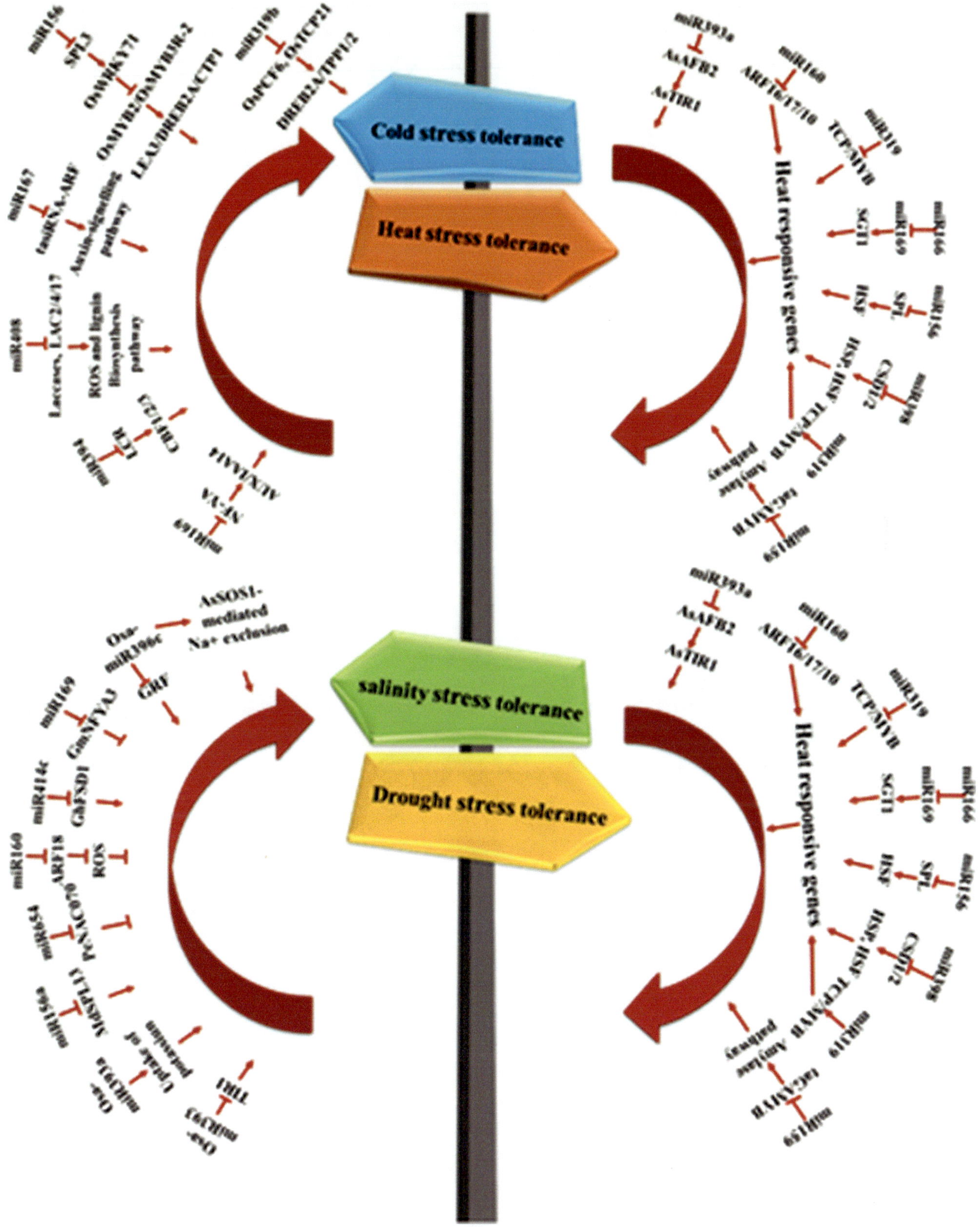

FIGURE 16.4 Schematic representation of the role of non-coding RNAs (ncRNAs) in mediating plant stress tolerance. The figure highlights the interplay between various stress responses, including cold, heat, salinity and drought stress. Arrows indicate the interactions among ncRNAs and their respective target genes associated with each stress tolerance mechanism. This illustrates the complex regulatory networks whereby ncRNAs contribute to enhancing plant resilience against environmental stressors.

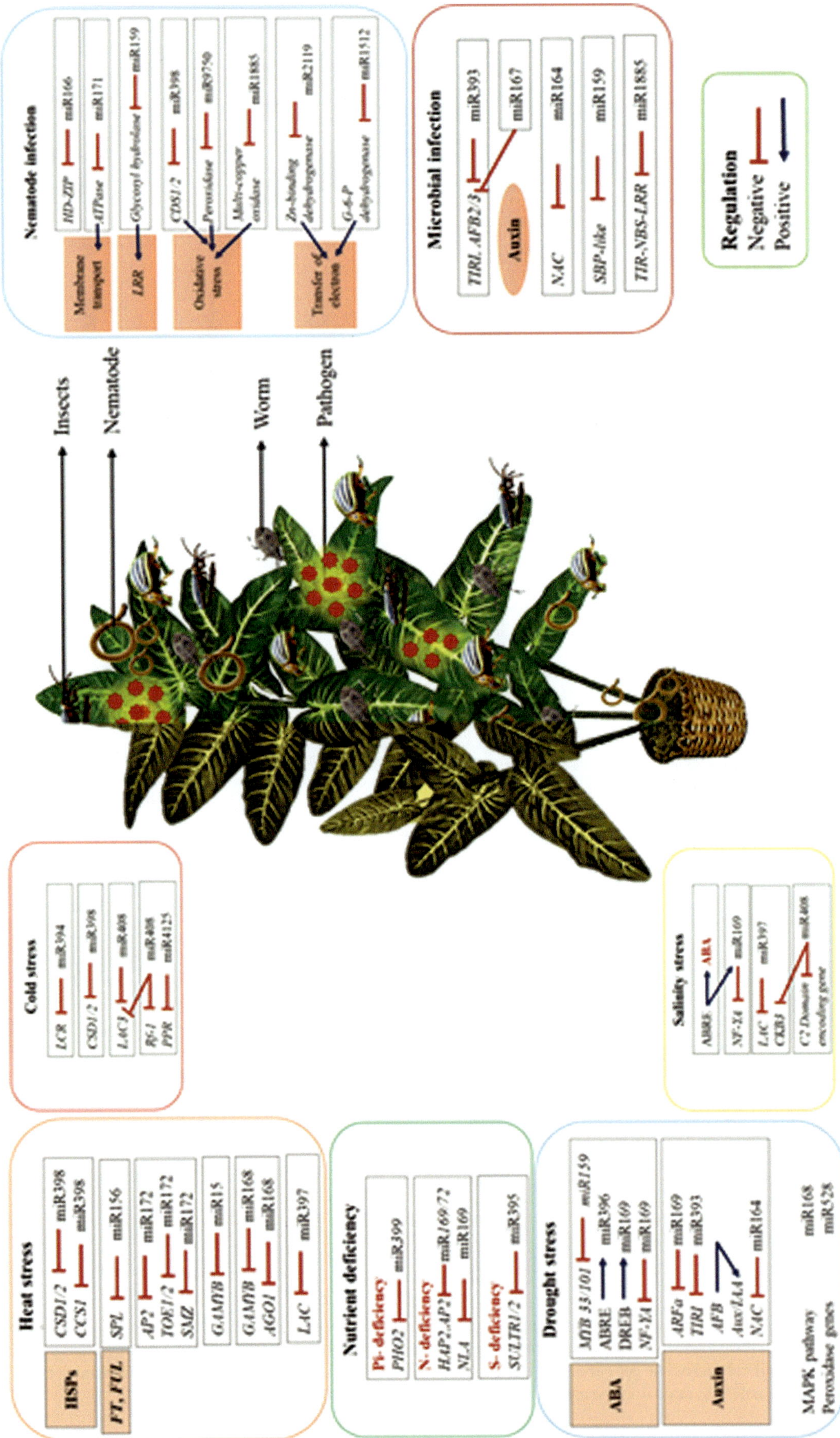

FIGURE 16.5 Mechanisms of plant responses to various stresses. This illustration depicts a plant exhibiting responses to heat stress, cold stress, nutrient deficiencies, drought stress, salinity stress and pathogen infection. Each stress category is associated with specific components, such as heat shock proteins (HSPs), various target genes and regulatory miRNAs. Arrows indicate the nature of regulation (negative or positive) of these miRNAs on the target genes, emphasising the intricate network of gene regulation involved in plant stress response mechanisms.

threats such as drought and salt stress and extreme temperature stress. Multiple plant stress response pathways follow the dominance of ncRNAs because these RNA molecules control various stress response mechanisms through their functional interactions. Plants can adapt to stress through combined activities of ncRNAs together with other stress response systems, as per Nawaz et al. (2023) and Umer et al. (2021). The three active ncRNA components that mediate plant stress responses include miRNAs, lncRNAs and siRNAs. Stress response functions receive their main inputs through the operation of transcription factors together with hormone signalling pathways and epigenetic changes as well as other stress-related pathway components (Jha et al., 2020; Mattick et al., 2023).

The fundamental regulatory function of miRNAs exists through their small RNA molecule status that operates between 20 and 24 nucleotides long and silences or destroys messenger RNAs (mRNAs). Stress reactions by miRNAs occur when these molecules select their target factors from among transcription elements, stress-responsive genes and hormone pathway molecule components, as described by Valinezhad Orang et al. (2014) and Catalanotto et al. (2016). PHO2 is suppressed, and phosphate uptake is enhanced with increased expression of miR-399 under phosphate starvation (Li et al., 2016; Yang et al., 2024). Notably, hormone signalling pathways and miRNAs communicate. One of the key hormones in abiotic stress responses, abscisic acid (ABA), controls stress tolerance through targeting multiple miRNAs. For example, miR-160 modulates auxin-response factors (ARFs) that affect ABA sensitivity, and miR-159 targets MYB transcription factors participating in ABA signalling. This kind of interaction between hormone signalling and miRNAs offers precise control over plant stress responses (Figure 16.5) (Singh et al., 2021; Zhang et al., 2022).

16.6 TECHNIQUES FOR STUDYING NON-CODING RNAs IN PLANT STRESS TOLERANCE

ncRNAs are the central regulators of the processes involved in stress tolerance in plants, and numerous methods have been developed to analyse their activity, expression pattern and interaction. The research methods exist in two categories which include research that depends on computation and experimental work. Each research method creates distinct knowledge regarding the defence mechanisms of plant cells against stress through ncRNAs. Through complex automated RNA-Seq methods, researchers gain access to study the transcriptome that contains ncRNAs. The research method enables the detection of new ncRNAs by producing extensive data about RNA expression levels. This process is described by Nejat and Mantri (2018) and Jha et al. (2020) together with Luk et al. (2014) as using bioinformatics software to examine RNA-Seq data for identifying ncRNA functions along with probable targets. Short RNA sequencing (sRNA-Seq) which operates as an improved form of RNA-Seq allows scientists to detect and quantify miRNAs and siRNAs as two examples of ncRNAs. The investigation of ncRNA regulatory control in plant stress responses depends particularly on this method, according to Vivek and Kumar (2021) and Liu et al. (2021). The sRNA sequences reveal newly discovered stress-tolerant small ncRNAs together with stress-specific expression patterns and the complete processing and origin pathways for small ncRNAs. Begum (2022) and Pei et al. (2017) confirmed that qRT-PCR establishes an accurate and specific methodology for ncRNA expression level measurement.

The stress-inducible ncRNA molecules from high-throughput screens undergo qRT-PCR verification followed by measurement of expression changes in these ncRNAs under different plant tissues and stress or time-dependent stress responses. Research on plant ncRNA tissue distribution and cellular localisation depends on in situ hybridisation methods as described by Peng et al. (2019) and Kupper et al. (2007). Under stress circumstances researchers can observe ncRNA production in various cell types and track the physical localisation of ncRNA molecules in plant tissues through the application of tagged probes which specifically detect ncRNAs. Epigenetic stress investigation has become possible through ChIP and ChIP-Seq technology to study ncRNA–chromatin associations, therefore advancing stress response comprehension (Mishra & Kanduri, 2019; Ounzain et al., 2014; Ye et al., 2020). The method utilises stress-induced chromatin remodelling mechanics together with chromatin modification enzymes to study both ncRNA-related chromatin transformations and the direct networking of ncRNAs. The study of ncRNA stress tolerance requires investigation of RNA–protein relationships through the implementation of RNA immunoprecipitation (RIP) combined with RIP-Seq methods according to Nickerson and Momen-Heravi (2024) and Bure et al. (2022).

These exploratory methods could help scientists determine the following: (1) how ncRNAs bind with proteins, (2) how ribonucleoprotein complexes form during stress responses as well as (3) ncRNA contributions to stress signalling pathways. GWAS studies can verify the genetic variations of stress tolerance characteristics alongside ncRNAs in plants. The approach defines which regions regulate stress-responsive regulatory networks while also exposing genetic stress factors that ncRNAs control but also detects regions related to stress tolerance in ncRNAs. Functional genomics research have undergone considerable advancement in studying ncRNAs using the CRISPR–Cas9 technology. This technique allows investigators to produce overexpression lines along with knockout/knockdown lines of individual ncRNAs and regulatory region modifications for stress tolerance analysis (Gill et al., 2024; Goyal et al., 2017; Javid et al., 2022). The artificial miRNA (amiRNA) technology produces effective gene and ncRNA silencing. Researches achieve stress-resistant plants through alterations in ncRNA which enables studies into stress-related contributions of select ncRNAs while revealing their gene silencing mechanisms under stress conditions (Bravo-Vázquez et al., 2025; Ossowski et al., 2008). This method enables structural predictions for ncRNAs and functional sequence identification as well as target identification along with interaction network detection which leads to stress response model development through ncRNA functional application.

The increasing application of Oxford Nanopore and PacBio in long-read sequencing technology research enables scientists to explore new possibilities of studying ncRNA (Roychowdhury et al., 2023; Zhong et al., 2021). lncRNA discovery and delineation together with annotation of new splice variants and isoforms become possible through these research approaches that additionally help explain complex ncRNA structures. Multiple sequencing approaches developed a systematic method to find out how ncRNAs control plant stress tolerance while creating opportunities for scientists to discover plant stress adaptation mechanisms (Chillón and Marcia, 2020; Faraz et al., 2024, Jovic et al., 2022).

16.7 THE APPLICATIONS OF ncRNA IN CROP DEVELOPMENT FIELD

The regulatory function of ncRNAs makes them suitable candidates for crop improvement work due to their important role in gene expression regulation. Science professionals have developed innovative approaches to harness ncRNA capabilities after scientists gained better understanding of its biological properties. This section identifies two major approaches for crop retention through ncRNA utilisation, including marker-assisted selection (MAS) as well as genetic engineering techniques.2055

16.7.1 ncRNA Genetic Engineering Techniques

Through genetic engineering methods, researchers can develop excellent crop properties, including yield production enhancement together with stress resistance attributes and improved nutrient production by modifying the expression patterns and activities of ncRNA in plants. Adjusting the expression of particular ncRNAs represents an essential method to develop superior crops guided by ncRNAs. The desired ncRNA sequence requires transgenes for manufacturing, while RNA interference (RNAi) serves to decrease ncRNA amounts. The addition of overexpressed miR-399 in tomatoes produced better phosphate handling for plant growth under low-phosphate conditions. The suppression of miR-156 in switchgrass plants enhanced biomass production and ethanol generation, according to Jovic et al. (2022) and Zhou et al. (2020).

The genome editing field has evolved through practical and economical methods that CRISPR/Cas9 can use to modify ncRNA sequences and their regulatory elements. Scientists utilise this method to build better manufacturing approaches while also setting the target specificity for nucleic acid regulators back to normal. Awareness about expression manipulation and trait selection was achieved through target gene miRNA-binding site redesign using CRISPR/Cas9 systems.

Scientists altered the miR-156 binding site of the rice IPA1 gene to enhance both plant structure and grain production in specific plants (Bravo-Vázquez et al., 2024; Li et al., 2018; Yang et al., 2018). The creation of synthetic ncRNA molecules creates a new method to advance crop development. Research tools developed from synthetic ncRNAs enable the generation of gene-targeting entities through defined activities that can produce desired phenotypes. Research has shown that using transgenic artificial miRNAs (amiRNAs) creates new possibilities to establish viral resistance in multiple crops. Scientific research proves that artificial miRNAs targeting the coat protein gene of *Turnip mosaic virus* (TuMV) can provide resistance to Brassica and Arabidopsis crops (Bravo-Vázquez et al., 2024; Carbonell, 2019; Wekesa et al., 2019). The epigenetic controlling phenomenon made possible by ncRNAs utilises DNA methylation and histone modification as its main mechanisms.

The modification of ncRNAs operating in these molecular processes creates heritable epigenetic modifications that do not modify DNA sequences. Research shows that DNA methylation frequencies change at specific loci when scientists overexpress particular lncRNAs, leading to modifications in gene expression levels and subsequent behavioural changes. Aquatic biologists have utilised this method to develop crops by changing developmental patterns and increasing stress tolerance, according to Davalos et al. (2019) and Prandi et al. (2022).

The developed research enables scientists to construct genome-editing methods that could function alongside ncRNA technology management systems through the implementation of rnADirected Nucleases operated by Cas13 to adjust problematic ncRNA sequences found in RNA transcripts. Plant scientists develop nutrient-rich resistant plants through CRISPR/Cas technology, according to Praveen et al. (2023) and Ma et al. (2022).

16.7.2 Marker-aided ncRNA Selection

Organisations use the MAS technique as a robust method for selecting plants based on markers for desirable features. Through inclusion of ncRNA-related markers in MAS methods, researchers can enhance crop development by increasing programme precision and efficacy (Bhandawat et al., 2020; Singh et al., 2023). The field of plant breeding and genetics achieves supplemental research from markers derived from ncRNAs. Differences in ncRNA sequence components or regulatory elements serve as these markers that link with particular desirable characteristics. The biological processes of growth and gene control and stress response function through the vital role of significant RNA molecules known as ncRNAs. The application of ncRNA-based markers during breeding programmes leads to multiple benefits in selecting desired traits. Among the most common molecular markers derived from ncRNAs are single nucleotide polymorphisms (SNPs) together with insertions/deletions (indels). SNPs in DNA consist of sequence modifications that involve one base pair, while insertions/deletions (indels) describe DNA modification through both addition and deletion of multiple base pairs. The development of PCR markers from sequence variations allows affordable and quick screening of breeding populations where phenotypic changes affect ncRNA genes and regulatory elements. Marking DNA by converting ncRNA-associated SNPs and indels requires implementation of multiple steps. The first technique for detecting sequence variations involves whole-genome sequencing or focused sequencing of ncRNA genes. Primers were designed specifically after researchers identified sequence variations for amplification of regions that encompassed the SNP or indel. Genetic analysis using DNA primers with selected breeders enables identification of desired traits related to particular ncRNA sequence variations when testing a population through PCR methods (Davalos et al., 2019; Jaiswal et al., 2020).

A range of ncRNA molecular markers proves effective at improving crops, including specifically rice cultivation through breeding programmes. The genetic variations in miRNA sequences link to key agricultural traits such as grain dimensions together with quality attributes. The expression of genes after transcriptional completion depends on miRNA regulation because these tiny ncRNAs participate as miRNA regulators. Plant development and growth-related gene expression seems to be under sequence control of miRNA elements and their targets. By analysing these miRNA-associated SNPs, breeders gain the ability to select rice varieties which have greater yields and improved quality content. NcRNA-based molecular markers include different types of ncRNAs besides miRNAs. Different plant biological processes rely on lncRNAs for their regulation, including blooming time control and plant stress response. Scientific research demonstrates that plant selection regarding stress tolerance and flowering timing characteristics becomes attainable when using sequencing markers of lncRNA along with regulatory markers (Bhogireddy et al., 2021; Ku et al., 2022; Sarfraz et al., 2025).

The research of complex phenotypes controlled by multiple genes or regulatory networks can benefit significantly from ncRNA-based molecular markers. These markers serve to understand complex phenotype genetics due to ncRNA regulatory functions and cellular process involvement (Frías-Lasserre & Villagra, 2017; Kaikkonen et al., 2011). The information regarding ncRNA-based markers helps breeders who wish to improve difficult-to-identify traits using traditional phenotypic selection practices. Scientific groups can apply sequence changes from ncRNA to create both affordable and efficient markers that aid trait identification. These genetic markers improve selection processes because they show information about how important agricultural traits are genetically controlled. Crop breeding applications of ncRNA-based markers will expand because we gain better insights into ncRNA regulation and function (Asghar et al., 2024; Liu and Zhu, 2014). Expression-based markers serve plants and breeders as valuable tools to discover and pick beneficial crop features in genetics and breeding programmes. The technological foundation of these markers relies on specific ncRNAs because their expression levels influence gene expression and cell function despite their inability to produce proteins. Scientists along with breeders can improve their understanding of molecular mechanisms which affect stress tolerance and disease resistance and yield potential through the utilisation of expression-based markers (Michelmore et al., 2017; De Los Reyes, 2019).

Complex molecular methods enable researchers to determine the expression levels of particular ncRNAs through expression-based markers. qPCR adopts fluorescent dyes or probes to track the real-time amplification of target RNA sequences, allowing scientists to precisely determine sample RNA molecule concentrations. RNA sequencing technologies provide researchers with extensive assessment of ncRNA expression throughout the entire transcriptome. Researchers can detect and measure numerous ncRNAs through high-throughput technologies in one operation which provides them a basic understanding of gene expression patterns (Abbas et al., 2016; Jorge et al., 2012). The expression marker technology proved valuable in agricultural development particularly through its applications in drought-resistant maize line development. The essentiality of enhancing water usage efficiency stands clear for maintaining food security because drought stress constitutes a major obstacle to global agricultural productivity. Scientists achieved vital findings about drought-resistant genetic markers through their research on short ncRNA levels of miRNAs during stress-induced drought in maize plants. miRNAs demonstrate important value as selection and breeding indicators for improved water usage efficiency and drought-resistant maize line breeding because of their significant associations (Costa et al., 2013; Mishra et al., 2016; Palande et al., 2025).

Expression-based markers supply a superior package of advantages when compared to traditional DNA-based markers. Expression markers help researchers establish straight links between genetic expressions and observable characteristics of phenotypes, thus aiding prediction of plant behaviours under particular environmental situations. Generation of expression-based indices allows researchers to calculate gene dynamics by monitoring plant responses to stimuli across development since expression-based indices detect plant reactions at different times. Static DNA markers lose their ability to detect spatial and time-related characteristics following their identification process. Such evaluation produces a robust method to uncover how ncRNAs regulate complex phenotypes in order to interpret genetic control mechanisms of essential agronomic characteristics (Figure 16.6) (Kumar et al., 2022; Lang et al., 2015; Yin et al., 2003).

Various genetic science and breeding operations benefit from implementing ncRNA target site markers for their

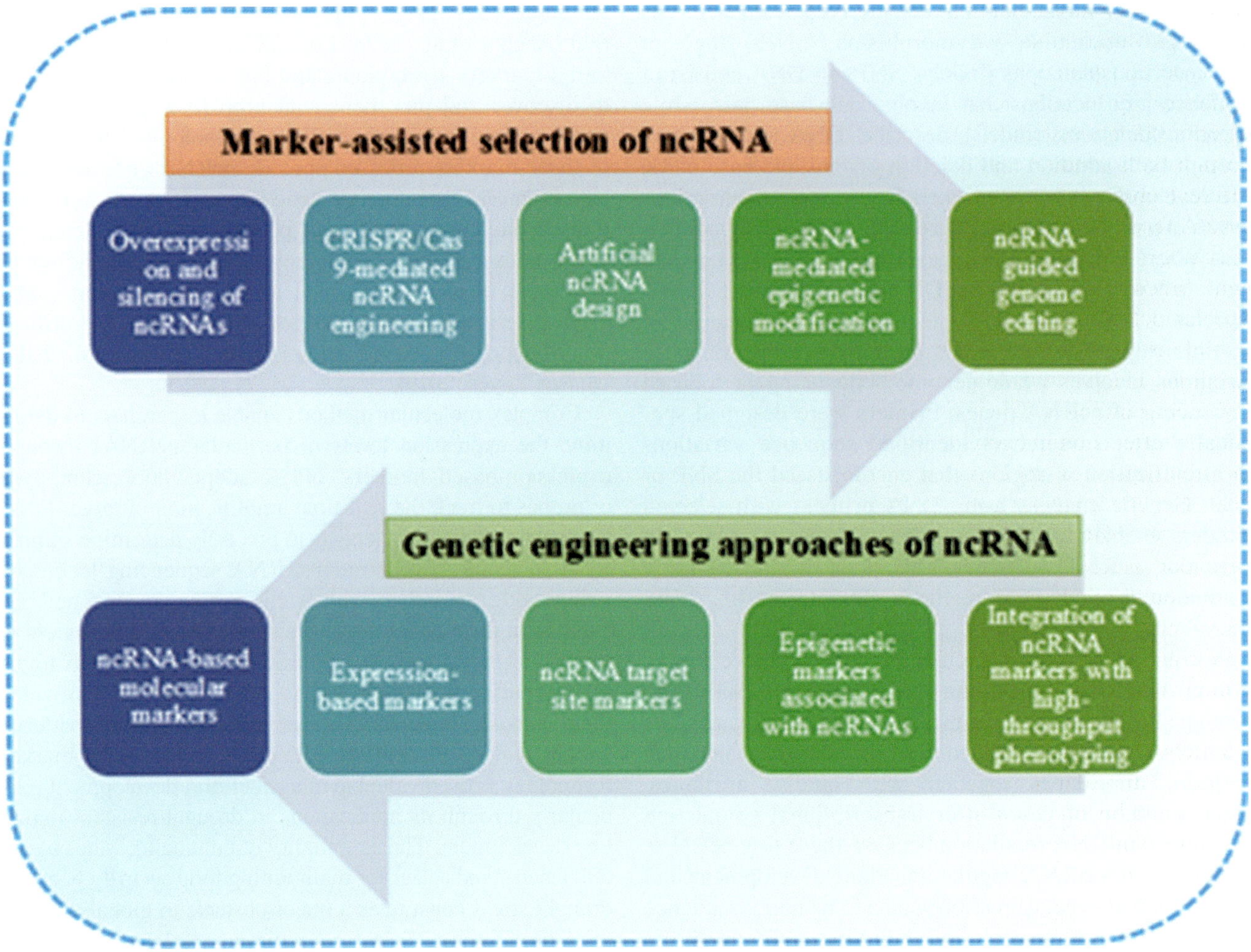

FIGURE 16.6 The paper presents two fundamental methods of applying ncRNA strategies in crop improvement which include marker-assisted selection of ncRNA and genetic engineering approaches of ncRNA. The first approach includes both overexpression and CRISPR/Cas9-guided engineering together with various ncRNA design and engineering methods. The application section of the article focuses on molecular markers while discussing expression-based markers together with high-throughput phenotyping integration with ncRNA markers.

capacity as a powerful tool. The markers were established by scientists who studied multiple targets in non-coding ribonucleic acids (ncRNAs) that control gene expression but lack protein production abilities. The high usefulness for plant breeders exists in these markers because they reveal modifications of plant traits triggered by ncRNA regulatory systems. Alternative genotypes in the sequence of ncRNA target sites significantly influence the binding performance of ncRNA between its target site and the relevant gene. Evolutionary changes in plant populations as well as different breeding techniques generate these sequence variants. Researchers and breeders can establish useful selection markers based on these polymorphisms by determining their specific traits (Budak et al., 2020; Davalos et al., 2019; Yi et al., 2015).

This method showcases outstanding advantages because it enables direct associations to form between genetic regulatory changes and target site markers for ncRNA. The identification of true causal trait variants which affect expression becomes more feasible through these ncRNA target site markers because they act differently from other markers linked to traits but without direct causality. Breeding plans achieve more effective selection possibilities through markers because of their direct link to functional changes. The wheat breeding programme demonstrates how effective ncRNA target site markers can be when examining the wheat miR172 binding site variance of the Q gene that impacts spike density and threshability attributes. Selection of wheat cultivars with enhanced grain quality became possible through breeder identification of markers linked to this specific polymorphism which leads to improved yields and quality products (Bhandawat et al., 2020; Gupta et al., 1999; Rasheed and Xia, 2019).

Many researchers have employed ncRNA target site markers for various crop species after their original development in wheat. Research on rice has established a connection between miRNA target site polymorphisms and grain size variation as well as resistance against diseases. The type of ripening that occurs in tomato fruits shows a relationship to polymorphisms found at miRNA target regions. Plant breeders use marker identification of polymorphisms to conduct easy selection of desired traits across different crops. The development and implementation of ncRNA target site markers became possible when molecular biology united with high-end genomics and traditional plant breeding methods. The approach for developing superior ncRNA-based markers appears promising as we expand our knowledge about gene regulations and ncRNA functions. The method strengthens fundamental plant biology understanding together with better insights into genetic diversity effects on phenotype manifestation and enhances the effectiveness of plant breeding operations (Budak et al., 2020; Fu, 2014; Hemasai et al., 2024; Singh et al., 2023; Tang et al., 2016). ncRNAs act as regulatory elements in epigenetic processes that refer to heritable changes of gene expression originating from DNA sequence modifications. The maintenance of all epigenetic modifications including DNA methylation and histone modifications happens through ncRNAs. Dynamic epigenetic markers that link with ncRNA functioning could act as valuable identification tools in plant breeding programmes for MAS selection methods. Particular genetic markers like DNA methylation patterns and histone modifications allow breeders to select desirable plant features according to Deniz and Erman (2017) and Heo et al. (2013).

Plant breeders are expected to enhance the selection process together with the creation of improved plant varieties through better traits because the epigenetic markers link to ncRNA functions. Selection of stress-tolerant plants is the key advantage from using epigenetic markers associated with ncRNAs in MAS programmes. Most plants defend themselves from environmental stresses through epigenetic mechanisms facilitated by ncRNA processes. Plant breeders can develop crops which show enhanced resistance to major environmental stressors through selecting proper epigenetic markers linked to stress tolerance, according to Karlik et al. (2019), Sun et al. (2021) and Singroha et al. (2022). Plants with various developmental patterns become selectable through epigenetic markers associated with ncRNAs which control intricate epigenetic processes typically managing plant development. The evaluation of specific markers enables plant breeders to locate plants featuring desired characteristics for development such as plant growth traits and blooming patterns.

The approach depends on ncRNA markers that scientists extract from genomic regions where researchers identified baseline control elements for cellular processes and gene expression without requiring coding proteins. The combination between high-throughput phenotyping technology and the ncRNA markers delivers specific genetic data that scientists and plant breeders use to connect plant characteristics to accelerate better plant-type development. Plant biologists achieve widespread alterations in their plant evaluations through high-throughput phenotyping technologies. Sophisticated imaging techniques, spectroscopic approaches and computer data acquisition systems enable rapid gathering of large phenotypic information on large plant populations. Plant growth, architectural morphology, photosynthetic function, stress response and other morphological and physiological characteristics are all measured in this dataset. Thanks to the simplicity and speed of gathering such large-scale phenotypic data, researchers can identify minute functional variations in plants that may go undetected with conventional observation methods (Li et al., 2021; Xiao et al., 2022; Yang et al., 2020).

Researchers may build strong correlations between genetic markers and particular plant characteristics by cross-referencing ncRNA marker information with the vast phenotypic data obtained by high-throughput phenotyping. This association simplifies the search for ncRNA markers strongly related with desirable traits like greater yield, disease resistance and stress tolerance. The integration of the two strategies provides a more complete knowledge of the hereditary processes underlying complex traits and their expression in the observable characteristics of the plant. By allowing breeders to select more promising breeding lines based on a greater knowledge of breeding techniques, this combinative approach significantly enhances the efficacy of crop development initiatives. By integrating phenotyping data with ncRNA markers, scientists may more reliably predict the performance of several plant lines than by combining subjective evaluations or protracted conventional breeding techniques (Fu & Jiang, 2022; Li et al., 2021; Xiao et al., 2022). This predictive ability enables breeders to save time and money when developing new crop kinds with enhanced characteristics by guiding them to select the most promising candidates.

16.8 CONCLUSION

Plants' primary regulators of stress tolerance are ncRNAs, which provide extra insight on their adaptation to surroundings stimuli. In response to stress, lncRNAs, miRNAs and siRNAs all serve different purposes in modulating gene expression and cellular activities. By targeting particular messenger RNAs, miRNAs help plants to respond more accurately to various signals, therefore speeding up their adaptation. LncRNAs change stress responses by means of chromatin remodelling, ceRNA activity and control of stress-responsive gene expression. SiRNAs promote stress resistance through RNA-directed DNA methylation and post-transcriptional gene silencing, which are essential for transposon silencing and stress memory formation. These molecules' changing expression patterns reveal how ncRNAs control complex response networks in stressful surroundings. ncRNAs are excellent candidates for producing stress-resistant varieties of crops given their adaptability, hence assisting agriculture in overcoming the challenges caused by climate change. Future advances in this field will most likely increase food security and agricultural production. As ncRNAs become more understood, so are their applications in agriculture and biotechnology, and fresh paths for increasing crop productivity and stress tolerance under harsh environmental conditions are revealed.

REFERENCES

(2004). cytoplasmic localization of a nuclear RNA binding protein in *Medicago truncatula. Plant Cell*, *16*, 1047–1059.

Abbas, Q., Raza, S. M., Biyabani, A. A., & Jaffar, M. A. (2016). A review of computational methods for finding non-coding RNA genes. *Genes*, *7*(12), 113.

Abubakar, M., Hajjaj, M., Naqvi, Z. E. Z., Shanawaz, H., Naeem, A., Padakanti, S. S. N., & Faraz, M. A. (2024). Non-coding RNA-mediated gene regulation in cardiovascular disorders: Current insights and future directions. *Journal of Cardiovascular Translational Research*, *17*(4), 739–767.

Ahmad, Y., Haider, S., Iqbal, J., Abbasi, B. A., Yaseen, T., & Mahmood, T. (2022). The mechanisms of genome editing technologies in crop plants. In *Principles and practices of OMICS and genome editing for crop improvement* (pp. 295–313). Cham: Springer International Publishing.

Ahmad, Y., Haider, S., Iqbal, J., Naseer, S., Attia, K. A., Mohammed, A. A., & Mahmood, T. (2024). In-silico analysis and transformation of OsMYB48 transcription factor driven by CaMV35S promoter in model plant–*Nicotiana tabacum* L. conferring abiotic stress tolerance. *GM Crops & Food*, *15*(1), 130–149.

Ahmed, H. G. M. D., Fatima, T., & Khaliq, N. (2024). Non-coding RNA-Mediated Regulation of Genes Expression in Response to Abiotic Stresses. *ncRNAs: Mediated Regulation: The Unarguable Biological Function to Environmental Constrains*, 235–264.

Ali, I., Salah, K. B. H., Sher, H., Ali, H., Ullah, Z., Ali, A., & Mahmood, T. (2022). Drought stress enhances the efficiency of floral dip method of Agrobacterium-mediated transformation in *Arabidopsis thaliana. Brazilian Journal of Biology*, *84*, e259326.

Ariel, F., Romero-Barrios, N., Jégu, T., Benhamed, M., & Crespi, M. (2015). Battles and hijacks: Noncoding transcription in plants. *Trends in Plant Science*, *20*, 362–371.

Asghar, A., Ullah, Z., Ali, T., Iqbal, J., Abbasi, B. A., Ijaz, S., & Mahmood, T. (2024). Introduction to nc-RNA and their crucial role in biological regulation under environmental stress. In *ncRNAs: Mediated regulation: The unarguable biological function to environmental constrains* (pp. 1–46). Springer Nature Switzerland.

Bandameedi, A., Rama Devi, S. J. S., Mamidala, P., & Vutharadhi, S. (2022). Roles of small interfering RNA (siRNA) in modulating plant defense responses. In *Molecular response and genetic engineering for stress in plants, volume 2: Biotic stress* (pp. 8–1). IOP Publishing.

Basso, M. F., Ferreira, P. C. G., Kobayashi, A. K., Harmon, F. G., Nepomuceno, A. L., Molinari, H. B. C., & Grossi-de-Sa, M. F. (2019). Micro RNA s and new biotechnological tools for its modulation and improving stress tolerance in plants. *Plant Biotechnology Journal*, *17*(8), 1482–1500.

Begum, Y. (2022). Regulatory role of microRNAs (miRNAs) in the recent development of abiotic stress tolerance of plants. *Gene*, *821*, 146283.

Beňačka, R., Szabóová, D., Guľašová, Z., & Hertelyová, Z. (2024). Non-coding RNAs in breast cancer: Diagnostic and therapeutic implications. *International Journal of Molecular Sciences*, *26*(1), 127.

Betancur, J. G. (2016). Pervasive lncRNA binding by epigenetic modifying complexes—The challenges ahead. *Biochimica et Biophysica Acta (BBA)-Gene Regulatory Mechanisms*, *1859*(1), 93–101.

Bhandawat, A., Sharma, H., Pundir, N., Madhawan, A., & Roy, J. (2020). Genome-wide identification and characterization of novel non-coding RNA-derived SSRs in wheat. *Molecular Biology Reports*, *47*, 6111–6125.

Bhogireddy, S., Mangrauthia, S. K., Kumar, R., Pandey, A. K., Singh, S., Jain, A., ... and Kudapa, H. (2021). Regulatory non-coding RNAs: A new frontier in regulation of plant biology. *Functional and Integrative Genomics*, *21*(3), 313–330.

Borsani, O., Zhu, J., Verslues, P. E., Sunkar, R., & Zhu, J. K. (2005). Endogenous siRNAs derived from a pair of natural cis-antisense transcripts regulate salt tolerance in Arabidopsis. *Cell*, *123*(7), 1279–1291.

Bravo-Vázquez, L. A., Castro-Pacheco, A. M., Pérez-Vargas, R., Velázquez-Jiménez, J. F., & Paul, S. (2025). The emerging applications of artificial MicroRNA-mediated Gene silencing in plant biotechnology. *Non-Coding RNA*, *11*(2), 19.

Bravo-Vázquez, L. A., Méndez-García, A., Chamu-García, V., Rodríguez, A. L., Bandyopadhyay, A., & Paul, S. (2024). The applications of CRISPR/Cas-mediated microRNA and lncRNA editing in plant biology: Shaping the future of plant non-coding RNA research. *Planta*, *259*(2), 32.

Brosnan, C. A., & Voinnet, O. (2009). The long and the short of noncoding RNAs. *Current Opinion in Cell Biology*, *21*, 416–425.

Budak, H., Kaya, S. B., & Cagirici, H. B. (2020). Long non-coding RNA in plants in the era of reference sequences. *Frontiers in Plant Science*, *11*, 276.

Bure, I. V., Nemtsova, M. V., & Kuznetsova, E. B. (2022). Histone modifications and non-coding RNAs: Mutual epigenetic regulation and role in pathogenesis. *International Journal of Molecular Sciences*, *23*(10), 5801.

Campalans, A., Kondorosi, A., & Crespi, M. (2004). *Enod40, a short open reading frame-containing mRNA, induces.*

Carbonell, A. (2019). Design and high-throughput generation of artificial small RNA constructs for plants. *Plant MicroRNAs: Methods and Protocols*, 247–260.

Catalanotto, C., Cogoni, C., & Zardo, G. (2016). MicroRNA in control of gene expression: An overview of nuclear functions. *International Journal of Molecular Sciences*, *17*(10), 1712.

Cech, T. R., & Steitz, J. A. (2014). The noncoding RNA revolution-trashing old rules to forge new ones. *Cell*, *157*, 77–94.

Chillón, I., & Marcia, M. (2020). The molecular structure of long non-coding RNAs: Emerging patterns and functional implications. *Critical Reviews in Biochemistry and Molecular Biology*, *55*(6), 662–690.

Chinnusamy, V., Zhu, J., Zhou, T., & Zhu, J. K. (2007). Small RNAs: big role in abiotic stress tolerance of plants. *Advances in molecular breeding toward drought and salt tolerant crops*, 223–260.

Cohu, C. M., Abdel-Ghany, S. E., Reynolds, K. A. G., Onofrio, A. M., Bodecker, J. R., Kimbrel, J. A., ..., & Pilon, M. (2009). Copper delivery by the copper chaperone for chloroplast and cytosolic copper/zinc-superoxide dismutases: Regulation and unexpected phenotypes in an Arabidopsis mutant. *Molecular Plant*, *2*(6), 1336–1350.

Costa, M., Nobre, M. S., Becker, J. D., Masiero, S., Amorim, M. I., Pereira, L. G., & Coimbra, S. (2013). Expression-based and co-localization detection of arabinogalactan protein 6 and arabinogalactan protein 11 interactors in Arabidopsis pollen and pollen tubes. *BMC Plant Biology*, *13*, 1–19.

Cui, L., Zheng, J., Lin, Y., Lin, P., Lu, Y., Zheng, Y., …, & Zhao, X. (2024). Decoding the ribosome's hidden language: rRNA modifications as key players in cancer dynamics and targeted therapies. *Clinical and Translational Medicine*, *14*(5), e1705.

D'Ario, M., Griffiths-Jones, S., & Kim, M. (2017). Small RNAs: Big impact on plant development. *Trends in Plant Science*, *22*, 1056–1068.

Davalos, A., Henriques, R., Latasa, M. J., Laparra, M., & Coca, M. (2019). Literature review of baseline information on non-coding RNA (ncRNA) to support the risk assessment of ncRNA-based genetically modified plants for food and feed. *EFSA Supporting Publications*, *16*(8), 1688E.

Decatur, W. A., & Fournier, M. J. (2003). RNA-guided nucleotide modification of ribosomal and other RNAs. *Journal of Biological Chemistry*, *278*(2), 695–698.

De Los Reyes, B. G. (2019). Genomic and epigenomic bases of transgressive segregation—New breeding paradigm for novel plant phenotypes. *Plant Science*, *288*, 110213.

Deniz, E., & Erman, B. (2017). Long noncoding RNA (lincRNA), a new paradigm in gene expression control. *Functional and Integrative Genomics*, *17*, 135–143.

Deogharia, M., & Gurha, P. (2022). The "guiding" principles of noncoding RNA function. *Wiley Interdisciplinary Reviews: RNA*, *13*(4), e1704.

Diallo, I., & Provost, P. (2020). RNA-sequencing analyses of small bacterial RNAs and their emergence as virulence factors in host-pathogen interactions. *International journal of molecular sciences*, *21*(5), 1627.

Eddy, S. R. (2001). Non-coding RNA genes and the modern RNA world. *Nature Reviews Genetics*, *2*, 919–929.

Erdmann, R. M., & Picard, C. L. (2020). RNA-directed DNA methylation. *PLoS Genetics*, *16*(10), e1009034.

Frías-Lasserre, D., & Villagra, C. A. (2017). The importance of ncRNAs as epigenetic mechanisms in phenotypic variation and organic evolution. *Frontiers in Microbiology*, *8*, 2483.

Fu, X., & Jiang, D. (2022). High-throughput phenotyping: The latest research tool for sustainable crop production under global climate change scenarios. In *Sustainable crop productivity and quality under climate change* (pp. 313–381). Academic Press.

Fu, X. D. (2014). Non-coding RNA: a new frontier in regulatory biology. *National Science Review*, *1*(2), 190–204.

George, T. P., Subramanian, S., & Supriya, M. H. (2024). A brief review of noncoding RNA. *Egyptian Journal of Medical Human Genetics*, *25*(1), 98.

Gill, S. S., Khan, N. A., Agarwala, N., Singh, K., Sunkar, R., & Tuteja, N. (2024). ncRNAs in plant development and stress responses. *Plant Physiology and Biochemistry*, 108950.

Goyal, A., Myacheva, K., Groß, M., Klingenberg, M., Duran Arqué, B., & Diederichs, S. (2017). Challenges of CRISPR/Cas9 applications for long non-coding RNA genes. *Nucleic acids research*, *45*(3), e12–e12.

Gupta, P. K., Varshney, R. K., Sharma, P. C., & Ramesh, B. J. P. B. (1999). Molecular markers and their applications in wheat breeding. *Plant Breeding*, *118*(5), 369–390.

Haider, S., Iqbal, J., Naseer, S., Yaseen, T., Shaukat, M., Bibi, H., & Mahmood, T. (2021). Molecular mechanisms of plant tolerance to heat stress: Current landscape and future perspectives. *Plant Cell Reports*, 1–25.

Haider, S., Raza, A., Iqbal, J., Shaukat, M., & Mahmood, T. (2022). Analyzing the regulatory role of heat shock transcription factors in plant heat stress tolerance: A brief appraisal. *Molecular Biology Reports*, *49*(6), 5771–5785.

Hemasai, B., Kumbha, D. K., Modem, V. N., Gannavarapu, S. K., Bommaka, R. R., Mallapuram, S., … & Vemireddy, L. R. (2024). Development of miRNA-SSR and target-SSR markers from yield-associate genes and their applicability in the assessment of genetic diversity and association mapping in rice (*Oryza sativa* L.). *Molecular Breeding*, *44*(4), 30.

Heo, J. B., Lee, Y. S., & Sung, S. (2013). Epigenetic regulation by long noncoding RNAs in plants. *Chromosome Research*, *21*, 685–693.

Hernández, Y., & Sanan-Mishra, N. (2017). miRNA mediated regulation of NAC transcription factors in plant development and environment stress response. *Plant Gene*, *11*, 190–198.

Hirayama, T., & Shinozaki, K. (2010). Research on plant abiotic stress responses in the post genome era: Past, present and future. *The Plant Journal*, *61*, 1041–1052.

Ijaz, S., Iqbal, J., Abbasi, B. A., Yaseen, T., Ullah, Z., Alqahtani, T., …, & Mahmood, T. (2024). nc-RNA: A key player in stress response pathways, a molecular perspective. *In ncRNAs: Mediated regulation: The unarguable biological function to environmental constrains* (pp. 139–163). Springer Nature Switzerland.

Jaiswal, V., Rawoof, A., Dubey, M., Chhapekar, S. S., Sharma, V., & Ramchiary, N. (2020). Development and characterization of non-coding RNA based simple sequence repeat markers in Capsicum species. *Genomics*, *112*(2), 1554–1564.

Javid, S., Bihamta, M. R., Omidi, M., Abbasi, A. R., Alipour, H., & Ingvarsson, P. K. (2022). Genome-Wide Association Study (GWAS) and genome prediction of seedling salt tolerance in bread wheat (*Triticum aestivum* L.). *BMC Plant Biology*, *22*(1), 581.

Jha, U. C., Nayyar, H., Jha, R., Khurshid, M., Zhou, M., Mantri, N., & Siddique, K. H. (2020). Long non-coding RNAs: Emerging players regulating plant abiotic stress response and adaptation. *BMC Plant Biology*, *20*, 1–20.

Jiao, L., Liu, Y., Yu, X. Y., Pan, X., Zhang, Y., Tu, J., …, & Li, Y. (2023). Ribosome biogenesis in disease: new players and therapeutic targets. *Signal Transduction and Targeted Therapy*, *8*(1), 15.

Jin, Q., Chachar, M., Ali, A., Chachar, Z., Zhang, P., Riaz, A., Ahmed, N., & Chachar, S. (2024). Epigenetic regulation for heat stress adaptation in plants: new horizons for crop improvement under climate change. *Agronomy*, *14*(9), 2105.

Jorge, N. A. N., Ferreira, C. G., & Passetti, F. (2012). Bioinformatics of cancer ncRNA in high throughput sequencing: present state and challenges. *Frontiers in genetics*, *3*, 287.

Jouravleva, K., & Zamore, P. D. (2025). A guide to the biogenesis and functions of endogenous small non-coding RNAs in animals. *Nature Reviews Molecular Cell Biology*, 1–24.

Jovic, D., Liang, X., Zeng, H., Lin, L., Xu, F., & Luo, Y. (2022). Single-cell RNA sequencing technologies and applications: A brief overview. *Clinical and Translational Medicine*, *12*(3), e694.

Kacso, T. P., Zahu, R., Tirpe, A., Paslari, E. V., Nuțu, A., & Berindan-Neagoe, I. (2022). Reactive oxygen species and long non-coding RNAs, an unexpected crossroad in cancer cells. *International Journal of Molecular Sciences*, *23*(17), 10133.

Kaikkonen, M. U., Lam, M. T., & Glass, C. K. (2011). Non-coding RNAs as regulators of gene expression and epigenetics. *Cardiovascular Research*, *90*(3), 430–440.

Karlik, E., Ari, S., & Gozukirmizi, N. (2019). LncRNAs: genetic and epigenetic effects in plants. *Biotechnology and Biotechnological Equipment*, *33*(1), 429–439.

Koroban, N. V., Kudryavtseva, A. V., Krasnov, G. S., Sadritdinova, A. F., Fedorova, M. S., Snezhkina, A. V., …, & Melnikova, N. V. (2016). The role of microRNA in abiotic stress response in plants. *Molecular Biology*, *50*, 337–343.

Kryovrysanaki, N., James, A., Tselika, M., Bardani, E., & Kalantidis, K. (2021). RNA silencing pathways in plant development and defense. *The International Journal of Developmental Biology*, *66*(1–2–3), 163–175.

Ku, Y.-S., Cheung, M.-Y., Cheng, S.-S., Nadeem, M. A., Chung, G., & Lam, H.-M. (2022). "Using the Knowledge of Post-transcriptional Regulations to Guide Gene Selections for Molecular Breeding in Soybean. *Frontiers in Plant Science*, *13*, 867731.

Kumar, S., Kamboj, D., Mishra, C. N., & Singh, G. P. (2022). Prospects of molecular markers for wheat improvement in postgenomic era. In *Bioinformatics in agriculture* (pp. 323–340). Academic Press.

Kumar, V., Srivastava, A. K., & AbdElgawad, H. (2025). Transcriptional and post-transcriptional regulation of plant growth, development, and stress responses. *Journal of Plant Growth Regulation*, 1–6.

Kupper, H., Seib, L. O., Sivaguru, M., Hoekenga, O. A., & Kochian, L. V. (2007). A method for cellular localization of gene expression via quantitative in situ hybridization in plants. *The Plant Journal*, *50*(1), 159–187.

Lang, J. E., Wecsler, J. S., Press, M. F., & Tripathy, D. (2015). Molecular markers for breast cancer diagnosis, prognosis and targeted therapy. *Journal of Surgical Oncology*, *111*(1), 81–90.

Lee, Y., & Rio, D. C. (2015). Mechanisms and regulation of alternative pre-mRNA splicing. *Annual Review of Biochemistry*, *84*(1), 291–323.

Li, B., Qu, L., & Yang, J. (2023). RNA-guided RNA modifications: Biogenesis, functions, and applications. *Accounts of Chemical Research*, *56*(22), 3198–3210.

Li, D., Quan, C., Song, Z., Li, X., Yu, G., Li, C., & Muhammad, A. (2021). High-throughput plant phenotyping platform (HT3P) as a novel tool for estimating agronomic traits from the lab to the field. *Frontiers in Bioengineering and Biotechnology*, *8*, 623705.

Li, L. J., Leng, R. X., Fan, Y. G., Pan, H. F., & Ye, D. Q. (2017). Translation of noncoding RNAs: Focus on lncRNAs, pri-miRNAs, and circRNAs. *Experimental Cell Research*, *361*(1), 1–8.

Li, R., Fu, D., Zhu, B., Luo, Y., & Zhu, H. (2018). CRISPR/Cas9-mediated mutagenesis of lncRNA1459 alters tomato fruit ripening. *The Plant Journal*, *94*(3), 513–524.

Li, Z., Zhang, X., Liu, X., Zhao, Y., Wang, B., & Zhang, J. (2016). miRNA alterations are important mechanism in maize adaptations to low-phosphate environments. *Plant Science*, *252*, 103–117.

Liang, Z., Geng, Y., & Gu, X. (2018). Adenine methylation: new epigenetic marker of DNA and mRNA. *Molecular Plant*, *11*(10), 1219–1221.

Liu, Q., Ding, C., Lang, X., Guo, G., Chen, J., & Su, X. (2021). Small noncoding RNA discovery and profiling with sRNA-tools based on high-throughput sequencing. *Briefings in Bioinformatics*, *22*(1), 463–473.

Liu, R., & Zhu, J. K. (2014). Non-coding RNAs as potent tools for crop improvement. *National Science Review*, *1*(2), 186–189.

Liu, Y., Teng, C., Xia, R., & Meyers, B. C. (2020). PhasiRNAs in plants: Their biogenesis, genic sources, and roles in stress responses, development, and reproduction. *Plant Cell*, *32*(10), 3059–3080.

Ludwig, N., Leidinger, P., Becker, K., Backes, C., Fehlmann, T., Pallasch, C., … , & Keller, A. (2016). Distribution of miRNA expression across human tissues. *Nucleic Acids Research*, *44*(8), 3865–3877.

Luk, A. C., Chan, W. Y., Rennert, O. M., & Lee, T. L. (2014). Long noncoding RNAs in spermatogenesis: insights from recent high-throughput transcriptome studies. *Reproduction*, *147*(5), R131–R141.

Luo, X., & He, Y. (2020). Experiencing winter for spring flowering: A molecular epigenetic perspective on vernalization. *Journal of Integrative Plant Biology*, *62*(1), 104–117.

Ma, X., Zhao, F., & Zhou, B. (2022). The characters of non-coding RNAs and their biological roles in plant development and abiotic stress response. *International Journal of Molecular Sciences*, *23*(8), 4124.

Mann, C. W., Sawyer, A., Gardiner, D. M., Mitter, N., Carroll, B. J., & Eamens, A. L. (2023). RNA-based control of fungal pathogens in plants. *International Journal of Molecular Sciences*, *24*(15), 12391.

Mapuranga, J., Chang, J., Zhang, L., Zhang, N., & Yang, W. (2022). Fungal secondary metabolites and small RNAs enhance pathogenicity during plant-fungal pathogen interactions. *Journal of Fungi*, *9*(1), 4.

Marques, T. M., & Gama-Carvalho, M. (2022). Network approaches to study endogenous RNA competition and its impact on tissue-specific microRNA functions. *Biomolecules*, *12*(2), 332.

Mattick, J. S., Amaral, P. P., Carninci, P., Carpenter, S., Chang, H. Y., Chen, L. L., …, & Wu, M. (2023). Long non-coding RNAs: definitions, functions, challenges and recommendations. *Nature Reviews Molecular Cell Biology*, *24*(6), 430–447.

Michelmore, R., Coaker, G., Bart, R., Beattie, G., Bent, A., Bruce, T., …, & Walsh, J. (2017). Foundational and translational research opportunities to improve plant health. *Molecular Plant-Microbe Interactions*, *30*(7), 515–516.

Mishra, K., & Kanduri, C. (2019). Understanding long noncoding RNA and chromatin interactions: what we know so far. *Non-Coding RNA*, *5*(4), 54.

Mishra, S., Bansal, S., Mishra, B., Sangwan, R. S., Asha, J. S., Jadaun, N. S., & Sangwan. (2016). RNAi and homologous over-expression based functional approaches reveal triterpenoid synthase gene-cycloartenol synthase is involved in downstream withanolide biosynthesis in *Withania somnifera*. *PLoS One*, *11*(2), e0149691.

Mittler, R. (2006). Abiotic stress, the field environment and stress combination. *Trends in Plant Science*, *11*, 15–19.

Morris, K. V., & Mattick, J. S. (2014). The rise of regulatory RNA. *Nature Reviews Genetics*, *15*, 423–437.

Motamedi, H., Ari, M. M., Alvandi, A., & Abiri, R. (2024). Principle, application and challenges of development siRNA-based therapeutics against bacterial and viral infections: A comprehensive review. *Frontiers in Microbiology*, *15*, 1393646.

Nawaz, M., Sun, J., Shabbir, S., Khattak, W. A., Ren, G., Nie, X., …, & Sonne, C. (2023). A review of plants strategies to resist biotic and abiotic environmental stressors. *Science of The Total Environment*, *900*, 165832.

Nejat, N., & Mantri, N. (2018). Emerging roles of long non-coding RNAs in plant response to biotic and abiotic stresses. *Critical Reviews in Biotechnology*, *38*(1), 93–105.

Nickerson, J. A., & Momen-Heravi, F. (2024). Long non-coding RNAs: roles in cellular stress responses and epigenetic mechanisms regulating chromatin. *Nucleus, 15*(1), 2350180.

Olejniczak, M., Kotowska-Zimmer, A., & Krzyzosiak, W. (2018). Stress-induced changes in miRNA biogenesis and functioning. *Cellular and Molecular Life Sciences, 75*, 177–191.

Ossowski, S., Schwab, R., & Weigel, D. (2008). Gene silencing in plants using artificial microRNAs and other small RNAs. *The Plant Journal, 53*(4), 674–690.

Ounzain, S., Pezzuto, I., Micheletti, R., Burdet, F., Sheta, R., Nemir, M., …, & Pedrazzini, T. (2014). Functional importance of cardiac enhancer-associated noncoding RNAs in heart development and disease. *Journal of Molecular and Cellular Cardiology, 76*, 55–70.

Palande, S., Arsenault, J., Basurto-Lozada, P., Bleich, A., Brown, B. N., Buysse, S. F., & Rougon-Cardoso, A. (2025). Expression-based machine learning models for predicting plant tissue identity. *Applications in Plant Sciences, e11621.*

Pandey, R., Bhardwaj, A. R., & Katiyar-Agarwal, S. (2012). Endogenous Small RNAs and Antibacterial Resistance in Plants. *MicroRNAs in Plant Development and Stress Responses*, 233–259.

Pang, J., Zhang, X., Ma, X., & Zhao, J. (2019). Spatio-temporal transcriptional dynamics of maize long non-coding RNAs responsive to drought stress. *Genes, 10*(2), 138.

Pauli, A., Rinn, J. L., & Schier, A. F. (2011). Non-coding RNAs as regulators of embryogenesis. *Nature Reviews Genetics, 12*, 136–149.

Pecoraro, V., Rosina, A., & Polacek, N. (2022). Ribosome-associated ncRNAs (rancRNAs) adjust translation and shape proteomes. *Non-coding RNA, 8*(2), 22.

Pei, G., Sun, T., Chen, S., Chen, L., & Zhang, W. (2017). Systematic and functional identification of small non-coding RNAs associated with exogenous biofuel stress in cyanobacterium *Synechocystis* sp. PCC 6803. *Biotechnology for Biofuels, 10*(1), 17.

Peschansky, V. J., & Wahlestedt, C. W. C. (2014). Non-coding RNAs as direct and indirect modulators of epigenetic regulation. *Epigenetics, 9*, 3–12.

Pierouli, K., Papakonstantinou, E., Papageorgiou, L., Diakou, I., Mitsis, T., Dragoumani, K., …, & Vlachakis, D. (2022). Long non-coding RNAs and microRNAs as regulators of stress in cancer. *Molecular Medicine Reports, 26*(6), 361.

Ponjavic, J., Ponting, C. P., & Lunter, G. (2007). Functionality or transcriptional noise? Evidence for selection within long noncoding RNAs. *Genome Research, 17*, 556–565.

Prandi, F. R., Lecis, D., Illuminato, F., Milite, M., Celotto, R., Lerakis, S., …, & Barillà, F. (2022). Epigenetic modifications and non-coding RNA in diabetes-mellitus-induced coronary artery disease: pathophysiological link and new therapeutic frontiers. *International Journal of Molecular Sciences, 23*(9), 4589.

Praveen, A., Dubey, S., Singh, S., & Sharma, V. K. (2023). Abiotic stress tolerance in plants: a fascinating action of defense mechanisms. *3 Biotech, 13*(3), 102.

Puchta-Jasińska, M., Bolc, P., Pietrusińska-Radzio, A., Motor, A., & Boczkowska, M. (2025). Small interfering RNAs as critical regulators of plant life process: New perspectives on regulating the transcriptomic machinery. *International Journal of Molecular Sciences, 26*(4), 1624.

Rai, M. I., Alam, M., Lightfoot, D. A., Gurha, P., & Afzal, A. J. (2019). Classification and experimental identification of plant long non-coding RNAs. *Genomics, 111*, 997–1005.

Rajeevkumar, S., Anunanthini, P., & Sathishkumar, R. (2015). Epigenetic silencing in transgenic plants. *Frontiers in Plant Science, 6*, 693.

Ramakrishnan, M., Zhang, Z., Mullasseri, S., Kalendar, R., Ahmad, Z., Sharma, A., …, & Wei, Q. (2022). Epigenetic stress memory: A new approach to study cold and heat stress responses in plants. *Frontiers in Plant Science, 13*, 1075279.

Rasheed, A., & Xia, X. (2019). From markers to genome-based breeding in wheat. *Theoretical and Applied Genetics, 132*, 767–784.

Reddy, A. S. N. (2001). Nuclear pre-mRNA splicing in plants. *Critical Reviews in Plant Sciences, 20*(6), 523–571.

Roychowdhury, R., Das, S. P., Gupta, A., Parihar, P., Chandrasekhar, K., Sarker, U., …, & Sudhakar, C. (2023). Multi-omics pipeline and omics-integration approach to decipher plant's abiotic stress tolerance responses. *Genes, 14*(6), 1281.

Ruiz-Ferrer, V., & Voinnet, O. (2009). Roles of plant small RNAs in biotic stress responses. *Annual Review of Plant Biology, 60*(1), 485–510.

Sabana, A. A., Rajesh, M. K., & Antony, G. (2020). Dynamic changes in the expression pattern of miRNAs and associated target genes during coconut somatic embryogenesis. *Planta, 251*, 1–18.

Sahebi, M., Hanafi, M. M., van Wijnen, A. J., Azizi, P., Abiri, R., Ashkani, S., & Taheri, S. (2016). Towards understanding pre-mRNA splicing mechanisms and the role of SR proteins. *Gene, 587*(2), 107–119.

Samad, A. F., Sajad, M., Nazaruddin, N., Fauzi, I. A., Murad, A. M., Zainal, Z., & Ismail, I. (2017). MicroRNA and transcription factor: Key players in plant regulatory network. *Frontiers in Plant Science, 8*, 565.

Sanan-Mishra, N., Abdul Kader Jailani, A., Mandal, B., & Mukherjee, S. K. (2021). Secondary siRNAs in plants: Biosynthesis, various functions, and applications in virology. *Frontiers in Plant Science, 12*, 610283.

Santosh, B., Varshney, A., & Yadava, P. K. (2015). Non-coding RNAs: Biological functions and applications. *Cell Biochemistry and Function, 33*(1), 14–22.

Sarfraz, Z., Zarlashat, Y., Ambreen, A., Mujahid, M., Iqbal, M. S., Fatima, S. A., …, & Fiaz, S. (2025). Plant biochemistry in the era of omics: Integrated omics approaches to unravel the genetic basis of plant stress tolerance. *Plant Breeding.*

Shimoni, Y., Friedlander, G., Hetzroni, G., Niv, G., Altuvia, S., Biham, O., & Margalit, H. (2007). Regulation of gene expression by small-noncoding RNAs: A quantitative view. *Molecular Systems Biology, 3*, 138.

Shukla, L. I., Chinnusamy, V., & Sunkar, R. (2008). The role of microRNAs and other endogenous small RNAs in plant stress responses. *Biochimica et Biophysica Acta (BBA)-Gene Regulatory Mechanisms, 1779*(11), 743–748.

Singh, P., Dutta, P., & Chakrabarty, D. (2021). miRNAs play critical roles in response to abiotic stress by modulating cross-talk of phytohormone signaling. *Plant Cell Reports, 40*(9), 1617–1630.

Singh, R. S., Singh, P., Sinha, S., Kumar, U., Kumari, R., & Kumar, S. (2023). Non-coding RNA based marker: A new weapon in armory of molecular markers. In *Molecular marker techniques: A potential approach of crop improvement* (pp. 117–125). Springer Nature Singapore.

Singh, S., Assumi, R., & Bhadrecha, P. (2023). Crosstalk of MicroRNAs with phytohormone signalling pathways. In *Plant MicroRNAs and stress response* (pp. 257–276). CRC Press.

Singroha, G., Kumar, S., Gupta, O. P., Singh, G. P., & Sharma, P. (2022). Uncovering the epigenetic marks involved in mediating salt stress tolerance in plants. *Frontiers in Genetics*, *13*, 811732.

Singroha, G., & Sharma, P. (2023). miRNAs perspective in mitigating waterlogging stress in plants. In *Plant small RNA in food crops* (pp. 347–364). Academic Press.

Song, L., Fang, Y., Chen, L., Wang, J., & Chen, X. (2021). Role of non-coding RNAs in plant immunity. *Plant Communications*, *2*(3).

Sun, C., Ali, K., Yan, K., Fiaz, S., Dormatey, R., Bi, Z., & Bai, J. (2021). Exploration of epigenetics for improvement of drought and other stress resistance in crops: A review. *Plants*, *10*(6), 1226.

Sunkar, R., Kapoor, A., & Zhu, J. K. (2006). Posttranscriptional induction of two Cu/Zn superoxide dismutase genes in Arabidopsis is mediated by downregulation of miR398 and important for oxidative stress tolerance. *The Plant Cell*, *18*(8), 2051–2065.

Sunkar, R., & Zhu, J. K. (2004). Novel and stress-regulated microRNAs and other small RNAs from Arabidopsis. *The Plant Cell*, *16*, 2001–2019.

Surdonja, K., Eggert, K., Hajirezaei, M. R., Harshavardhan, V. T., Seiler, C., Von Wirén, N., …, & Kuhlmann, M. (2017). Increase of DNA methylation at the HvCKX2. 1 promoter by terminal drought stress in barley. *Epigenomes*, *1*(2), 9.

Tang, W., Wu, T., Ye, J., Sun, J., Jiang, Y., Yu, J., …, & Wan, J. (2016). SNP-based analysis of genetic diversity reveals important alleles associated with seed size in rice. *BMC Plant Biology*, *16*, 1–11.

Umar, O. B., Ranti, L. A., Abdulbaki, A. S., Bola, A. U. L., Abdulhamid, A. K., Biola, M. R., & Victor, K. O. (2021). Stresses in plants: Biotic and abiotic. In *Current trends in wheat research*. IntechOpen.

Valinezhad Orang, A., Safaralizadeh, R., & Kazemzadeh-Bavili, M. (2014). Mechanisms of miRNA-mediated gene regulation from common downregulation to mRNA-specific upregulation. *International Journal of Genomics*, *2014*(1), 970607.

Vivek, A. T., & Kumar, S. (2021). Computational methods for annotation of plant regulatory non-coding RNAs using RNA-seq. *Briefings in Bioinformatics*, *22*(4), bbaa322.

Wang, J., Meng, X., Dobrovolskaya, O. B., Orlov, Y. L., & Chen, M. (2017). Non-coding RNAs and their roles in stress response in plants. *Genomics, Proteomics and Bioinformatics*, *15*(5), 301–312.

Wekesa, J. S., Luan, Y., Chen, M., & Meng, J. (2019). A hybrid prediction method for plant lncRNA-protein interaction. *Cells*, *8*(6), 521.

Wen, M., Lin, X., Xie, M., Wang, Y., Shen, X., Liufu, Z., …, & Tang, T. (2016). Small RNA transcriptomes of mangroves evolve adaptively in extreme environments. *Scientific Reports*, *6*(1), 27551.

Xiao, Q., Bai, X., Zhang, C., & He, Y. (2022). Advanced high-throughput plant phenotyping techniques for genome-wide association studies: A review. *Journal of Advanced Research*, *35*, 215–230.

Yadav, A., Mathan, J., Dubey, A. K., & Singh, A. (2024). The emerging role of non-coding RNAs (ncRNAs) in plant growth, development, and stress response signaling. *Non-Coding RNA*, *10*(1), 13.

Yang, J., Meng, X., Pan, J., Jiang, N., Zhou, C., Wu, Z., & Gong, Z. (2018). CRISPR/Cas9-mediated noncoding RNA editing in human cancers. *RNA Biology*, *15*(1), 35–43.

Yang, W., Feng, H., Zhang, X., Zhang, J., Doonan, J. H., Batchelor, W. D., …, & Yan, J. (2020). Crop phenomics and high-throughput phenotyping: Past decades, current challenges, and future perspectives. *Molecular Plant*, *13*(2), 187–214.

Yang, Y., Liang, Y., Wang, C., & Wang, Y. (2024). MicroRNAs as potent regulators in nitrogen and phosphorus signaling transduction and their applications. *Stress Biology*, *4*(1), 38.

Ye, M., Xie, L., Zhang, J., Liu, B., Liu, X., He, J., …, & Dong, K. (2020). Determination of long non-coding RNAs associated with EZH2 in neuroblastoma by RIP-seq, RNA-seq and ChIP-seq. *Oncology Letters*, *20*(4), 1.

Yi, X., Zhang, Z., Ling, Y., Xu, W., & Su, Z. (2015). PNRD: A plant non-coding RNA database. *Nucleic Acids Research*, *43*(D1), D982–D989.

Yin, X., Stam, P., Kropff, M. J., & Schapendonk, A. H. (2003). Crop modeling, QTL mapping, and their complementary role in plant breeding. *Agronomy Journal*, *95*(1), 90–98.

Zeng, J., Gupta, V. K., Jiang, Y., Yang, B., Gong, L., & Zhu, H. (2019). Cross-kingdom small RNAs among animals, plants and microbes. *Cells*, *8*(4), 371.

Zhang, F., Yang, J., Zhang, N., Wu, J., & Si, H. (2022). Roles of microRNAs in abiotic stress response and characteristics regulation of plant. *Frontiers in Plant Science*, *13*, 919243.

Zhang, L., Lin, T., Zhu, G., Wu, B., Zhang, C., & Zhu, H. (2023). LncRNAs exert indispensable roles in orchestrating the interaction among diverse noncoding RNAs and enrich the regulatory network of plant growth and its adaptive environmental stress response. *Horticulture Research*, *10*(12), uhad234.

Zhong, L., Zhen, M., Sun, J., & Zhao, Q. (2021). Recent advances on the machine learning methods in predicting ncRNA-protein interactions. *Molecular Genetics and Genomics*, *296*(2), 243–258.

Zhou, X., Cui, J., Meng, J., & Luan, Y. (2020). Interactions and links among the noncoding RNAs in plants under stresses. *Theoretical and Applied Genetics*, *133*, 3235–3248.

Zuo, J., Zhang, Z., Li, M., Yang, Y., Zheng, B., Wang, P., & Zhou, S. (2022). The crosstalk between reactive oxygen species and noncoding RNAs: From cancer code to drug role. *Molecular Cancer*, *21*(1), 30.

17 Innovations in Plant Biology
The Contribution of ncRNAs to Abiotic Stress Tolerance

Nadia Riaz, Afifa Younas, Zubaida Yousaf, Nimra Tariq and Muhammad Kaleem

17.1 NON-CODING RNAs (ncRNAs)

Non-coding RNAs (ncRNAs) have emerged significantly in recent years as critical bioactive molecules contributing to genomic and phenotypic variation. Although up to 90% of the eukaryotic genome is transcribed into RNA, only about 2% of these transcripts result in protein production (Pauli et al., 2011; Rai et al., 2019). The remainder of the transcriptome consists of ncRNAs, which were previously thought to be transcriptional noise due to their lack of protein-coding potential and poorly conserved sequences (Ariel et al., 2015; Pauli et al., 2011). However, advancements in high-throughput sequencing and experimental studies have revealed the crucial roles of ncRNAs in regulating genes at various levels, including epigenetic, transcriptional and post-transcriptional processes (Cech & Steitz, 2014; Peschansky & Wahlestedt, 2014). Consequently, the development of advanced next-generation sequencing (NGS) technologies, combined with homology-based or experimental approaches, has been instrumental in uncovering ncRNAs in plants (Waititu et al., 2020).

Plants are exposed to several types of biotic and abiotic stresses. Concerned with the abiotic stresses limiting plant growth and development are salt, drought, heat, cold and heavy metal stresses. To cope with this stress, plants have adopted several types of homeostasis regulatory mechanisms (Figure 17.1). The advancement in sequencing technology such as next-generation DNA and RNA sequencing and bioinformatics tools have revealed potential of ncRNA. ncRNAs are capable of regulating plant homeostasis through DNA- and RNA-associated interactions. These ncRNAs interact with the genetic material in cytoplasm and nucleus, thus altering the biological processes.

17.2 BIOGENESIS OF ncRNAs

ncRNAs exhibit a wide range of origins and functions (Contreras-Cubas et al., 2012). Various theories explain their origin, including evolution from prevailing transposable elements (TEs), gene duplication, random hairpin structures, replication of RNA viruses, pseudogenisation of protein-coding sequences, synthesis of double-stranded RNAs (dsRNAs) from heterochromatin regions, as well as DNA repeat sequences. These processes contribute to the synthesis of different ncRNAs such as small interfering RNAs (siRNAs), microRNAs (miRNAs), long ncRNAs (lncRNAs) and piwi-interacting RNAs (piRNAs) (Waititu et al., 2020).

Based on their origin, biogenesis and functions, ncRNAs were categorised into many distinct classes since their discovery. Depending on their molecular structure, ncRNAs in plants are divided into two categories, circular ncRNAs and linear ncRNAs. Circular ncRNAs are categorised into intronic circRNAs, exonic circRNAs, intergenic circRNAs, untranslated region (UTR) circRNAs and many other circular RNAs (circRNAs) which were derived from two or more genes. On the other hand, linear ncRNAs are classified into regulatory ncRNAs and housekeeping ncRNAs (Liu et al., 2017; Ma et al., 2022). Housekeeping ncRNAs are mainly involved in cellular and ribosomal functions, consisting of transfer RNAs (tRNAs), ribosomal RNAs (rRNAs), small nuclear RNAs (snRNAs) and small nucleolar RNAs (snoRNAs), while regulatory ncRNAs include siRNAs, miRNAs, lncRNAs and piRNAs (Morris & Mattick, 2014). The regulatory ncRNAs can be transcribed from DNA but unable to translate into proteins; they play crucial role in plant growth, development and responses to abiotic stresses through transcriptional and post-transcriptional regulation (D'Ario et al., 2017).

17.3 MicroRNAs (miRNAs)

These are small endogenous ncRNAs, generally 20–24 nucleotides in length. miRNA genes are transcribed into primary miRNA transcripts, and then these primary transcripts are transcribed by DICER-LIKE (DCL) proteins along with miRNA. Furthermore, these mature miRNAs after reacting with ARGONAUTE (AGO) proteins form a miRNA-induced silencing complex (miRISC). And eventually this complex works to downstream mRNAs through targeted complementary sequences (Song et al., 2019).

miRNAs play a crucial role in plant growth and development through regulation of post-translational gene expression (Wu, 2013). They target a wide range of genes as well as transcription factors to regulate plant development. Several researchers observed that miRNA family or even a single miRNA can regulate similar or dissimilar members of a gene family (Dong et al., 2022). Furthermore, it

DOI: 10.1201/9781003583257-17

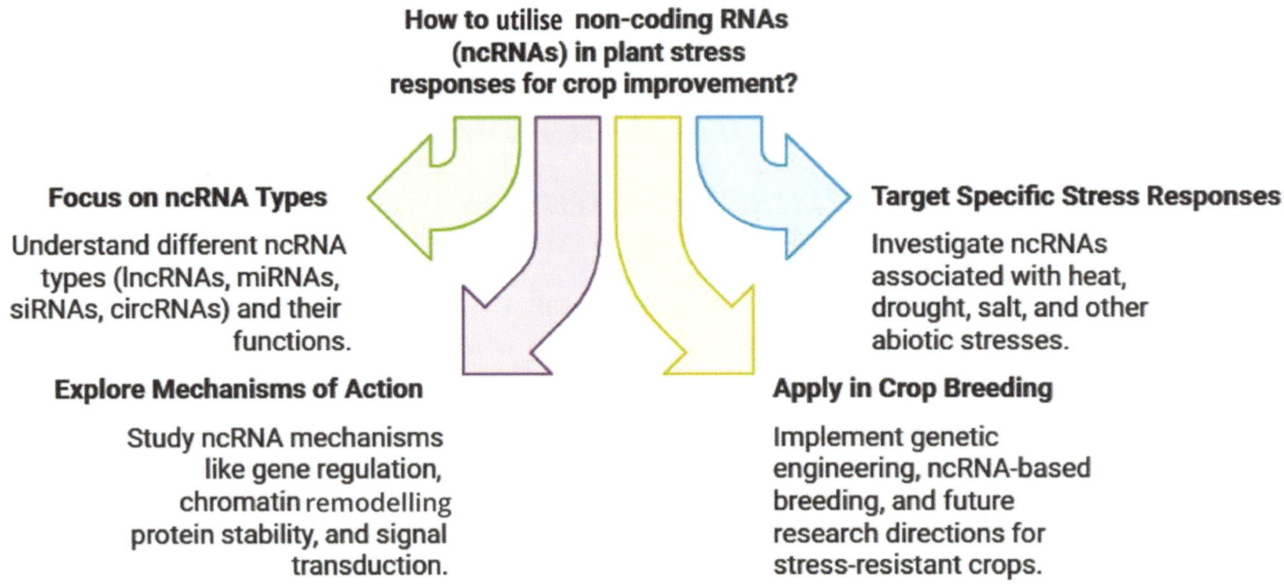

FIGURE 17.1 Functional mechanisms of ncRNAs in stress response.

was also observed that a conserved plant species possesses miRNA along with its targets (Jones-Rhoades, 2012). For example, in many flowering plants, a conserved miRNA (miR156/miR157) family plays an important role in regulation of both vegetative and reproductive transition phases through regulation of squamosa promoter-binding protein-like (SPL) transcription factors (Zheng et al., 2019; Zhou et al., 2021). In *Arabidopsis*, miR172 regulates flowering time, floral development and somatic embryogenesis by targeting AP2 (Apetala2) and WUS (WUSCHEL) transcription factors (Nowak et al., 2022; O'Maoileidigh et al., 2021).

Likewise, miR172 regulates blooming phase in maize, rice and barley (Smoczynska & Szweykowska-Kulinska, 2016). Another miRNA family, miR164, regulates lateral root formation in maize and *Arabidopsis* by regulating NAC1 transcription factor. Similarly, in *Medicago sativa* (alfalfa) it was reported that conserved miR164 targets MtNAC1, and its overexpression reduces nodule formation in alfalfa roots. Interestingly, miRNAs and their targets often have diverse roles beyond their conserved functions. For example, the miR156-SPL module regulates grain size in rice, panicle branching, ear development as well as seed setting in maize, and tiller formation in switchgrass (Cai et al., 2022; D'Ario et al., 2017).

Moreover, some biological processes can be regulated by diverse miRNAs associated with performing this task. For instance, miR165 or miR166 along with miR394 family uphold the shoot apical meristem (SAM) in *Arabidopsis* by targeting Leaf Curling Responsiveness (LCR) genes as well as homeodomain leucine zipper III (HD-ZIP III) transcription factor (D'Ario et al., 2017). MiR165 or miR166 along with miR390 family target genesand control leaf polarity in plants. Likewise, two miRNA families such as miRNA157 and miR172 mutually regulate different processes in *Arabidopsis* like transition from vegetative to reproductive stage and flowering process. Tang and Chu (2017) reported that rice grain size is regulated by miR156, miR396 and miR397. Additionally, unrelated miRNA (miR156 and miR319) modules regulate flower induction and floral development in tomato in response to gibberellins (Silva et al., 2019).

Moreover, species-specific miRNAs also play a vital part in plant growth and development. For instance, the miR171-SCL module activates the miR156-SPL pathway to mediate vegetative phase transition in barley. In rice, both miR393 and miR444a are involved in nitrate signalling mediated processes like root development and tiller formation. miR444 regulates tiller formation in rice by targeting OsMADS 23/27a/57 transcription factors through the repression of RNA-Dependent RNA Polymerase 1 (Li et al., 2016; Yan et al., 2014). These findings emphasise the functional conservation and versatility of miRNAs in regulating various plant growth stages. They also highlight their potential for improving plant growth and yield. Further studies on miRNA-target modules will enhance our understanding of plant development and offer insights into agricultural applications.

17.4 SMALL INTERFERING RNAs (siRNAs)

siRNAs are a class of small ncRNAs (sRNAs) derived from various dsRNAs, approximately 22 nucleotides long, and their biogenesis is initiated through the action of DCL proteins. This biogenesis process begins with synthesis of dsRNAs in an RNA-dependent manner by RNA polymerases. These dsRNAs are then processed by specific AGO proteins to produce different types of siRNAs (Axtell, 2013). Unlike miRNAs, which have a heteroduplex structure, siRNAs are homoduplex structure. miRNAs evolved from specific genes, while siRNAs are typically formed from transposons or heterochromatic regions. A key distinction between the miRNA and siRNA pathways is their usage of AGO proteins. SiRNAs predominantly use AGO_2, along with other AGO proteins, whereas miRNA-mediated gene silencing can involve any AGO protein from AGO_1 to AGO_{10}. SiRNAs are highly conserved across species and play essential roles in mRNA target repression and DNA methylation (Jha et al., 2021). SiRNAs are categorised based on their origin into trans-acting siRNAs (tasiRNAs), heterochromatic siRNAs (hc-siRNA), TE-derived siRNAs, repeat-associated siRNAs and long small interfering siRNAs.

SiRNAs control plant growth by mediating transcriptional gene silencing (TGS) through DNA/histone methylation or both by RdDM (RNA-directed DNA methylation) pathway. These siRNAs regulate reproductive processes in flowering plants. For example, in the course of development of pollen grain (male gametophyte), some TEs undergo demethylation in vegetative nucleus and produce siRNAs of 21, 22 and 23 nucleotides that later move into sperm cells (Martinez et al., 2016). These sperm cells show lower methylation at CHH sites, suggesting reduced induction of the RdDM pathway. But methylation at CHH sites increases during embryogenesis. The siRNAs involved in regulating methylation in embryonic cells come from sperm or endosperm cells, facilitating link between the parental genomes. TE-derived siRNAs in tomatoes play a vital role in determination of hybrid phenotype (Liu et al., 2020).

ta-siRNAs are a family of phasiRNAs derived from TAS loci and play a role in plant development. For example, the miR390–AGO7 target transcript generates TAS3 tasiRNA, which targets ARF family genes to regulate the transition from vegetative to reproductive phase, somatic embryogenesis, flower architecture, root architecture, leaf morphogenesis and phytohormone crosstalk in plant species such as potato, tomato, *Arabidopsis*, maize and rice (Ding et al., 2020; Liu et al., 2020). SiRNAs produced from NLR (nucleotide-binding leucine-rich repeat) control nodule development in alfalfa by targeting NLR/TAS5 transcripts and LTR (long terminal repeat) elements. However, phasiRNAs control reproductive development in soybean by targeting retrotransposons (Arikit et al., 2014). Overall, siRNAs and phasiRNAs are critical during the reproductive phase and in the transmission of information across generations via DNA or histone modifications along with other roles played in the developmental stages being explored, offering new insights into their diverse functions.

17.5 LONG NON-CODING RNAs (lncRNA)

lncRNAs are basically RNA transcripts having size of more than 200 nt and don't encode proteins. Early research questioned their significance in plants. However, over time, these molecules, once thought to be 'transcriptional noise', have proven to be vital regulators of gene expression across eukaryotes, from single-celled organisms like yeast to plants and mammals (van Werven et al., 2012). lncRNAs are typically transcribed by RNA polymerase II, but RNA polymerase IV and V have also been involved. lncRNAs transcribed by RNA polymerase IV often act as precursors for synthesis of siRNA. In many plants, lncRNAs are regulated at specific stages of development and in response to environmental changes and attributing to the functional components of transcriptome. Based on their origin, they are classified into several types: incRNAs (intronic ncRNAs), lincRNAs (long intergenic ncRNAs), circRNAs (circular lncRNAs) and NATs (natural antisense transcripts) (Mattick & Rinn, 2015). Most lncRNAs act as decoys/backbones/sponges, as well as guides. Moreover, they can act as endogenous target mimics (eTMs) for various miRNAs and regulate alternative splicing (Jha et al., 2021).

lncRNAs play a significant role in plant developmental processes by modulating epigenetic changes, as well as the regulation at transcriptional and translational level of target genes through both cis- (adjacent genes) and trans- (distant genes) mechanisms (Chen et al., 2020). Some lncRNAs can encode small peptides under 100 amino acids in length. For example, ENOD40 (Early Nodulin 40) peptide is encoded by the ENOD40 gene and interacts with the nodulin-100 protein to facilitate sucrose uptake during nodule development and nodule organogenesis in roots of alfalfa and soybean. Interestingly, ENOD40 also encodes NAT (N-acetyltransferase) and DONE40 genes, which can serve as a molecular scaffold or decoy, as seen in the nodule initiation process in peanuts (*Arachis hypogea*) (Ganguly et al., 2021).

In addition, the lncRNA COOLAIR (cold-inducible long antisense intragenic RNA) regulates the FLC (Flowering Locus C) gene, triggering flowering in *Arabidopsis* when exposed to cold temperatures (Csorba et al., 2014). LDMAR (long-day-specific male fertility-associated RNA) controls pollen development in rice under long-day conditions (Ding et al., 2012). Npc 356 (Non-Protein Coding 356) has been shown to overexpress *Arabidopsis* under salt-stressed conditions and promotes the growth of both primary and secondary roots in *Arabidopsis*. NATs, which make up about 70% of *Arabidopsis* mRNAs, are thought to facilitate histone modifications in their corresponding genes, thereby influencing gene expression. For example, LAIR (LRK intergenic antisense RNA) regulates the expression of LRK (leucine-rich repeat receptor kinase) genes via histone modifications and affects grain yield traits in rice (Wang et al., 2018). Overall, these findings highlight the diverse roles of lncRNAs in various stages of plant development, especially in response to environmental stimuli, significantly impacting plant growth and yield. Further research on lncRNAs, their encoded peptides and their target genes will provide deeper insights into plant development, survival and yield improvement.

17.6 REGULATORY ROLE OF NON-CODING RNAs IN DROUGHT STRESS TOLERANCE IN PLANTS

Drought stress is one of the most difficult abiotic stressors on plants, negatively impacting growth, productivity and survival. In response to such environmental challenges, plants deploy a variety of adaptation mechanisms to maintain cellular homeostasis and maximise water utilisation. Emerging data has emphasised the critical functions of ncRNAs in regulating gene expression during drought stress. These ncRNAs, which have miRNAs, small interfering RNAs (siRNAs) and lncRNAs, are important regulatory molecules that improve drought resilience by modulating stress-responsive gene networks at transcriptional, post-transcriptional and epigenetic levels (Yadav et al., 2024). This chapter investigates the regulatory roles of ncRNAs in drought tolerance, with an emphasis on their interactions with particular gene networks and stress-response mechanisms (Figure 17.2). The regulatory network of ncRNAs in drought tolerance frequently includes complicated interactions between miRNAs, lncRNAs, and mRNAs. In mustard, lncRNAs have been found to interact with miRNAs targeting aquaporin genes, which are required for water transport and cellular hydration during drought stress. lncRNAs limit the degradation of mRNAs producing aquaporins by sequestering miRNAs, hence improving water retention capacity and cellular osmotic balance under drought circumstances (Wei et al., 2024).

In addition to direct gene regulation, ncRNAs interact with hormonal pathways, particularly those involving abscisic acid (ABA), an important hormone in drought response. miRNAs like miR160 and miR167 affect auxin signalling pathways, which impact root development and architecture, which is critical for maximising water intake during drought (Yadav et al., 2024). Furthermore, lncRNAs influence ABA-responsive genes, therefore strengthening drought-responsive transcription factor networks that lead to stomatal control and decreased water loss.

miRNAs are tiny ncRNAs that predominantly influence gene expression by mRNA cleavage or translational inhibition. Under drought circumstances, miRNAs influence a diverse set of stress-responsive genes, notably those involved in water retention, antioxidant defence, and hormone signalling pathways. miRNAs like miR159, miR169 and miR398 have been linked to drought tolerance by targeting genes that control stomatal closure, root architecture and reactive oxygen species (ROS) scavenging. For example, miR159 inhibits MYB transcription factors involved in ABA signalling, which improves stomatal closure and reduces water loss during drought (Wei et al., 2024). miR398 controls antioxidant enzymes like superoxide dismutase, which reduces oxidative stress during drought-induced ROS buildup and protects cellular structures from harm (Balhara et al., 2024).

In *Brassica juncea*, miRNAs have been demonstrated to influence the expression of MYB transcription factors and aquaporin genes, both of which are essential for water transport and cellular osmotic equilibrium during drought. This control includes the direct silencing of certain mRNAs, which reduces water loss and improves drought resilience (Ackah et al., 2022).

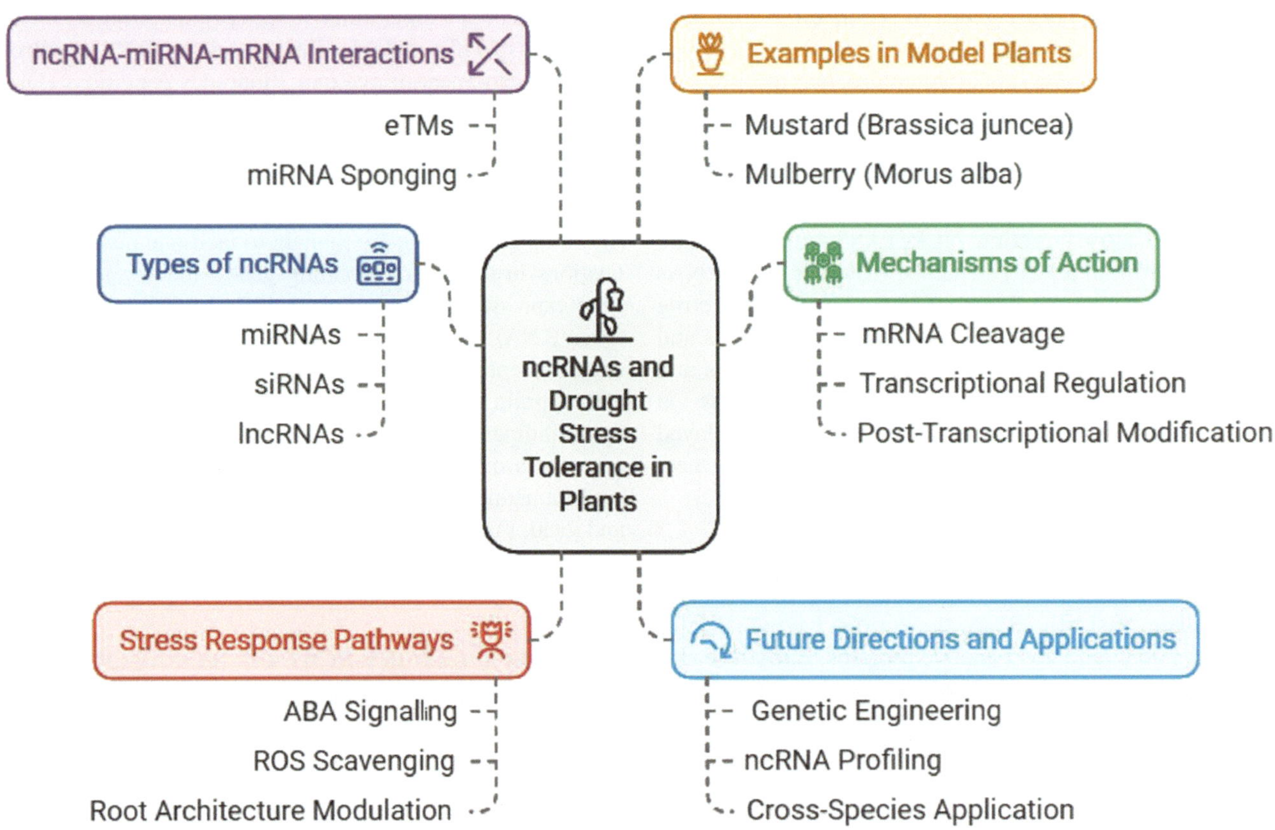

FIGURE 17.2 A flow chart indicating regulatory roles of ncRNAs in drought tolerance.

lncRNAs, which are generally longer than 200 nucleotides, play a variety of regulatory activities, including acting as molecular sponges for miRNAs, modifying chromatin states and impacting transcription and post-transcriptional processes. Recent research has highlighted their significance as key mediators in drought stress tolerance through interactions with miRNAs and other ncRNAs.

lncRNAs contribute to drought tolerance by acting as eTMs for miRNAs, thus freeing target mRNAs from miRNA-mediated repression. In drought-tolerant mustard (*B. juncea*), certain lncRNAs function as eTMs for miRNAs that regulate MYB transcription factors involved in drought response. By sequestering these miRNAs, lncRNAs enhance the expression of MYB genes, supporting osmotic adjustment and other physiological adaptations under drought (Balhara et al., 2024).

In another study on *Morus alba*, lncRNAs were identified as key players in activating secondary metabolism pathways, which are crucial for the biosynthesis of osmoprotectants and antioxidants. These compounds help stabilise cellular structures and reduce oxidative damage caused by drought-induced ROS (Figure 17.2) (Ackah et al., 2022).

17.7 CASE STUDIES: ncRNA-DRIVEN DROUGHT TOLERANCE IN MODEL PLANTS

17.7.1 Mustard (*Brassica juncea*)

ncRNAs in *B. juncea* have been intensively investigated for their functions in drought tolerance. Specific lncRNAs and miRNAs have been discovered as modulating the MYB and aquaporin gene families, hence increasing drought resilience. The connection between ncRNAs and their target genes promotes physiological processes like osmotic balance, ROS scavenging and root growth, all of which are essential for survival in water-stressed environments (Wei et al., 2024).

17.7.2 Mulberry (*Morus alba*)

Studies on drought-stressed mulberry plants show that ncRNAs, particularly lncRNAs, control secondary metabolic pathways that are critical for antioxidant synthesis and cellular defence against oxidative damage. This control allows mulberry to retain metabolic activity and structural integrity throughout extended drought periods (Ackah et al., 2022).

The study of ncRNAs has helped us understand how plants respond to drought stress, with miRNAs and lncRNAs emerging as important modulators in stress tolerance networks. These ncRNAs do not function in isolation; rather, they are part of complex networks that include feedback loops and crosstalk with critical drought-responsive genes and hormone pathways. Advances in high-throughput sequencing and analytics have enabled precise ncRNA profiling, allowing for the targeting of ncRNAs for genetic engineering to generate drought-tolerant crops. Future study might further unravel ncRNA interactions with epigenetic and proteomic alterations to build holistic methods for improving drought resilience in crops.

17.8 REGULATORY ROLE OF ncRNAs IN SALT STRESS TOLERANCE

Salt stress is an important abiotic element that influences plant development and production, especially in agricultural environments where elevated salinity can lead to lower crop yields. Plants have developed complex ways to cope with salt stress, including the use of ncRNAs, which modulate gene expression under stress conditions (Jin et al., 2024). This chapter investigates the functions of various ncRNAs, with a particular emphasis on lncRNAs and small interfering RNAs (siRNAs), in salt tolerance processes.

ncRNAs, which include miRNAs, small interfering RNAs (siRNAs) and lncRNAs, regulate gene expression at both the transcriptional and post-transcriptional stages. They are largely classified by length, with lncRNAs often more than 200 nucleotides and siRNAs typically shorter, approximately 20–25 nucleotides (Wang et al., 2024; Zhao et al., 2024).

lncRNAs have emerged as important regulatory molecules in salt stress settings. For example, in *Arabidopsis thaliana*, the lncRNA DRIR (drought-induced lncRNA) is activated by salt stress, which improves plant resilience by regulating multiple stress-responsive genes (Wang et al., 2024). Similarly, the lncRNA LAL in *Medicago truncatula* improves salinity tolerance by regulating the expression of light-harvesting chlorophyll a/b binding proteins, which are involved in photosynthesis and ROS control (Table 17.1) (Zhao et al., 2024).

lncRNAs frequently function as competitive endogenous RNAs (ceRNAs), binding to miRNAs and inhibiting the degradation of their target mRNAs (Jin et al., 2024). For example, particular lncRNAs in wheat have been found to

TABLE 17.1
Case Studies of ncRNAs in Salt Tolerance

Species	ncRNA	Function	References
Arabidopsis thaliana	DRIR	Enhances drought and salt tolerance by regulating ABA signalling	Wang et al. (2024)
Medicago truncatula	LAL	Modulates chlorophyll-binding proteins, enhancing ROS regulation	Zhao et al. (2024)
Oryza sativa (rice)	NAT-derived lncRNAs	Increases ROS scavenging, contributes to salt tolerance	Jin et al. (2024)

TABLE 17.2
Functions of Key Heat-Responsive lncRNAs in Plants

lncRNA	Target Genes/Proteins	Role in Heat Stress Tolerance	References
lncRNA-heat-induced	HSF, HSPs	Enhances protein stability by upregulating HSP expression	Babaei et al. (2024)
ELENA1	MED19a, PR1	Increases heat tolerance by modifying chromatin and transcription	Jin et al. (2024)
AUXIN-REGULATED	WRKY42, APOLO	Modifies chromatin structure, inducing root elongation under heat	Yadav et al. (2024)
MISSEN	Tubulin, HeFP	Prevents abnormal cell expansion during early heat stress stages	Imaduwage and Hewadikaram (2024)

function as ceRNAs by interacting with miRNAs involved in ROS homeostasis, which is crucial under salt stress conditions (Zhao et al., 2024).

NATs are a distinct type of lncRNA that is transcribed in the opposite direction as protein-coding genes. Certain NATs in *Arabidopsis* have been demonstrated to boost ROS scavenging ability under salt stress conditions, further improving plant tolerance to high salinity settings (Table 17.1) (Wang et al., 2024).

17.9 REGULATORY ROLE OF ncRNAs IN HEAT STRESS TOLERANCE IN PLANTS

Global climate change has increased the frequency of severe temperature occurrences, which have a detrimental impact on agricultural yield and food security. Heat stress is the most harmful of these abiotic stressors, influencing a variety of physiological and biochemical processes inside plants (Babaei et al., 2024). ncRNAs, particularly lncRNAs and miRNAs, play critical regulatory functions in plants' adaptive responses to heat stress by influencing gene expression, protein stability and cellular signalling pathways (Tables 17.2 and 17.3) (Jin et al., 2024; Yadav et al., 2024). ncRNAs regulate gene expression by many methods, such as transcription, chromatin remodelling, mRNA splicing and post-transcriptional modification (Yadav et al., 2024).

lncRNAs are longer than 200 nucleotides and serve a variety of functions, including molecular scaffolding, decoys and guides for protein complexes implicated in stress signalling (Babaei et al., 2024). Heat-responsive lncRNAs, such as lncRNA-heat stress factor (HSF) binding lncRNAs, target heat-shock proteins (HSPs) and transcription factors to stabilise cellular structures and prevent protein denaturation (Imaduwage & Hewadikaram, 2024).

miRNAs are usually 20–24 nucleotides long and work by targeting mRNAs for cleavage or translational inhibition. Heat stress induces miRNAs such as miR156 and miR172, which regulate transcription factors involved in heat-responsive pathways (Jin et al., 2024). ncRNAs influence stress responses through regulatory networks that include gene silencing, RNA processing and ceRNA networks, all of which are critical for heat stress resistance (Babaei et al., 2024; Jin et al., 2024). Certain lncRNAs bind transcription factors to chromatin areas associated with heat-responsive genes. For example, the lncRNA ELENA1 binds to the mediator complex member MED19a, facilitating the activation of heat stress-related genes, including PATHOGENESIS-RELATED GENE 1 (PR1) (Jin et al., 2024). This connection promotes chromatin remodelling and transcription in response to high temperatures (Imaduwage & Hewadikaram, 2024). In ceRNA networks, lncRNAs and mRNAs share similar miRNA binding sites, functioning as decoys to reduce miRNA effects on heat-responsive genes. For example, the lncRNA that competes for miR156 binding might stabilise transcription factors required for heat tolerance by lowering miRNA repression on target mRNAs (Babaei et al., 2024).

The discovery and functional confirmation of ncRNAs in heat stress response pave the way for genetic engineering and breeding initiatives aiming at producing heat-tolerant crops. The CRISPR/Cas9 system has the capacity to precisely modify ncRNA genes, allowing the regulation of

TABLE 17.3
Heat Stress-Responsive miRNAs and Their Functions

miRNA	Target mRNA	Function in Heat Stress Tolerance	References
miR156	SPL transcription factors	Regulates downstream heat-responsive genes	Jin et al. (2024)
miR159	GAMYB	Modulates hormone signalling pathways under high temperatures	Babaei et al. (2024)
miR398	CSD1, CSD2	Enhances antioxidant response, reducing oxidative damage	Yadav et al. (2024)
miR408	Copper-binding proteins	Promotes cellular resilience by stabilising electron transport	Imaduwage and Hewadikaram (2024)

stress-responsive pathways (Tables 17.2 and 17.3) (Yadav et al., 2024).

17.10 REGULATORY ROLE OF ncRNAs IN COLD STRESS TOLERANCE IN PLANTS

The intensification of global warming and climate variability has led to more frequent and severe extreme weather events, including both high- and low-temperature stress, creating significant challenges for crop production. Although global warming is often associated with increased heat, abrupt episodes of cold stress have also become a prominent environmental factor, imposing additional challenges on agricultural productivity. Cold stress disrupts critical cellular processes, lowering metabolic rates and limiting nutrient uptake of plants, collectively impacting their growth, yield and survival. Enhancing cold tolerance in plants, particularly in major crops, is essential to ensure food security as environmental conditions continue to fluctuate. Addressing this challenge requires a deeper understanding of the genetic and molecular mechanisms that plants utilise to endure cold stress.

Historically, research on plant adaptation has centred on protein-coding genes, but ncRNAs have emerged as critical regulators of gene expression and plant resilience. Unlike messenger RNAs (mRNAs), which encode proteins, ncRNAs function independently of protein translation, influencing gene regulation at multiple levels and facilitating plants' adaptation to changing environments. These ncRNAs are classified into various types based on their size and function, including miRNAs, small interfering RNAs (siRNAs), lncRNAs and circRNAs. They play essential regulatory roles across numerous biological processes, from transcriptional control to epigenetic modifications, and are integral to plant responses to both biotic and abiotic stresses. Recent studies emphasise the critical role ncRNAs play in enhancing plants' cold stress tolerance by modulating pathways and genes necessary for survival in low-temperature conditions (Cao et al., 2024; Yadav et al., 2024).

Jha et al. (2023) examined how unusual daily temperature fluctuations driven by climate change and variability adversely impact crop production. Given that plants are stationary and continuously exposed to external environmental stimuli, such as extreme high (heat) and low (cold) temperatures, they have evolved sophisticated molecular regulatory mechanisms to sustain growth under stress. Central to these mechanisms are ncRNAs, particularly miRNAs, small interfering RNAs (siRNAs) and lncRNAs, which collectively play an essential role in regulating heat and cold stress tolerance. Jha et al. emphasise significant findings on the regulatory functions of these ncRNAs in supporting plant adaptation to extreme temperatures. Additionally, this review also addresses the challenges involved in identifying ncRNAs and comprehending their roles in enhancing plant resilience to temperature stress, thus highlighting the potential of ncRNA research in improving agricultural sustainability.

A specific subset of ncRNAs, lncRNAs, has shown remarkable potential in regulating plant responses to cold stress. These lncRNAs are longer than other ncRNAs (usually over 200 nucleotides) and have been identified as crucial players in modulating gene expression under stress conditions. For example, in *Arabidopsis thaliana*, a well-studied model organism, the lncRNA COLD INDUCED lncRNA 1 (CIL1) was found out to be a positive regulator of cold stress response, significantly increasing in expression when the plant is exposed to cold stress. This discovery paved the way for studying similar lncRNAs in other crop species to enhance their cold tolerance (Liu et al., 2022).

In another example, *Ammopiptanthus nanus*, a plant species that survives in extremely cold conditions, researchers identified over 1,300 cold-responsive lncRNAs, including TCONS00065739. This lncRNA was found to interact with miRNAs and mRNAs to regulate stress response pathways, illustrating the complex role that lncRNAs play in cold stress tolerance (Zhu et al., 2023).

In their 2023 review, Zuo et al. investigated how plants, which frequently encounter a range of adverse environmental conditions, have evolved sophisticated mechanisms to cope with these stressors. A notable mechanism discussed is the development of transcriptional memory, a process that enables plants to remember past stress encounters and adapt more effectively when faced with similar or even different stressors in the future. This stress-induced memory can enhance a plant's resilience and adaptability, potentially reducing damage from recurrent stress events. Understanding the molecular basis of transcriptional memory in plants has significant implications for breeding stress-tolerant crops, as it provides a theoretical foundation for improving agricultural resilience under changing climatic conditions. A comprehensive grasp of transcriptional memory mechanisms is essential for agronomic practices and breeding strategies aimed at enhancing agricultural productivity and quality.

In a related review, Hazra et al. (2023) focused on how abiotic stressors often intensified by climate change impact plants at a cellular level, particularly through the generation of ROS, which can impair vital cellular functions. To counteract ROS-induced damage, plants have evolved sophisticated antioxidant systems. However, the increasing frequency and intensity of abiotic stressors pose a substantial risk to crop yields and global food security. ncRNAs have emerged as crucial regulators in the plant response to various abiotic and biotic stresses. Key types of ncRNAs, such as miRNAs and lncRNAs, play crucial roles in regulating stress-related gene expression across transcriptional, post-transcriptional and translational levels. Additionally, lncRNAs contribute to epigenetic regulation by altering histone modification patterns and chromatin organisation, affecting the expression of target genes. Their review highlights the potential application of ncRNAs in maintaining ROS homeostasis, emphasising their promising role in enhancing crop resilience against abiotic stressors through molecular and genetic approaches.

The molecular mechanisms by which ncRNAs contribute to cold tolerance are diverse and multifaceted. At the transcriptional level, ncRNAs such as miRNAs and siRNAs can downregulate or silence specific target genes that may be sensitive to cold. By silencing these genes, ncRNAs prevent processes that would otherwise weaken the plant under cold conditions. In contrast, lncRNAs often function at the post-transcriptional and translational levels by acting as molecular decoys, sponges or scaffolds, thereby enhancing the stability and expression of genes necessary for cold tolerance (Jin et al., 2024). For instance, some lncRNAs bind to miRNAs as 'sponges', inhibiting miRNA activity and allowing target genes to be expressed. In *A. nanus*, TCONS00065739 acts as a competitive target of miR530, enabling the expression of the TZP gene, which is associated with enhanced cold tolerance (Zhu et al., 2023). Additionally, lncRNAs play a role in epigenetic modifications, such as histone modifications and DNA methylation, to 'memorize' cold responses, effectively priming plants to better respond to subsequent cold events (Zuo et al., 2023).

The study of ncRNAs in plants has advanced significantly with the development of high-throughput sequencing and bioinformatics tools, which have enabled the discovery of numerous ncRNAs involved in stress responses. Understanding the regulatory roles of ncRNAs, particularly lncRNAs, offers promising strategies for crop improvement. By identifying and manipulating cold-responsive ncRNAs in staple crops like rice, wheat and maize, researchers can potentially enhance these plants' resilience to cold stress, contributing to stable yields under fluctuating climates (Cao et al., 2024). ncRNAs' capacity to control stress-responsive gene networks positions them as valuable targets for developing new biotechnological approaches, such as CRISPR-based genome editing, to modulate ncRNA expression and improve stress tolerance in crops.

Despite these promising advancements, there are still challenges in fully understanding ncRNAs' roles in cold stress tolerance. Due to the complex and overlapping pathways in which ncRNAs operate, it can be difficult to isolate specific ncRNAs and determine their exact functions. For example, lncRNAs often interact with other ncRNAs, such as miRNAs, and with various regulatory proteins, forming intricate regulatory networks that can vary between plant species. Additionally, much remains unknown about the precise mechanisms by which ncRNAs influence epigenetic changes under cold stress, a critical area for understanding how plants 'remember' cold exposure and adapt to subsequent stresses (Hazra et al., 2023; Jha et al., 2023). Future research efforts could benefit from integrative approaches that combine genetic, molecular and computational methods to map out ncRNA interactions and identify targets for genetic engineering. Moreover, interdisciplinary research that merges plant biology with climate science and agronomy can enhance our understanding of ncRNAs' roles in crop resilience and contribute to sustainable agricultural solutions for a changing climate.

lncRNAs are essential regulators in plant development and stress response, while miRNAs function in transcriptional and post-transcriptional gene regulation. The specific role of lncRNA-mediated cold stress response in plants, and the complex cooperation among lncRNAs, miRNAs and target mRNAs as part of the competing endogenous RNA (ceRNA) network, remains to be fully elucidated. In recent research on the winter wheat cultivar Dongnongdongmai 1 (Dn1), 9970 putative lncRNAs were identified through high-throughput sequencing across three temperature conditions (5°C, −10°C and −25°C). Of the 14,626 genes detected via weighted gene co-expression network analysis, 7435 lncRNAs were found to be co-expressed with 7191 mRNAs, revealing six modules within the lncRNA–mRNA weighted co-expression network associated with cold resistance, where antioxidant systems and hormones were prominent in low-temperature response (Lu et al., 2020).

Further analysis highlighted the significance of ceRNA interactions in cold stress tolerance, involving 384 lncRNAs and 70 miRNAs, and the formation of complex regulatory networks with 225 lncRNAs, 60 miRNAs and 621 target mRNAs. Interestingly, a conserved region was identified in the complementary sequences of lncRNAs and miR164/408, although they exhibited opposing expression trends within the ceRNA network. These findings suggest key roles for lncRNAs and mRNAs in cold resistance, presenting valuable insights for genomic strategies in wheat breeding to enhance cold tolerance.

ncRNAs have emerged as vital regulators of cold stress tolerance in plants, influencing gene expression networks and enabling plants to adapt to extreme temperature fluctuations. Among ncRNAs, lncRNAs play a particularly significant role in modulating cold-responsive genes, as demonstrated by studies in *Arabidopsis*, *A. nanus* and other plant species. By acting at various levels of gene regulation, from transcriptional silencing to epigenetic modifications, ncRNAs enable plants to withstand cold stress and maintain productivity. As global climate change continues to challenge agricultural systems, understanding and harnessing ncRNAs in crop improvement holds substantial promise for enhancing cold resilience in economically important crops. Continued advancements in this field, coupled with new biotechnological tools, could pave the way for novel, sustainable strategies to ensure crop stability and food security under future climate conditions.

17.11 REGULATORY ROLE OF ncRNAs IN CONTROLLING HEAVY METAL TOXICITY IN PLANTS

Globally, heavy metals are affecting the environmental health, ecological surroundings and all living beings. Persistent pollutants, which do not degrade through natural processes and, therefore, deposit within soil, water and living organisms over time, are heavy metals like lead (Pb), cadmium (Cd), mercury (Hg) and arsenic (As). The presence of these compounds affects cellular function and evokes detrimental health effects across biological systems. Heavy metals disrupt biological pathways and, with long-time exposure, can cause neurotoxicity, carcinogenicity and chronic diseases. Hence, further knowledge of molecular mechanisms related to responses to HM exposure is urgently needed to address the effective mitigation strategies and their negative impacts both on environmental and human health (Yu et al., 2024). The new findings have significantly been concentrated on the ncRNAs' role in heavy metal stress responses, newly discovered insights into gene regulation, epigenetic modulation and resilience pathways within organisms.

Epigenetics, or gene modifications in gene expression that don't alter the DNA sequence, has emerged as a key mechanism in understanding how organisms adapt themselves to environmental challenges such as heavy metal toxicity. Epigenetic modifications include DNA methylation, histone modification and actions of ncRNAs that are necessary for the accurate regulation of genes associated with stress responses. These epigenetic changes are not only critical for maintaining cellular health but are also reversible, hence making them an attractive target for therapeutic and biotechnological interventions. There are so many heavy metals that can interfere with the epigenome: they affect gene expression patterns needed for environmental and human resilience (Adegbola & Adetutu, 2024). In contrast to organic compounds, heavy metals are thermodynamically stable and persist in the environment, thereby forming long-lasting risks to ecosystems, agriculture and human diets.

Emerging evidence indicates that once considered as "transcriptional noise", ncRNAs are actually key mediators to mediate adaptive responses under environmental stressors, which includes heavy metals, and modified ncRNAs are potential agent target in the management of toxicity imposed by HMs, moving both the environmental and biomedical fields. Heavy metals primarily affect cell health through the introduction of oxidative stress. Increased levels of cellular production of ROS lead to oxidative damage to proteins, lipids and DNA; it characterises the mechanisms of HM exposure. If oxidative stress is not adequately regulated, it disrupts cellular homeostasis, leading to apoptosis. Plants make use of ncRNAs for regulating ROS levels while coordinating antioxidant responses as parts of the mechanisms of resilience to heavy metal stress. Plants are typically sessile in nature. They have, therefore, been exposed to variations in environmental conditions such as heavy metal toxicity over long periods. As such, plants developed complex molecular networks to resist these adverse environmental conditions over time. NcRNAs, including the miRNAs, siRNAs and lncRNAs, have been shown to be integral regulatory molecules in plants counteracting the toxic effects from heavy metals to maintain cell homeostasis and reconstitute damaged tissue (Bhattacharya & Roychoudhury, 2023).

In the context of agricultural productivity, knowing how ncRNAs regulate plant responses to HMs can be the basis for crop variety development that is more resistant to environmental stress, an essential adaptation in regions affected by metal pollution. Exposure to heavy metals also brings with it significant health risks to humans. These pollutants may cross biological barriers, settle into tissues and have interaction with critical cellular systems while causing toxicity in multiple organ systems. Neurotoxicity includes interference with neurodevelopment via metals, such as lead and mercury, and contributes towards neurodegenerative disease. The epigenetic mechanisms are involved in which ncRNAs regulate those pathways involved in neuronal survival, stress response or oxidative defence. Heavy metals also induce alterations in the expressions of ncRNAs through regulation of genes of essential neurodevelopmental and neuroprotective pathways. This newly found knowledge is therapeutically significant in suggesting that the modulation of ncRNA may assist in decreasing neurotoxicity induced by heavy metals. For instance, the investigation for epigenetic reprogramming techniques that could be employed to prevent or reverse HM-induced neurotoxicity points towards ncRNAs as promising therapeutic targets (Sharma et al., 2024). Besides this, the role of epigenetic changes in HM-induced neurotoxicity also throws a pointer to the idea that toxicity caused by heavy metals might be reversed and the damage induced by HM could be repaired.

The functions of ncRNAs go beyond the heavy metal responses and range from abiotic stresses such as salinity, drought and extreme temperatures. Heavy metals are abiotic stresses requiring complex regulation, and their responses are coordinated by the expression of stress-related genes where ncRNAs play central roles. Some examples of miRNAs include short RNA sequences that bind target mRNAs and affect their expression. It specifically upregulates certain detoxification and antioxidant-related gene targets for creating a more robust defence in an HM-stressed plant, although lncRNAs generally act as molecular scaffolds or ceRNAs in sponging miRNAs for fine-tuning the cellular response to heavy metal stress. These complex regulatory networks expose mechanisms by which plants manage toxicity of heavy metals, and these offer new ways through which phytoremediation – which involves cleaning up metal-polluted environments with plants – can be improved. It is on the basis of such a mechanism that researchers may have the opportunity to engineer crops, not only resilient to such environmental stressors but able to detoxify soils, for instance, by means of increased uptake and accumulation of heavy metals. With the advent of high-throughput sequencing and bioinformatics, the study of ncRNAs has been expanded rapidly to provide new tools for analysing gene expression and identifying the complex roles of these molecules in environmental stress responses. For example, in the model plant *Arabidopsis thaliana*, N6-methyladenosine (m6A) RNA modifications have been shown to influence transcript stability and translation under HM-induced oxidative stress, further emphasising the regulatory role of ncRNAs over cellular stress responses (Jha et al., 2023).

Epitranscriptomic modifications hence add another layer to our understanding of gene regulation, specifically regarding how plants dynamically respond to heavy metal stress. The results of ncRNA studies in *Arabidopsis* reveal how they can also play a role in other crops and therefore be potentially useful in developing agriculturally resilient systems that can recover from the effects of HMs, among other abiotic stressors. ncRNAs are also in humans linked to the modulating of responses to stressors, including environmental exposure like pollution and heavy metals.

Exposure to HMs relates to the development of numerous chronic diseases, including cancers. These ncRNAs regulate multiple cancer-related pathways through RNA methylation, histone modifications and direct mRNA targeting. Research into these mechanisms, especially in m6A methylation pathways involving pollution-induced tumorigenesis, has a good possibility of being explored as biomarkers for early diagnosis and targets for preventive measures. Changes mediated by protein factors, including METTL3 and YTHDF1, can affect key pathways which are related to cell survival, immune response and carcinogenesis, suggesting that ncRNAs play central roles in human responses to HM-induced pathologies. The field of ncRNAs may provide a new horizon for therapeutic development targeting epigenetic regulators of gene expression, especially in pollution-exacerbated diseases, such as lung, skin and bladder cancer. ncRNAs are an invaluable part of understanding and managing impacts of heavy metals across organisms. They influence gene expression, epigenetic modulation and adaptive responses as promising pathways for mitigating adverse effects of environmental pollutants. As the functional complexities of ncRNAs continue to be discovered, their use in environmental management and therapeutic interventions is becoming increasingly possible. Whether it is to enhance phytoremediation or to develop ncRNA-targeted therapies in humans, the understanding of ncRNAs and their role in heavy metal stress responses is an essential framework for addressing the environmental and health challenges posed by heavy metal pollution. A deep understanding of mechanisms mediated by ncRNAs is thus necessary for devising strategies to protect ecosystems, improve agricultural resilience and promote public health in the face of persistent environmental pollution.

17.12 REGULATORY ROLE OF ncRNAs IN DEVELOPMENT AND NUTRIENT DEFICIENCY OF PLANTS

Plants depend on many genetic and molecular processes for improved growth, development and responses to changes in their surroundings. Among the newly described mechanisms are the ncRNAs, which are considered some of the most important regulators in plants. Once only known as messengers and structural components, such as ribosomal and tRNAs, now ncRNAs form a diverse category of molecules that regulate expression and regulation at multi-levels, including transcriptional, post-transcriptional and epigenetic systems. Various ncRNA classes, including small, medium-sized and lncRNAs, and circular RNAs, play significant roles in regulating plant growth and stress response. lncRNAs have diverse regulatory mechanisms at various gene regions, including promoters, 5′UTRs, introns, exons and 3′UTR regions. They also participate in phase separation processes and serve as sponges, precursors, spliceosomes, stabilisation scaffolds or interaction factors (Zhang et al., 2023).

In plants, ncRNAs play a significant role in various developmental stages, which influence gene expression in organs such as roots, leaves, seeds and endosperms. They function as molecular switches, orchestrating complex regulatory networks. For instance, Yadav et al. (2024) highlight the importance of ncRNAs in modulating growth across root, leaf and seed tissues. Through intricate signalling pathways, these ncRNAs guide cellular differentiation, division and expansion, ultimately shaping plant morphology and organogenesis. This diversity in the kinds of ncRNAs makes for a broad toolkit in regulation, each type contributing uniquely to developmental and growth modulation. For example, siRNAs mainly focus on the degradation of specific mRNA transcripts, hence lowering the level of gene expression post-transcriptionally, while miRNAs take part in fine-tuning gene expression by binding to complementary mRNAs and inhibiting translation.

Another class of interest is lncRNAs, which have been shown to play a crucial role in the regulation of gene expression during plant development. Interconnected regulatory mechanisms mediate the fine-tuning of gene-expression-governing interactions with soil biota, nutrient acquisition and heavy metal tolerance in plants. Recently, with high-throughput sequencing, it is possible to identify plant lncRNAs that respond to rhizosphere biotic and abiotic signals. Here, the mechanisms of lncRNA biogenesis, classification and mode of action explain the functions of lncRNAs in mediating the adaptation of plants to a variety of rhizosphere factors. Few studies have also revealed expression of the target genes of lncRNAs during developmental plasticity and stress responses to rhizobium. lncRNAs have recently been reported to interact with many transcription factors and chromatin-modifying proteins to create tissue-specific patterns of gene expression. Fahad et al. (2024) stated that environment-responsive lncRNAs were identified through high-throughput sequencing, indicating their role in developmental plasticity. lncRNAs act as molecular scaffolds, sponges, and signalling molecules, thus helping the plants adapt to changes in their environment and handle internal signalling related to growth.

Macronutrients such as N and Pi are important for the normal growth of plants, which influence many metabolic and physiological processes of survival and growth. According to Yang et al. (2024), 'miRNAs play intricate regulatory roles in N and Pi signaling transduction by regulating nutrient uptake with metabolic need'. In this process, miRNAs interact with various classes of ncRNAs for fine-tuning N and Pi uptake and mobilisation by targeting and modulating specific genes involved in these pathways.

For instance, in response to nitrogen deficiency, miRNAs regulate nitrogen metabolism and root architecture, thereby improving nitrogen use efficiency (Yang et al., 2024). Qiu et al. (2023) also show that some lncRNAs, such as lncRNA24320.6, improve nitrogen uptake and enhance growth in rice seedlings under low-nitrogen conditions. The lncRNA enhances the expression of flavonoid 3′-hydroxylase 5 (OsF3′H5), which is essential for nitrogen metabolism and plant growth in a nitrogen-deficient environment.

Even though trace amounts are needed, micronutrients are extremely crucial in plant health, allowing many enzymatic and biochemical reactions to take place. According to Nayak et al. (2023), ncRNAs regulate the uptake and transport of the necessary micronutrients such as iron, zinc and manganese through gene modulation, thus optimising the absorption and distribution of micronutrients according to the nutrient availability in the soil. Micronutrient uptake involves controlling the expression of nutrient-transporting genes as well as genes encoding for transport enzymes and is further fine-tuned with tsRNAs, including siRNAs and miRNAs, to ensure homeostasis even under dynamic conditions. These fine-tunings allow for optimised nutrition for plant growth, increased resistance to biotic and abiotic stress and higher yields.

Environmental stresses, including drought, salinity and extreme temperatures, disrupt metabolic functions and impact plant productivity. Yadav et al. (2024) state that ncRNAs significantly enable the plant to cope with environmental stresses. A few of the ncRNAs, including miRNAs and lncRNAs, activate stress response pathways and initiate protective mechanisms for improved resilience under stress conditions.

For example, this study uses whole-transcriptome sequencing to compare cucumber leaves with varying sink strength. It reveals that the growth process in cucumber can be regulated by ncRNAs and alternative polyadenylation. The analysis reveals enriched pathways for photosynthesis, energy production, hormone signal transduction, starch metabolism and protein synthesis. lncRNAs also regulate plant stress responses that are mediated through hormonal responses. They regulate transduction of signal from ABA, ethylene and salicylic acid – all of these are considered to be pivotal components for the response in stressful conditions (Wang et al., 2024).

Moreover, the relationships among lncRNAs, miRNAs and siRNAs render it an intricate network of regulatory interactions responding to environmental clues in the regulation of gene expression involving stress responses. For instance, Wang et al. postulated a model for the ceRNA hypothesis wherein lncRNAs function like miRNA sponges to eliminate their target mRNA from repression, thereby enhancing their quick response to stressors.

One of the more interesting functions of lncRNAs is that they act as molecular scaffolds to recruit RNA-binding proteins and chromatin modifiers to target gene loci. According to Zhang et al. (2023), lncRNAs are critical for chromatin remodelling through interactions with histone modification complexes that open up transcription factor accessibility and gene activation. lncRNAs can modulate the chromatin architecture to initiate or suppress gene expression responses in response to developmental cues or environmental signals, thus providing another layer of regulation for plant adaptive responses. The study identified 1628 lncRNAs in rice roots under low-nitrogen treatment. After 4 hours, 50 lncRNAs and 373 mRNAs were significantly upregulated, while 43 lncRNAs and 536 mRNAs were downregulated. One lncRNA, lncRNA24320.6, was found to positively regulate the expression of a flavonoid 3′-hydroxylase 5 gene and improve nitrogen absorption and growth in rice seedlings under LN conditions (Qiu et al., 2023).

Plant hormones are critical determinants of growth, stress response and nutrient allocation. Indirectly through their regulation of hormone biosynthesis and signalling, ncRNAs affect the physiological responses required for survival under stress conditions. Yadav et al. (2024) suggest that ncRNAs mediate the pathways of phytohormones, especially under nutrient starvation and under stress conditions. ncRNAs regulate the synthesis and transport of ABA and gibberellins. Thus, they control the pattern of plant growth under various external stimuli.

The new discoveries show sophisticated roles of ncRNAs in plant development, nutrient uptake, and tolerance towards different stresses. This diverse tool, therefore, associates with other biomolecules in complex relationships that are highly diverse. This aids in the regulation as well as the expression at levels such as transcription and epigenetic. An added advantage of the ncRNA endowments shall be keeping the plants alive in more variable environments, explaining how regulatory networks based on ncRNA play an important role in agricultural sustainability.

17.13 FUTURE RECOMMENDATIONS

ncRNAs, particularly lncRNAs, are being recognised as critical regulators of plant responses to salt stress. lncRNAs act through ceRNA networks, NAT functions and other regulatory mechanisms to offer a layer of resistance against environmental stresses. Understanding these pathways will be critical for creating genetically altered crops with improved salt tolerance. The CRISPR/Cas9 system has the capacity to precisely modify ncRNA genes, allowing the regulation of stress-responsive pathways. Moreover, understanding the mechanisms underlying ncRNA function not only opens new avenues for agricultural biotechnology but also advances our fundamental knowledge of plant biology, offering potential solutions to pressing challenges such as food security and climate resilience.

REFERENCES

Ackah, M., Jin, X., Zhang, Q., Amoako, F. K., Wang, L., et al. (2022). Long noncoding RNA transcriptome analysis reveals novel lncRNAs in *Morus alba* response to drought stress. *The Plant Genome*, *17*, e20273. https://doi.org/10.1002/tpg2.20273

Adegbola, P. I., & Adetutu, A. (2024). Genetic and epigenetic modulations in toxicity: The two-sided roles of heavy metals and polycyclic aromatic hydrocarbons from the environment. *Toxicology Reports*, 12, 502.

Ariel, F., Romero-Barrios, N., Jégu, T., Benhamed, M., & Crespi, M. (2015). Battles and hijacks: Noncoding transcription in plants. *Trends in Plant Science*, *20*(6), 362–371.

Arikit, S., Xia, R., Kakrana, A., Huang, K., Zhai, J., Yan, Z., …, & Meyers, B. C. (2014). An atlas of soybean small RNAs identifies phased siRNAs from hundreds of coding genes. *The Plant Cell*, *26*(12), 4584–4601.

Axtell, M. J. (2013). Classification and comparison of small RNAs from plants. *Annual Review of Plant Biology*, *64*, 137–159.

Babaei, S., Bhalla, P. L., & Singh, M. B. (2024). Identifying long non-coding RNAs involved in heat stress response during wheat pollen development. *Frontiers in Plant Science*, *15*, 1344928. https://doi.org/10.3389/fpls.2024.1344928

Balhara, R., Verma, D., Kaur, R., & Singh, K. (2024). MYB transcription factors, their regulation and interactions with non-coding RNAs during drought stress in *Brassica juncea*. *BMC Plant Biology*, *24*, 999. https://doi.org/10.1186/s12870-024-05736-8

Bhattacharya, A., & Roychoudhury, A. (2023). Plant non-coding RNAs and their regulation in salinity and heavy metal stress. In *Non-coding RNAs* (pp. 128–147). CRC Press.

Cai, J., Liu, W., Li, W., Zhao, L., Chen, G., Bai, Y., …, & Zhang, X. (2022). Downregulation of miR156-targeted PvSPL6 in switchgrass delays flowering and increases biomass yield. *Frontiers in Plant Science*, *13*, 834431.

Cao, W., Yang, L., Zhuang, M., Lv, H., Wang, Y., Zhang, Y., & Ji, J. (2024). Plant non-coding RNAs: The new frontier for the regulation of plant development and adaptation to stress. *Plant Physiology and Biochemistry*, *208*, 108435.

Cech, T. R., & Steitz, J. A. (2014). The noncoding RNA revolution-trashing old rules to forge new ones. *Cell*, *157*(1), 77–94.

Chen, L., Zhu, Q. H., & Kaufmann, K. (2020). Long non-coding RNAs in plants: emerging modulators of gene activity in development and stress responses. *Planta*, *252*, 92.

Contreras-Cubas, C., Palomar, M., Arteaga-Vazquez, M., Reyes, J. L., & Covarrubias, A. A. (2012). Non-coding RNAs in the plant response to abiotic stress. *Planta*, *236*, 943–958.

Csorba, T., Questa, J. I., Sun, Q., & Dean, C. (2014). Antisense COOLAIR mediates the coordinated switching of chromatin states at FLC during vernalization. *Proceedings of the National Academy of Sciences*, *111*(45), 16160–16165.

D'Ario, M., Griffiths-Jones, S., & Kim, M. (2017). Small RNAs: Big impact on plant development. *Trends in Plant Science*, *22*(12), 1056–1068.

Ding, B., Xia, R., Lin, Q., Gurung, V., Sagawa, J. M., Stanley, L. E., …, & Yuan, Y. W. (2020). Developmental genetics of corolla tube formation: role of the tasiRNA-ARF pathway and a conceptual model. *Plant Cell*, *32*(11), 3452–3468.

Ding, J., Lu, Q., Ouyang, Y., Mao, H., Zhang, P., Yao, J., …, & Zhang, Q. (2012). A long noncoding RNA regulates photoperiod-sensitive male sterility, an essential component of hybrid rice. *Proceedings of the National Academy of Sciences*, *109*(7), 2654–2659.

Dong, Q., Hu, B., & Zhang, C. (2022). microRNAs and their roles in plant development. *Frontiers in Plant Science*, *13*, 824240.

Fahad, M., Tariq, L., Muhammad, S., & Wu, L. (2024). Underground communication: Long noncoding RNA signaling in the plant rhizosphere. *Plant Communications*, 5(7).

Ganguly, P., Roy, D., Das, T., Kundu, A., Cartieaux, F., Ghosh, Z., & DasGupta, M. (2021). The natural antisense transcript DONE40 derived from the lncRNA ENOD40 locus interacts with SET domain protein ASHR3 during inception of symbiosis in *Arachis hypogaea*. *Molecular Plant-Microbe Interactions*, *34*(9), 1057–1070.

Hazra, S., Moulick, D., Mukherjee, A., Sahib, S., Chowardhara, B., Majumdar, A., …, & Dey, A. (2023). Evaluation of efficacy of non-coding RNA in abiotic stress management of field crops: Current status and future prospective. *Plant Physiology and Biochemistry*, *203*, 107940.

Imaduwage, I. U. H., & Hewadikaram, M. (2024). Predicted roles of long non-coding RNAs in abiotic stress tolerance responses of plants. *Molecular Horticulture*, *4 (1)*, 20. https://doi.org/10.1186/s43897-024-00094-3

Jha, U. C., Nayyar, H., Roychowdhury, R., Prasad, P. V., Parida, S. K., & Siddique, K. H. (2023). Non-coding RNAs (ncRNAs) in plants: Master regulators for adapting to extreme temperature conditions. *Plant Physiology and Biochemistry*, *205*, 108164.

Jha, V., Narjala, A., Basu, D., Tn, S., Pachamuthu, K., Chenna, S., …, & Shivaprasad, P. V. (2021). Essential role of γ-clade RNA-dependent RNA polymerases in rice development and yield-related traits is linked to their atypical polymerase activities regulating specific genomic regions. *New Phytologist*, *232*(4), 1674–1691.

Jin, X., Wang, Z., Li, X., Ai, Q., Wong, D. C. J., Zhang, F., Yang, J., Zhang, N., & Si, H. (2024). Current perspectives of lncRNAs in abiotic and biotic stress tolerance in plants. *Frontiers in Plant Science*, *14*, 1334620. https://doi.org/10.3389/fpls.2023.1334620

Jones-Rhoades, M. W. (2012). Conservation and divergence in plant microRNAs. *Plant Molecular Biology*, *80*, 3–16.

Li, X., Xia, K., Liang, Z., Chen, K., Gao, C., & Zhang, M., 2016. MicroRNA393 is involved in nitrogen-promoted rice tillering through regulation of auxin signal transduction in axillary buds. *Scientific Reports*, *6*, 32158.

Liu, D., Mewalal, R., Hu, R., Tuskan, G. A., & Yang, X. (2017). New technologies accelerate the exploration of non-coding RNAs in horticultural plants. *Horticulture Research*, *4*.

Liu, G., Liu, F., Wang, Y., & Liu, X. (2022). A novel long non-coding RNA CIL1 enhances cold stress tolerance in Arabidopsis. *Plant Science*, *323*, 111370.

Liu, Y., Teng, C., Xia, R., & Meyers, B. C. (2020). PhasiRNAs in plants: their biogenesis, genic sources, and roles in stress responses, development, and reproduction. *Plant Cell*, *32*, 3059–3080.

Lu, Q., Xu, Q., Guo, F., Lv, Y., Song, C., Feng, M., …, & Cang, J. (2020). Identification and characterization of long non-coding RNAs as competing endogenous RNAs in the cold stress response of *Triticum aestivum*. *Plant Biology*, *22*(4), 635–645.

Ma, X., Zhao, F., & Zhou, B. (2022). The characters of non-coding RNAs and their biological roles in plant development and abiotic stress response. *International Journal of Molecular Sciences*, *23*(8), 4124.

Martinez, G., Panda, K., Köhler, C., & Slotkin, R. K. (2016). Silencing in sperm cells is directed by RNA movement from the surrounding nurse cell. *Nature Plants*, *2*(4), 1–8.

Mattick, J. S., & Rinn, J. L. (2015). Discovery and annotation of long noncoding RNAs. *Nature Structural & Molecular Biology*, *22*, 5–7.

Morris, K. V., & Mattick, J. S. (2014). The rise of regulatory RNA. *Nature Reviews Genetics*, *15*(6), 423–437.

Nayak, J. K., Chauhan, R., Kumar, S., & Prabhakaran, S. (2023). *Whispers of nourishment: Unveiling the role of non-coding RNA in plant nutrient availability*.

Nowak, K., Moronczyk, J., Grzyb, M., Szczygieł-Sommer, A., & Gaj, M. D. (2022). miR172 regulates WUS during somatic embryogenesis in *Arabidopsis* via AP2. *Cells*, *11*(4), 718.

O'Maoileidigh, D. S., van Driel, A. D., Singh, A., Sang, Q., Le Bec, N., Vincent, C., …, & Coupland, G. (2021). Systematic analyses of the MIR172 family members of *Arabidopsis* define their distinct roles in regulation of APETALA2 during floral transition. *PLoS Biology*, *19*(2), e3001043.

Pauli, A., Rinn, J. L., & Schier, A. F. (2011). Non-coding RNAs as regulators of embryogenesis. *Nature Reviews Genetics*, *12*(2), 136–149.

Peschansky, V. J., & Wahlestedt, C. (2014). Non-coding RNAs as direct and indirect modulators of epigenetic regulation. *Epigenetics*, *9*(1), 3–12.

Qiu, D., Wu, Y., Xia, K., Zhang, M., Zhang, Z., & Tian, Z. (2023). Genome-wide identification and characterization of long non-coding RNAs in roots of rice seedlings under nitrogen deficiency. *Plants*, *12*(23), 4047.

Rai, M. I., Alam, M., Lightfoot, D. A., Gurha, P., & Afzal, A. J. (2019). Classification and experimental identification of plant long non-coding RNAs. *Genomics*, *111*(5), 997–1005.

Sharma, B., Govindan, G., Li, Y., Sunkar, R., & Gregory, B. D. (2024). RNA N6-methyladenosine affects copper-induced oxidative stress response in *Arabidopsis thaliana*. *Non-Coding RNA*, *10*(1), 8.

Shin, S. Y., Jeong, J. S., Lim, J. Y., Kim, T., Park, J. H., Kim, J. K., & Shin, C. (2018). Transcriptomic analyses of rice (*Oryza sativa*) genes and non-coding RNAs under nitrogen starvation using multiple omics technologies. *BMC Genomics*, *19*, 1–20.

Silva, G. F., Silva, E. M., Correa, J. P., Vicente, M. H., Jiang, N., Notini, M. M., …, & Nogueira, F. T. (2019). Tomato floral induction and flower development are orchestrated by the interplay between gibberellin and two unrelated micro RNA-controlled modules. *New Phytologist*, *221*(3), 1328–1344.

Smoczynska, A., & Szweykowska-Kulinska, Z. (2016). MicroRNA-mediated regulation of flower development in grasses. *Acta Biochimica Polonica*, *63*, 687–692.

Song, X., Li, Y., Cao, X., & Qi, Y. (2019). MicroRNAs and their regulatory roles in plant-environment interactions. *Annual Review of Plant Biology*, *70*(1), 489–525.

Tang, J., & Chu, C. (2017). MicroRNAs in crop improvement: fine-tuners for complex traits. *Nature Plants*, *3*, 17077.

van Werven, F. J., Neuert, G., Hendrick, N., Lardenois, A., Buratowski, S., van Oudenaarden, A., … & Amon, A. (2012). Transcription of two long noncoding RNAs mediates mating-type control of gametogenesis in budding yeast. *Cell*, *150*(6), 1170–1181.

Waititu, J. K., Zhang, C., Liu, J., & Wang, H. (2020). Plant non-coding RNAs: Origin, biogenesis, mode of action and their roles in abiotic stress. *International Journal of Molecular Sciences*, *21*(21), 8401.

Wang, L., Fu, Y., Yuan, Z., et al. (2024). Identification and analysis of short-term and long-term salt-associated lncRNAs in the leaf of *Avicennia marina*. *BMC Plant Biology*, *24*, 500. https://doi.org/10.1186/s12870-024-05216-z

Wang, M., Zhao, W., Gao, L., & Zhao, L. (2018). Genome-wide profiling of long non-coding RNAs from tomato and a comparison with mRNAs associated with the regulation of fruit ripening. *BMC Plant Biology*, *18*, 1–10.

Wang, Y., Zhang, H., Zhang, Z., Hua, B., Liu, J., & Miao, M. (2024). Source leaves are regulated by sink strengths through non-coding RNAs and alternative polyadenylation in cucumber (*Cucumis sativus* L.). *BMC Plant Biology*, *24*(1), 812.

Wei, J., Li, H., Huang, X., Zhao, Y., Ouyang, L., et al. (2024). Elucidating the regulatory role of long non-coding RNAs in drought stress response during seed germination in leaf mustard. *PeerJ*, *12*, e17661. https://doi.org/10.7717/peerj.17661

Wu, G. (2013). Plant microRNAs and development. *Journal of Genetics and Genomics*, *40*(5), 217–230.

Xu, F., Yan, T. T., & Luo, Y. L. (2013). Thermosensitive poly (N-isopropyl acrylamide-co-N, N-dimethyl acryl amide)-block-poly (d, l-lactide) amphiphilic block copolymer micelles for prednisone drug release. *Journal of Bioactive and Compatible Polymers*, *28*(1), 66–85.

Yadav, A., Mathan, J., Dubey, A. K., & Singh, A. (2024). The emerging role of non-coding RNAs (ncRNAs) in plant growth, development, and stress response signaling. *Non-Coding RNA*, *10(1)*, 13. https://doi.org/10.3390/ncrna10010013

Yan, Y., Wang, H., Hamera, S., Chen, X., & Fang, R. (2014). MiR444a has multiple functions in the rice nitrate-signaling pathway. *The Plant Journal*, *78*, 44–55.

Yang, Y., Liang, Y., Wang, C., & Wang, Y. (2024). MicroRNAs as potent regulators in nitrogen and phosphorus signaling transduction and their applications. *Stress Biology*, *4*(1), 38.

Yu, G., Wu, L., Su, Q., Ji, X., Zhou, J., Wu, S., …, & Li, H. (2024). Neurotoxic effects of heavy metal pollutants in the environment: Focusing on epigenetic mechanisms. *Environmental Pollution*, *345*, 123563.

Zhang, F., Yang, J., Zhang, N., Wu, J., & Si, H. (2022). Roles of microRNAs in abiotic stress response and characteristics regulation of plant. *Frontiers in Plant Science*, *13*, 919243.

Zhang, L., Lin, T., Zhu, G., Wu, B., Zhang, C., & Zhu, H. (2023). LncRNAs exert indispensable roles in orchestrating the interaction among diverse noncoding RNAs and enrich the regulatory network of plant growth and its adaptive environmental stress response. *Horticulture Research*, *10*(12), uhad234.

Zhao, Y., Liu, Y., Zhang, F., et al. (2024). The long noncoding RNA LAL contributes to salinity tolerance by modulating LHCB1s' expression in *Medicago truncatula*. *Communications Biology*, *7*, 289. https://doi.org/10.1038/s42003-024-05953

Zheng, C., Ye, M., Sang, M., & Wu, R. (2019). A regulatory network for miR156-SPL module in *Arabidopsis thaliana*. *International Journal of Molecular Sciences*, *20*(24), 6166.

Zhou, Q., Shi, J., Li, Z., Zhang, S., Zhang, S., Zhang, J., …, & Liu, G. (2021). miR156/157 targets SPLs to regulate flowering transition, plant architecture and flower organ size in petunia. *Plant and Cell Physiology*, *62*(5), 839–857.

Zhu, M., Dong, Q., Bing, J., Songbuerbatu, Zheng, L., Dorjee, T., …, & Gao, F. (2023). Combined lncRNA and mRNA expression profiles identified the lncRNA–miRNA–mRNA modules regulating the cold stress response in *Ammopiptanthus nanus*. *International Journal of Molecular Sciences*, *24*(7), 6502.

Zuo, D. D., Ahammed, G. J., & Guo, D. L. (2023). Plant transcriptional memory and associated mechanism of abiotic stress tolerance. *Plant Physiology and Biochemistry*, *201*, 107917.

Index

Note: Locators in *italics* represent figures and **bold** indicate tables in the text.

D

E

F

G

N

O

P

Q

R

S